PERGAMON MATERIALS SERIES
VOLUME 5

The Coming of Materials Science

PERGAMON MATERIALS SERIES

Series Editor: **Robert W. Cahn** FRS
Department of Materials Science and Metallurgy, University of Cambridge, Cambridge, UK

Vol. 1 **CALPHAD** by N. Saunders and A. P. Miodownik
Vol. 2 **Non-Equilibrium Processing of Materials** edited by C. Suryanarayana
Vol. 3 **Wettability at High Temperatures** by N. Eustathopoulos, M. G. Nicholas and B. Drevet
Vol. 4 **Structural Biological Materials** edited by M. Elices
Vol. 5 **The Coming of Materials Science** by R. W. Cahn
Vol. 6 **Multinuclear Solid State NMR of Inorganic Materials** by K. J. D. Mackenzie and M. E. Smith
Vol. 7 **Underneath the Bragg Peaks: Structural Analysis of Complex Materials** by T. Egami and S. L. J. Billinge
Vol. 8 **Thermally Activated Mechanisms in Crystal Plasticity** by D. Caillard and J.-L. Martin

A selection of forthcoming titles in this series:

Phase Transformations in Titanium- and Zirconium-Based Alloys by S. Banerjee and P. Mukhopadhyay
Nucleation by A. L. Greer and K. F. Kelton
Non-Equilibrium Solidification of Metastable Materials from Undercooled Melts by D. M. Herlach and B. Wei
The Local Chemical Analysis of Materials by J.-W. Martin
Synthesis of Metal Extractants by C. K. Gupta

PERGAMON MATERIALS SERIES

The Coming of Materials Science

by

Robert W. Cahn, FRS

Department of Materials Science and Metallurgy,
University of Cambridge,
Cambridge, UK

PERGAMON
An Imprint of Elsevier Science
Amsterdam – London – New York – Oxford – Paris – Shannon – Tokyo

ELSEVIER SCIENCE Ltd
The Boulevard, Langford Lane
Kidlington, Oxford OX5 1GB, UK

First edition 2001
Second impression 2003

Library of Congress Cataloging in Publication Data
A catalog record from the Library of Congress has been applied for.

British Library Cataloguing in Publication Data
A catalogue record from the British Library has been applied for.

ISBN: 0-08-042679-4

∝ The paper used in this publication meets the requirements of ANSI/NISO Z39.48-1992 (Permanence of Paper).

Transferred to Digital Printing 2009

This book is dedicated to the memory of
Professor DANIEL HANSON (1892–1953)
of Birmingham University
who played a major role in modernising the teaching of Metallurgy
and thereby helped clear the ground for the emergence of Materials Science

Preface

My objective in writing this book, which has been many years in preparation, has been twofold. The discipline of materials science and engineering emerged from small beginnings during my professional life, and I became closely involved with its development; accordingly, I wanted to place on record the historical stages of that development, as well as premonitory things that happened long ago. My second objective, inseparable from the first, was to draw an impressionistic map of the present state of the subject, for readers coming new to it as well as for those well ensconced in research on materials. My subject-matter is the science, not the craft that preceded it, which has been well treated in a number of major texts. My book is meant primarily for working scientists and engineers, and also for students with an interest in the origins of their subject; but if some professional historians of science also find the contents to be of interest, I shall be particularly pleased.

The first chapter examines the emergence of the materials science concept, in both academe and industry, while the second and third chapters delve back into the prehistory of materials science (examining the growth of such concepts as atoms, crystals and thermodynamics) and also examine the evolution of a number of neighbouring disciplines, to see what helpful parallels might emerge. Thereafter, I pursue different aspects of the subject in varying depth. The book is in no sense a textbook of materials science; it should rather be regarded as a pointilliste portrait of the discipline, to be viewed from a slight distance. The space devoted to a particular topic is not to be regarded as a measure of the importance I attach to it, neither is the omission of a theme meant to express any kind of value judgment. I sought merely to achieve a reasonable balance between many kinds of themes within an acceptable overall length, and to focus on a few of the multitude of men and women who together have constructed materials science and engineering.

The numerous literature references are directed to two distinct ends: many refer to the earliest key papers and books, while others are to sources, often books, that paint a picture of the present state of a topic. In the early parts of the book, most references are to the distant past, but later on, as I treat the more modern parts of my subject, I refer to more recent sources.

There has been some dispute among professional historians of science as to who should be entitled to write a history such as this. Those trained as historians are understandably apt to resent the presumption of working scientists, in the evening of their days, in trying to take the bread from the historians' mouths. We, the superannuated scientists, are decried by some historians as 'whigs', mere uncritical celebrants of a perpetually advancing and improving insight into and control over nature. (A.R. Hall has called Whiggism "the writing of history as the story of an ascent to a splendid and virtuous climax"). There is some justice in this criticism, although not as much as its proponents are apt to claim. Another dispute, which has erupted recently into the so-called 'science wars', is between externalists who perceive science as an approach conditioned largely by social pressures (generally not recognized by the scientific practitioners themselves) and those, like myself, who take a mostly internalist stance and see scientific research as being primarily conditioned by the questions which flow directly from developing knowledge and from technological imperatives. The internalist/externalist dispute will never be finally resolved but the reader should at least be aware of its existence. At any rate, I have striven to be critical about the history of my own discipline, and to draw general conclusions about scientific practice from what I have discovered about the evolution of materials science.

One other set of issues runs through the book like a leitmotif: What *is* a scientific discipline? How do disciplines emerge and differentiate? Can a discipline also be interdisciplinary? Is materials science a real discipline? These questions are not just an exercise in lexicography and, looking back, it is perhaps the last of these questions which gave me the impetus to embark on the book.

A huge range of themes is presented here and I am bound to have got some matters wrong. Any reader who spots an error will be doing me a favor by kindly writing in and telling me about it at: rwc12@cam.ac.uk. Then, if by any chance there is a further edition, I can include corrections.

ROBERT CAHN
Cambridge, August 2000

Preface to Second Printing

The first printing being disposed of, the time has come to prepare a second printing. I am taking this opportunity to correct a substantial number of typographic mistakes and other small errors, which had escaped repeated critical read-throughs before the first printing. In addition, a small number of more substantial matters, such as inaccurate claims for priority of discovery, need to be put right, and these matters are dealt with in a *Corrigenda* at the very end of the book.

I am grateful to several reviewers and commentators for uncovering misprints, omissions and factual errors which I have been able to correct in this printing. My thanks go especially to Masahiro Koiwa in Japan, Jean-Paul Poirier and Jean Philibert in France, Jack Westbrook and Arne Hessenbruch in the United States.

ROBERT CAHN
Cambridge, October 2002

Acknowledgments

My thanks go first of all to Professor Sir Alan Cottrell, metallurgist, my friend and mentor for more than half a century, who has given me sage advice almost since I emerged from swaddling clothes. He has also very kindly read this book in typescript and offered his comments, helpful as always.

Next, I want to acknowledge my deep debt to the late Professor Cyril Stanley Smith, metallurgist and historian, who taught me much of what I know about the proper approach to the history of a technological discipline and gave me copies of many of his incomparable books, which are repeatedly cited in mine.

Professor Sir Brian Pippard gave me the opportunity, in 1993, to prepare a book chapter on the history of the physics of materials for a book, *Twentieth Century Physics*, that he was editing and which appeared in 1995; this chapter was a useful 'dry run' for the present work. I have also found his own contributions to that book a valuable source.

A book published in 1992, *Out of the Crystal Maze*, edited by Lillian Hoddeson and others, was also a particularly valuable source of information about the physics of materials, shading into materials science.

Dr. Frederick Seitz, doyen of solid-state physicists, has given me much helpful information, about the history of semiconductors in particular, and has provided an invaluable exemplar (as has Sir Alan Cottrell) of what a scientist can achieve in retirement.

Professor Colin Russell, historian of science and emeritus professor at the Open University, gave me helpful counsel on the history of chemistry and showed me how to take a philosophical attitude to the disagreements that beset the relation between practising scientists and historians of science. I am grateful to him.

The facilities of the Science Periodicals Library of Cambridge University, an unequalled source of information recent and ancient, and its helpful staff, together with those of the Whipple Library of the History and Philosophy of Science and the Library of the Department of Materials Science and Metallurgy, have been an indispensable resource.

Professors Derek Hull, Colin Humphreys and Alan Windle of my Department in Cambridge have successively provided ideal facilities that have enabled me to devote myself to the preparation of this book. My thanks go to them.

Hundreds of friends and colleagues all over the world, far too many to name, have sent me preprints and reprints, often spontaneously. The following have provided specific information, comments or illustrations, or given me interviews:

Kelly Anderson, V.S. Arunachalam, Bell Laboratory Archives, Yann le Bouar (who kindly provided Fig. 12.3(f) used on the cover), Stephen Bragg, Ernest Braun, Paul D. Bristowe, Joseph E. Burke, the late Hendrik B.G. Casimir, Leo Clarebrough, Clive Cohen, Peter Day, Anne Smith Denman, Cyril Domb, Peter Duncumb, Peter Edwards, Morris Fine, Joan Fitch, Jacques Friedel, Robert L. Fullman, Stefano Gialanella, Jon Gjønnes, Herbert Gleiter, Gerhard Goldbeck-Wood, Charles D. Graham, Martin L. Green, A. Lindsay Greer, Karl A. Gschneidner Jr, the late Peter Haasen, Richard H.J. Hannink, Jack Harris, Sir David Harrison, Peter W. Hawkes, Mats Hillert, Sir Peter Hirsch, Michael Hoare, Gerald Holton, the late John P. Howe, Archibald Howie, Paley Johnson, Stephen Keith, the late Andrew Keller, Peter Keller, the late David Kingery, Reiner Kirchheim, Ernest Kirkendall, Ole Kleppa, Masahiro Koiwa, Gero Kostorz, Eduard V. Kozlov, Edward Kramer, Kehsin Kuo, Vladislav G. Kurdyumov, Elisabeth Leedham-Green, Lionel M. Levinson, Eric Lifshin, James Livingston, John W. Martin, Thaddeus Massalski, David Melford, the late Sir Harry Melville, Peter Morris, Jennifer Moss, William W. Mullins, John Mundy, Frank Nabarro, Hideo Nakajima, the late Louis Néel, Arthur S. Nowick, Kazuhiro Otsuka, Ronald Ottewill, David Pettifor, Jean-Paul Poirier, G.D. Price, Eugen Rabkin, Srinivasa Ranganathan, C.N.R. Rao, Percy Reboul, M.Wyn Roberts, John H. Rodgers, Rustum Roy, Derek W. Saunders, Peter Paul Schepp, Hermann Schmalzried, Changxu Shi, K. Shimizu, Frans Spaepen, Hein Stüwe, Robb Thomson, Victor Trefilov, C. Tuijn, David Turnbull, Ruslan Valiev, Ajit Ram Verma, Jeffrey Wadsworth, Sir Frederick (Ned) Warner, James A. Warren, Robert C. Weast, Jack H. Westbrook, Guy White, Robert J. Young, Xiao-Dong Xiang. I apologise for any inadvertent omissions from this list.

Erik Oosterwijk and Lorna Canderton of Elsevier have efficiently seen to the minutiae of book production and I thank them for all they have done.

My son Andrew has steadfastly encouraged me in the writing of this book, and I thank him for this filial support. My dear wife, Pat, has commented on various passages. Moreover, she has made this whole enterprise feasible, not only by her confidence in her eccentric husband's successive pursuits but by always providing an affectionate domestic environment; I cannot possibly ever thank her enough.

ROBERT CAHN

Contents

CHAPTER 6
CHARACTERIZATION   213

CHAPTER 7
FUNCTIONAL MATERIALS   253

CHAPTER 8
THE POLYMER REVOLUTION   307

CHAPTER 9
CRAFT TURNED INTO SCIENCE   343

CHAPTER 10
MATERIALS IN EXTREME STATES   393

Chapter 1
Introduction

Chapter 1
Introduction

1.1. GENESIS OF A CONCEPT

Materials science emerged in USA, some time in the early 1950s. That phrase denoted a new scientific concept, born out of metallurgy, and this book is devoted to the emergence, development and consequences of that concept, in the US and elsewhere. Just who first coined the phrase is not known, but it is clear that by 1956 a number of senior research scientists had acquired the habit of using it. In 1958 and 1959 the new concept began to stimulate two developments in America: the beginnings of a change in the nature of undergraduate and graduate teaching in universities, and a radically new way of organising academic research on materials. The concept also changed the way industrial research was conceived, in a few important laboratories at least.

In this introductory chapter, I shall focus on the institutional beginnings of materials science, and materials engineering as well; indeed, "MSE" became an accepted abbreviation at quite an early stage. Following an examination, in Chapter 2, of the earlier emergence of some related disciplines, the intellectual antecedents to and development of materials science in its early stages are treated in Chapter 3. The field made its first appearance in USA, and for a number of years developed only in that country. Its development elsewhere was delayed by at least a decade.

1.1.1 Materials science and engineering in universities

Northwestern University, in Illinois not far from Chicago, was the first university to adopt materials science as part of a department title. That grew out of a department of metallurgy. Morris Fine, who was head of the department at the time, has documented the stages of the change (Fine 1990, 1994, 1996). He was a metallurgist, doing research at Bell Laboratories, when in early 1954 he was invited to visit Northwestern University to discuss plans to create a new graduate department of metallurgy there. (It is common at the leading American universities to organise departments primarily for work at graduate level, and in contrast to many other countries, the graduate students are exposed to extensive compulsory lecture courses.) In the autumn of 1954 Fine started at the University as a member of the new metallurgy department. In his letter of acceptance he had already mooted his wish to start a materials science programme in cooperation with other departments.

3

In spite of its graduate status, the new department did offer some undergraduate courses, initially for students in other departments. One of the members of faculty was Jack Frankel, who "was a disciple of Daniel Rosenthal at the University of California, Los Angeles... who had developed such a course there". Frankel worked out some of the implications of this precursor by developing a broadly based undergraduate lecture course at Northwestern and, on the basis of this, writing a book entitled *Principles of the Properties of Materials* (Frankel 1957). Fine remarks that "this course and Jack's thinking were key elements in developing materials science at Northwestern". Various other departments accepted this as a service course. According to the minutes of a faculty meeting in May 1956, it was resolved to publish in the next University Bulletin a paragraph which included the statement: "A student who has satisfactorily completed a programme of study which includes most of these (undergraduate) courses will be adequately prepared for professional work or graduate study in metallurgy and *materials science*". So, from 1957, undergraduates could undertake a broad study of materials in a course provided by what was still a metallurgy department. In February of 1958, a memorandum was submitted to the responsible academic dean, with the title *The Importance of Materials Science and Engineering*. One sentence in this document, which was received with favour by the dean, reads: "Traditionally the field of material science (even at this early stage, the final 's' in the adjective, 'materials', was toggled on and off) has developed along somewhat separate channels – solid state physics, metallurgy, polymer chemistry, inorganic chemistry, mineralogy, glass and ceramic technology. Advance in materials science and technology is hampered by this artificial division of the whole science into separate parts." The document went on to emphasise "the advantages of bringing together a group of specialists in the various types of materials and allowing and *encouraging* their cooperation and free interchange of ideas". Clearly this proposal was approved at a high level, for at a meeting a few months later, in December 1958, the metallurgy faculty meeting resolved, nemine contradicente, to change the name of the Graduate Department of Metallurgy to Graduate Department of Materials Science, and in January 1959 the university trustees approved this change.

At almost the same time as the 1958 faculty meeting, the US President's Science Advisory Committee referred to universities' attempts to "establish a new materials science and engineering" and claimed that they needed government help (Psaras and Langford 1987, p. 23).

The dean told the head of the department that various senior metallurgists around America had warned that the new department might "lose out in attracting students" by not having 'metallurgy' as part of its title. That issue was left open, but the department clearly did not allow itself to be intimidated and *Materials Science* became its unqualified name (although '*and Engineering*' was soon afterwards added to that name, to "better recognise the character of the department that had been formed"). The department did not lose out. Other departments in the English-speaking world have been more cautious: thus, my own department in Cambridge University began as "Metallurgy", eventually became "Metallurgy and Materials Science" and finally, greatly daring, changed to "Materials Science and Metallurgy". The final step cannot be more than a few decades off. The administrators of Oxford University, true to their reputation for pernicketiness, raised their collective eyebrows at the use of a plural noun, 'materials', in adjectival function. The department of materials science there, incensed, changed its name simply to 'Department of Materials', and some other universities followed suit.

Fine, who as we have seen played a major part in willing the new structure into existence, had (Fine 1996) "studied solid-state quantum mechanics and statistical mechanics as a graduate student in metallurgy (at the University of Minnesota)". It is striking that, as long ago as the 1940s, it was possible for an American student of metallurgy to work on such topics in his graduate years: it must have been this early breadth of outlook that caused materials science education, which is centred on the pursuit of breadth, to begin in that part of the world.

From 1959, then, the department of materials science at Northwestern University taught graduates the new, broad discipline, and an undergraduate course for materials science and engineering majors followed in due course. The idea of that discipline spread fast through American universities, though some eminent metal-lurgists such as Robert F. Mehl fiercely defended the orthodox approach to physical metallurgy. Nevertheless, by 1969 (Harwood 1970) some 30% of America's many university departments of metallurgy carried a title involving combinations of the words 'materials science' and 'metallurgy'. We are not told how quickly the 'materials engineering' part of the nomenclature was brought in. By 1974, the COSMAT Report (COSMAT 1974), on the status of MSE, remarked that America had some 90 "materials-designated" baccalaureate degree courses, ≈60 of them accredited, and that ≈50 institutions in America by then offered graduate degrees in materials. Today, not many departments of metallurgy remain in America; they have almost all changed to MSE. Different observers give somewhat inconsistent figures; thus, Table 1.1 gives statistics assembled by Lyle Schwartz in 1987, from American Society of Metals sources.

Henceforth, 'materials science' will normally be used as the name of the field with which this book is concerned; when the context makes it particularly appropiate to include 'and engineering' in the name, I shall use the abbreviation "MSE", and occasionally I shall be discussing materials engineering by itself.

There were also universities which did not set up departments of materials science but instead developed graduate programmes as an interdepartmental venture, usually but not always within a 'College of Engineering'. An early

Table 1.1. Trends in titles of materials departments at US universities, 1964–1985, after Lyle, in Psaras and Langford 1987.

Department title	Number of departments, by year		
	1964	1970	1985
Minerals and mining	9	7	5
Metallurgy	31	21	17
Materials	11	29	51
Other	18	21	17
Total	69	78	90

example of this approach was in the University of Texas at Austin, and this is described in some detail by Fine (1994). At the time he wrote his overview, 38 fulltime faculty members and 90 students were involved in this graduate programme: the students gain higher degrees in MSE even though there is no department of that name. "Faculty expertise and graduate student research efforts are concentrated in the areas of materials processing, solid-state chemistry, polymer engineering and science, X-ray crystallography, biomaterials, structural materials, theory of materials (whatever that means!) and solid-state (electronic?) materials and devices". Fine discusses the pros and cons of the two distinct ways of running graduate programmes in MSE. It may well be that the Texas way is a more effective way of forcing polymers into the curriculum; that has always proved to be a difficult undertaking. I return to this issue in Chapter 14. The philosophy underlying such interdepartmental programmes is closely akin to that which led in 1960 to the interdisciplinary materials research laboratories in the USA (Section 1.1.3).

To give a little more balance to this story, it is desirable to outline events at another American university, the Massachusetts Institute of Technology. A good account of the very gradual conversion from metallurgy to MSE has been provided in a book (Bever 1988) written to celebrate the centenary of the first course in which metallurgy was taught there (in combination with mining); this has been usefully supplemented by an unpublished text supplied by David Kingery (1927–2000), an eminent ceramist (Kingery 1999). As is common in American universities, a number of specialities first featured at graduate level and by stages filtered through to undergraduate teaching. One of these specialities was ceramics, introduced at MIT by Frederick H. Norton who joined the faculty in 1933 and taught there for 29 years. Norton, a physicist by training, wrote the definitive text on refractory materials. His field expanded as mineral engineering declined and was in due course sloughed off to another department. Kingery, a chemist by background, did his doctoral research with Norton and joined the faculty himself in 1950. He says: "Materials science, ceramic science and most of what we think of as advanced technology did not exist in 1950, but the seeds had been sown in previous decades and were ready to sprout. The Metallurgy Department had interests in process metallurgy, physical metallurgy, chemical metallurgy and corrosion, but, in truth, *the properties and uses of metals are not very exciting* (my italics). The ceramics activity was one division of the Metallurgy Department, and from 1949 onwards, higher degrees in ceramic engineering could be earned. During the 1950s, we developed a ceramics program as a fully interdisciplinary activity." He goes on to list the topics of courses taken by (presumably graduate) students at that time, in colloid science, spectroscopy, thermodynamics and surface chemistry, crystal structure and X-ray diffraction, dielectric and ferroelectric materials and quantum physics. The words in italics, above, show what we all know, that to succeed in a new endeavour it is necessary to focus one's enthusiasm intensely. For Kingery, who has been extremely influential in the evolution of ceramics as a constituent of MSE, ceramics constitute the heart and soul of MSE. With two colleagues, he wrote a standard undergraduate text on ceramics (Kingery 1976). By stages, he refocused on the truly modern aspects of ceramics, such as the role of chemically modified grain boundaries in determining the behaviour of electronic devices (Kingery 1981).

In 1967, the department's name (after much discussion) was changed to 'Metallurgy and Materials Science' and not long after that, a greatly broadened undergraduate syllabus was introduced. By that time, 9 years after the Northwestern initiative, MIT took the view that the change of name would actually enhance the department's attractiveness to prospective students. In 1974, after further somewhat acrimonious debates, the department's name changed again to 'Materials Science and Engineering'. It is to be noted, however, that the reality changed well before the name did. Shakespeare's Juliet had, perhaps, the essence of the matter:

"What's in a name? That which we call a rose By any other name would smell as sweet".

All the foregoing has been about American universities. Materials science was not introduced in European universities until well into the 1960s. I was in fact the first professor to teach and organise research in MSE in Britain – first as professor of materials technology at the University College of North Wales, 1962–1964, and then as professor of materials science at the University of Sussex, 1965–1981. But before any of this came about, the Department of Physical Metallurgy at the University of Birmingham, in central England, under the visionary leadership of Professor Daniel Hanson and starting in 1946, transformed the teaching of that hitherto rather qualitative subject into a quantitative and rigorous approach. With the essential cooperation of Alan Cottrell and Geoffrey Raynor, John Eshelby and Frank Nabarro, that Department laid the foundation of what was to come later in

America and then the world. This book is dedicated to the memory of Daniel Hanson.

1.1.2 MSE in industry

A few industrial research and development laboratories were already applying the ideas of MSE before those ideas had acquired a name. This was true in particular of William Shockley's group at the Bell Telephone Laboratories in New Jersey and also of General Electric's Corporate Laboratory in Schenectady, New York State. At Bell, physicists, chemists and metallurgists all worked together on the processing of the semiconductors, germanium and silicon, required for the manufacture of transistors and diodes: William Pfann, the man who invented zone-refining, without which it would have been impossible in the 1950s to make semiconductors pure enough for devices to operate at all (Riordan and Hoddeson 1997), was trained as a chemical engineer and inspired by his contact with a famous academic metallurgist. Later, Bell's interdisciplinary scientists led the way in developing hard metallic superconductors. Such a broad approach was not restricted to inorganic materials; the DuPont Research Station in Delaware, as early as the 1930s, had enabled an organic chemist, Carothers, and a physical chemist, Flory, both scientists of genius, to create the scientific backing that eventually brought nylon to market (Morawetz 1985, Hounshell and Smith 1988, Furukawa 1998); the two of them, though both chemists, made quite distinct contributions.

The General Electric Laboratory has a special place in the history of industrial research in America: initially directed by the chemist Willis Whitney from 1900, it was the first American industrial laboratory to advance beyond the status of a troubleshooting addendum to a factory (Wise 1985). The renowned GE scientists, William Coolidge and Irving Langmuir (the latter a Nobel prizewinner for the work he did at GE) first made themselves indispensable by perfecting the techniques of manufacturing ductile tungsten for incandescent light bulbs, turning it into coiled filaments to reduce heat loss and using inert gases to inhibit blackening of the light bulb (Cox 1979). Langmuir's painstaking research on the interaction of gases and metal surfaces not only turned the incandescent light bulb into a practical reality but also provided a vital contribution to the understanding of heterogeneous catalysis (Gaines and Wise 1983). A steady stream of scientifically intriguing and commercially valuable discoveries and inventions continued to come from the Schenectady laboratory, many of them relating to materials: as to the tungsten episode, a book published for Coolidge's 100th birthday and presenting the stages of the tungsten story in chronological detail (including a succession of happy accidents that were promptly exploited) claims that an investment of just $116,000 produced astronomical profits for GE (Liebhafsky 1974).

In 1946, a metallurgist of great vision joined this laboratory in order to form a new metallurgy research group. He was J.H. (Herbert) Hollomon (1919–1985). One of the first researchers he recruited was David Turnbull, a physical chemist by background. I quote some comments by Turnbull about his remarkable boss, taken from an unpublished autobiography (Turnbull 1986): "Holloman, then a trim young man aged 26, was a most unusual person with quite an overpowering personality. He was brash, intense, completely self-assured and overflowing with enthusiasm about prospects for the new group. He described the fascinating, but poorly understood, responses of metals to mechanical and thermal treatments and his plans to form an interdisciplinary team, with representation from metallurgy, applied mechanics, chemistry and physics, to attack the problems posed by this behaviour. He was certain that these researches would lead to greatly improved ability to design and synthesise new materials that would find important technological uses and expressed the view that equipment performance was becoming more materials- than design-limited... Hollomon was like no other manager. He was rarely neutral about anything and had very strong likes and dislikes of people and ideas. These were expressed openly and vehemently and often changed dramatically from time to time. Those closely associated with him usually were welcomed to his inner sanctum or consigned to his outer doghouse. Most of us made, I think, several circuits between the sanctum and the doghouse. Hollomon would advocate an idea or model vociferously and stubbornly but, if confronted with contrary evidence of a convincing nature, would quickly and completely reverse his position without the slightest show of embarrassment and then uphold the contrary view with as much vigour as he did the former one..." In an internal GE obituary, Charles Bean comments: "Once here, he quickly assembled an interdisciplinary team that led the transformation of metallurgy from an empirical art to a field of study based on principles of physics and chemistry". This transformation is the subject-matter of Chapter 5 in this book.

Hollomon's ethos, combined with his ferocious energy and determination, and his sustained determination to recruit only the best researchers to join his group, over the next 15 years led to a sequence of remarkable innovations related to materials, including man-made diamond, high-quality thermal insulation, a vacuum circuit-breaker, products based on etched particle tracks in irradiated solids, polycarbonate plastic and, particularly, the "Lucalox" alumina envelope for a metal–vapour lamp. (Of course many managers besides Hollomon were involved.) A brilliant, detailed account of these innovations and the arrangements that made them possible was later written by Guy Suits and his successor as director, Arthur Bueche (Suits and Bueche 1967). Some of these specific episodes will feature later in this book, but it helps to reinforce the points made here about Hollomon's conception of broad research on materials if I point out that the invention of translucent alumina tubes for lamps was a direct result of untrammelled research by R.L. Coble on the mechanism of densification during the sintering of a ceramic powder. There have been too few such published case-histories of industrial innovation in materials; many years ago, I put the case for pursuing this approach to gaining insight (Cahn 1970).

The projects outlined by Suits and Bueche involved collaborations between many distinct disciplines (names and scientific backgrounds are punctiliously listed), and it was around this time that some of the protagonists began to think of themselves as materials scientists. Hollomon outlined his own conception of "Materials Science and Engineering"; this indeed was the title of an essay he brought out some years after he had joined GE (Hollomon 1958), and here he explains what kind of creatures he conceived materials scientists and materials engineers to be. John Howe, who worked in the neighbouring Knolls Atomic Power Laboratory at that time, has told me that in the 1950s, he and Hollomon frequently discussed "the need for a broader term as more fundamental concepts were developed" (Howe 1987), and it is quite possible that the new terminology in fact evolved from these discussions at GE. Hollomon concluded his essay: "The professional societies must recognise this new alignment and arrange for its stimulation and for the association of those who practice both the science and engineering of materials. We might even need an American Materials Society with divisions of science and engineering. Metallurgical engineering will become materials engineering. OUT OF METALLURGY, BY PHYSICS, COMES *MATERIALS SCIENCE* (my capitals)." It was to be many years before this prescient advice was heeded; I return to this issue in Chapter 14. Westbrook and Fleischer, two luminaries of the GE Laboratory's golden days, recently dedicated a major book to Hollomon, with the words: "Wise, vigorous, effective advocate of the relevance and value of scientific research in industry" (Westbrook and Fleischer 1995); but a little later still, Fleischer in another book (Fleischer 1998) remarked drily that when Hollomon left the Research Center to take up the directorship of GE's General Engineering Laboratory, he suddenly began saying in public: "Well, we know as much about science as we need. Now is the time to go out and use it". Circumstances alter cases. It is not surprising that as he grew older, Hollomon polarised observers into fierce devotees and implacable opponents, just as though he had been a politician.

Suits and Bueche conclude their case-histories with a superb analysis of the sources, tactics and uses of applied research, and make the comment: "The case histories just summarised show, first of all, the futility of trying to label various elements of the research and development process as 'basic', 'applied' or 'development'. Given almost any definition of these terms, one can find variations or exceptions among the examples."

Hollomon's standing in the national industrial community was recognised in 1955 when the US National Chamber of Commerce chose him as one of the ten outstanding young men in the country. Seven years later, President Kennedy brought Hollomon to Washington as the first Assistant Secretary of Commerce for Science and Technology, where he did such notable things as setting up a President's Commission on the Patent System in order to provide better incentives for overcoming problems in innovation. He showed his scientific background in his habit of answering the question: "What is the problem?" with "90% of the problem is in understanding the problem" (Christenson 1985).

1.1.3 The materials research laboratories

As we have seen, the concept of MSE emerged early in the 1950s and by 1960, it had become firmly established, as the result of a number of decisions in academe and in industry. In that year, as the result of a sustained period of intense discussion and political lobbying in Washington, another major decision was taken, this time by agencies of the US Government. The Interdisciplinary Laboratories were born.

According to recent memoirs by Frederick Seitz (1994) and Sproull (1987), the tortuous negotiations that led to this outcome began in 1954, when the great mathematician and computer theorist, John von Neumann, became '*the* scientist commissioner' of the five-member Atomic Energy Commission (AEC). (This remark presumably means that the other four commissioners were not scientists.) He thereupon invited Seitz to visit him (he had witnessed Seitz's researches in materials science – indeed, Seitz is one of the most eminent progenitors of materials science – during his frequent visits to the University of Illinois) and explained that he "was especially upset that time and time again what he wanted to do was prevented by an inadequate science of materials. When he asked what limited the growth of that science, he was told 'Lack of people'." According to Seitz, von Neumann worried that MSE was being treated as a side issue by the Government, and he proposed that federal agencies, starting with the AEC, join in funding a number of interdisciplinary materials research laboratories at universities. He then asked Seitz to join him in specifically developing a proposal for the protoype laboratory to be set up at the University of Illinois, to be funded at that stage just by the AEC. Clearly in view of his complaint, what von Neumann had in mind was *both* a place where interdisciplinary research on materials would be fostered *and* one where large numbers of new experts would be nurtured.

A formal proposal was developed and submitted, early in 1957, but before this could result in a contract, von Neumann was taken ill and died. Things were then held in abeyance until the launch of the Soviet Sputnik satellite in October 1957 changed everything. Two things then happened: a proposal to fund 12 laboratories emerged in Washington and Charles Yost of the Air Force's Office of Air Research was put in charge of making this happen. Thereupon Donald Stevens, head of the

Metallurgy and Materials Branch of the AEC, who remembered von Neumann's visionary plan for the University of Illinois specifically, set about putting this into effect. Seitz (1994) recounts the almost surrealistic difficulties put in the way of this project by a succession of pork-barrelling Senators; Illinois failed to become one of the three (not twelve, as initially proposed) initial Materials Research Laboratories chosen out of numerous applicants (the first ones were set up at Cornell, Pennsylvania and Northwestern), but in 1962 Illinois did finally acquire an MRL. Sproull (1987) goes into considerable detail concerning the many Government agencies that, under a steady push from Dr. Stevens and Commissioner Willard Libby of the AEC, collaborated in getting the project under way. Amusingly, a formal proposal from Hollomon, in early 1958, that a National Materials Laboratory should be created instead, quickly united everyone behind the original proposal; they all recognised that Hollomon's proposed laboratory would do nothing to enhance the supply of trained materials scientists and engineers.

Some 20 years after the pressure for the creation of the new interdisciplinary laboratories was first felt, one of the academics who became involved very early on, Prof. Rustum Roy of Pennsylvania State University, wrote eloquently about the underlying ideal of interdisciplinarity (Roy 1977). He also emphasised the supportive role played by some influential industrial scientists in that creation, notably Dr. Guy Suits of GE, whom we have already encountered, and Dr. William Baker of Bell Laboratories who was a major force in pushing for interdisciplinary materials research in industry and academe alike. A magisterial survey by Baker (1967), under the title *Solid State Science and Materials Development*, indicates the breadth and scope of his scientific interests.

Administratively, the genesis of these Laboratories, which initially were called Interdisciplinary Research Laboratories and later, Materials Research Laboratories, involved many complications, most of them in Washington, not least when in 1972 responsibility for them was successfully transferred to the National Science Foundation (NSF). As Sproull cynically remarks: "To those unfamiliar with the workings of federal government (and especially Capitol Hill), transfer of a program sounds simple, but it is simple only if the purpose of transfer is to kill the program".

Lyle, in a multiauthor book published by the two National Academies to celebrate the 25th birthday of the MRLs (Psaras and Langford, 1987), gives a great deal of information about their achievements and modus operandi. By then, 17 MRLs had been created, and 7 had either been closed down or were in process of termination. The essential feature of the laboratories was, and is, the close proximity and consequent cooperation between members of many different academic departments, by constructing dedicated buildings in which the participating faculty members had offices as well as laboratories. This did not impede the faculty members' continuing close involvement with their own departments' activities. At the time of the transfer to the NSF, according to Lyle, in 12 MRLs, some 35% were physicists, 25% were chemists, 19% were metallurgists *or members of MSE departments*, 16% were from other engineering disciplines (mainly electrical), and 5% from other departments such as mathematics or earth sciences. In my view, the most significant feature of these statistics is the large percentage of physicists who in this way became intimately involved in the study of materials. This is to be viewed in relation to Sproull's remark (Sproull 1987) that in 1910, "chemistry and metallurgy had already hailed many centuries of contributions to the understanding of materials... but physics' contribution had been nearly zero".

The COSMAT Report of 1974 (a major examination of every aspect of MSE, national and international, organised by the National Academy of Sciences, itself reviewed in 1976 in some depth by Cahn (reprinted 1992), was somewhat critical of the MRLs in that the rate of increase of higher degrees in the traditional metallurgy/materials department was no faster than that of engineering degrees overall. Lyle counters this criticism by concluding that "much of the interdisciplinarity sought in the original... concept was realised through evolutionary changes in the traditional materials departments rather than by dramatic changes in interactions across university departmental lines. This cross-departmental interaction would come only with the group research concept introduced by NSF." The point here is that teaching in the 'traditional' departments, even at undergraduate levels, was deeply influenced by the research done in the MRLs. From the perspective of today, the 37 years, to date, of MRLs can be considered an undiluted good.

1.1.4 Precursors, definitions and terminology

This book is primarily directed at professional materials scientists and engineers, and they have no urgent need to see themselves defined. Indeed, it would be perfectly reasonable to say about materials science what Aaron Katchalsky used to say about his new discipline, biophysics: "Biophysics is like my wife. I know her, but I cannot define her" (Markl 1998). Nevertheless, in this preliminary canter through the early history of MSE, it is instructive to examine briefly how various eminent practitioners have perceived their changing domain.

David Turnbull, in his illuminating *Commentary on the Emergence and Evolution of "Materials Science"* (Turnbull 1983), defined materials science "broadly" as "the characterisation, understanding, and control of the structure of matter at the ultramolecular level and the relating of this structure to properties (mechanical, magnetic, electrical, etc.). That is, it is 'Ultramolecular Science'." In professional and educational practice, however, he says that materials science focuses on the more complex features of behaviour, and especially those aspects controlled by crystal defects. His definition at once betrays Turnbull's origin as a physical chemist. Only a chemist, or possibly a polymer physicist, would focus on molecules when so many important materials have no molecules, as distinct from atoms or ions. Nomenclature in our field is sometimes highly confusing: thus in 1995 a journal began publication under the name *Supramolecular Science*, by which the editor-in-chief means "supramolecular aggregates, assemblies and nanoscopic materials"; that last adjective seems to be a neologism.

The COSMAT Report of 1974, with all its unique group authority, defines MSE as being "concerned with the generation and application of knowledge relating the composition, structure, and processing of materials to their properties and uses". It is probably a fair comment on this simple definition that in the early days of MSE the chief emphasis was on structure and especially structural defects (as evidenced by a famous early symposium proceedings entitled *Imperfections in Nearly Perfect Crystals* (Shockley et al. 1952), while in recent years more and more attention has been paid to the influence of processing variables.

As mentioned above, Sproull (1987) claimed that physics had contributed almost nothing to the understanding of materials before 1910, but went on to say that in the 1930s, books such as Hume-Rothery's *The Structure of Metals and Alloys*, Mott and Jones's *Properties of Metals and Alloys*, and especially Seitz's extremely influential *The Modern Theory of Solids* of 1940, rapidly advanced the science of the solid state and gave investigators a common language and common concepts. Sproull's emphasis was a strongly physical one. Indeed, the statistics given above of disciplinary affiliations in the MRLs show that physicists, after a long period of disdain, eventually leapt into the study of materials with great enthusiasm. Solid-state physics itself had a hard birth in the face of much scepticism from the rest of the physics profession (Mott 1980, Hoddeson et al. 1992). But now, physics has become so closely linked with MSE that at times there have been academic takeover bids from physicists for the entire MSE enterprise... unsuccessful up to now.

Names of disciplines, one might think, are not particularly important: it is the reality that matters. I have already quoted Shakespeare to that effect. But it is not really as simple as that, as the following story from China (Kuo 1996) illustrates. In 1956, my correspondent, an electron microscopist, returned to China after a period in the West and was asked to help in formulating a Twelve-Year Plan of Scientific and Technological Development. At that time, China was overrun by thousands of Soviet experts who were not backward in making suggestions. They advised the"Chinese authorities to educate a large number of scientists in *metallovedenie*, a Russian term which means 'metal-knowledge', close to metallography, itself an antiquated German concept (*Metallographie*) which later converted into *Metallkunde* (what we today call *physical metallurgy* in English). The Russians translated *metallovedenie* into the Chinese term for *metal physics*, since Chinese does not have a word for physical metallurgy. The end-result of this misunderstanding was that in the mid-1960s, the Chinese found that they had far too many metal physicists, all educated in metal physics divisions of physics departments in 17 universities, and a bad lack of "engineers who understand alloys and their heat-treatment", yet it was this last which the Soviet experts had really meant. By that time, Mao had become hostile to the Soviet Union and the Soviet experts were gone. By 1980, only 3 of the original 17 metal physics divisions remained in the universities. An attempt was later made to train students in materials science. In the days when all graduates were still directed to their places of work in China, the "gentleman in the State Planning Department" did not really understand what materials science meant, and was inclined to give materials science graduates "a post in the materials depot".

Although almost the whole of this introductory chapter has been focused on the American experience, because this is where MSE began, later the 'superdiscipline' spread to many countries. In the later chapters of this book, I have been careful to avoid any kind of exclusive focus on the US. The Chinese anecdote shows, albeit in an extreme form, that other countries also were forced to learn from experience and change their modes of education and research. In fact, in most of the rest of this book, the emphasis is on topics and approaches in research, and not on particular places. One thing which is entirely clear is that the pessimists, always among us, who assert that all the really important discoveries in MSE have been made, are wrong: in Turnbull's words at a symposium (Turnbull 1980), "10 or 15 years from now there will be a conference similar to this one where many young enthusiasts, too naive to realize that all the important discoveries have been made, will be describing materials and processes that we, at present, have no inkling of". Indeed, there was and they did.

REFERENCES

Baker, W.O. (1967) *J. Mater.* **2**, 917.
Bever, M.B. (1988) *Metallurgy and Materials Science and Engineering at MIT: 1865–1988* (privately published by the MSE Department).
Cahn, R.W. (1970) *Nature* **225**, 693.
Cahn, R.W. (1992) *Artifice and Artefacts: 100 Essays in Materials Science* (Institute of Physics Publishing, Bristol and Philadelphia) p. 314.
Christenson, G.A. (1985) Address at memorial service for Herbet Hollomon, Boston, 18 May.
COSMAT (1974) *Materials and Man's Needs: Materials Science and Engineering. Summary Report of the Committee on the Survey of Materials Science and Engineering* (National Academy of Sciences, Washington, DC) pp. 1, 39.
Cox, J.A. (1979) *A Century of Light* (Benjamin Company for The General Electric Company, New York).

Fine, M.E. (1990) The First Thirty Years, in *Tech, The Early Years: a History of the Technological Institute at Northwestern University from 1939 to 1969* (privately published by Northwestern University) p. 121.

Fine, M.E. (1994) *Annu. Rev. Mater. Sci.* **24**, 1.

Fine, M.E. (1996) Letter to the author dated 20 March 1996.

Fleischer R.L. (1998) *Tracks to Innovation* (Springer, New York) p. 31.

Frankel, J.P. (1957) *Principles of the Properties of Materials* (McGraw-Hill, New York).

Furukawa, Y. (1998) *Inventing Polymer Science* (University of Pennsylvania Press, Philadelphia).

Gaines, G.L. and Wise, G. (1983) in: *Heterogeneous Catalysis: Selected American Histories. ACS Symposium Series 222* (American Chemical Society, Washington, DC) p. 13.

Harwood, J.J. (1970) Emergence of the field and early hopes, in *Materials Science and Engineering in the United States*, ed. Roy, R. (Pennsylvania State University Press) p. 1.

Hoddeson, L., Braun, E., Teichmann, J. and Weart, S. (editors) (1992) *Out of the Crystal Maze* (Oxford University Press, Oxford).

Hollomon, J.H. (1958) *J. Metals (AIME)*, **10**, 796.

Hounshell, D.A. and Smith, J.K. (1988) *Science and Corporate Strategy: Du Pont R&D, 1902–1980* (Cambridge University Press, Cambridge) pp. 229, 245, 249.

Howe, J.P. (1987) Letters to the author dated 6 January and 24 June 1987.

Kingery, W.D., Bowen, H.K. and Uhlmann, D.R. (1976) *Introduction to Ceramics*, 2nd edition (Wiley, New York).

Kingery, W.D. (1981) in *Grain Boundary Phenomena in Electronic Ceramics*, ed. Levinson, L.M. (American Ceramic Society, Columbus, OH) p. 1.

Kingery, W.D. (1999) Text of an unpublished lecture, *The Changing World of Ceramics 1949–1999*, communicated by the author.

Kuo, K.H. (1996) Letter to the author dated 30 April 1996.

Liebhafsky, H.A. (1974) *William David Coolidge: A Centenarian and his Work* (Wiley-Interscience, New York).

Markl, H. (1998) *European Review* **6**, 333.

Morawetz, H. (1985) *Polymers: The Origins and Growth of a Science* (Wiley-Interscience, New York; republished in a Dover edition, 1995).

Mott, N.F. (organizer) (1980) The Beginnings of Solid State Physics, *Proc. Roy. Soc. (Lond.)* **371**, 1.

Psaras, P.A. and Langford, H.D. (eds.) (1987) *Advancing Materials Research* (National Academy Press, Washington DC) p. 35.

Riordan, M. and Hoddeson, L. (1997) *Crystal Fire: The Birth of the Information Age* (W.W. Norton, New York).

Roy, R. (1977) Interdisciplinary Science on Campus – the Elusive Dream, *Chemical Engineering News*, August.

Seitz, F. (1994) *MRS Bulletin* **19/3**, 60.

Shockley, W., Hollomon, J.H., Maurer, R. and Seitz, F. (editors) (1952) *Imperfections in Nearly Perfect Crystals* (Wiley, New York).

Sproull, R.L. (1987) *Annu. Rev. Mater. Sci.* **17**, 1.

Suits, C.G. and Bueche, A.M. (1967) in *Applied Science and Technological Progress: A Report to the Committee on Science and Astronautics, US House of Representatives, by the National Academy of Sciences* (US Government Printing Office, Washington, DC) p. 297.

Turnbull, D. (1980) in *Laser and Electron Beam Processing of Materials*, ed. White, C.W. and Peercy, P.S. (Academic Press, New York) p. 1.

Turnbull, D. (1983) *Annu. Rev. Mater. Sci.* **13**, 1.

Turnbull, D. (1986) *Autobiography*, unpublished typescript.

Westbrook, J.H. and Fleischer, R.L. (1995) *Intermetallic Compounds: Principles and Practice* (Wiley, Chichester, UK).

Wise, G. (1985) *Willis R. Whitney, General Electric, and the Origins of US Industrial Research* (Columbia University Press, New York).

Chapter 2
The Emergence of Disciplines

Chapter 2
The Emergence of Disciplines

2.1. DRAWING PARALLELS

This entire book is about the emergence, nature and cultivation of a new discipline, materials science and engineering. To draw together the strings of this story, it helps to be clear about what a scientific discipline actually is; that, in turn, becomes clearer if one looks at the emergence of some earlier disciplines which have had more time to reach a condition of maturity. Comparisons can help in definition; we can narrow a vague concept by examining what apparently diverse examples have in common.

John Ziman is a renowned theoretical solid-state physicist who has turned himself into a distinguished metascientist (one who examines the nature and institutions of scientific research in general). In fact, he has successfully switched disciplines. In a lecture delivered in 1995 to the Royal Society of London (Ziman 1996), he has this to say: "Academic science could not function without some sort of internal social structure. This structure is provided by subject specialisation. Academic science is divided into disciplines, each of which is a recognised domain of organised teaching and research. It is practically impossible to be an academic scientist without locating oneself initially in an established discipline. *The fact that disciplines are usually very loosely organised* (my italics) does not make them ineffective. An academic discipline is much more than a conglomerate of university departments, learned societies and scientific journals. It is an 'invisible college', *whose members share a particular research tradition* (my italics). This is where academic scientists acquire the various theoretical paradigms, codes of practice and technical methods that are considered 'good science' in their particular disciplines... A recognised discipline or sub-discipline provides an academic scientist with a home base, a tribal identity, a social stage on which to perform as a researcher." Another attempt to define the concept of a scientific discipline, by the science historian Servos (1990, Preface), is fairly similar, but focuses more on intellectual concerns: "By a discipline, I mean a family-like grouping of individuals sharing intellectual ancestry and united at any given time by an interest in common or overlapping problems, techniques and institutions". These two wordings are probably as close as we can get to the definition of a scientific discipline in general.

The concept of an 'invisible college', mentioned by Ziman, is the creation of Derek de Solla Price, an influential historian of science and "herald of scientometrics" (Yagi et al. 1996), who wrote at length about such colleges and their role in the scientific enterprise (Price 1963, 1986). Price was one of the first to apply quantitative

21

methods to the analysis of publication, reading, citation, preprint distribution and other forms of personal communication among scientists, including 'conference-crawling'. These activities define groups, the members of which, he explains, "seem to have mastered the art of attracting invitations from centres where they can work along with several members of the group for a short time. This done, they move to the next centre and other members. Then they return to home base, but always their allegiance is to the group rather than to the institution which supports them, unless it happens to be a station on such a circuit. For each group there exists a sort of commuting circuit of institutions, research centres, and summer schools giving them an opportunity to meet piecemeal, so that over an interval of a few years everybody who is anybody has worked with everybody else in the same category. Such groups constitute an *invisible college*, in the same sense as did those first unofficial pioneers who later banded together to found the Royal Society in 1660." An invisible college, as Price paints it, is apt to define, not a mature discipline but rather an emergent grouping which may or may not later ripen into a fully blown discipline, and this may happen at breakneck speed, as it did for molecular biology after the nature of DNA had been discovered in 1953, or slowly and deliberately, as has happened with materials science.

There are two particularly difficult problems associated with attempts to map the nature of a new discipline and the timing of its emergence. One is the fierce reluctance of many traditional scientists to accept that a new scientific grouping has any validity, just as within a discipline, a revolutionary new scientific paradigm (Kuhn 1970) meets hostility from the adherents of the established model. The other difficulty is more specific: a new discipline may either be a highly specific breakaway from an established broad field, or it may on the contrary represent a broad synthesis from a number of older, narrower fields: the splitting of physical chemistry away from synthetic organic chemistry in the nineteenth century is an instance of the former, the emergence of materials science as a kind of synthesis from metallurgy, solid-state physics and physical chemistry exemplifies the latter. For brevity, we might name these two alternatives *emergence by splitting* and *emergence by integration*. The objections that are raised against these two kinds of disciplinary creation are apt to be different: emergence by splitting is criticised for breaking up a hard-won intellectual unity, while emergence by integration is criticised as a woolly bridging of hitherto clearcut intellectual distinctions.

Materials science has in its time suffered a great deal of the second type of criticism. Thus Calvert (1997) asserts that "metallurgy remains a proper discipline, with fundamental theories, methods and boundaries. Things fell apart when the subject extended to become materials science, with the growing use of polymers, ceramics, glasses and composites in engineering. The problem is that all materials are different and we no longer have a discipline."

Materials science was, however, not alone in its integrationist ambitions. Thus, Montgomery (1996) recently described his own science, geology, in these terms: "Geology is a magnificent science; a great many phenomenologies of the world fall under its purview. It is unique in defining a realm all its own yet drawing within its borders the knowledge and discourse of so many other fields – physics, chemistry, botany, zoology, astronomy, various types of engineering and more (geologists are at once true 'experts' and hopeless 'generalists')." Just one of these assertions is erroneous: geology is not unique in this respect... materials scientists are both true experts and hopeless generalists in much the same way.

However a new discipline may arrive at its identity, once it has become properly established the corresponding scientific community becomes "extraordinarily tight", in the words of Passmore (1978). He goes on to cite the philosopher Feyerabend, who compared science to a church, closing its ranks against heretics, and substituting for the traditional "outside the church there is no salvation" the new motto "outside my particular science there is no knowledge". The most famous specific example of this is Rutherford's arrogant assertion early in this century: "There's physics... and there's stamp-collecting". This intense pressure towards exclusivity among the devotees of an established discipline has led to a counter-pressure for the emergence of broad, inclusive disciplines by the process of integration, and this has played a major part in the coming of materials science.

In this chapter, I shall try to set the stage for the story of the emergence of materials science by looking at case-histories of some related disciplines. They were all formed by splitting but in due course matured by a process of integration. So, perhaps, the distinction between the two kinds of emergence will prove not to be absolute. My examples are: physical chemistry, chemical engineering and polymer science, with brief asides about colloid science, solid-state physics and chemistry, and mechanics in its various forms.

2.1.1 The emergence of physical chemistry

In the middle of the nineteenth century, there was no such concept as *physical chemistry*. There had long been a discipline of inorganic chemistry (the French call it 'mineral chemistry'), concerned with the formation and properties of a great variety of acids, bases and salts. Concepts such as equivalent weights and, in due course, valency very slowly developed. In distinction to (and increasingly in opposition to) inorganic chemistry was the burgeoning discipline of organic chemistry. The very name implied the early belief that compounds of interest to organic chemists, made up of carbon, hydrogen and oxygen primarily, were the exclusive domain of living matter, in the sense that such compounds could only be synthesised by living organisms. This notion was eventually disproved by the celebrated synthesis of urea, but by this time the name, organic chemistry, was firmly established. In fact, the term has been in use for nearly two centuries.

Organic and inorganic chemists came into ever increasing conflict throughout the nineteenth century, and indeed as recently as 1969 an eminent British chemist was quoted as asserting that "inorganic chemistry is a ridiculous field". This quotation comes from an admirably clear historical treatment, by Colin Russell, of the progress of the conflict, in the form of a teaching unit of the Open University in England (Russell 1976). The organic chemists became ever more firmly focused on the synthesis of new compounds and their compositional analysis. Understanding of what was going on was bedevilled by a number of confusions, for instance, between gaseous atoms and molecules, the absence of such concepts as stereochemistry and isomerism, and a lack of understanding of the nature of chemical affinity. More important, there was no agreed atomic theory, and even more serious, there was uncertainty surrounding atomic weights, especially those of 'inorganic' elements. In 1860, what may have been the first international scientific conference was organised in Karlsruhe by the German chemist August Kekulé (1829–1896 – he who later, in 1865, conceived the benzene ring); some 140 chemists came, and spent most of their time quarrelling. One participant was an Italian chemist, Stanislao Cannizzaro (1826–1910) who had rediscovered his countryman Avogadro's Hypothesis (originally proposed in 1811 and promptly forgotten); that Hypothesis (it deserves its capital letter!) cleared the way for a clear distinction between, for instance, H and H_2. Cannizzaro eloquently pleaded Avogadro's cause at the Karlsruhe conference and distributed a pamphlet he had brought with him (the first scattering of reprints at a scientific conference, perhaps); this pamphlet finally convinced the numerous waverers of the rightness of Avogadro's ideas, ideas which we all learn in school nowadays.

This thumbnail sketch of where chemistry had got to by 1860 is offered here to indicate that chemists were mostly incurious about such matters as the nature and strength of the chemical bond or how quickly reactions happened; all their efforts went into methods of synthesis and the tricky attempts to determine the numbers of different atoms in a newly synthesised compound. The standoff between organic and inorganic chemistry did not help the development of the subject, although by the time of the Karlsruhe Conference in 1860, in Germany at least, the organic synthetic chemists ruled the roost.

Early in the 19th century, there were giants of natural philosophy, such as Dalton, Davy and most especially Faraday, who would have defied attempts to categorise them as physicists or chemists, but by the late century, the sheer mass of accumulated information was such that chemists felt they could not afford to dabble in physics, or vice versa, for fear of being thought dilettantes.

In 1877, a man graduated in chemistry who was not afraid of being thought a dilettante. This was the German Wilhelm Ostwald (1853–1932). He graduated with a master's degree in chemistry in Dorpat, a "remote outpost of German scholarship in Russia's Baltic provinces", to quote a superb historical survey by Servos (1990); Dorpat, now called Tartu, is in what has become Latvia, and its disproportionate role in 19th-century science has recently been surveyed (Siilivask 1998). Ostwald was a man of broad interests, and as a student of chemistry, he devoted much time to literature, music and painting – an ideal student, many would say today. During his master's examination, Ostwald asserted that "modern chemistry is in need of reform". Again, in Servos's words, "Ostwald's blunt assertion... appears as an early sign of the urgent and driving desire to reshape his environment, intellectual and institutional, that ran as an extended motif through his career... He sought to redirect chemists' attention from the substances participating in chemical reactions to the reactions themselves. Ostwald thought that chemists had long overemphasised the taxonomic aspects of their science by focusing too narrowly upon the composition, structure and properties of the species involved in chemical processes... For all its success, the taxonomic approach to chemistry left questions relating to the rate, direction and yield of chemical reactions unanswered. To resolve these questions and to promote chemistry from the ranks of the descriptive to the company of the analytical sciences, Ostwald believed chemists would have to study the conditions under which compounds formed and decomposed and pay attention to the problems of chemical affinity and equilibrium, mass action and reaction velocity. The arrow or equal sign in chemical equations must, he thought, become chemists' principal object of investigation."

For some years he remained in his remote outpost, tinkering with ideas of chemical affinity, and with only a single research student to assist him. Then, in 1887, at the young age of 34, he was offered a chair in chemistry at the University of Leipzig, one of the powerhouses of German research, and his life changed utterly. He called his institute (as the Germans call academic departments) by the name of 'general chemistry' initially; the name 'physical chemistry' came a little later, and by the late 1890s was in very widespread use. Ostwald's was however only the Second Institute of Chemistry in Leipzig; the First Institute was devoted to organic chemistry, Ostwald's bête noire. Physics was required for the realisation of his objectives because, as Ostwald perceived matters, physics had developed beyond the descriptive stage to the stage of determining the general laws to which phenomena were subject; chemistry, he thought, had not yet attained this crucial stage. Ostwald would have sympathised with Rutherford's gibe about physics and stamp-collecting. It is ironic that Rutherford received a Nobel Prize in *Chemistry* for his researches on radioactivity. Ostwald himself also received the Nobel Prize for Chemistry, in 1909, nominally at least for his work in catalysis, although his founding work in physical chemistry was on the law of mass action. (It would be a while before the Swedish

Academy of Sciences felt confident enough to award a chemistry prize overtly for prowess in physical chemistry, upstart that it was.)

Servos gives a beautifully clear explanation of the subject-matter of physical chemistry, as Ostwald pursued it. Another excellent recent book on the evolution of physical chemistry, by Laidler (1993) is more guarded in its attempts at definition. He says that "it can be defined as that part of chemistry that is done using the methods of physics, or that part of physics that is concerned with chemistry, i.e., with specific chemical substances", and goes on to say that it cannot be precisely defined, but that he can recognise it when he sees it! Laidler's attempt at a definition is not entirely satisfactory, since Ostwald's objective was to get away from insights which were specific to individual substances and to attempt to establish laws which were general.

About the time that Ostwald moved to Leipzig, he established contact with two scientists who are regarded today as the other founding fathers of physical chemistry: a Dutchman, Jacobus van 't Hoff (1852–1911) and a Swede, Svante Arrhenius (1859–1927). Some historians would include Robert Bunsen (1811–1899) among the founding fathers, but he was really concerned with experimental techniques, not with chemical theory.

Van't Hoff began as an organic chemist. By the time he had obtained his doctorate, in 1874, he had already published what became a very famous pamphlet on the 'tetrahedral carbon atom' which gave rise to modern organic stereochemistry. After this he moved, first to Utrecht, then to Amsterdam and later to Berlin; from 1878, he embarked on researches in physical chemistry, specifically on reaction dynamics, on osmotic pressure in solutions and on polymorphism (van't Hoff 1901), and in 1901 he was awarded the first Nobel Prize in chemistry. The fact that he was the first of the trio to receive the Nobel Prize accords with the general judgment today that he was the most distinguished and original scientist of the three.

Arrhenius, insofar as his profession could be defined at all, began as a physicist. He worked with a physics professor in Stockholm and presented a thesis on the electrical conductivities of aqueous solutions of salts. A recent biography (Crawford 1996) presents in detail the humiliating treatment of Arrhenius by his sceptical examiners in 1884, which nearly put an end to his scientific career; he was not adjudged fit for a university career. He was not the last innovator to have trouble with examiners. Yet, a bare 19 years later, in 1903, he received the Nobel Prize for Chemistry. It shows the unusual attitude of this founder of physical chemistry that he was distinctly surprised not to receive the Physics Prize, because he thought of himself as a physicist.

Arrhenius's great achievement in his youth was the recognition and proof of the notion that the constituent atoms of salts, when dissolved in water, dissociated into charged forms which duly came to be called *ions*. This insight emerged from laborious and systematic work on the electrical conductivity of such solutions as they were progressively diluted: it was a measure of the 'physical' approach of this research that although the absolute conductivity decreases on dilution, the molecular conductivity goes up... i.e., each dissolved atom or ion becomes more efficient on average in conducting electricity. Arrhenius also recognised that no current was needed to promote ionic dissociation. These insights, obvious as they seem to us now, required enormous originality at the time.

It was Arrhenius's work on ionic dissociation that brought him into close association with Ostwald, and made his name; Ostwald at once accepted his ideas and fostered his career. Arrhenius and Ostwald together founded what an amused German chemist called "the wild army of ionists"; they were so named because (Crawford 1996) "they believed that chemical reactions in solution involve only ions and not dissociated molecules", and thereby the ionists became "the Cossacks of the movement to reform German chemistry, making it more analytical and scientific". The ionists generated extensive hostility among some – but by no means all – chemists, both in Europe and later in America, when Ostwald's ideas migrated there in the brains of his many American research students (many of whom had been attracted to him in the first place by his influential textbook, *Lehrbuch der Allgemeinen Chemie*).

Later, in the 1890s, Arrhenius moved to quite different concerns, but it is intriguing that materials scientists today do not think of him in terms of the concept of ions (which are so familiar that few are concerned about who first thought up the concept), but rather venerate him for the *Arrhenius equation* for the rate of a chemical reaction (Arrhenius 1889), with its universally familiar exponential temperature dependence. That equation was in fact first proposed by van't Hoff, but Arrhenius claimed that van't Hoff's derivation was not watertight and so it is now called after Arrhenius rather than van't Hoff (who was in any case an almost pathologically modest and retiring man).

Another notable scientist who embraced the study of ions in solution – he oscillated so much between physics and chemistry that it is hard to say where his prime loyalty belonged – was Walther Nernst, who in the way typical of German students in the 19th century wandered from university to university (Zürich, Berlin, Graz, Würzburg), picking up Boltzmann's ideas about statistical mechanics and chemical thermodynamics on the way, until he fell, in 1887, under Ostwald's spell and was invited to join him in Leipzig. Nernst fastened on the theory of electrochemistry as the key theme for his research and in due course he brought out a precocious book entitled *Theoretische Chemie*. His world is painted, together with acute sketch-portraits of Ostwald, Arrhenius, Boltzmann and other key figures of physical chemistry, by Mendelssohn (1973). We shall meet Nernst again in Section 9.3.2.

During the early years of physical chemistry, Ostwald did not believe in the existence of atoms... and yet he was somehow included in the wild army of ionists. He was resolute in his scepticism and in the 1890s he sustained an obscure theory of 'energetics' to take the place of the atomic hypothesis. How ions could be formed in a solution containing no atoms was not altogether clear. Finally, in 1905, when Einstein had shown in rigorous detail how the Brownian motion studied by Perrin could be interpreted in terms of the collision of dust motes with moving molecules (Chapter 3, Section 3.1.1), Ostwald relented and publicly embraced the existence of atoms.

In Britain, the teaching of the ionists was met with furious opposition among both chemists and physicists, as recounted by Dolby (1976a) in an article entitled "Debate on the Theory of Solutions – A Study of Dissent" and also in a book chapter (Dolby 1976b). A rearguard action continued for a long time. Thus, Dolby (1976a) cites an eminent British chemist, Henry Armstrong (1848–1937) as declaring, as late as 4 years after Ostwald's death (Armstrong 1936), that "the fact is, there has been a split of chemists into two schools since the intrusion of the Arrhenian faith... a new class of workers into our profession – people without knowledge of the laboratory and with sufficient mathematics at their command to be led astray by curvilinear agreements." It had been nearly 50 years before, in 1888–1898, that Armstrong first tangled with the ionists' ideas and, as Dolby comments, he was "an extreme individualist, who would never yield to the social pressures of a scientific community or follow scientific trends". The British physicist F.G. Fitzgerald, according to Servos, "suspected the ionists of practising physics without a licence". Every new discipline encounters resolute foes like Armstrong and Fitzgerald; materials science was no exception.

In the United States, physical chemistry grew directly through the influence of Ostwald's 44 American students, such as Willis Whitney who founded America's first industrial research laboratory for General Electric (Wise 1985) and, in the same laboratory, the Nobel prizewinner Irving Langmuir (who began his education as a metallurgist and went on to undertake research in the physical chemistry of gases and surfaces which was to have a profound effect on industrial innovation, especially of incandescent lamps). The influence of these two and others at GE was also outlined by the industrial historian Wise (1983) in an essay entitled "Ionists in Industry: Physical Chemistry at General Electric, 1900–1915". In passing, Wise here remarks: "Ionists could accept the atomic hypothesis, and some did; but they did not have to". According to Wise, "to these pioneers, an ion was not a mere incomplete atom, as it later became for scientists". The path to understanding is usually long and tortuous. The stages of American acceptance of the new discipline is also a main theme of Servos's (1990) historical study.

Two marks of the acceptance of the new discipline, physical chemistry, in the early 20th century were the Nobel prizes for its three founders and enthusiastic industrial approval in America. A third test is of course the recognition of a discipline in universities. Ostwald's institute carried the name of physical chemistry well before the end of the 19th century. In America, the great chemist William Noyes (1866–1936), yet another of Ostwald's students, battled hard for many years to establish physical chemistry at MIT which at the turn of the century was not greatly noted for its interest in fundamental research. As Servos recounts in considerable detail, Noyes had to inject his own money into MIT to get a graduate school of physical chemistry established. In the end, exhausted by his struggle, in 1919 he left MIT and moved west to California to establish physical chemistry there, jointly with such giants as Gilbert Lewis (1875–1946). When Noyes moved to Pasadena, as Servos puts it, California was as well known for its science as New England was for growing oranges; this did not take long to change. In America, the name of an academic department is secondary; it is the creation of a research (graduate) school that defines the acceptance of a discipline. In Europe, departmental names are more important, and physical chemistry departments were created in a number of major universities such as for instance Cambridge and Bristol; in others, chemistry departments were divided into a number of subdepartments, physical chemistry included. By the interwar period, physical chemistry was firmly established in European as well as American universities.

Another test of the acceptance of a new discipline is the successful establishment of new journals devoted to it, following the gradual incursion of that discipline into existing journals. The leading American chemical journal has long been the *Journal of the American Chemical Society*. According to Servos, in the key year 1896 only 5% of the articles in *JACS* were devoted to physical chemistry; 10 years later this had increased to 15% and by the mid 1920s, to more than 25%. The first journal devoted to physical chemistry was founded in Germany by Ostwald in 1887, the year he moved to his power base in Leipzig. The journal's initial title was *Zeitschrift für physikalische Chemie, Stöchiometrie und Verwandtschaftslehre* (the last word means 'lore of relationships'), and a portrait of Bunsen decorated its first title page.

Nine years later, the *Zeitschrift für physikalische Chemie* was followed by the *Journal of Physical Chemistry*, founded in the USA by Wilder Bancroft (1867–1953), one of Ostwald's American students. The 'chequered career' of this journal is instructively analysed by both Laidler (1993) and Servos (1990). Bancroft (who spent more than half a century at Cornell University) seems to have been a difficult man, with an eccentric sense of humour; thus at a Ph.D. oral examination he asked the candidate "What in water puts out fires?", and after rejecting some of the answers the student gave with increasing desperation, Bancroft revealed that the right answer was 'a fireboat'. Any scientific author will recognize that this is not the ideal way for a journal editor to behave, let alone an examiner. There is no space here to go into the vagaries of Bancroft's personality (Laidler can be consulted about this), but many American physical chemists, Noyes among them, were so incensed by him and his editorial judgment that they boycotted his journal. It ran into financial problems; for a while it was supported from Bancroft's own ample means, but the end of the financial road was reached in 1932 when he had to resign as editor and the journal was taken over by the American Chemical Society. In Laidler's words, "the various negotiations and discussions that led to the wresting of the editorship from Bancroft also led to the founding of an important new journal, the *Journal of Chemical Physics*, which appeared in 1933". It was initially edited by Harold Urey (1893–1981) who promptly received the Nobel Prize for Chemistry in 1934 for his isolation of deuterium (it might just as well have been the physics prize). Urey remarked at the time that publication in the *Journal of Physical Chemistry* was "burial without a tombstone" since so few *physicists* read it. The new journal also received strong support from the ACS, in spite of (or because of?) the fact that it was aimed at physicists.

These two journals, devoted to *physical chemistry* and *chemical physics*, have continued to flourish peaceably side by side until the present day. I have asked expert colleagues to define for me the difference in the reach of these two fields, but most of them asked to be excused. One believes that chemical physics was introduced when quantum theory first began to influence the understanding of the chemical bond and of chemical processes, as a means of ensuring proper attention to quantum mechanics among chemists. It is clear that many eminent practitioners read and publish impartially in both journals. The evidence suggests that *JCP* was founded in 1933 because of despair about the declining standards of *JPC*. Those standards soon recovered after the change of editor, but a new journal launched with hope and fanfare does not readily disappear and so *JCP* sailed on. The inside front page of *JCP* carries this message: "The purpose of the *JCP* is to bridge a gap between the journals of physics and journals of chemistry. The artificial boundaries between physics and chemistry have now been in actual fact completely eliminated, and a large and active group is engaged in research which is as much the one as the other. It is to this group that the journal is rendering its principal service...".

One of the papers published in the first issue of *JCP*, by F.G. Foote and E.R. Jette, was devoted to the defect structure of FeO and is widely regarded as a classic. Frank Foote (1906–1998), a metallurgist, later became renowned for his contribution to the Manhattan Project and to nuclear metallurgy generally; so chemical physics certainly did not exclude metallurgy.

It is to be noted that 'chemical physics', its own journal apart, does not carry most of the other trappings of a recognised discipline, such as university departments bearing that name. It is probably enough to suggest that those who want to be thought of as chemists publish in *JPC* and those who prefer to be regarded as physicists, in *JCP* (together with a few who are neither physicists nor chemists).

But I am informed that theoretical *chemists* tend to prefer *JCP*. The path of the generaliser is a difficult one.

The final stage in the strange history of physical chemistry and chemical physics is the emergence of a new journal in 1999. This is called *PCCP*, and its subtitle is: *Physical Chemistry Chemical Physics: A Journal of the European Chemical Societies*. *PCCP*, we are told "represents the fusion of two long-established journals, *Faraday Transactions* and *Berichte der Bunsen-Gesellschaft* – the respective physical chemistry journals of the Royal Society of Chemistry (UK) and the Deutsche Bunsen-Gesellschaft für Physikalische Chemie...". Several other European chemical societies are also involved in the new journal. There is a 'college' of 12 editors. This development appears to herald the re-uniting of two sisterly disciplines after 66 years of separation.

One other journal which has played a key part in the recognition and development of physical chemistry needs to be mentioned; in fact, it is one of the precursors of the new *PCCP*. In 1903, the Faraday Society was founded in London. Its stated object was to "promote the study of electrochemistry, electrometallurgy, chemical physics, metallography and kindred subjects". In 1905, the *Transactions of the Faraday Society* began publication. Although 'physical chemistry' was not mentioned in the quoted objective, yet the Transactions have always carried a hefty dose of physical chemistry. The journal included the occasional reports of 'Faraday Discussions', special occasions for which all the papers are published in advance so that the meeting can concentrate wholly on intensive debate. From 1947, these *Faraday Discussions* have been published as a separate series; some have become famous in their own right, such as the 1949 and 1993 *Discussions on Crystal Growth*. Recently, the 100th volume (Faraday Division 1995) was devoted to a *Celebration of Physical Chemistry*, including a riveting account by John Polanyi of "How discoveries are made, and why it matters".

Servos had this to say about the emergence of physical chemistry: "Born out of revolt against the disciplinary structure of the physical sciences in the late 19th century, it (physical chemistry) soon acquired all the trappings of a discipline itself. Taking form in the 1880s, it grew explosively until, by 1930, it had given rise to a half-dozen or more specialities. . ." – the perfect illustration of *emergence by splitting*, twice over. Yet none of these subsidiary specialities have achieved the status of fullblown disciplines, and physical chemistry – with chemical physics, its alter ego – has become an umbrella field taking under its shelter a great variety of scientific activities.

There is yet another test of the acceptance of a would-be new discipline, and that is the publication of textbooks devoted to the subject. By this test, physical chemistry took a long time to 'arrive'. One distinguished physical chemist has written an autobiography (Johnson 1996) in which he says of his final year's study for a chemistry degree in Cambridge in 1937: "Unfortunately at this time, there was no textbook (in English) in general physical chemistry available so that to a large extent it was necessary to look up the original scientific papers referred to in the lectures. In many ways this was good practice though it was time-consuming." In 1940 this lack was at last rectified; it took more than half a century after the founding of the first journal in physical chemistry before the new discipline was codified in a comprehensive English-language text (Glasstone 1940).

So, physical chemistry has developed far beyond the vision of its three famous founders. But then, the great mathematician A.N. Whitehead once remarked that "a science which hesitates to forget its founders is lost"; he meant that it is dangerous to refuse to venture in new directions. Neither physical chemistry nor materials science has ever been guilty of such a refusal.

2.1.2 The origins of chemical engineering

Chemical engineering, as a tentative discipline, began at about the same time as did physical chemistry, in the 1880s, but it took rather longer to become properly established. In fact, the earliest systematic attempt to develop a branch of engineering focused on the large-scale manufacture of industrial chemicals took place at Boston Tech, the precursor of the Massachusetts Institute of Technology, MIT. According to a recent account of the early history of chemical engineering (Cohen 1996), the earliest course in the United States to be given the title 'chemical engineering' was organized and offered by Lewis Norton at Boston Tech in 1888. Norton, like so many other Americans, had taken a doctorate in chemistry in Germany. It is noteworthy that the first hints of the new discipline came in the form of a university teaching course and not, as with physical chemistry, in the form of a research programme. In that difference lay the source of an increasingly bitter quarrel between the chemical engineers and the physical chemists at Boston Tech, just about the time it became MIT.

Norton's course combined a "rather thorough curriculum in mechanical engineering with a fair background in general, theoretical and applied chemistry". Norton died young and the struggling chemical engineering course, which was under the tutelage of the chemistry department until 1921, came in due course under the aegis of William Walker, yet another German-trained American chemist who had established a lucrative sideline as a consulting chemist to industry. From the beginning of the 1900s, an irreconcilable difference in objectives built up in the Chemistry Department, between two factions headed by Arthur Noyes (see Section 2.1.1) and William Walker. Their quarrels are memorably described in Servos's book (1990). The issue was put by Servos in these words: "Should MIT broaden its goals by becoming a science-based university (which it scarcely was in 1900) with a graduate school oriented towards basic research and an undergraduate curriculum rooted in the fundamental sciences? Or should it reaffirm its heritage by focusing on the training of engineers and cultivating work in the applied sciences? Was basic science to be a means towards an end, or should it become an end in itself?" This neatly encapsulates an undying dispute in the academic world; it is one that cannot be ultimately resolved because right is on both sides, but the passage of time gradually attenuates the disagreement.

Noyes struggled to build up research in physical chemistry, even, as we have seen, putting his own personal funds into the endeavour, and Walker's insistence on focusing on industrial case-histories, cost analyses and, more generally, enabling students to master production by the ton rather than by the test tube, was wormwood and gall to Noyes. Nevertheless, Walker's resolute industry-centred approach brought ever-increasing student numbers to the chemical engineering programme (there was a sevenfold increase over 20 years), and so Noyes's influence waned and Walker's grew, until in desperation, as we have seen, Noyes went off to the California Institute of Technology. That was another academic institution which had begun as an obscure local 'Tech' and under the leadership of a succession of pure scientists it forged ahead in the art of merging the fundamental with the practical. The founders of MSE had to cope with the same kinds of forceful disagreements as did Noyes and Walker.

The peculiar innovation which characterised university courses from an early stage was the concept of *unit operations*, coined by Arthur Little at MIT in 1916. In Cohen's (1996) words, these are "specific processes (usually involving physical, rather than chemical change) which were common throughout the chemical industry. Examples are heating and cooling of fluids, distillation, crystallisation, filtration, pulverisation and so forth." Walker introduced unit operations into his course at MIT in 1905 (though not yet under that name), and later he, with coauthors, presented them in an influential textbook. Of the several advantages of this concept listed by Cohen, the most intriguing is the idea that, because unit operations were so general, they constituted a system which a consultant could use throughout the chemical industry without breaking his clients' confidences. Walker, and other chemical engineers in universities, introduced unit operations because of their practical orientation, but as Cohen explains, over the years a largely empirical treatment of processes was replaced by an ever more analytical and science-based approach. The force of circumstance and the advance in insight set at naught the vicious quarrel between the practical men and the worshippers of fundamental science.

Chemical engineering, like every other new discipline, also encountered discord as to its name: terms like 'industrial chemistry' or 'chemical technology' were widely used and this in turn led to serious objections from existing bodies when the need arose to establish new professional organisations. For instance, in Britain the Society for Chemical Industry powerfully opposed the creation of a specialised institution for chemical engineers. There is no space to detail here the involved minuets which took place in connection with the British and American Institutes of Chemical Engineering; Cohen's essay should be consulted for particulars.

The science/engineering standoff in connection with chemical engineering education was moderated in Britain because of a remarkable initiative that took place in Cambridge, England. Just after the War, in 1945, Shell, the oil and petrochemicals giant, gave a generous benefaction to Cambridge University to create a department of chemical engineering. The department was headed by a perfectionist mechanical engineer, Terence Fox (1912–1962)[1], who brought in many chemists, physical chemists in particular. One physical chemist, Peter Danckwerts (1916–1984), was sent away to MIT to learn some chemical engineering and later, in 1959, became a famous department head in his turn. (This was an echo of an early Cambridge professor of chemistry in the unregenerate days of the university in the 18th century, a priest who was sent off to the Continent to learn a little chemistry.) The unusual feature in Cambridge chemical engineering was that students could enter the department either after 2 years' initial study in engineering or alternatively after 2 years study in the natural sciences, including chemistry. Either way, they received the same specialist tuition once they started chemical engineering. This has worked well; according to an early staff member (Harrison 1996), 80–90% of chemical engineering students have always come by the 'science route'. This experience shows that science and engineering outlooks can coexist in fruitful harmony.

It is significant that the Cambridge benefaction came from the petroleum industry. In the early days of chemical engineering education, pioneered in Britain in Imperial College and University College in London, graduates had great difficulty in finding acceptance in the heavy chemicals industry, especially Imperial Chemical Industries, which reckoned that chemists could do everything needful. Chemical engineering graduates were however readily accepted by the oil industry, especially when refineries began at last to be built in Britain from 1953 onwards (Warner 1996). Indeed, one British university (Birmingham) created a department of oil engineering and later converted it to chemical engineering. Warner (1996) believes that chemists held in contempt the forcible breakdown of petroleum constituents before they were put together again into larger molecules, because this was so different from the classical methods of synthesis of complex organic molecules. So the standoff between

---

[1] Fox's perfectionism is illustrated by an anecdote: At a meeting held at ICI (his previous employer), Fox presented his final design for a two-mile cable transporter. Suddenly he clapped his hand to his head and exclaimed: "How *could* I have made such an error!" Then he explained to his alarmed colleagues: "I forgot to allow for the curvature of the Earth".

organic and physical chemists finds an echo in the early hostility between organic chemists and petroleum technologists. Other early chemical engineers went into the explosives industry and, especially, into atomic energy.

It took much longer for chemical engineering, as a technological profession, to find general acceptance, than it took for physical chemistry to become accepted as a valid field of research. Finally it was achieved. The second edition of the great Oxford English Dictionary, which is constructed on historical principles, cites an article in a technical journal published in 1957: "Chemical engineering is now recognized as one of the four primary technologies, alongside civil, mechanical and electrical engineering".

2.1.3 Polymer science

In 1980, Alexander Todd, at that time President of the Royal Society of Chemistry in London, was asked what had been chemistry's biggest contribution to society. He thought that despite all the marvellous medical advances, chemistry's biggest contribution was the development of polymerisation, according to the preface of a recent book devoted to the history of high-technology polymers (Seymour and Kirshenbaum 1986). I turn now to the stages of that development and the scientific insights that accompanied it.

During the 19th century chemists concentrated hard on the global composition of compounds and slowly felt their way towards the concepts of stereochemistry and one of its consequences, optical isomerism. It was van 't Hoff in 1874, at the age of 22, who proposed that a carbon atom carries its 4 valencies (the existence of which had been recognized by August Kekulé (1829–1896) in a famous 1858 paper) directed towards the vertices of a regular tetrahedron, and it was that recognition which really stimulated chemists to propose *structural* formulae for organic compounds. But well before this very major step had been taken, the great Swedish chemist Jöns Jacob Berzelius (1779–1848), stimulated by some comparative compositional analyses of butene and ethylene published by Michael Faraday, had proposed in 1832 that "substances of equal composition but different properties be called *isomers*". The following year he suggested that when two compounds had the same relative composition but different absolute numbers of atoms in each molecule, the larger one be called *polymeric*. These two terms are constructed from the Greek roots *mer* (a part), *iso* (same) and *poly* (many).

The term 'polymer' was slow in finding acceptance, and the concept it represented, even slower. The French chemist Marcellin Berthelot (1827–1907) used it in the 1860s for what we would now call an *oligomer* (*oligo* = few), a molecule made by assembling just 2 or 3 monomers into a slightly larger molecule; the use of the term to denote long-chain (macro-) molecules was delayed by many years. In a lecture he delivered in 1863, Berthelot was the first to discuss polymerisation (actually, oligomerisation) in some chemical detail.

Van't Hoff's genial insight showed that a carbon atom bonded to chemically distinct groups would be asymmetric and, depending on how the groups were disposed in space, the consequent compound should show optical activity – that is, when dissolved in a liquid it would rotate the plane of polarisation of plane-polarised light. Louis Pasteur (1822–1895), in a famously precocious study, had discovered such optical activity in tartrates as early as 1850, but it took another 24 years before van't Hoff recognized the causal linkage between optical rotation and molecular structure, and showed that laevorotary and dextrorotary tartrates were *stereoisomers*: they had structures related by reflection. Three-dimensional molecular structure interested very few chemists in this period, and indeed van't Hoff had to put up with some virulent attacks from sceptical colleagues, notably from Berthelot who, as well as being a scientist of great energy and ingenuity, was also something of an intellectual tyrant who could never admit to being wrong (Jacques 1987). It was thus natural that he spent some years in politics as foreign minister and minister of education.

These early studies opened the path to the later recognition of steroisomerism in polymers, which proved to be an absolutely central concept in the science of polymers.

These historical stages are derived from a brilliant historical study of polymer science, by Morawetz (1985, 1995). This is focused strongly on the organic and physical chemistry of macromolecules. The corresponding technology, and its close linkage to the chemistry and stereochemistry of polymerisation, is treated in other books, for instance those by McMillan (1979), Liebhafsky et al. (1978), and Mossman and Morris (1994), as well as the previously mentioned book by Seymour and Kirshenbaum (1986).

Once stereochemistry had become orthodox, the chemistry of monomers, oligomers and polymers could at length move ahead. This happened very slowly in the remainder of the 19th century, although the first industrial plastics (based on natural products which were already polymerised), like celluloid and viscose rayon, were produced in the closing years of the century without benefit of detailed chemical understanding (Mossman and Morris 1994). Much effort went into attempts to understand the structure of natural rubber, especially after the discovery of vulcanisation by Charles Goodyear in 1855: rubber was broken down into constituents (devulcanised, in effect) and then many attempted to re-polymerise the monomer isoprene, with very indifferent success until O. Wallach, in 1887, succeeded in doing so with the aid of illumination – photopolymerisation. It was not till 1897 that a German chemist, C. Engler, recognised that "one need not assume that only similar molecules assemble" – the first hint that copolymers (like future synthetic rubbers) were a possibility in principle.

Rubber was only one of the many natural macromolecules which were first studied in the nineteenth century. This study was accompanied by a growing revolt among organic chemists against the notion that polymerised products really consisted of long chains with (inevitably) varying molecular weights. For the organic chemists, the holy grail was a well defined molecule of known and constant composition, molecular weight, melting-point, etc., usually purified by distillation or crystallisation, and those processes could not usually be applied to polymers. Since there were at that time no *reliable* methods for determining large molecular weights, it was hard to counter this resolute scepticism. One chemist, O. Zinoffsky, in 1886 found a highly ingenious way of proving that molecular weights of several thousands did after all exist. He determined an empirical formula of $C_{712}H_{1130}N_{214}S_2Fe_1O_{245}$ for haemoglobin. Since a molecule could not very well contain only a fraction of one iron atom, this empirical formula also represented the smallest possible size of the haemoglobin molecule, of weight 16,700. A molecule like haemoglobin was one thing, and just about acceptable to sceptical organic chemists: after all, it had a constant molecular weight, unlike the situation that the new chemists were suggesting for synthetic long-chain molecules.

At the end of the nineteenth century, there was one active branch of chemistry, the study of colloids, which stood in the way of the development of polymer chemistry. Colloid science will feature in Section 2.1.4; suffice it to say here that students of colloids, a family of materials like the glues which gave colloids their name, perceived them as small particles or micelles each consisting of several molecules. Such particles were supposed to be held together internally by weak, "secondary valences" (today we would call these van der Waals forces), and it became an article of orthodoxy that supposed macromolecules were actually micelles held together by weak forces and were called 'association colloids'. (Another view was that some polymers consisted of short closed-ring structures.) As Morawetz puts it, "there was almost universal conviction that large particles must be considered aggregates"; even the great physical chemist Arthur Noyes publicly endorsed this view in 1904. Wolfgang Ostwald (1886–1943), the son of Wilhelm Ostwald, was the leading exponent of colloid science and the ringleader of the many who scoffed at the idea that any long-chain molecules existed. Much of the early work on polymers was published in the *Kolloid-Zeitschrift*.

There was one German chemist, Hermann Staudinger (1881–1965), at one time a colleague of the above-mentioned Engler who had predicted copolymerisation, who was the central and obstinate proponent of the reality of long-chain molecules held together by covalent bonds. He first announced this conviction in a lecture in 1917 to the Swiss Chemical Society. He referred to "high-molecular compounds" from which later the term "high polymers" was coined to denote very long chains. Until he was 39, Staudinger practised conventional organic chemistry. Then he switched universities, returning from Switzerland to Freiburg in Germany, and resolved to devote the rest of his long active scientific life to macromolecules, especially to synthetic ones. As Flory puts it in the historical introduction to his celebrated polymer textbook of 1953, Staudinger showed that "in contrast to association colloids, high polymers exhibit colloidal properties in all solvents in which they dissolve" – in other words, they had *stable* molecules of large size.

At the end of the 1920s, Staudinger also joined a group of other scientists in Germany who began to apply the new technique of X-ray diffraction to polymers, notably Herman Mark (1895–1992) who was to achieve great fame as one of the fathers of modern polymer science (he was an Austrian who made his greatest contributions in America and anglicised his first name). One of the great achievements of this group was to show that natural rubber (which was amorphous or glasslike) could be crystallised by stretching; so polymers were after all not incapable of crystallising, which made rubber slightly more respectable in the eyes of the opponents of long chains. Staudinger devoted much time to the study of poly(oxymethylenes), and showed that it was possible to crystallise some of them (one of the organic chemists' criteria for 'real' chemical compounds). He showed that his crystalline poly(oxymethylene) chains, and other polymers too, were far too long to fit into one unit cell of the crystal structures revealed by X-ray diffraction, and concluded that the chains could terminate anywhere in a crystal after meandering through several unit cells. This, once again, was a red rag to the organic bulls, but finally in 1930, a meeting of the Kolloid-Gesellschaft, in Morawetz's words, "clearly signified the victory of the concept of long-chain molecules". *The consensus is that this fruitless battle, between the proponents of long-chain molecules and those who insisted that polymers were simply colloidal aggregates, delayed the arrival of large-scale synthetic polymers by a decade or more.*

Just how long-chain molecules can in fact be incorporated in regular crystal lattices, when the molecules are bound to extend through many unit cells, took a long time to explain. Finally, in 1957, three experimental teams found the answer; this episode is presented in Chapter 8.

The story of Staudinger's researches and struggles against opposition, and also of the contributions of Carothers who is introduced in the next paragraph, is brilliantly told in a very recent historical study (Furukawa 1998).

There are two great families of synthetic polymers, those made by addition methods (notably, polyethylene and other polyolefines), in which successive monomers simply become attached to a long chain, and those made by condensation reactions (polyesters, polyamides, etc.) in which a monomer becomes attached to the end of a chain with the generation of a small by-product molecule, such as water. The first sustained programme of research directed specifically to finding new synthetic macromolecules involved mostly condensation reactions and was masterminded by Wallace Carothers (1896–1937) an organic chemist of genius who in 1928 was recruited by the Du Pont company in America and the next year (just before the colloid scientists threw in the towel) started his brilliant series of investigations that resulted notably in the discovery and commercialisation, just before the War, of nylon. In Flory's words, Carothers's investigations "were singularly successful in establishing the molecular viewpoint and in dispelling the attitude of mysticism then prevailing in the field". Another major distinction which needs to be made is between polymers made from bifunctional monomers (i.e., those with just two reactive sites) and monomers with three or more reactive sites. The former can form unbranched chains, the latter form branched, three-dimensional macromolecules. What follows refers to the first kind.

The first big step in making addition polymers came in 1933 when ICI, in England, decided to apply high-pressure methods to the search, inspired by the great American physicist Percy Bridgman (1882–1961) who devoted his life as an experimentalist to determining the changes in materials wrought by large hydrostatic pressures (see Section 4.2.3). ICI found that in the presence of traces of oxygen, ethylene gas under high pressure and at somewhat raised temperature would polymerise (Mossman and Morris 1994). Finally, after many problems had been overcome, on the day in 1939 that Germany invaded Poland, the process was successfully scaled up to a production level. Nothing was announced, because it turned out that this high-pressure polyethylene was ideal as an insulator in radar circuits, with excellent dielectric properties. The Germans did not have this product, because Staudinger did not believe that ethylene could be polymerised. Correspondingly, nylon was not made publicly available during the War, being used to make parachutes instead.

The ICI process, though it played a key part in winning the Battle of Britain, was difficult and expensive and it was hard to find markets after the War for such a costly product. It was therefore profoundly exciting to the world of polymers when, in 1953, it became known that a 'stereoactive' polymerisation catalyst (aluminium triethyl plus titanium tetrachloride) had been discovered by the German chemist Karl Ziegler (1898–1973) that was able to polymerise ethylene to yield crystallisable ('high-density') polyethylene. This consisted of unbranched chains with a regular (trans) spatial arrangement of the CH_2 groups. It was 'high-density' because the regularly constructed chains can pack more densely than the partly amorphous ('semicrystalline') low-density material made by ICI's process.

Ziegler's success was followed shortly afterwards by the corresponding achievement by the Italian chemist Giulio Natta (1903–1979), who used a similar catalyst to produce stereoregular (isotactic) polypropylene in crystalline form. That in turn was followed in short order by the use of a similar catalyst in America to produce stereoregular polyisoprene, what came to be called by the oxymoron 'synthetic natural rubber'. These three products, polyethylene, polypropylene and polyisoprene and their many derivatives, were instantly taken up by industry around the world and transformed the markets for polymers, because (for instance) high-density polyethylene was very much cheaper to make than the low-density form and moreover its properties and physical form could be tailor-made for particular end-uses. Through the canny drafting of contracts, Ziegler was one of the few innovators who has actually made a good deal of money from his discovery.

This entire huge development was dependent on two scientific insights and one improvement in technique. The insights were the recognition of the chain nature of high polymers and of the role of the stereotactic nature of those chains. These insights were not generally accepted until after 1930. The technique (or better, battery of techniques) was the collection of gradually improved methods to determine average molecular weight and of molecular weight distribution. These methods included osmometry and viscometry (early methods) and moved on to use of the ultracentrifuge, light-scattering and finally, gel-permeation chromatography. A lively eyewitness account of some of these developments is provided by two of the pioneers, Stockmayer and Zimm (1984), under the title "When polymer science looked easy".

Up to about 1930, polymer science was the exclusive province of experimental chemists. Thereafter, there was an ever-growing input from theoretical chemists and also physicists, who applied the methods of statistical mechanics to understanding the thermodynamics of assemblies of long-chain molecules, and in particular to the elucidation of rubber elasticity, which was perhaps *the* characteristic topic in polymer science. The most distinguished contributor to the statistical mechanics was Paul Flory (1910–1985), who learnt his polymer science while working with Carothers at Du Pont. His textbook of polymer chemistry (Flory 1953) is perhaps the most distinguished overview of the entire field and is still much cited, 48 years after publication.

The input of physicists has become ever greater: two of the most active have been Samuel Edwards in Cambridge and Pierre-Gilles de Gennes in Paris; the latter introduced the method of the renormalisation group (invented by particle physicists) to the statistics of polymer chains (de Gennes 1979) and also, jointly with Edwards, came to an understanding of diffusion in polymers. The physics of polymers (chain statistics, rubber elasticity, crystallisation mechanisms, viscoelasticity and plasticity, dielectric behaviour) has gradually become an identifiable subfield and has been systematised in a recent textbook (Strobl 1996).

Physical chemistry, as we have seen, after its founding quickly acquired dedicated scientific journals, but was very slow in acquiring textbooks. Polymer science was slow on both counts. Flory's text of 1953 was the first major book devoted to the field, though Staudinger made an early first attempt (Staudinger 1932). Many of the early papers appeared in the *Kolloid-Zeitschrift*; this was founded in 1906 and continued under that name until 1973, when it was converted into *Colloid and Polymer Science*. In spite of the uneasy coexistence of colloid science and polymer science in the 1920s, the journal still today mixes papers in the two disciplines, though polymer papers predominate. As late as 1960, only four journals were devoted exclusively to polymers – two in English, one in German and one in Russian. Now, however, the field is saturated: a survey in 1994 came up with 57 journal titles devoted to polymers that could be found in the Science Citation Index, and this does not include minor journals that were not cited. One major publisher, alone, publishes 9 polymer journals! *Macromolecules*, *Polymer* and *Journal of Applied Polymer Science* are the most extensively cited titles. One journal (*Journal of Polymer Science: Polymer Physics*) has 'physics' in its title. Many of the 57 journals have an engineering or applied science flavour, and the field of polymer science is by no means now coterminous with polymer chemistry, as it was half a century ago.

So, although the discipline had a very slow and hesitant emergence, there is no doubt that polymer science is now an autonomous and thoroughly recognised field. It has had its share of Nobel Prizes – Staudinger, Ziegler, Natta, Flory and de Gennes spring to mind. The 1994 *Metallurgy/Materials Education Yearbook* published by ASM International lists 15 university departments in North America specialising in polymer science, with names like Polymer Science and Engineering, Macromolecular Science and Plastics Engineering. Many observers of the MSE field judge that polymers are on their way to becoming, before long, the most widespread and important of all classes of materials.

More about polymers will be found in Chapter 8.

2.1.4 Colloids

The concept of a colloid goes back to an Englishman, Thomas Graham (1805–1869) (Graham 1848). He made a comprehensive study of the diffusion kinetics of a number of liquids, including notably solutions of a variety of substances. Some substances, he found, are characterised by ultraslow diffusion (solutions of starch or dextrin, and albumin, for instance) and are moreover unable to crystallise out of solution: he called these *colloids* (i.e., glue-like). The term, apparently, could apply either to the solution or just to the solute. Ordinary solutions (of salts, for instance), in which diffusion was rapid, were named *crystalloids*. Graham also proposed the nomenclature of *sols* (highly fluid solutions of colloids) and *gels* (gelatinous solutions). What Graham did not realise (he did not have the techniques needed to arrive at such a conclusion) was that what his colloids had in common was a large particle size – large, that is, compared to the size of atoms or molecules, but generally too small to be seen in optical microscopes. That recognition came a little later.

What was recognised from the beginning was that colloidal solutions are two-phase materials.

The study of colloids accelerated rapidly after Graham's initial researches, and much attention was focused on the properties of interfaces, adsorption behaviour in particular. Because of this, 'colloid chemistry' expanded to cover emulsions and foams, as well as aerosols. It took quite a long time to reach the recognition that though a sol (like the gold sol earlier studied by Faraday, for instance) had to be regarded as a suspension of tiny particles of one phase (solid gold) in another phase, water, yet such a two-phase solution behaved identically, in regard to such properties as osmotic pressure, to a true (crystalloid) solution. This was established by Perrin's elegant experiments in 1908 which showed that the equilibrium distribution in a gravitational field of suspended colloid particles large enough to be observed in a microscope follows the same law as the distribution of gas molecules in the atmosphere, and thereby, a century after John Dalton, at last convinced residual sceptics of the reality of atoms and molecules (Nye 1972) (see also Chapter 3, Section 3.1.1).

As Morawetz puts the matter, "an acceptance of the validity of the laws governing colligative properties (i.e., properties such as osmotic pressure) for polymer solutions had no bearing on the question whether the osmotically active particle is a molecule or a molecular aggregate". The colloid chemists, as we have seen, in regard to polymer solutions came to favour the second alternative, and hence created the standoff with the proponents of macromolecular status outlined above.

What concerns us here is the attempt by the champions of colloid chemistry to establish it as a distinct discipline. There was something of an argument about its name; for a while, the term 'capillarity' favoured by Herbert Freundlich (1881–1941), a former assistant of Wilhelm Ostwald, held pride of place. The field has long had its own journals (e.g., the *Kolloid-Zeitschrift* already referred to) and a number of substantial texts have been published. An introduction to colloid chemistry by Wolfgang Ostwald, which originally appeared in 1914, went through numerous editions (Ostwald 1914). Its title, in translation, means "the world of neglected dimensions", and as this suggests, his book has a messianic air about it. Other important texts were those by the American chemist Weiser (1939) and especially a major overview by the Cambridge physical chemists Alexander and Johnson (1949). The last of these was entitled *Colloid Science* (not colloid chemistry) and the authors indicate in their preface that the main reason for this choice of title was that this was the name of an academic department in Cambridge in which they had worked for some years.

That department, the Department of Colloid Science in Cambridge University, was the creation and brainchild of Eric Rideal (1890–1974). In his own words, writing in 1947, "some twenty years ago it was my duty to attempt to build up a laboratory for teaching and research which would serve as a bridge between the biological sciences and physics and chemistry". As a physical chemistry lecturer in Cambridge in 1920, he was intensely interested in surfaces and interfaces and he collaborated with an extraordinary range of Cambridge scientists, with interests in photochemistry, electrochemistry, corrosion (metallurgy) and the statistical mechanics of gases. A wellwisher secured an endowment from the International Education Board, a charity, and a chair in Colloidal Physics was created in 1930. Rideal was appointed to it and moved into exiguous quarters to build up the department. Soon, a further charitable donation materialised, specifically intended for the setting up of chairs in 'bridging subjects', and so the chair in Colloidal Physics was allowed to lapse and Rideal became Professor of Colloid Science instead. As Rideal remarked much later (Rideal 1970), "Not having the remotest idea what colloidal physics were, I naturally accepted it (the chair).... (Later) I was asked whether I would resign my chair and be appointed the first Plummer Professor of Colloid Science, a name which I coined because I thought it was much more suitable than Colloidal Physics. It sounded better and meant just as little." On such accidents do the names of disciplines, or would-be disciplines, depend. At first, the new department was actually a subdepartment of the Chemistry department, but in 1943 Rideal was able to force independence for his fief, on the grounds that in this way collaboration with biologists would be easier. Rideal's interest in interfaces was both literal and metaphorical.

Much of this outline history comes from Johnson's unpublished autobiography (1996). This, and Rideal's obituary for the Royal Society (Eley 1976) show that in research terms the department was a great success, with excellent staff and a horde of research students. Rideal was one of those research supervisors who throw out an endless stream of bright ideas and indications of relevant literature, and then leaves the student to work out all the details; this worked. It did not always work, however; the young Charles Snow was one of his collaborators; Snow and another young man thought that they had discovered a new vitamin and celebrated the discovery with Rideal in a local pub. As Rideal remarked later (Rideal 1970): "It was all wrong unfortunately... C.P. Snow... went off to Sicily, or maybe Sardinia, and thought he was going to die and started to write. He came back with a book; and this book, 'The Search', he presented to me, and that started him on his literary career." One never knows what an unsuccessful piece of research will lead to. Unfortunately, Snow disliked his mentor and is reputed to have used him as raw material for one of his less sympathetic fictional characters.

The department's input to undergraduate teaching was slight, and moreover it was geographically separated from the rest of Cambridge chemistry. In 1946, Rideal accepted an invitation to become director of the Royal Institution in London, taking some of his staff with him, and another professor of colloid science (Francis

Roughton) was appointed to succeed him in Cambridge. In due course the university set up a highly secret committee to consider the future of the department, and it was only years later that its decision to wind up the department leaked out, to the fury of many in the university (Johnson 1996). Nevertheless, the committee members were more effective politicians than were the friends of colloid science, and when the second professor retired in 1966, the department vanished from the scene. (An organisation that is cataloguing Roughton's personal archives has recently commented (NCUACS 2000) that Roughton "presided over a rather disparate group in the Department whose interests ranged from physical chemistry of proteins to ore flotation. During the latter part of his tenure he attempted to redirect the work of the Department towards the study of membranes and biological surface effects. *However, such were the doubts about the existence of a definable subject called Colloid Science* (my emphasis) that on his retirement in 1966 the title of the department was extinguished in favour of Biophysics.")

One of the Department's luminaries, Ronald Ottewill, went off to Bristol University, where he became first professor of colloid science and then professor of physical chemistry, both in the Department of Physical Chemistry. The Bristol department has been one of the most distinguished exponents of colloid science in recent years, but Ottewill considers that it is best practised under the umbrella of physical chemistry.

It is perhaps appropriate that the old premises of the Department of Colloid Science are now occupied by the Department of the History and Philosophy of Science. To the best of my knowledge, there has never been another department of colloid science anywhere in the academic world.

This episode has been displayed in some detail because colloid science is a clear instance of a major field of research which has never quite succeeded in gaining recognition as a distinct discipline, in spite of determined attempts by a number of its practitioners. The one feature that most distinguishes colloid science from physical chemistry, polymer science and chemical engineering is that universities have not awarded degrees in colloid science. That is, perhaps, what counts most for fields with ambitions to become fullblown disciplines.

Lest I leave the erroneous impression here that colloid science, in spite of the impossibility of defining it, is not a vigorous branch of research, I shall conclude by explaining that in the last few years, an entire subspeciality has sprung up around the topic of *colloidal (pseudo-) crystals*. These are regular arrays that are formed when a suspension (sol) of polymeric (e.g., latex) spheres around half a micrometre in diameter is allowed to settle out under gravity. The suspension can include spheres of one size only, or there may be two populations of different sizes, and the radius ratio as well as the quantity proportions of the two sizes are both controllable variables. 'Crystals' such as AB_2, AB_4 and AB_{13} can form (Bartlett et al. 1992, Bartlett and van

Megen 1993, Grier 1998, Pusey 2001); there is an entire new crystallography in play. The field has in fact emerged from a study of natural opal, which consists of tiny silica spheres in the form of colloidal crystals. Such colloidal crystals are in fact stabilised by subtle entropic factors (Frankel 1993) combined with a weak repulsion provided by electrostatic charges at the particle surfaces. In fact, the kind of colloidal supension used in this work was designed some years ago by colloid chemists as a medium for paints, and now they are used by physicists to study (in slow motion, because of the weak interactions) phase transitions, 'melting' in particular (Larsen and Grier 1996). This growing body of research makes copious use of colloid ideas but is carried out in departments of physical chemistry and physics. An inchoate field of 'colloid engineering' is emerging; colloidal crystals can be used to confine and control light, analogously to bandgap engineering in semiconductors; photons with energies lying in the bandgap cannot propagate through the medium. Such 'photonic band gap' materials have recently been discussed by Joannopoulos et al. (1997) and by Berger (1999); a particularly clear explanation is by Pendry (1999).

The broader field of colloid science continues to attract overviews, the most recent being a book entitled *The Colloidal Domain*, *Where Physics*, *Chemistry and Biology Meet* (Evans and Wenneström 1999).

2.1.5 Solid-state physics and chemistry

Both of these crucial fields of research will surface repeatedly later in this book; here they are briefly discussed only as fields which by at least one of the criteria I have examined do not appear to qualify as fully blown disciplines. Both have emerged only in this century, because a knowledge of crystal structure is indispensable to both and that only emerged after 1912, when X-ray diffraction from crystals was discovered.

The beginnings of the enormous field of solid-state physics were concisely set out in a fascinating series of recollections by some of the pioneers at a Royal Society Symposium (Mott 1980), with the participation of a number of professional historians of science, and in much greater detail in a large, impressive book by a number of historians (Hoddeson et al. 1992), dealing in depth with such histories as the roots of solid-state physics in the years before quantum mechanics, the quantum theory of metals and band theory, point defects and colour centres, magnetism, mechanical behaviour of solids, semiconductor physics and critical statistical theory.

As for solid-state chemistry, that began in the form of 'crystal chemistry', the systematic study of the chemical (and physical) factors that govern the structures in which specific chemicals and chemical families crystallise, and many books on this topic were published from the 1930s onwards. The most important addition to straight crystal chemistry from the 1940s onwards was the examination of *crystal defects* – point, line and planar defects, including grain boundaries and interphase boundaries. In fact, crystal defects were first studied by the solid-state physicists; the first compilation of insights relating to crystal defects was a symposium proceedings organised by a group of (mostly) American physicists (Shockley et al. 1952). This was followed after some years by a classic book by the Dutch chemist Kroeger (1974), again focused entirely on crystal defects and their linkage to non-stoichiometry, and an excellent book on disorder in crystals (Parsonage and Staveley 1979). The current status is surveyed in an excellent overview (Rao and Gopalakrishnan 1986, 1997). It will clarify the present status of solid-state chemistry to list the chapter headings in this book: Structure of solids – old and new facets; new and improved methods of characterisation; preparative strategies; phase transitions; new light on an old problem – defects and non-stoichiometry; structure-property relations; fashioning solids for specific purposes – aspects of materials design; reactivity of solids. The linkage with materials science is clear enough.

The enormous amount of research at the interface between physical and structural chemistry has been expertly reviewed recently by Schmalzried in a book about chemical kinetics of solids (Schmalzried 1995), dealing with matters such as morphology and reactions at evolving interfaces, oxidation specifically, internal reactions (such as internal oxidation), reactions under irradiation, etc.

Both fields are very well supplied with journals, some even combining physics with chemistry (e.g., *Journal of Physics and Chemistry of Solids*). Some are venerable journals now focusing on solid-state physics without indicating this in the title, such as *Philosophical Magazine*. The *Journal of Solid-State Chemistry* has recently been complemented by several journals with 'materials chemistry' in the title, but I know of no journals devoted explicitly to the physics of materials: indeed that phrase has only just entered use, though it was the title of a historical piece I wrote recently (Cahn 1995), and the term has been used in the titles of multiauthor books (e.g., Fujita 1994, 1998). 'Applied physics', which overlaps extensively with the concept of physics of materials, appears in the title of numerous journals. (Some mathematicians eschew the term "applied mathematics" and prefer to use "applicable mathematics", as being more logical; "applicable physics" would be a good term, but it has never been used.) Many papers in both solid-state physics and solid-state chemistry are of course published in general physics and chemistry journals.

An eminent researcher at the boundaries between physics and chemistry, Howard Reiss, some years ago explained the difference between a solid-state chemist and a solid-state physicist. The first thinks in configuration space, the second in momentum space; so, one is the Fourier transform of the other.

It is striking that in the English-speaking world, where academic 'departments' are normal, no departments of either solid-state physics or of solid-state chemistry are to be found. These vast fields have been kept securely tethered to their respective parent disciplines, without any visible ill consequences for either; students are given a broad background in physics or in chemistry, and in the later parts of their courses they are given the chance to choose emphasis on solids if they so wish... but their degrees are simply in physics or in chemistry (or, indeed, in physical chemistry). In continental Europe, where specialised 'institutes' take the place of departments, there are many institutes devoted to subfields of solid-state physics and solid-state chemistry, and a few large ones, as in the University of Paris-Sud and in the University of Bordeaux, cover these respective fields in their entirety.

2.1.6 Continuum mechanics and atomistic mechanics of solids

My objective here is to exemplify the stability of some scientific fields in the face of developments which might have been expected to lead to mergers with newer fields which have developed alongside.

Most materials scientists at an early stage in their university courses learn some elementary aspects of what is still miscalled "strength of materials". This field incorporates elementary treatments of problems such as the elastic response of beams to continuous or localised loading, the distribution of torque across a shaft under torsion, or the elastic stresses in the components of a simple girder. 'Materials' come into it only insofar as the specific elastic properties of a particular metal or timber determine the numerical values for some of the symbols in the algebraic treatment. This kind of simple theory is an example of *continuum mechanics*, and its derivation does not require any knowledge of the crystal structure or crystal properties of simple materials or of the microstructure of more complex materials. The specific aim is to design simple structures that will not exceed their elastic limit under load.

From 'strength of materials' one can move two ways. On the one hand, mechanical and civil engineers and applied mathematicians shift towards more elaborate situations, such as "plastic shakedown" in elaborate roof trusses; here some transient plastic deformation is planned for. Other problems involve very complex elastic situations. This kind of continuum mechanics is a huge field with a large literature of its own (an example is the celebrated book by Timoshenko 1934), and it has essentially nothing to do with materials science or engineering because it is not specific to any material or even family of materials.

From this kind of continuum mechanics one can move further towards the domain of almost pure mathematics until one reaches the field of *rational mechanics*, which harks back to Joseph Lagrange's (1736–1813) mechanics of rigid bodies and to earlier mathematicians such as Leonhard Euler (1707–1783) and later ones such as Augustin Cauchy (1789–1857), who developed the mechanics of deformable bodies. The preeminent exponent of this kind of continuum mechanics was probably Clifford Truesdell in Baltimore. An example of his extensive writings is *A First Course in*

Rational Continuum Mechanics (1977, 1991); this is a volume in a series devoted to pure and applied mathematics, and the author makes it very clear that rational continuum mechanics is to be regarded as almost pure mathematics; at one point in his preface, he remarks that "physicists should be able to understand it, should they wish to". His initial quotations are, first, from a metaphysician and, second, from the pure mathematician David Hilbert on the subject of rigorous proofs. Truesdell's (1977, 1991) book contains no illustrations; in this he explicitly follows the model of Lagrange, who considered that a good algebraist had no need of that kind of support. I should perhaps add that Dr. Truesdell wrote many of his books in the study of his renaissance-style home, called *Il Palazzetto*, reportedly using a quill pen. I do not know why the adjective 'rational' is thought necessary to denote this branch of mathematics; one would have thought it tautological.

I cannot judge whether Truesdell's kind of continuum mechanics is of use to mechanical engineers who have to design structures to withstand specific demands, but the total absence of diagrams causes me to wonder. In any case, I understand (Walters 1998, Tanner and Walters 1998) that rational mechanics was effectively Truesdell's invention and is likely to end with him. The birth and death of would-be disciplines go on all the time.

At the other extreme from rational continuum mechanics we have the study of elastic and plastic behaviour of single crystals. Crystal elasticity is a specialised field of its own, going back to the mineralogists of the nineteenth century, and involving tensor mathematics and a detailed understanding of the effects of different crystal symmetries; the aforementioned Cauchy had a hand in this too. Crystal elasticity is of considerable practical use, for instance in connection with the oscillating slivers of quartz used in electronic watches; these slivers must be cut to precisely the right orientation to ensure that the relevant elastic modulus of the sliver is invariant with temperature over a limited temperature range. The plastic behaviour of metal crystals has been studied since the beginning of the present century, when Walter Rosenhain (see Chapter 3, Section 3.2.1) first saw slip lines on the surface of polished polycrystalline metal after bending and recognised that plasticity involved shear ('slip') along particular lattice planes and vectors. Crystal plasticity was studied intensely from the early 1920s onwards, and understanding was codified in two important experimental texts (Schmid and Boas 1935, Elam 1935); crucial laws such as the critical shear stress law for the start of plastic deformation were established. In the 1930s a start was also made with the study of plastic deformation in polycrystalline metals in terms of slip in the constituent grains. This required a combination of continuum mechanics and the physics of single-crystal plasticity. This branch of mechanics has developed fruitfully as a joint venture between mechanical engineers, applied (applicable) mathematicians, metallurgists and solid-state physicists. The leading spirit in this venture was Geoffrey (G.I.) Taylor

(1886–1975), a remarkable English fluid dynamics expert who became interested in plasticity of solids when in 1922 he heard a lecture at the Royal Society about the work of Dr. Constance Elam (the author of one of the above-mentioned books). Elam and Taylor worked together on single-crystal plasticity for some 10 years and this research led Taylor to the co-invention of the dislocation concept in 1934, and then on to a classic paper on polycrystal plasticity (Taylor 1938). This paper is still frequently cited: for instance, Taylor's theory concerning the minimum number (5) of distinct slip elements needed to ensure an arbitrary shape change of a grain embedded in a polycrystal has been enormously influential in the understanding of plastic deformability. Taylor's collected papers include no fewer than 41 papers on the solid state (Batchelor 1958). His profound influence on the field of plasticity is vividly analysed in a recent biography (Batchelor 1996).

A good picture of the present state of understanding of polycrystal plasticity can be gleaned from a textbook by Khan and Huang (1995). Criteria for plastic yield, for instance, are developed both for a purely continuum type of medium and for a polycrystal undergoing slip. This book contains numerous figures and represents a successful attempt to meet crystal plasticity experts at a halfway point. A corresponding treatment by metallurgists is entitled "deformation and texture of metals at large strains" and discusses the rotation of individual crystallites during plastic deformation, which is of industrial importance (Aernoudt et al. 1993).

In 1934, a new kind of crystal defect, the dislocation, was invented (independently by three scientists) and its existence was confirmed some years later. The dislocation can be briefly described as the normal (but not exclusive) vector of plastic deformation in crystals. This transformed the understanding of such deformation, especially once the elastic theory of dislocation interaction had been developed (Cottrell 1953). Cottrell went on to write a splendid student text in which he contrived to marry continuum mechanics and 'crystal mechanics' into an almost seamless whole (Cottrell 1964). From that point on, the understanding, in terms of the interaction of point, line and planar defects, of both fast and slow plastic deformation in single and polycrystals developed rapidly. A fine example of what modern theory can achieve is the creation of *deformation-mechanism maps* by Frost and Ashby (1982); such maps plot normalised stress and normalised temperature on a double-log plot, for particular metals or ceramics with a particular grain (crystal) size, and using theoretically derived *constitutive relations*, the domain of the graph is divided into areas corresponding to different deformation mechanisms (some further details are in Section 5.1.2.2). This kind of map has proved very useful both to materials engineers who develop new materials, and to mechanical engineers who use them.

The upshot of all this is that the mechanics of elastic and plastic types of deformation spans a spectrum from the uncompromising and highly general rational mechanics to the study of crystal slip in single crystals and its interpretation in terms of the elastic theory of interaction between defects, leading to insights that are specific to particular materials. There is some degree of a meeting of minds in the middle between the mathematicians and mechanical engineers on the one side and the metallurgists, physicists and materials scientists on the other, but it is also true to say that continuum mechanics and what might (for want of a better term) be called *atomistic mechanics* have remained substantially divergent approaches to the same set of problems. One is a part of mechanical engineering or more rarefied applied mathematics, the other has become an undisputed component of materials science and engineering, and the two kinds of specialists rarely meet and converse. This is not likely to change.

Another subsidiary domain of mechanics which has grown in stature and importance in parallel with the evolution of polymer science is *rheology*, the science of flow, which applies to fluids, gels and soft solids. It is an engaging mix of advanced mathematics and experimental ingenuity and provides a good deal of insight specific to particular materials, polymers in particular. A historical outline of rheology, with concise biographical sketches of many of its pioneers, has been published by Tanner and Walters (1998).

Very recently, people who engage in computer simulation of crystals that contain dislocations have begun attempts to bridge the continuum/atomistic divide, now that extremely powerful computers have become available. It is now possible to model a variety of aspects of dislocation mechanics in terms of the atomic structure of the lattice around dislocations, instead of simply treating them as lines with 'macroscopic' properties (Schiøtz et al. 1998, Gumbsch 1998). What this amounts to is 'linking computational methods across different length scales' (Bulatov et al. 1996). We will return to this briefly in Chapter 12.

2.2. THE NATURAL HISTORY OF DISCIPLINES

At this stage of my enquiry I can draw only a few tentative conclusions from the case-histories presented above. I shall return at the end of the book to the issue of how disciplines evolve and when, to adopt biological parlance, a new discipline becomes self-fertile.

We have seen that physical chemistry evolved from a deep dissatisfaction in the minds of a few pioneers with the current state of chemistry as a whole – one could say that its emergence was research-driven and spread across the world by hordes of new Ph.Ds. Chemical engineering was driven by industrial needs and the corresponding changes that were required in undergraduate education. Polymer science started from a wish to understand certain natural products and moved by slow stages, once the key concept had been admitted, to the design, production and understanding of synthetic materials. One could say that it was a synthesis-driven discipline. Colloid science (the one that 'got away' and never reached the full status of a discipline) emerged from a quasi-mystic beginning as a branch of very applied chemistry. Solid-state physics and chemistry are of crucial importance to the development of modern materials science but have remained fixed by firm anchors to their parent disciplines, of which they remain undisputed parts. Finally, the mechanics of elastic and plastic deformation is a field which has always been, and remains, split down the middle, and neither half is in any sense a recognisable discipline. The mechanics of flow, rheology, is closer to being an accepted discipline in its own right.

Different fields, we have seen, differ in the speed at which journals and textbooks have appeared; the development of professional associations is an aspect that I have not considered at this stage. What seems best to distinguish recognized disciplines from other fields is academic organisation. Disciplines have their own distinct university departments and, even more important perhaps, those departments have earned the right to award degrees in their disciplines. Perhaps it is through the harsh trial of academic infighting that disciplines win their spurs.

REFERENCES

Aernoudt, K., van Houtte, P. and Leffers, T. (1993) in *Plastic Deformation and Fracture of Materials*, edited by H. Mughrabi, Volume 6 of *Materials Science and Technology*, ed. R.W., Cahn, P. Haasen and E.J. Kramer (VCH, Weinheim) p. 89.

Alexander, A.E. and Johnson, P. (1949) *Colloid Science*, 2 volumes (Clarendon Press, Oxford).

Armstrong, H.E. (1936) *Chem. Indus.* **14**, 917.

Arrhenius, S. (1889) Z. *Phys. Chem.* **4**, 226.

Bartlett, P., Ottewill, R.H. and Pusey, P.N. (1992) *Phys. Rev. Lett.* **68**, 3801.

Bartlett, P. and van Megen, W. (1993) in *Granular Matter*, ed. A. Mehta (Springer, Berlin) p. 195.

Batchelor, G.K. (1958) *G.I. Taylor, Scientific Papers, Volume 1, Mechanics of Solids* (Cambridge University Press, Cambridge).

Batchelor, G.K. (1996) *The Life and Legacy of G.I. Taylor*, Chapter 11 (Cambridge University Press, Cambridge).

Berger, V. (1999) *Curr. Opi. Solid State Mater. Sci.* **4**, 209.

Bulatov, V.V., Yip, Si. and Arias, T. (1996) *J. Computer-Aided Mater. Design* **3**, 61.

Cahn, R.W. (1995) in *Twentieth Century Physics*, ed. L.M. Brown, A. Pais and B. Pippard, vol. 3 (Institute of Physics Publishing, Bristol and American Institute of Physics Press, New York) p. 1505.

Calvert, P. (1997) (book review) *Nature* **388**, 242.

Cohen, C. (1996) *British Journal of the History of Science* **29**, 171.

Cottrell, A.H. (1953) *Dislocations and Plastic Flow in Crystals* (Clarendon Press, Oxford).

Cottrell, A.H. (1964) *The Mechanical Properties of Matter* (Wiley, New York).

Crawford, E. (1996) *Arrhenius: From Ionic Theory to Greenhouse Effect* (Science History Publications/USA, Canton, MA).

De Gennes, P.-G. (1979) *Scaling Concepts in Polymer Physics* (Cornell University Press, Ithaca, NY).

Dolby, R.G.A. (1976a) *Hist. Stud. Phys. Sci.* **7**, 297.

Dolby, R.G.A. (1976b) in *Perspectives on the Emergence of Scientific Disciplines*, eds. G. Lemaine, R. MacLeod, M. Mulkay and P. Weingart (The Hague, Mouton) p. 63.

Elam, C.F. (1935) *Distortion of Metal Crystals* (Clarendon Press, Oxford).

Eley, D.D. (1976) Memoir of Eric Rideal, *Biogr. Mem. Fellows Roy. Soc.* **22**, 381.

Evans, D.F. and Wenneström, H. (1999) *The Colloidal Domain, Where Physics, Chemistry and Biology Meet* (Wiley-VCH, Weinheim).

Faraday Division, Roy. Soc. of Chem., London (1995) A celebration of physical chemistry, *Faraday Discussions*, No. 100.

Flory, P.J. (1953) *Principles of Polymer Chemistry* (Cornell University Press, Ithaca, NY).

Frankel, D. (1993) *Physics World*, February, p. 24.

Frost, H.J. and Ashby, M.F. (1982) *Deformation-Mechanism Maps: The Plasticity and Creep of Metals and Ceramics* (Pergamon Press, Oxford).

Fujita, F.E. (editor) (1994, 1998) *Physics of New Materials* (Springer, Berlin).

Furukawa, Y. (1998) *Inventing Polymer Science: Staudinger, Carothers and the Emergence of Macromolecular Chemistry* (Pennsylvania University Press, Philadelphia).

Glasstone, S. (1940) *Textbook of Physical Chemistry* (Macmillan, London).

Graham, T. (1848) *Phil. Trans. Roy. Soc. Lond.* **151**, 183.

Grier, D.G. (editor) (1998) A series of papers on colloidal crystals, in *MRS Bulletin*, October 1998.

Gumbsch, P. (1998) *Science* **279**, 1489.

Harrison, D. (1996) Interview.

Hoddeson, L., Braun, E., Teichmann, J. and Weart, S. (editors) (1992) *Out of the Crystal Maze: Chapters from the History of Solid-State Physics* (Oxford University Press, Oxford).

Jacques, J. (1987) *Berthelot: Autopsie d'un Mythe* (Belin, Paris).

Joannopoulos, J.D., Villeneuve, P.R. and Fan, S. (1997) *Nature* **386**, 143.

Johnson, P. (1996) *Unpublished autobiography*.

Khan, A.S. and Huang, S. (1995) *Continuum Theory of Plasticity* (Wiley, New York).

Kroeger, F.A. (1974) *The Chemistry of Imperfect Crystals*, 2 volumes (North Holland, Amsterdam).

Kuhn, T. (1970) *The Structure of Scientific Revolutions*, 2nd revised edition (Chicago University Press).

Laidler, K.J. (1993) *The World of Physical Chemistry* (Oxford University Press, Oxford).

Larsen, A.E. and Grier, D.G. (1996) *Phys. Rev. Lett.* **76**, 3862.

Liebhafsky, H.A., Liebhafsky, S.S. and Wise, G. (1978) *Silicones under the Monogram: A Story of Industrial Research* (Wiley-Interscience, New York).

McMillan, F.M. (1979) *The Chain Straighteners – Fruitful Innovation: the Discovery of Linear and Stereoregular Synthetic Polymers* (Macmillan, London).

Mendelssohn, K. (1973) *The World of Walther Nernst* (Macmillan, London). A German translation published 1976 by Physik-Verlag, Weinheim, as *Walther Nernst und seine Zeit*.

Montgomery, S.L. (1996) *The Scientific Voice* (The Guilford Press, New York) p. viii.

Morawetz, H. (1985) *Polymers: The Origins and Growth of a Science* (Wiley, New York) (Reprinted (1995) as a Dover, Mineola, NY edition).

Mossman, S.T.E and Morris, P.J.T. (1994) *The Development of Plastics* (Royal Society of Chemistry, London).

Mott, N.F. (editor) (1980) *Proc. Roy. Soc. Lond.* **371A**, 1.

NCUACS (2000) Annual Report of the National Cataloguing Unit for the Archives of Contemporary Scientists, University of Bath, UK, p. 10.

Nye, M.J. (1972) *Molecular Reality: A Perspective on the Scientific Work of Jean Perrin* (Macdonald, London and, American Elsevier, New York).

Ostwald, W. (1914) *Die Welt der Vernachlässigten Dimensionen: Eine Einführung in die Kolloidchemie* (Steinkopff, Dresden and Leipzig).

Parsonage, N.G. and Staveley, L.A.K. (1979) *Disorder in Crystals* (Oxford University Press, Oxford).

Passmore, J. (1978) *Science and Its Critics* (Duckworth, London) p. 56.

Pendry, J.B. (1999) *Current Science (India)* **76**, 1311.

Price, I. de Solla J. (1963) *Little Science, Big Science*, Chapter 3. (Reprinted in (1986) *Little Science, Big Science... and Beyond*) (Columbia University Press, New York).

Pusey, P.N. (2001) Colloidal Crystals, in *Encyclopedia of Materials* ed. K.H.J. Buschee et al. (Pergamon, Oxford) in press.

Rao, C.N.R. and Gopalakrishnan, J. (1986, 1997) *New Directions in Solid State Chemistry* (Cambridge University Press, Cambridge).

Rideal, E. (1970) Text of a talk, "Sixty Years of Chemistry", presented on the occasion of the official opening of the West Wing, Unilever Research Laboratory, Port Sunlight, 20 July, 1970 (privately printed).

Russell, C.A. (1976) *The Structure of Chemistry – A Third-Level Course* (The Open University Press, Milton Keynes, UK).

Schiøtz, J., DiTolla, F.D. and Jacobsen, K.W. (1998) *Nature* **391**, 561.

Schmalzried, H. (1995) *Chemical Kinetics of Solids* (VCH, Weinheim).

Schmid, E. and Boas, W. (1935) *Kristallplastizität* (Springer, Berlin).

Servos, J.W. (1990) *Physical Chemistry from Ostwald to Pauling: The Making of a Science in America* (Princeton University Press, Princeton, NJ).

Seymour, R.B. and Kirshenbaum, G.S. (1986) *High Performance Polymers: Their Origin and Development* (Elsevier, New York).

Shockley, W., Hollomon, J.H., Maurer, R. and Seitz, F. (editors) (1952) *Imperfections in Nearly Perfect Crystals* (Wiley, New York, and Chapman and Hall, London).

Siilivask, K. (1998) Europe, Science and the Baltic Sea, in *Euroscientia Forum* (European Commission, Brussels) p. 29.

Staudinger, H. (1932) *Die Hochmolekularen Organischen Verbindungen* (Springer, Berlin).

Stockmayer, W.H. and Zimm, B.H. (1984) *Annu. Rev. Phys. Chem.* **35**, 1.

Strobl, G. (1996) *The Physics of Polymers* (Springer, Berlin).

Taylor, G.I. (1938) *J. Inst. Metals* **62**, 307.

Tanner, R.I. and Walters, K. (1998) *Rheology: An Historical Perspective* (Elsevier Amsterdam).

Timoshenko, S. (1934) *Introduction to the Theory of Elasticity for Engineers and Physicists* (Oxford University Press, London).

Truesdell, C.A. (1977, 1991) *A First Course in Rational Continuum Mechanics* (Academic Press, Boston).

van 't Hoff, J.H. (1901) *Zinn, Gips und Stahl vom physikalisch-chemischen Standpunkt* (Oldenbourg, München and Berlin).

Walters, K. (1998) private communication.

Warner, F. (1996) Interview.

Weiser, H.B. (1939) *A Textbook of Colloid Chemistry*, 2nd edition (Wiley, New York).

Wise, G. (1983) *Isis* **74**, 7.

Wise, G. (1985) *Willis R. Whitney, General Electric and the Origins of the US Industrial Revolution* (Columbia University Press, New York).

Yagi, E., Badash, L. and Beaver, D. de B. (1996) *Interdiscip. Sci. Rev.* **21**, 64.

Ziman, J. (1996) *Sci. Stud.* **9**, 67.

Chapter 3
Precursors of Materials Science

Chapter 3
Precursors of Materials Science

3.1. THE LEGS OF THE TRIPOD

In Cambridge University, the final examination for a bachelor's degree, irrespective of subject, is called a 'tripos'. This word is the Latin for a three-legged stool, or tripod, because in the old days, when examinations were conducted orally, one of the participants sat on such a stool. Materials science is examined as one option in the Natural Sciences Tripos, which itself was not instituted until 1848; metallurgy was introduced as late as 1932, and this was progressively replaced by materials science in the 1960s. In earlier days, it was neither the nervous candidate, nor the severe examiner, who sat on the 'tripos'; this was occupied by a man sometimes called the 'prevaricator' who, from the 14th century, if not earlier, was present in order to inject some light relief into the proceedings: when things became too tense, he would crack a joke or two and then invite the examiner to proceed. I believe this system is still sometimes used for doctoral examinations in Sweden.

The tripod and its occupant, then, through the centuries helped students of classics, philosophy, mathematics and eventually natural science to maintain a sense of proportion. One might say that the three prerequisites for doing well in such an examination were (and remain) knowledge, judgment and good humour, three preconditions of a good life. By analogy, I suggest that there were three preconditions of the emergence of materials science, constituting another tripod: those preconditions were an understanding of (1) atoms and crystals, (2) phase equilibria, and (3) microstructure. These three forms of understanding were the crucial precursors of our modern understanding and control of materials. For a beginning, I shall outline how these forms of understanding developed.

3.1.1 Atoms and crystals
The very gradual recognition that matter consists of atoms stretched over more than two millennia, and that recognition was linked for several centuries with the struggles of successive generations of scientists to understand the nature of crystals. This is why I am here combining sketches of the history of atoms and of the history of crystals, two huge subjects.

The notion that matter had ultimate constituents which could not be further subdivided goes back to the Greeks (atom = Greek *a-tomos*, not capable of being cut). Democritus (circa 460 BC – circa 370 BC), probably leaning on the ideas of

57

Epicurus, was a very early proponent of this idea; from the beginning, the amount of empty space associated with atoms and the question whether neighbouring atoms could actually be in contact was a source of difficulty, and Democritus suggested that solids with more circumatomic void space were in consequence softer. A century later, Aristotle praised Democritus and continued speculating about atoms, in connection with the problem of explaining how materials can change by combining with each other... *mixtion*, as the process came to be called (Emerton 1984).

Even though Democritus and his contemporaries were only able to speculate about the nature of material reality, yet their role in the creation of modern science is more crucial than is generally recognised. That eminent physicist, Erwin Schrödinger, who in his little book on *Nature and the Greeks* (Schrödinger 1954, 1996) has an illuminating chapter about *The Atomists*, put the matter like this: "The grand idea that informed these men was that the world around them was something that *could be understood*, if only one took the trouble to observe it properly; that it was not the playground of gods and ghosts and spirits who acted on the spur of the moment and more or less arbitrarily, who were moved by passions, by wrath and love and desire for revenge, who vented their hatred, and could be propitiated by pious offerings. These men had freed themselves of superstition, they would have none of all this. They saw the world as a rather complicated mechanism, according to eternal innate laws, which they were curious to find out. This is of course the fundamental attitude of science to this day." In this sense, materials science and all other modern disciplines owe their origin to the great Greek philosophers.

The next major atomist was the Roman Lucretius (95 BC – circa 55 BC), who is best known for his great poem, *De rerum natura* (Of the Nature of Things), in which the author presents a comprehensive atomic hypothesis, involving such aspects as the ceaseless motion of atoms through the associated void (Furley 1973). Lucretius thought that atoms were characterised by their shape, size and weight, and he dealt with the problem of their mutual attraction by visualising them as bearing hooks and eyes... a kind of primordial 'Velcro'. He was probably the last to set forth a detailed scientific position in the form of verse.

After this there was a long pause until the time of the 'schoolmen' in the Middle Ages (roughly 1100–1500). People like Roger Bacon (1220–1292), Albertus Magnus (1200–1280) and also some Arab/Moorish scholars such as Averroes (1126–1198) took up the issue; some of them, notably Albertus, at this time already grappled with the problem of the nature of crystalline minerals. Averroes asserted that "the natural minimum...is that ultimate state in which the form is preserved in the division of a natural body". Thus, the smallest part of, say, alum would be a particle which in some sense had the form of alum. The alternative view, atomism proper, was that alum and all other substances are made up of a few basic building units none of which is specific to alum or to any other single chemical compound. This difference of opinion (in modern terms, the distinction between a molecule and an atom) ran through the centuries and the balance of dogma swung backwards and forwards. The notion of molecules as distinct from atoms was only revived seriously in the 17th century, by such scientists as the Dutchman Isaac Beeckman (1588–1637) (see Emerton 1984, p. 112). Another early atomist, who was inspired by Democritus and proposed a detailed model according to which atoms were in perpetual and intrinsic motion and because of this were able to collide and form molecules, was the French philosopher Pierre Gassendi (1592–1655). For the extremely involved history of these ideas in antiquity, the Middle Ages and the early scientific period, Emerton's excellent book should be consulted.

From an early stage, as already mentioned, scholars grappled with the nature of crystals, which mostly meant naturally occurring minerals. This aspect of the history of science can be looked at from two distinct perspectives – one involves a focus on the appearance, classification and explanation of the forms of crystals (i.e., crystallography), the other, the role of mineralogy in giving birth to a proper science of the earth (i.e., geology). The first approach was taken, for instance, by Burke (1966) in an outstanding short account of the origins of crystallography, the second, in a more recent study by Laudan (1987).

As the era of modern science approached and chemical analysis improved, some observers classified minerals in terms of their compositions, others in terms of their external appearance. The 'externalists' began by measuring angles between crystal faces; soon, crystal symmetry also began to be analysed. An influential early student of minerals – i.e., crystals – was the Dane Nicolaus Stenonius, generally known as Steno (1638–1686), who early recognised the constancy of interfacial angles and set out his observations in his book, *The Podromus, A Dissertation on Solids Naturally Contained within Solids* (see English translation in Scherz 1969). Here he also examines the juxtaposition of different minerals, hence the title. Steno accepted the possibility of the existence of atoms, as one of a number of rival hypotheses. The Swedish biologist Carolus Linnaeus (1707–1778) somewhat later attempted to extend his taxonomic system from plants and animals to minerals, basing himself on crystal shape; his classification also involved a theory of the genesis of minerals with a sexual component; his near-contemporaries, Romé de l'Isle and Haüy (see below) credited Linnaeus with being the true founder of crystallography, because of his many careful measurements of crystals; but his system did not last long, and he was not interested in speculations about atoms or molecules.

From quite an early stage, some scientists realised that the existence of flat crystal faces could be interpreted in terms of the regular piling together of spherical or ellipsoidal atoms. Figure 3.1 shows some 17th-century drawings of postulated crystal structures due to the Englishman Robert Hooke (1635–1703) and the Dutchman Christiaan Huygens (1629–1695). The great astronomer, Johannes

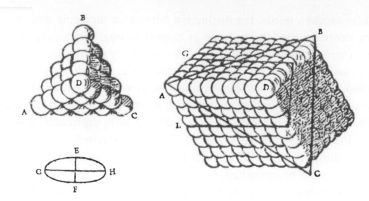

Figure 3.1. (from Emerson, p. 134) Possible arrangements of spherical particles, according to Hooke (left, from a republication in *Micrographia Restaurata*, London 1745) and Huygens (right, from *Traité de la Lumière*, Leiden 1690).

Kepler (1571–1630) had made similar suggestions some decades earlier. Both Kepler and Huygens were early analysts of crystal symmetries in terms of atomic packing. This use of undifferentiated atoms in regular arrays was very different from the influential corpuscular models of René Descartes (1596–1650), as outlined by Emerton (1984, p. 131 et seq.): Descartes proposed that crystals were built up of complicated units (star- or flower-shaped, for instance) in irregular packing; according to Emerton, this neglect of regularity was due to Descartes's emphasis on the motion of particles and partly because of his devotion to Lucretius's unsymmetrical hook-and-eye atoms.

In the 18th century, the role of simple, spherical atoms was once more in retreat. An eminent historian of metallurgy, Cyril Stanley Smith, in his review of Emerton's book (Smith 1985) comments: "...corpuscular thinking disappeared in the 18th century under the impact of Newtonian anti-Cartesianism. The new math was so useful because its smoothed functions could use empirical constants without attention to substructure, while simple symmetry sufficed for externals. Even the models of Kepler, Hooke and Huygens showing how the polyhedral form of crystals could arise from the stacking of spherical or spheroidal parts were forgotten." The great French crystallographers of that century, Romé de l'Isle and Haüy, thought once again in terms of non-spherical 'molecules' shaped like diminutive crystals, and not in terms of atoms.

Jean-Baptiste Romé de l'Isle (1736–1790) and René Haüy (1743–1822), while they, as remarked, credited Linnaeus with the creation of quantitative crystallography, themselves really deserve this accolade. Romé de l'Isle was essentially a chemist and much concerned with the genesis of different sorts of crystal, but his real claim to fame is that he first clearly established the principle that the interfacial angles of a particular species of crystal were always the same, however different the shape of individual specimens might be, tabular, elongated or equiaxed – a principle foreshadowed a hundred years earlier by Steno. This insight was based on very exact measurements using contact goniometers; the even more exact optical goniometer was not invented until 1809 by William Wollaston (1766–1826). (Wollaston, incidentally, was yet another scientist who showed how the stacking of spherical atoms could generate crystal forms. He was also an early scientific metallurgist, who found out how to make malleable platinum and also discovered palladium and rhodium.)

Haüy, a cleric turned experimental mineralogist, built on Romé's findings: he was the first to analyse in quantitative detail the relationship between the arrangement of building-blocks (which he called 'integrant molecules') and the position of crystal faces: he formulated what is now known as the law of rational intercepts, which is the mathematical expression of the regular pattern of 'treads and steps' illustrated in Figure 3.2(a), reproduced from his *Traité de Cristallographie* of 1822. The tale is often told how he was led to the idea of a crystal made up of integrant molecules shaped like the crystal itself, by an accident when he dropped a crystal of iceland spar and found that the small cleavage fragments all had the same shape as the original large crystal. "Tout est trouvé!" he is reputed to have exclaimed in triumph.

From the 19th century onwards, chemists made much of the running in studying the relationship between atoms and crystals. The role of a German chemist, Eilhardt Mitscherlich (1794–1863, Figure 3.2(b)) was crucial (for a biography, see Schütt 1997). He was a man of unusual breadth who had studied oriental philology and history, became 'disillusioned with these disciplines' in the words of Burke (1966) and turned to medicine, and finally from that to chemistry. It was Mitscherlich who discovered, first, the phenomenon of isomorphism and, second, that of polymorphism. Many salts of related compositions, say, sodium carbonate and calcium carbonate, turned out to have similar crystal symmetries and axial ratios, and sometimes it was even possible to use the crystals of one species as nuclei for the growth of another species. It soon proved possible to use such *isomorphous* crystals for the determination of atomic weights: thus Mitscherlich used potassium selenite, isomorphous with potassium sulphate, to determine the atomic weight of selenium from the already known atomic weight of sulphur. Later, Mitscherlich established firmly that one and the same compound might have two or even more distinct crystal structures, stable (as was eventually recognised) in different ranges of temperature. (Calcite and aragonite, two quite different *polymorphs* of calcium carbonate, were for mineralogists the most important and puzzling example.) Finally, Wollaston and the French chemist François Beudant, at about the same time, established the existence of mixed crystals, what today we would in English call *solid solutions* (though *Mischkristall* is a term still used in German).

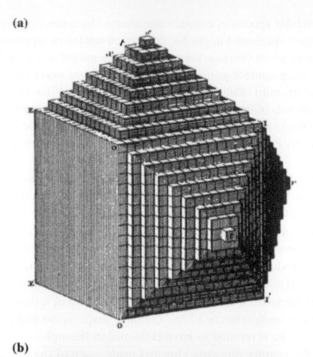

(a)

(b)

These three findings – isomorphism, polymorphism, mixed crystals – spelled the doom of Haüy's central idea that each compound had one – and one only – integrant molecule the shape of which determined the shape of the consequent crystal and, again according to Cyril Smith (Smith 1960, p. 190), it was the molecule as the combination of atoms in fixed proportions – rather than the atoms themselves, or any integrant molecules – which now became the centre of chemical interest. When John Dalton (1766–1844) enunciated his atomic hypothesis in 1808, he did touch on the role of regularly combined and arranged atoms in generating crystals, but he was too modest to speculate about the constitution of molecules; he thought that "it seems premature to form any theory on this subject till we have discovered *from other principles* (my italics) the number and order of the primary elements" (Dalton 1808).

The great Swedish chemist Jöns Berzelius (1779–1848) considered the findings of Mitscherlich, together with Dulong and Petit's discovery in 1819 that the specific heats of solids varied inversely as their atomic weights, to be the most important empirical proofs of the atomic hypothesis at that time. It is to be noted that one of these two cornerstones was based on crystallography, which thus became one of the foundations of modern atomic theory.

Another 19th century scientist is one we have met before, in Chapter 2, Section 2.1.4. Thomas Graham (1805–1869), the originator of the concept of colloids, made a reputation by studying the diffusion of fluids (both gases and liquids) in each other in a quantitative way. As one recent commentator (Barr 1997) has put it, "the crucial point about Graham's law (of diffusion) is its quantitative nature and that it could be understood, if not completely explained, by the kinetic theory of gases developed by Maxwell and Clausius shortly after the middle of the nineteenth century. In this way the ideas of diffusion being connected with the random motion of molecules over a characteristic distance, the mean free path, entered science." Jean Perrin, whose crucial researches we examine next, could be said to be the inheritor of Graham's insights. Many years later, in 1900, William Roberts-Austen (1843–1905), a disciple of Graham, remarked of him (Barr 1997): "I doubt whether he would have wished any other recognition than that so universally accorded to him of being the leading atomist of his age".

We move now to the late 19th century and the beginning of the 20th, a period during which a number of eminent chemists and some physicists were still resolutely sceptical concerning the existence of atoms, as late as hundred years after John Dalton's flowering. Ostwald's scepticism was briefly discussed in Section 2.1.1, as

Figure 3.2. (a) Treads and risers forming crystal faces of various kinds, starting from a cubic primitive form (after Haüy 1822). (b) Eilhardt Mitscherlich (1794–1863) (courtesy Deutsches Museum, Munich).

was his final conversion by Einstein's successful quantitative interpretation of Brownian motion in 1905 in terms of the collisions between molecules and small suspended particles, taken together with Jean Perrin's painstaking measurements of the Brownian motion of suspended colloidal gamboge particles, which together actually produced a good estimate of Avogadro's number. Perrin's remarkable experimental tour de force is the subject of an excellent historical book (Nye 1972); it is not unreasonable to give Perrin the credit for finally establishing the atomic hypothesis beyond cavil, and Nye even makes a case for Perrin as having preceded Rutherford in his recognition of the necessity of a compound atom. Perrin published his results in detail, first in a long paper (Perrin 1909) and then in a book (Perrin 1913). The scientific essayist Morowitz (1993) laments that "one of the truly great scientific books of this century gathers dust on library shelves and is missing from all libraries established after 1930". Morowitz shows a table from Perrin's 1913 book, reproduced here in the earlier form presented by Nye (1972), which gives values of Avogadro's number from 15 distinct kinds of experiment; given the experimental difficulties involved, these values cluster impressively just above the value accepted today, 60.22×10^{22}. If no atoms...then no Avogadro's number. Perrin received the Nobel Prize for Physics in 1926.

Phenomena observed	N (Avogadro's Number)/10^{22}
Viscosity of gases (kinetic theory)	62
Vertical distribution in dilute emulsions	68
Vertical distribution in concentrated emulsions	60
Brownian movement (Perrin)	
Displacements	69
Rotations	65
Diffusion	69
Density fluctuation in concentrated emulsions	60
Critical opalescence	75
Blueness of the sky	65
Diffusion of light in argon	69
Blackbody spectrum	61
Charge on microscopic particles	62
Radioactivity	
Helium produced	64
Radium lost	71
Energy radiated	60

The detailed reasons for Ostwald's atomic scepticism when he gave a major lecture in Germany in 1895 are set out systematically in a book by Stehle (1994), who remarks: "The great obstacle faced by those trying to convince the sceptics of the reality of atoms and molecules was the lack of phenomena making apparent the graininess of matter. It was only by seeing individual constituents, either directly or indirectly through the observation of fluctuations about the mean behaviour predicted by kinetic theory, that the existence of these particles could be shown unambiguously. Nothing of the kind had been seen as yet, as Ostwald so forcefully pointed out...". In fact, Johann Loschmidt (1821–1895) in 1866 had used Maxwell's kinetic theory of gases (which of course presupposes the reality of atoms, or rather molecules) together with a reasonable estimate of an atomic cross-section, to calculate a good value for Avogadro's Number, that longterm criterion of atomic respectability. Ostwald's resolute negation of the existence of atoms distressed some eminent scientists; thus, Ludwig Boltzmann's statistical version of thermodynamics (see Section 3.3.2), which was rooted in the reality of molecules, was attacked by opponents of atomism such as Ostwald, and it has been asserted by some historians that this (together with Ernst Mach's similarly implacable hostility) drove Boltzmann into a depression which in turn led to his suicide in 1906. Even today, the essential link between the atomic hypothesis and statistical thermodynamics provokes elaborate historical analyses such as a recent book by Diu (1997).

Just after Ostwald made his sceptical speech in 1895, the avalanche of experiments that peaked a decade later made his doubts untenable. In the 4th (1908) edition of his textbook, *Grundriss der physikalischen Chemie*, he finally accepted, exactly a hundred years after Dalton enunciated his atomic theory and two years after Boltzmann's despairing suicide, that Thomson's discovery of the electron as well as Perrin's work on Brownian motion meant that "we arrived a short time ago at the possession of experimental proof for the discrete or particulate nature of matter – proof which the atomic hypothesis has vainly sought for a hundred years, even a thousand years" (Nye 1972, p. 151). Not only Einstein's 1905 paper and Perrin's 1909 overview of his researches (Perrin 1909), but the discovery of the electron by J.J. Thomson in 1897 and thereafter the photographs taken with Wilson's cloud-chamber (the 'grainiest' of experiments), Rutherford's long programme of experiments on radioactive atoms, scattering of subatomic projectiles and the consequent establishment of the planetary atom, followed by Moseley's measurement of atomic X-ray spectra in 1913 and the deductions that Bohr drew from these... all this established the atom to the satisfaction of most of the dyed-in-the-wool disbelievers. The early stages, centred around the electron, are beautifully set out in a very recent book (Dahl 1997). The physicist's modern atom in due course led to the chemist's modern atom, as perfected by Linus Pauling in his hugely influential book, *The Nature of the Chemical Bond and the Structure of Molecules and Crystals*, first published in 1939. Both the physicist's and the chemist's atoms were necessary precursors of modern materials science.

Nevertheless, a very few eminent scientists held out to the end. Perhaps the most famous of these was the Austrian Ernst Mach (1838–1916), one of those who inspired Albert Einstein in his development of special relativity. As one brief biography puts it (Daintith *et al.* 1994), "he hoped to eliminate metaphysics – all those purely 'thought-things' which cannot be pointed to in experience – from science". Atoms, for him, were "economical ways of symbolising experience. But we have as little right to expect from them, as from the symbols of algebra, more than we have put into them". Not all, it is clear, accepted the legacy of the Greek philosophers, but it is appropriate to conclude with the words (Andrade 1923) of Edward Andrade (1887–1971): "The triumph of the atomic hypothesis is the epitome of modern physics".

3.1.1.1 X-ray diffraction. The most important episode of all in the history of crystallography was yet to come: the discovery that crystals can diffract X-rays and that this allows the investigator to establish just where the atoms are situated in the crystalline unit cell. But before that episode is outlined, it is necessary to mention the most remarkable episode in crystallographic theory – the working out of the 230 space groups. In the mid-19th century, and based on external appearances, the entire crystal kingdom was divided into 7 systems, 14 space lattices and 32 point-groups (the last being all the self-consistent ways of disposing a collection of symmetry elements passing through a single point), but none of these exhausted all the intrinsically different ways in which a motif (a repeated group of atoms) can in principle be distributed within a crystal's unit cell. This is far more complicated than the point-groups, because (1) new symmetry elements are possible which combine rotation or reflection with translation and (2) the various symmetry elements, including those just mentioned, can be situated in various positions within a unit cell and generally do not all pass through one point in the unit cell. This was recognised and analysed by three mathematically gifted theorists: E. Fedorov in Russia (in 1891), A. Schoenfliess in Germany (in 1891) and W. Barlow in England (in 1894). All the three independently established the existence of 230 distinct space groups (of symmetry elements in space), although there was some delay in settling the last three groups. Fedorov's work was not published in German until 1895 (Fedorov 1895), though it appeared in Russian in 1891, shortly before the other two published their versions. Fedorov found no comprehension in the Russia of his time, and so his priority is sometimes forgotten. Accounts of the circumstances as they affected Fedorov and Schoenfliess were published in 1962, in *Fifty Years of X-ray Diffraction* (Ewald 1962, pp. 341, 351), and a number of the earliest papers related to this theory are reprinted by Bijvoet *et al.* (1972). The remarkable feature of this piece of triplicated pure theory is that it was perfected 20 years before an experimental method was discovered for the analysis of actual crystal structures, and when such a method at length appeared, the theory of space groups turned out to be an indispensable aid to the process of interpreting the diffraction patterns, since it means that when one atom has been located in a unit cell, then many others are automatically located as well if the space group has been identified (which is not difficult to do from the diffraction pattern itself). The Swiss crystallographer P. Niggli asserted in 1928 that "every scientific structure analysis must begin with a determination of the space group", and indeed it had been Niggli (1917) who was the first to work out the systematics that would allow a space group to be identified from systematic absences in X-ray diffractograms.

In 1912 Max von Laue (1879–1960), in Munich, instructed two assistants, Paul Knipping and Walter Friedrich, to send a beam of (polychromatic) X-rays through a crystal of copper sulphate and on to a photographic plate, and immediately afterwards they did the same with a zincblende crystal: they observed the first diffraction spots from a crystal. Laue had been inspired to set up this experiment by a conversation with Paul Ewald, who pointed out to him that atoms in a crystal had to be not only periodically arranged but much more closely spaced than a light wavelength. (This followed simply from a knowledge of Avogadro's Number and the measured density of a crystal.) At the time, no one knew whether X-rays were waves or particles, and certainly no one suspected that they were both. As he says in his posthumous autobiography (Von Laue 1962), he was impressed by the calculations of Arnold Sommerfeld, also in Munich, which were based on some recent experiments on the diffraction of X-rays at a wedge-shaped slit; it was this set of calculations, published earlier in 1912, that led von Laue to the idea that X-rays had a short wavelength and that crystals might work better than slits. So the experiments with copper sulphate and zincblende showed to von Laue's (and most other people's) satisfaction that X-rays were indeed waves, with wavelengths of the order of 0.1 nm. The crucial experiment was almost aborted before it could begin because Sommerfeld forbade his assistants, Friedrich and Knipping, to get involved with von Laue; Sommerfeld's reason was that he estimated that thermal vibrations in crystals would be so large at room temperature that the essential periodicity would be completely destroyed. He proved to be wrong (the periodicity is not destroyed, only the intensity of diffraction is reduced by thermal motion). Friedrich and Knipping ignored their master (a hard thing to do in those days) and helped von Laue, who as a pure theorist could not do the experiment by himself. Sommerfeld was gracious: he at once perceived the importance of what had been discovered and forgave his errant assistants.

The crucial experiments that determined the structures of a number of very simple crystals, beginning with sodium chloride, were done, not by von Laue and his helpers, but by the Braggs, William (1862–1942) and Lawrence (1890–1971), father and son, over the following two years (Figure 3.3). The irony was that, as von Laue declares in his autobiographical essay, Bragg senior had only shortly before declared his conviction that X-rays were particles! It was his own son's work which led Bragg senior to declare at the end of 1912 that "the problem becomes...not to decide between two theories of X-rays, but to find...one theory which possesses the capabilities of both", a prescient conclusion indeed. At a meeting in London in 1952 to celebrate the 40th anniversary of his famous experiment, von Laue remarked in public how frustrated he had felt afterwards that he had left it to the Braggs to make these epoch-making determinations; he had not made them himself because he was focused, not on the nature of crystals but on the nature of X-rays. By the time he had shifted his focus, it was too late. It has repeatedly happened in the history of science that a fiercely focused discoverer of some major insight does not see the further consequences that stare him in the face. The Ewald volume already cited sets out the minutiae of the events of 1912 and includes a fascinating account of the sequence of events by Lawrence Bragg himself (pp. 59–63), while the subtle relations between Bragg père and Bragg fils are memorably described in Gwendolen Caroe's memoir of her father, William H. Bragg (Caroe 1978). Recent research by an Australian historian (Jenkin 1995), partly based on W.L. Bragg's unpublished autobiography,

Figure 3.3. Portraits of the two Braggs (courtesy Mr. Stephen Bragg).

has established that the six-year-old schoolboy Lawrence, in Adelaide, fell off his bicycle in 1896 and badly injured his elbow; his father, who had read about the discovery of X-rays by Wilhelm Röntgen at the end of 1895, had within a year of that discovery rigged up the first X-ray generator in Australia and so he was able to take a radiograph of his son's elbow – the first medical radiograph in Australia. This helped a surgeon to treat the boy's elbow properly over a period of time and thereby save its function. It is perhaps not so surprising that the thoughts of father and son turned to the use of X-rays in 1912.

Henry Lipson, a British crystallographer who knew both the Braggs has commented (Lipson 1990) that "W.H. and W.L. Bragg were quite different personalities. We can see how important the cooperation between people with different sorts of abilities is; W.H. was the good sound eminent physicist, whereas W.L. was the man with intuition. The idea of X-ray reflection came to him in the grounds of Trinity College, Cambridge, where he was a student of J.J. Thomson's and should not have been thinking of such things."

Lawrence Bragg continued for the next 59 years to make one innovation after another in the practice of crystal structure analysis; right at the end of his long and productive life he wrote a book about his lifetime's experiences, *The Development of X-ray Analysis* (Bragg 1975, 1992), published posthumously. In it he gives a striking insight into the beginnings of X-ray analysis. In 1912, he was still a very young researcher with J.J. Thomson in the Cavendish Laboratory in Cambridge, and he decided to use the Laue diffraction technique (using polychromatic X-rays) to study ZnS, NaCl and other ionic crystals. "When I achieved the first X-ray reflections, I worked the Rumkorff coil too hard in my excitement and burnt out the platinum contact. Lincoln, the mechanic, was very annoyed as a contact cost 10 shillings, and refused to provide me with another one for a month. In these days (i.e., ≈1970) a researcher who discovered an effect of such novelty and importance would have very different treatment. I could never have exploited my ideas about X-ray diffraction under such conditions... In my father's laboratory (in Leeds) the facilities were on quite a different scale." In 1913 he moved to Leeds and he and his father began to use a newly designed X-ray spectrometer with essentially monochromatic X-rays. A 1913 paper on the structure of diamond, in his own words "may be said to represent the start of X-ray crystallography". By the time he moved back to Cambridge as Cavendish professor in 1938, the facilities there had distinctly improved.

Though beaten in that race by the Braggs, von Laue received the Nobel Prize in 1914, one year before the Braggs did.

In spite of these prompt Nobel awards, it is striking how long it took for the new technique for determining atomic arrangements in crystals – crystal structures – to spread in the scientific community. This is demonstrated very clearly by an editorial written by the German mineralogist P. Groth in the *Zeitschrift für Kristallographie*, a journal which he had guided for many years. Groth, who also taught in Munich, was the most influential mineralogist of his generation and published a renowned textbook, *Chemische Kristallographie*. In his 1928 editorial he sets out the genesis and development of his journal and writes about many of the great crystallographers he had known. Though he refers to Federov, the creator of space groups (whom he hails as one of the two greatest geniuses of crystallography in the preceding 50 years), Groth has nothing whatever to say about X-ray diffraction and crystal structure analysis, 16 years after the original discovery. Indeed, in 1928, crystal structure analysis was only beginning to get into its stride, and mineralogists like Groth had as yet derived very few insights from it; in particular, the structure analysis of silicates was not to arrive till a few years later.[1]

Metallurgists, also, were slow to feel at ease with the new techniques, and did not begin to exploit X-ray diffraction in any significant way until 1923. Michael Polanyi (1891–1976), in an account of his early days in research (Polanyi 1962) describes how he and Herman Mark determined the crystal structure of white tin from a single crystal in 1923; just after they had done this, they received a visit from a Dutch colleague who had independently determined the same structure. The visitor vehemently maintained that Polanyi's structure was wrong; in Polanyi's words, "only after hours of discussion did it become apparent that his structure was actually the same as ours, but looked different because he represented it with axes turned by 45° relative to ours".

Even the originator was hesitant to blow his own trumpet. In 1917, the elder Bragg published an essay on "physical research and the way of its application", in a multiauthor book entitled "Science and the Nation" (Bragg 1917). Although he writes at some length on Röntgen and the discovery of X-rays, he includes not a word on X-ray diffraction, five years after the discoveries by his son and himself.

This slow diffusion of a crucial new technique can be compared with the invention of the scanning tunnelling microscope (STM) by Binnig and Rohrer, first made public in 1983, like X-ray diffraction rewarded with the Nobel Prize 3 years later, but unlike X-ray diffraction quickly adopted throughout the world. That invention, of comparable importance to the discoveries of 1912, now (2 decades later) has sprouted numerous variants and has virtually created a new branch of surface science. With it, investigators can not only see individual surface atoms but they can also manipulate atoms singly (Eigler and Schweitzer 1990). This rapid adoption of

---

[1] Yet when Max von Laue, in 1943, commemorated the centenary of Groth's birth, he praised him for keeping alive the hypothesis of the space lattice which was languishing everywhere else in Germany, and added that without this hypothesis it would have been unlikely that X-ray diffraction would have been discovered and even if it had been, it would have been quite impossible to make sense of it.

the STM is of course partly due to much better communications, but it is certainly in part to be attributed to the ability of so many scientists to recognise very rapidly what could be done with the new technique, in distinction to what happened in 1912.

In Sweden, a precocious school of crystallographic researchers developed who applied X-ray diffraction to the study of metallic phases. Their leaders were Arne Westgren and Gösta Phragmén. As early as 1922 (Westgren and Phragmén 1922) they performed a sophisticated analysis of the crystal structures of various phases in steels, and they were the first (from measurements of the changes of lattice parameter with solute concentration) to recognise that solutions of carbon in body-centred alpha-iron must be 'interstitial' – i.e., the carbon atoms take up positions between the regular lattice sites of iron. In a published discussion at the end of this paper, William Bragg pointed out that Sweden, having been spared the ravages of the War, was able to undertake these researches when the British could not, and appealed eloquently for investment in crystallography in Britain. The Swedish group also began to study intermetallic compounds, notably in alloy systems based on copper; Westgren found the unit cell dimensions of the compound Cu_5Zn_8 but could not work out the structure; that feat was left to one of Bragg's young research students, Albert Bradley, who was the first to determine such a complicated structure (with 52 atoms in the unit cell) from diffraction patterns made from a powder instead of a single crystal (Bradley and Thewlis 1926); this work was begun during a visit by Bradley to Sweden. This research was a direct precursor of the crucial researches of William Hume-Rothery in the 1920s and 1930s (see Section 3.3.1.1).

In spite of the slow development of crystal structure analysis, once it did 'take off' it involved a huge number of investigators: tens of thousands of crystal structures were determined, and as experimental and interpretational techniques became more sophisticated, the technique was extended to extremely complex biological molecules. The most notable early achievement was the structure analysis, in 1949, of crystalline penicillin by Dorothy Crowfoot-Hodgkin and Charles Bunn; this analysis achieved something that traditional chemical examination had not been able to do. By this time, the crystal structure, and crystal chemistry, of a huge variety of inorganic compounds had been established, and *that* was most certainly a prerequisite for the creation of modern materials science.

Crystallography is a very broad science, stretching from crystal-structure determination to crystal physics (especially the systematic study and mathematical analysis of anisotropy), crystal chemistry and the geometrical study of phase transitions in the solid state, and stretching to the prediction of crystal structures from first principles; this last is very active nowadays and is entirely dependent on recent advances in the electron theory of solids. There is also a flourishing field of applied crystallography, encompassing such skills as the determination of preferred orientations, alias textures, in polycrystalline assemblies. It would be fair to say that within this broad church, those who determine crystal structures regard themselves as being members of an aristocracy, and indeed they feature prominently among the recipients of the 26 Nobel Prizes that have gone to scientists best described as crystallographers; some of these double up as chemists, some as physicists, increasing numbers as biochemists, and the prizes were awarded in physics, chemistry or medicine. It is doubtful whether any of them would describe themselves as materials scientists![2]

Crystallography is one of those fields where physics and chemistry have become inextricably commingled; it is however also a field that has evinced more than its fair share of quarrelsomeness, since some physicists resolutely regard crystallography as a technique rather than as a science. (Thus an undergraduate specialisation in crystallography at Cambridge University was killed off some years ago, apparently at the instigation of physicists.) What all this shows is that scientists go on arguing about terminology as though this were an argument about the real world, and cannot it seems be cured of an urge to rank each other into categories of relative superiority and inferiority. Crystallography is further discussed below, in Section 4.2.4.

3.1.2 Phase equilibria and metastability

I come now to the second leg of our notional tripod – phase equilibria.

Until the 18th century, man-made materials such as bronze, steel and porcelain were not 'anatomised'; indeed, they were not usually perceived as having any 'anatomy', though a very few precocious natural philosophers did realise that such materials had structure at different scales. A notable exemplar was René de Réaumur (1683–1757) who deduced a good deal about the fine-scale structure of steels by closely examining fracture surfaces; in his splendid *History of Metallography*, Smith (1960) devotes an entire chapter to the study of fractures. This approach did not require the use of the microscope. The other macroscopic evidence for fine structure within an alloy came from the examination of metallic meteorites. An investigator of one collection of meteorites, the Austrian Aloys von Widmanstätten (1754–1849), had the happy inspiration to section and polish one meteorite and etch the polished section, and he observed the image shown in Figure 3.4, which was included in an atlas of illustrations of meteorites published by his assistant Carl von Schreibers in Vienna, in 1820 (see Smith 1960, p. 150). This 'micro'structure is very much coarser

---

[2] In a letter of unspecified date to a biologist, Linus Pauling is reported as writing (Anon 1998): "You refer to me as a biochemist, which is hardly correct. I can properly be called a chemist, or a physical chemist, or a physicist, or an X-ray crystallographer, or a mineralogist, or a molecular biologist, but not, I think, a biochemist."

Figure 3.4. (from Smith 1960, p. 151). The Elbogen iron meteorite, sectioned, polished and etched. The picture was made by inking the etched surface and using it as a printing plate. The picture is enlarged about twofold. From a book by Carl von Schreibers published in 1820, based upon the original observation by von Widmanstätten in 1808. (Reproduced from Smith 1960.) This kind of microstructure has since then been known as a Widmanstätten structure.

than anything in terrestrial alloys, and it is now known that the coarseness results from extremely slow cooling ($\approx$ one degree Celsius per one million years) of huge meteorites hurtling through space, and at some stage breaking up into smaller meteorites; the slow cooling permits the phase transformations during cooling to proceed on this very coarse scale. (This estimate results both from measurements of nickel distribution in the metallic part of the meteorite, and from an ingenious technique that involves measurement of damage tracks from plutonium fission fragments that only left residual traces – in mineral inclusions within the metallic body – once the meteorite had cooled below a critical temperature (Fleischer *et al.* 1968); a further estimate is that the meteorite during this slow-cooling stage in its life had a radius of 150–250 km.)

In the penultimate sentence, I have used the word 'phase'. This concept was unknown to von Widmanstätten, and neither was it familiar to Henry Sorby (1826–1908), an English amateur scientist who was the prime pioneer in the microscopic study of metallic structure. He began by studying mineralogical and petrographic sections under the microscope in transmitted polarised light, and is generally regarded as the originator of that approach to studying the microstructure of rocks; he was initially rewarded with contempt by such geologists as the Swiss de Saussure who cast ridicule on the notion that one could "look at mountains through a microscope". Living as he did in his native city of Sheffield, England, Sorby naturally moved on for some years, beginning in 1864, to look at polished sections of steels, adapting his microscope to operate by reflected light, and he showed, as one commentator later put it, that "it made sense to look at railway lines through a microscope". Sorby might be described as an intellectual descendant of the great mediaeval German craftsman Georgius Agricola (1494–1555), who became known as the father of geology as well as the recorder of metallurgical practice. Sorby went on to publish a range of observations on steels as well as description of his observational techniques, mostly in rather obscure publications; moreover, at that time it was not possible to publish micrographic photographs except by expensive engraving and his 1864 findings were published as a brief unillustrated abstract. The result was that few became aware of Sorby's pioneering work, although he did have a vital influence on the next generation of metallographers, Heycock and Neville in particular, as well as the French school of investigators such as Floris Osmond. Sorby's influence on the early scientific study of materials is analysed in a full chapter in Smith's (1960) book, and also in the proceedings of a symposium devoted to him (Smith 1965) on the occasion of the centenary of his first observations on steel. One thing he was the first to suggest (later, in 1887, when he published an overview of his ferrous researches) was that his micrographs indicated that a piece of steel consists of an array of separate small crystal grains.

Our next subject is a man who, in the opinion of some well-qualified observers, was the greatest native-born American man of science to date: Josiah Willard Gibbs (1839–1903, Figure 3.5). This genius began his university studies as a mechanical engineer before becoming professor of mathematical physics at Yale University in 1871, before he had even published any scientific papers. It is not clear why his chair had the title it did, since at the time of his appointment he had not yet turned to the theory of thermodynamics. Yale secured a remarkable bargain, especially as the university paid him no salary for many years and he lived from his family fortune. In passing, at this point, it is worth pointing out that a number of major pure scientists began their careers as engineers: the most notable example was Paul Dirac (electrical), another was John Cockroft (also electrical); Ludwig Wittgenstein, though hardly a scientist, began as an aeronautical engineer. Unlike these others, Gibbs continued to undertake such tasks as the design of a brake for railway cars and of the teeth for gearwheels, even while he was quietly revolutionising physical chemistry and metallurgy. He stayed at Yale all his life, working quietly by himself,

Figure 3.5. Portrait of Josiah Willard Gibbs (courtesy F. Seitz).

with a minimum of intellectual contacts outside. He did not marry. It is not perhaps too fanciful to compare Gibbs with another self-sufficient bachelor, Isaac Newton, in his quasi-monastic cell in Cambridge.

In the early 1870s, Gibbs turned his attention to the foundations of thermodynamics (a reasonable thing for a mechanical engineer to do), when through the work of Clausius and Carnot "it had achieved a measure of maturity", in the words of one of Gibbs' biographers (Klein 1970–1980). Gibbs sought to put the first and second laws of thermodynamics on as rigorous a basis as he could, and he focused on the role of entropy and its maximisation, publishing the first of his terse, masterly papers in 1873. He began by analysing the thermodynamics of fluids, but a little later went on to study systems in which different states of matter were present together. This situation caught his imagination and he moved on to his major opus, "On the equilibrium of heterogeneous substances", published in 1876 in the *Transactions of the Connecticut Academy of Arts and Sciences* (Gibbs 1875–1978). In the words of Klein, in this memoir of some 300 pages Gibbs hugely extended the reach of thermodynamics, including chemical, elastic, surface, electromagnetic and electrochemical phenomena in a single system. When Gibbs (1878) published a short memoir about this paper, he wrote as follows:

> "It is an inference naturally suggested by the general increase of entropy which accompanies the changes occurring in any isolated material system that when the entropy of the system has reached a maximum, the system will be in a state of equilibrium. Although this principle has by no means escaped the attention of physicists, its importance does not seem to have been duly appreciated. *Little has been done to develop the principle as a foundation for the general theory of thermodynamic equilibrium* (my italics)."

Gibbs focused on the concept of a *phase*. This concept is not altogether easy to define. Here are three definitions from important modern textbooks: (1) Darken and Gurry, in *Physical Chemistry of Metals* (1953) say: "Any homogeneous portion of a system is known as a phase. Different homogeneous portions at the same temperature, pressure and composition – such as droplets – are regarded as the same phase". (2) Ruoff, *Materials Science* (1973) says: "A phase is the material in a region of space which in principle can be mechanically separated from other phases". (3) Porter and Easterling, in *Phase Transformations in Metals and Alloys* (1981) say: "A phase can be defined as a portion of the system whose properties and composition are homogeneous and which is physically distinct from other parts of the system". A phase may contain one or more chemical *components*. The requirement for uniformity (homogeneity) of composition only applies so long as the system is required to be in equilibrium; metastable phases can have composition and property gradients; but then Gibbs was entirely concerned with the conditions for equilibrium to be attained.

In his 1878 abstract, Gibbs formulated two alternative but equivalent forms of the criterion for thermodynamic equilibrium: "For the equilibrium of any isolated system it is necessary and sufficient that in all possible variations of the state of the system which do not alter its energy (entropy), the variation of its entropy (energy) shall either vanish or be negative (positive)". Gibbs moved on immediately to apply this criterion to the issue of chemical equilibrium between phases. According to Klein, "the result of this work was described by Wilhelm Ostwald as determining the form and content of chemistry for a century to come, and by Henri Le Chatelier as comparable in its importance for chemistry with that of Antoine Lavoisier" (the co-discoverer of oxygen). From his criterion, Gibbs derived a corollary of general validity, the *phase rule*, formulated as $\delta = n + 2 - r$. This specifies the number of independent variations δ (usually called 'degrees of freedom') in a system of r coexistent phases containing n independent chemical components. The phase rule, when at last it became widely known, had a definitive effect on the understanding and determination of *phase*, or *equilibrium, diagrams*.

There are those who say nowadays that Gibbs's papers, including his immortal paper on heterogeneous equilibria, present no particular difficulties to the reader.

This was emphatically not the opinion of his contemporaries, to some of whom Gibbs circulated reprints since the *Connecticut Transactions* were hardly widely available in libraries. One of his most distinguished admirers was James Clerk Maxwell, who made it his business to alert his fellow British scientists to the importance of Gibbs's work, but in fact few of them were able to follow his meaning. According to Klein's memoir, "(Gibbs) rejected all suggestions that he write a treatise that would make his ideas easier to grasp. Even Lord Rayleigh (in a letter he wrote to Gibbs) thought the original paper 'too condensed and difficult for most, I might say all, readers'. Gibbs responded by saying that in his own view the memoir was instead 'too *long*' and showed a lack of 'sense of the value of time, of (his) own or others, when (he) wrote it'." In Germany, it was not till Ostwald translated Gibbs's papers in 1892 that his ideas filtered through.

The man who finally forced Gibbs's ideas, and the phase rule in particular, on the consciousness of his contemporaries was the Dutchman H.W. Bakhuis Roozeboom (1856–1907), a chemist who in 1886 succeeded van 't Hoff as professor of chemistry in the University of Amsterdam. Roozeboom heard of Gibbs's work through his Dutch colleague Johannes van der Waals and "saw it as a major breakthrough in chemical understanding" (Daintith *et al.* 1994). Roozeboom demonstrated in his own research the usefulness of the phase rule, in particular, in showing what topological features are thermodynamically possible or necessary in alloy equilibria – e.g., that single-phase regions must be separated by two-phase regions in an equilibrium diagram. In Cyril Smith's words, "it was left to Roozeboom (1900) to discuss constitution (equilibrium) diagrams in general, and by slightly adjusting (William) Roberts-Austen's constitution diagram, to show the great power of the phase rule". By 1900, others, such as Henri Le Chatelier (1850–1936) were using Gibbs's principles to clarify alloy equilibria; Le Chatelier's name is also immortalised by his Principle, deduced from Gibbs, which simply states that any change made to a system in equilibrium results in a shift in the equilibrium that minimises the change (see overview by Bever and Rocca 1951).

Roozeboom engaged in a long correspondence (outlined by Stockdale 1946) with two British researchers in Cambridge who had embarked on joint alloy studies, Charles Thomas Heycock (1858–1931) and Francis Henry Neville (1847–1915) (Figure 3.6), and thereby inspired them to determine the first really accurate non-ferrous equilibrium diagram, for the copper–tin binary system. Figure 3.7 reproduces this diagram, which has the status of a classic. Apart from the fact that in their work they respected the phase rule, they made two other major innovations. One was that they were able to measure high temperatures with great accuracy, for the first time, by carefully calibrating and employing the new platinum resistance thermometer, developed by Ernest Griffiths and Henry Callendar, both working in Cambridge (the former with Heycock and Neville and the latter in the Cavendish Laboratory); (at about the same time, in France, Le Chatelier perfected the platinum/platinum–

Figure 3.6. Charles Heycock and Francis Neville.

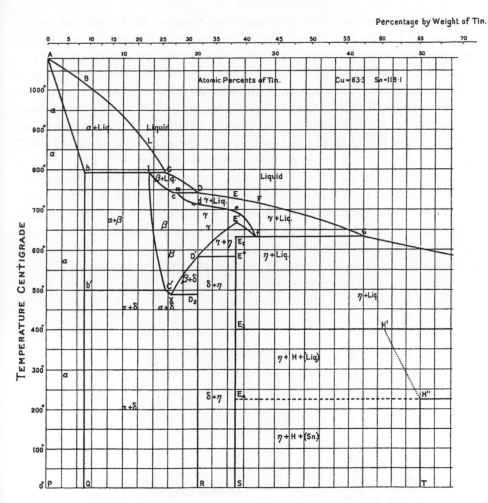

Figure 3.7. Part of Heycock and Neville's Cu–Sn phase diagram.

rhodium thermocouple). Heycock and Neville's other innovation was to use the microscope in the tradition of Sorby, but specifically to establish equilibria, notably those holding at high temperatures which involved quenching specimens from the relevant temperatures (see Section 3.1.3). Heycock and Neville set up their joint laboratory in the garden of one of the Cambridge colleges, Sidney Sussex, and there they studied alloy equilibria from 1884 until Neville's retirement in 1908. A full account of the circumstances leading to the operation of a research laboratory in a single college as distinct from a central university facility, and detailed information about the careers of Heycock and Neville, can be found in a book published in 1996 to mark the fourth centenary of Sidney Sussex College (Greer 1996).

The prehistory of the phase rule, the steps taken by Gibbs and the crucial importance of the rule in understanding phase equilibria, are outlined in an article published in a German journal to mark its centenary (Petzow and Henig 1977).

One other scientist played a major role in establishing the examination of equilibrium diagrams – alternatively called phase diagrams – as a major part of the study of materials. This was Gustav Tammann (1861–1938), born to a German-speaking member of the Russian nobility (Figure 3.8). One of his several forenames was Apollon and indeed he attained something of the aura of the sun god himself. Tammann is a hero to German-speaking metallurgists (Köster 1939, 1961) and he is also regarded by some as a co-founder of the discipline of physical chemistry; he knew Arrhenius and van 't Hoff well and interacted considerably with both; he knew Ostwald also but preferred to keep his distance: neither was a particularly easy man. He also came to know Roozeboom. As the biographical memoir by one of his descendants (Tammann 1970–1980) remarks, he first did research at the borders of chemistry and physics in 1883 when he began determining molecular weights from

Gustav Tammann [1861~1938]

Figure 3.8. Gustav Tammann.

the lowering of vapour pressures – and this was 4 years before Ostwald took up his chair in Leipzig. Influenced by Gibbs and Roozeboom, Tammann in his early base in Dorpat, in one of Russia's Baltic provinces (see Siilivask 1998) began in 1895 to study heterogeneous equilibria between liquid and vapour phases, and he also studied triple points. After some years, he reached the crucial conclusion (at variance with current opinion at the time) that all transitions from the crystalline state to other phases must be discontinuous, unlike the continuous liquid/vapour transition discovered by van der Waals. He published his researches leading to this conclusion in 1903.

A few years later he spent a time working with Nernst, before being invited in 1903 to occupy a newly established chair of inorganic chemistry in the University of Göttingen; when Nernst moved away from Göttingen in 1907, Tammann moved to his chair of physical chemistry and he held this until he retired in 1930. In Göttingen, Tammann worked with enormous energy (his biographer wrote that he was "a giant not only in stature but also in health and capacity for work; Tammann regularly worked in his laboratory for ten hours a day") and he directed a large stable of research students who were also expected to keep long hours, and provoked notorious outbursts of rage when they failed to live up to expectation. He generated some 500 publications, highly unusual for a materials scientist, including successive editions of his famous *Lehrbuch der Metallographie*. Initially he worked mostly on inorganic glasses, in which field he made major contributions, before shifting progressively towards metals and alloys. He then began a long series of approximate studies of binary alloy systems, setting out to study alloys of 20 common metallic elements mixed in varying proportions in steps of 10 at.%, requiring 1900 alloys altogether. Using mainly thermal analysis and micrographic study, he was able to identify which proportions of particular metals formed proper intermetallic compounds, and established that the familiar valence relationships and stoichiometric laws applicable to salts do not at all apply to intermetallic compounds. From these studies he also reached the precocious hypothesis that some intermetallic compounds must have a non-random distribution of the atomic species on the lattice... and this before X-ray diffraction was discovered. This inspired guess was confirmed experimentally, by X-ray diffraction, in 1925, by Swedish physicists stimulated by Tammann's hypothesis. Tammann's very numerous alloy phase diagrams were of necessity rough and ready and cannot be compared with Heycock and Neville's few, but ultraprecise, phase diagrams.

Later, after the War, Tammann moved further towards physics by becoming interested in the mechanism of plastic deformation and the repair of deformed metals by the process of recrystallisation (following in the footsteps of Ewing and Rosenhain in Cambridge at the turn of the century), paving the way for the very extensive studies of these topics that followed soon after. Tammann thus followed the dramatic shift of metallurgy away from chemical concerns towards physical aspects which had gathered momentum since 1900. In fact Tammann's chair was converted into a chair of physical metallurgy after his retirement, and this (after the next incumbent stepped down) eventually became a chair of metal physics.

The determination of equilibrium diagrams as a concern spread quite slowly at first, and it was only Tammann's extraordinary energy which made it a familiar concern. It took at least two decades after Roozeboom, and Heycock and Neville, at the turn of the century, to become widespread, but after that it became a central activity of metallurgists, ceramists and some kinds of chemists, sufficiently so that in 1936, as we shall see in Chapter 13, enough was known to permit the publication of a first handbook of binary metallic phase diagrams, and ternary diagrams followed in due course. In this slow start, the study of equilibrium diagrams resembled the determination of crystal structures after 1912.

As an indication of the central role that phase diagrams now play in the whole of materials science, the cumulative index for the whole of the 18-volume book series, *Materials Science and Technology* (Cahn *et al.* 1991–1998) can be cited in evidence. There are 89 entries under the heading "phase diagram", one of the most extensive of all listings in this 390-page index.

3.1.2.1 *Metastability*.

The emphasis in all of Gibbs's work, and in the students of phase diagrams who were inspired by him, was always on equilibrium conditions. A phase, or equilibrium, diagram denotes the phases (single or multiple), their compositions and ranges, stable for any alloy composition and any temperature. However, the long years of study of steels hardened by quenching into water, and the discovery in 1906 of age-hardening of aluminium alloys at room temperature, made it clear enough that the most interesting alloys are not in equilibrium, but find themselves, to coin a metaphor, in a state of suspended animation, waiting until conditions are right for them to approach true equilibrium at the temperature in question. This is possible because at sufficiently low temperatures, atomic movement (diffusion) in a crystalline material becomes so slow that all atoms are 'frozen' into their current positions. This kind of suspended animation is now seen to be a crucially important condition throughout materials science.

Wilhelm Ostwald was the first to recognise this state of affairs clearly. Indeed, he went further, and made an important distinction. In the second edition of his *Lehrbuch der Allgemeinen Chemie*, published in 1893, he introduced the concept of *metastability*, which he himself named. The simplest situation is just *instability*, which Ostwald likened to an inverted pyramid standing on its point. Once it begins to topple, it becomes ever more unstable until it has fallen on one of its sides, the new condition of stability. If, now, the tip is shaved off the pyramid, leaving a small flat surface parallel to the base where the tip had been, and the pyramid is again carefully inverted, it will stand metastably on this small surface, and if it is very slightly tilted, will return to its starting position. Only a larger imposed tilt will now topple the pyramid. Thus, departing from the analogy, Ostwald pointed out that each state of a material corresponds to a definite (free) energy: an unstable state has a local maximum in free energy, and as soon as the state is "unfrozen", it will slide down the free energy slope, so to speak. A metastable state, however, occupies a *local* minimum in free energy, and can remain there even if atomic mobility is reintroduced (typically, by warming); the state of true stability, which has a lower free energy, can only be attained by driving the state of the material over a neighbouring energy maximum. A water droplet supercooled below the thermodynamic freezing temperature is in metastable equilibrium so long as it is not cooled too far. A quenched aluminium alloy is initially in an unstable condition and, if the atoms can move, they form zones locally enriched in solute; such zones are then metastable against the nucleation of a transition phase which has a lower free energy than the starting state. Generally, whenever a process of *nucleation* is needed to create a more stable phase within a less stable phase, the latter can be maintained metastably; a tentative nucleus, or embryo, which is not large enough will redissolve and the metastable phase is locally restored. This is a very common situation in materials science.

The interpretation of metastable phases in terms of Gibbsian thermodynamics is set out simply in a paper by van den Broek and Dirks (1987).

3.1.2.2 Non-stoichiometry. One feature of phases that emerged clearly from the application of Gibbs's phase rule is that thermodynamics permit a phase to be not exactly stoichiometric; that is, a phase such as NiAl can depart from its ideal composition to, say, $Ni_{55}Al_{45}$ or $Ni_{45}Al_{55}$, without loss of its crystal structure; all that happens is that some atoms sit in locations meant for the other kind of atoms or, (in the special case of NiAl and a few other phases) some atomic sites remain vacant. The dawning recognition of the reality of non-stoichiometry, two centuries ago, convinced some chemists that atoms could not exist, otherwise, they supposed, strict stoichiometry would necessarily be enforced. One such sceptic was the famous French chemist Claude-Louis Berthollet (1748–1822); because of the observed non-stoichiometry of some compounds, he propounded a theory of *in*definite proportions in chemical combination, which the work of John Dalton (1766–1844) and others succeeded in refuting, early in the nineteenth century. For a century, compounds with a wide composition range in equilibrium were known as *berthollides* while those of a very narrow range round the stoichiometric composition were known as *daltonides*. This terminology has now, rather regrettably, fallen out of use; one of the last instances of its use was in a paper by the eminent Swedish crystallographer, Hägg (1950).

3.1.3 Microstructure

We come now to the third leg of the tripod, the third essential precursor of modern materials science. This is the study of microstructure in materials. When the practice of sectioning, polishing, etching and examining items such as steel ingots was first introduced, it was possible to see some features with the naked eye. Thus, Figure 3.9 shows the "macrostructure" of a cast ingot which has frozen rather slowly. The elongated, 'columnar' crystal grains stretching from the ingot surface into the interior can be clearly seen at actual (or even reduced) dimensions. But rapidly solidified metal has very fine grains which can only be seen under a microscope, as Henry Sorby came to recognise. The shape and size of grains in a single-phase metal or solid-solution alloy can thus fall within the province of either macro- or micro-structure.

At the turn of the century it was still widely believed that, while a metal in its 'natural' state is crystalline, after bending backwards and forwards (i.e., the process of fatigue damage), metal locally becomes amorphous (devoid of crystalline structure). Isolated observations (e.g., Percy 1864) showed that evidence of

Figure 3.9. Macrostructure in an ingot.

crystalline structure reappeared on heating, and it was thus supposed that the amorphous material re-crystallised. The man who first showed unambiguously that metals consist of small crystal grains was Walter Rosenhain (1875–1934), an engineer who in 1897 came from Australia to undertake research for his doctorate with an exceptional engineering professor, Alfred Ewing, at Cambridge. Ewing (1855–1935) had much broader interests than were common at the time, and was one of the early scientific students of ferromagnetism. He introduced the concept of hysteresis in connection with magnetic behaviour, and indeed coined the word. As professor of mechanism and applied mechanics at Cambridge University from 1890, he so effectively reformed engineering education that he reconciled traditionalists there to the presence of engineers on campus (Glazebrook 1932–1935), culminating in 1997 with the appointment of an engineer as permanent vice-chancellor (university president). Ewing may well have been the first engineering professor to study materials in their own right.

Ewing asked Rosenhain to find out how it was possible for a metal to undergo plastic deformation without losing its crystalline structure (which Ewing believed metals to have). Rosenhain began polishing sheets of a variety of metals, bending them slightly, and looking at them under a microscope. Figure 3.10 is an example of the kind of image he observed. This shows two things: plastic deformation entails displacement in shear along particular lattice planes, leaving 'slip bands', and those traces lie along different directions in neighboring regions... i.e., in neighboring crystal grains. The identification of these separate regions as distinct crystal grains was abetted by the fact that chemical attack produced crystallographic etch figures

Figure 3.10. Rosenhain's micrograph showing slip lines in lead grains.

of different shapes in the various regions. (Etching of polished metal sections duly became an art in its own right.) This work, published under the title *On the crystalline structure of metals* (Ewing and Rosenhain 1900), is one of the key publications in modern physical metallurgy. A byproduct of this piece of research, simple in approach but profound in implication, was the first clear recognition of recrystallisation after plastic deformation, which came soon after the work of 1900; it was shown that the boundaries between crystal grains can migrate at high temperatures. The very early observations on recrystallisation are summarised by Humphreys and Hatherly (1995).

It was ironic that a few years later, Rosenhain began to insist that the material inside the slip bands (i.e., between the layers of unaffected crystal) had become amorphous and that this accounted for the progressive hardening of metals as they were increasingly deformed: there was no instrument to test this hypothesis and so it was unfruitful, but none the less hotly defended!

In the first sentence of Ewing and Rosenhain's 1900 paper, the authors state that "The microscopic study of metals was initiated by Sorby, and has been pursued by Arnold, Behrens, Charpy, Chernoff, Howe, Martens, Osmond, Roberts-Austen, Sauveur, Stead, Wedding, Werth, and others". So, a range of British, French, German, Russian and American metallurgists had used the reflecting microscope (and Grignon in France in the 18th century had seen grains in iron even without benefit of a microscope, Smith 1960), but nevertheless it was not until 1900 that the crystalline nature of metals became unambiguously clear.

In the 1900 paper, there were also observations of deformation twinning in several metals such as cadmium. The authors referred to earlier observations in minerals by mineralogists of the German school; these had in fact also observed slip in non-metallic minerals, but that was not recognised by Ewing and Rosenhain. The repeated rediscovery of similar phenomena by scientists working with different categories of materials was a frequent feature of 19th-century research on materials.

As mentioned earlier, Heycock and Neville, at the same time as Ewing and Rosenhain were working on slip, pioneered the use of the metallurgical microscope to help in the determination of phase diagrams. In particular, the delineation of phase fields stable only at high temperatures, such as the β field in the Cu–Sn diagram (Figure 3.7) was made possible by the use of micrographs of alloys quenched from different temperatures, like those shown in Figure 3.11. The use of micrographs showing the identity, morphology and distribution of diverse phases in alloys and ceramic systems has continued ever since; after World War II this approach was immeasurably reinforced by the use of the electron microprobe to provide compositional analysis of individual phases in materials, with a resolution of a micrometre or so. An early text focused on the microstructure of steels was published by the American metallurgist Albert Sauveur (1863–1939), while an

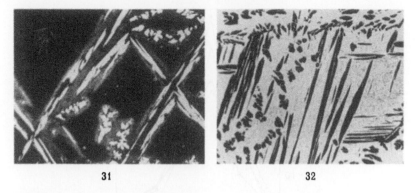

31   32

Figure 3.11. A selection of Heycock and Neville's micrographs of Cu–Sn alloys.

informative overview of the uses of microstructural examination in many branches of metallurgy, at a time before the electron microprobe was widely used, was published by Nutting and Baker (1965).

Ewing and Rosenhain pointed out that the shape of grains was initially determined simply by the chance collisions of separately nucleated crystallites growing in the melt. However, afterwards, when recrystallisation and grain growth began to be studied systematically, it was recognised that grain shapes by degrees approached metastable equilibrium – the ultimate equilibrium would be a single crystal, because any grain boundaries must raise the free energy. The notable English metallurgist Cyril Desch (1874–1958) (Desch 1919) first analysed the near-equilibrium shapes of metal grains in a polycrystal, and he made comparisons with the shapes of bubbles in a soapy water froth; but the proper topological analysis of grain shapes had to await the genius of Cyril Stanley Smith (1903–1992). His definitive work on this topic was published in 1952 and republished in fairly similar form, more accessibly, many years later (Smith 1952, 1981). Smith takes the comparison between metallic polycrystals and soap-bubble arrays under reduced air pressure further and demonstrates the similarity of form of grain-growth kinetics and bubble-growth kinetics. Grain boundaries are perceived as having an interface energy akin to the surface tension of soap films. He goes on to analyse in depth the topological relationships between numbers of faces, edges and corners of polyhedra in contact and the frequency distributions of polygonal faces with different numbers of edges as observed in metallic grains, biological cell assemblies and soap bubble arrays (Figure 3.12). This is an early example of a critical comparison between different categories of 'materials'. Cyril Smith was an exceptional man, whom we shall meet again in Chapter 14. Educated as a metallurgist in Birmingham University, he emigrated as a very young man to America where he became an industrial research metallurgist who published some key early papers on phase diagrams and phase

The Coming of Materials Science

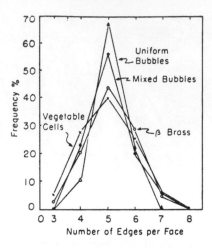

Figure 3.12. Frequency of various polygonal faces in grains, cells and bubbles (after C.S. Smith, *A Search for Structure*, 1981).

transformations, worked on the atom bomb at Los Alamos and then created the Institute for the Study of Metals at Chicago University (Section 14.4.1), before devoting himself wholly, at MIT, to the history of materials and to the relationship between the scientific and the artistic role of metals in particular. His books of 1960 and 1965 have already been mentioned.

The kind of quantitative shape comparisons published by Desch in 1919 and Smith in 1952 have since been taken much further and have given rise to a new science, first called quantitative metallography and later, *stereology*, which encompasses both materials science and anatomy. Using image analysers that apply computer software directly to micrographic images captured on computer screens, and working indifferently with single-phase and multiphase microstructures, quantities such as area fraction of phases, number density of particles, mean grain size and mean deviation of the distribution, mean free paths between phases, shape anisotropy, etc., can be determined together with an estimate of statistical reliability. A concise outline, with a listing of early texts, is by DeHoff (1986), while a more substantial recent overview is by Exner (1996). Figure 3.13, taken from Exner's treatment, shows examples of the ways in which quantities determined stereologically correlate with industrially important mechanical properties of materials. Stereology is further treated in Section 5.1.2.3.

A new technique, related to stereology, is *orientation-imaging*: here, the crystallographic orientations of a population of grains are determined and the misorientations between nearest neighbours are calculated and displayed graphically (Adams *et al.* 1993). Because properties of individual grain boundaries depend on

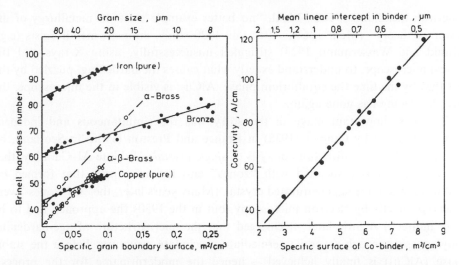

Figure 3.13. Simple relationships between properties and microstructural geometry: (a) hardness of some metals as a function of grain-boundary density; (b) coercivity of the cobalt phase in tungsten–carbide/cobalt 'hard metals' as a function of interface density (after Exner 1996).

the magnitude and nature of the misorientation, such a grain-boundary character distribution (gbcd) is linked to a number of macroscopic properties, corrosion resistance in particular; the life of the lead skeleton in an automobile battery has for instance been greatly extended by controlling the gbcd.

The study of phase transformations, another crucial aspect of modern materials science, is intimately linked with the examination of microstructure. Such matters as the crystallographic orientation of interfaces between two phases, the mutual orientation of the two neighbouring phase fields, the nature of ledges at the interface, the locations where a new phase can be nucleated (e.g., grain boundaries or lines where three grains meet), are examples of features which enter the modern understanding of phase transformations. A historically important aspect of this is *age-hardening*. This is the process of progressive hardening of an unstable (quenched) alloy, originally one based on Al–Cu, during storage at room temperature or slightly above. It was accidentally discovered by Alfred Wilm in Germany during 1906–1909; it remained a total mystery for more than a decade, until an American group, Merica *et al.* (1920) demonstrated that the solubility of copper in solid aluminium decreases sharply with falling temperature, so that an alloy consisting of a *stable* solid solution when hot becomes supersaturated when it has been quenched to room temperature, but can only approach equilibrium very slowly because of the low mobility of the atoms in the solid. This very important paper in the history of physical metallurgy at once supplied a basis for finding other alloy systems capable of age-hardening, on the basis of known phase diagrams of binary alloys. In the words of the eminent

American metallurgist, R.F. Mehl, "no better example exists in metallurgy of the power of theory" (Mehl 1967). After this 1920 study, eminent metallurgists (e.g., Schmid and Wassermann 1928) struggled unsuccessfully, using X-rays and the optical microscope, to understand exactly what causes the hardening, puzzled by the fact that by the time the equilibrium phase, $AlCu_2$, is visible in the microscope, the early hardening has gone again.

The next important stage in the story was the simultaneous and independent observation by Guinier (1938) in France and Preston (1938) in Scotland, by sophisticated X-ray diffraction analysis of *single crystals* of dilute Al–Cu alloy, that age-hardening was associated with "zones" enriched in copper that formed on {1 0 0} planes of the supersaturated crystal. (Many years later, the "GP zones" were observed directly by electron microscopy, but in the 1930s the approach had to be more indirect.) A little later, it emerged that the microstructure of age-hardening alloys passes through several intermediate precipitate structures before the stable phase ($AlCu_2$) is finally achieved – hence the modern name for the process, *precipitation-hardening*. Microstructural analysis by electron microscopy played a crucial part in all this, and dislocation theory has made possible a quantitative explanation for the increase of hardness as precipitates evolve in these alloys. After Guinier and Preston's pioneering research (published on successive pages of *Nature*), age-hardening in several other alloy systems was similarly analysed and a quarter century later, the field was largely researched out (Kelly and Nicholson 1963). One byproduct of all this was the recognition, by David Turnbull in America, that the whole process of age-hardening was only possible because the quenching process locked in a population of excess lattice vacancies, which greatly enhances atomic mobility. The entire story is very clearly summarised, with extracts from many classical papers, in a book by Martin (1968, 1998). It is worth emphasising here the fact that it was only when single crystals were used that it became possible to gain an understanding of the nature of age-hardening. Single crystals of metals are of no direct use in an industrial sense and so for many years no one thought of making them, but in the 1930s, their role in research began to blossom (Section 3.2.3 and Chapter 4, Section 4.2.1).

The sequence just outlined provides a salutary lesson in the nature of explanation in materials science. At first the process was a pure mystery. Then the relationship to the shape of the solid-solubility curve was uncovered; that was a *partial* explanation. Next it was found that the microstructural process that leads to age-hardening involves a succession of intermediate phases, none of them in equilibrium (a very common situation in materials science as we now know). An understanding of how these intermediate phases interact with dislocations was a further stage in explanation. Then came an understanding of the shape of the GP zones (planar in some alloys, globular in others). Next, the kinetics of the hardening needed to be understood in terms of excess vacancies and various short-circuit paths for diffusion. The holy grail of complete understanding recedes further and further as understanding deepens (so perhaps the field is after all *not* researched out).

The study of microstructures in relation to important properties of metals and alloys, especially mechanical properties, continues apace. A good overview of current concerns can be found in a multiauthor volume published in Germany (Anon. 1981), and many chapters in my own book on physical metallurgy (Cahn 1965) are devoted to the same issues.

Microstructural investigation affects not only an understanding of structural (load-bearing) materials like aluminium alloys, but also that of functional materials such as 'electronic ceramics', superconducting ceramics and that of materials subject to irradiation damage. Grain boundaries, their shape, composition and crystallographic nature, feature again and again. We shall encounter these cases later on. Even alloys which were once examined in the guise of structural materials have, years later, come to fresh life as functional materials: a striking example is Al–4wt%Cu, which is currently used to evaporate extremely fine metallic conducting 'interconnects' on microcircuits. Under the influence of a flowing current, such interconnects suffer a process called electromigration, which leads to the formation of voids and protuberances that can eventually create open circuits and thereby destroy the operation of the microcircuit. This process is being intensely studied by methods which involve a detailed examination of microstructure by electron microscopy and this, in turn, has led to strategies for bypassing the problem (e.g., Shi and Greer 1997).

3.1.3.1 Seeing is believing. To conclude this section, a broader observation is in order. In materials science as in particle physics, *seeing is believing*. This deep truth has not yet received a proper analysis where materials science is concerned, but it has been well analysed in connection with particle (nuclear) physics. The key event here was C.T.R. Wilson's invention in 1911 (on the basis of his observations of natural clouds while mountain-climbing) of the "cloud chamber", in which a sudden expansion and cooling of saturated water vapour in air through which high-energy particles are simultaneously passing causes water droplets to nucleate on air molecules ionised by the passing particles, revealing particle tracks. To say that this had a stimulating effect on particle physics would be a gross understatement, and indeed it is probably no accident (as radical politicians like to say) that Wilson's first cloud-chamber photographs were published at about the same time as the atomic hypothesis finally convinced most of the hardline sceptics, most of whom would certainly have agreed with Marcellin Berthelot's protest in 1877: "Who has ever seen, I repeat, a gaseous molecule or an atom?"

A research student in the history of science (Chaloner 1997) recently published an analysis of the impact of Wilson's innovation under the title "The most wonderful experiment in the world: a history of the cloud chamber", and the professor of the history of science at Harvard almost simultaneously published a fuller account of the same episode and its profound implications for the sources of scientific belief (Galison 1997). Chaloner at the outset of his article cites the great Lord Rutherford: "It may be argued that this new method of Mr. Wilson's has in the main only confirmed the deductions of the properties of the radiations made by other more indirect methods. While this is of course in some respects true, I would emphasize the importance to science of the gain in confidence of the accuracy of these deductions that followed from the publication of his beautiful photographs." There were those philosophers who questioned the credibility of a 'dummy' track, but as Galison tells us, no less an expert than the theoretical physicist Max Born made it clear that "there is something deeply valued about the visual character of evidence".

The study of microstructural change by micrographic techniques, applied to materials, has similarly, again and again, led to a "gain in confidence". This is the major reason for the importance of microstructure in materials science. A further consideration, not altogether incidental, is that micrographs can be objects of great beauty. As Chaloner points out, Wilson's cloud-chamber photographs were of exceptional technical perfection...they were beautiful (as Rutherford asserted), more so than those made by his successors, and because of that, they were reproduced again and again and their public impact thus accumulated. A medical scientist quoted by Chaloner remarked: "Perhaps it is more an article of faith for the morphologist, than a matter of demonstrated fact, that an image which is sharp, coherent, orderly, fine textured and *generally aesthetically pleasing* is more likely to be true than one which is coarse, disorderly and indistinct". Aesthetics are a touchstone for many: the great theoretical physicists Dirac and Chandrasekhar have recorded their conviction that mathematical beauty is a test of truth – as indeed did an eminent pure mathematician, Hardy.

It is not, then, an altogether superficial observation that metallographers, those who use microscopes to study metals (and other kinds of materials more recently), engage in frequent public competitions to determine who has made the most beautiful and striking images. The most remarkable micrographs, like Wilson's cloud-chamber photographs, are reproduced again and again over the years. A fine example is Figure 3.14 which was made about 1955 and is still frequently shown. It shows a dislocation source (see Section 3.2.3.2) in a thin slice of silicon. The silicon was 'decorated' with a small amount of copper at the surface of the slice; copper diffuses fast in silicon and makes a beeline for the dislocation where it is held fast by the elastic stress field surrounding any

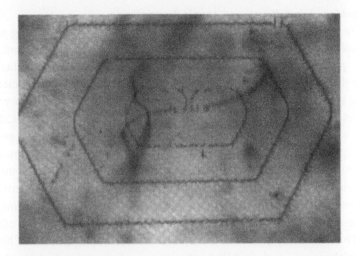

Figure 3.14. Optical micrograph of a dislocation source in silicon, decorated with copper (after W.C. Dash).

dislocation line. The sample has been photographed under a special microscope with optics transparent to infrared light; silicon is itself transparent to infrared, however, copper is not, and therefore the 'decorated' dislocation pattern shows up dark. This photograph was one of the very earliest direct observations of dislocations in a crystal; 'direct' here applies in the same sense in which it would apply to a track in one of Wilson's cloud-chambers. It is a ghost, but a very solid ghost.

3.2. SOME OTHER PRECURSORS

This chapter is entitled 'Precursors of Materials Science' and the foregoing major Sections have focused on the atomic hypothesis, crystallography, phase equilibria and microstructure, which I have presented as the main supports that made possible the emergence of modern materials science. In what follows, some other fields of study that made substantial contributions are more briefly discussed. It should be remembered that this is in no way a *textbook*; my task is not to explain the detailed nature of various phenomena and entitities, but only to outline how they came to be invented or recognised and how they have contributed to the edifice of modern materials science. The reader may well think that I have paid too much attention, up to now, to metals; that was inevitable, but I shall do my best to redress the balance in due course.

3.2.1 Old-fashioned metallurgy and physical metallurgy

Until the late 19th century metallurgy, while an exceedingly flourishing technology and the absolute precondition of material civilization, was a craft and neither a science nor, properly speaking, a technology. It is not part of my task here to examine the details of the slow evolution of metallurgy into a proper science, but it is instructive to outline a very few stages along that road, from the first widely read texts on metallurgical practice (Biringuccio 1540, 1945, Agricola 1556, 1912). Biringuccio was really the first craftsman to set down on paper the essentials of what was experimentally known in the 16th century about the preparation and working of metals and alloys. To quote from Cyril Smith's excellent introduction to the modern translation: "Biringuccio's approach is largely experimental: that is, he is concerned with operations that had been found to work without much regard to why. The state of chemical knowledge at the time permitted no other sound approach. Though Biringuccio has a number of working hypotheses, he does not follow the alchemists in their blind acceptance of theory which leads them to discard experimental evidence if it does not conform." Or as Smith remarked later (Smith 1977): "Despite their deep interest in manipulated changes in matter, the alchemists' overwhelming trust in theory blinded them to facts". The mutual, two-way interplay between theory and experiment which is the hallmark of modern science comes much later.

The lack of any independent methods to test such properties as "purity" could lead Biringuccio into reporting error. Thus, on page 60 of the 1945 translation, he writes: "That metal (i.e., tin) is known to be purer that shows its whiteness more or... if when some part of it is bent or squeezed by the teeth it gives its natural cracking noise...". That cracking noise, we now know, is caused by the rapid creation of deformation twins. When, in 1954, I was writing a review paper on twinning, I made up some intentionally very impure tin and bit it: it crackled merrily.

Reverting to the path from Biringuccio and Agricola towards modern scientific metallurgy, Cyril Smith, whom we have already met and who was the modern master of metallurgical history (though, by his own confession (Smith 1981), totally untrained in history), has analysed in great detail the gradual realisation that steel, known for centuries and used for weapons and armour, was in essence an alloy of iron *and carbon*. As he explained (Smith 1981), up to the late 18th century there was a popular phlogiston-based theory of the constitution of steel: the idea was that iron was but a stage *in the reduction to the purest state, which was steel*, and it was only a series of painstaking chemical analyses by eminent French scientists which finally revealed that the normal form of steel was a *less* pure form of iron, containing carbon and manganese in particular (by the time the existence of these elements was recognised around the time of the French revolution). The metallurgical historian Wertime (1961), who has mapped out in great detail the development of steel metallurgy and the understanding of the nature of steel, opines that "indeed, chemistry must in some degree attribute its very origins to iron and its makers".

This is an occasion for another aside. For millenia, it was fervently believed by natural philosophers that purity was the test of quality and utility, as well as of virtue, and all religions, Judaism prominent among them, aspire to purity in all things. The anthropologist Mary Douglas wrote a famous text vividly entitled *Purity and Danger*; this was about the dangers associated with *im*purity. In a curious but intriguing recent book (Hoffmann and Schmidt 1997), the two authors (one a famous chemist, the other an expert on the Mosaic laws of Judaism) devote a chapter to the subtleties of "Pure/Impure", prefacing it with an invocation by the prophet Jeremiah: "I have made you an assayer of My people – a refiner – You are to note and assay their ways. They are bronze and iron, they are all stubbornly defiant; they deal basely, all of them act corruptly." Metallurgy is a difficult craft: the authors note that US President Herbert Hoover (the modern translator of Agricola), who was a connoisseur of critically minded people, opined that Jeremiah was a metallurgist "which might account for his critical tenor of mind". The notion that *intentional* impurity (which is never called that – the name for it is 'alloying' or 'doping') is often highly beneficial took a very long time to be acceptable. Roald Hoffman, one of the authors of the above-mentioned book, heads one of his sections "Science and the Drive towards Impurity" and the reader quickly comes to appreciate the validity of the section title. So, a willing acceptance of intentional impurity is one of the hallmarks of modern materials science. However, all things go in cycles: once germanium and silicon began to be used as semiconductors, levels of purity never previously approached became indispensable, and before germanium or silicon could be usefully doped to make junctions and transistors, these metalloids had first to be ultrapurified. Purity necessarily precedes doping, or if you prefer, impurity comes before purity which leads to renewed impurity. That is today's orthodoxy.

Some of the first stirrings of a scientific, experimental approach to the study of metals and alloys are fully analysed in an interesting history by Aitchison (1960), in which such episodes as Sorby's precocious metallography and the discovery of age-hardening are gone into. Yet throughout the 19th century, and indeed later still, that scientific approach was habitually looked down upon by many of the most effective practical men. A good late example is a distinguished Englishman, Harry Brearley (1871–1948), who in 1913 invented (or should one say discovered?) stainless steel. He was very sceptical about the utility of 'metallographists', as scientific students of metals were known in his day. It is worth quoting *in extenso* what Brearley, undoubtedly a famous practical steelmaker, had to say in his (reissued) autobiography (Brearley 1995) about the conflict between the scientists and the practical men: "It would be foolish to deny the fruitfulness of the enormous labour, patient and often unrewarded, which has replaced the old cookery-book method of producing alloyed metals by an understanding intelligence which can be called scientific. But it would be hardly less foolish to imagine, because a subject can be talked about more intelligibly, that the words invariably will be words of wisdom. The operations of an old trade may not lend themselves to complete representations by symbols, and it is a grievous mistake to suppose that what the University Faculty does not know cannot be worth knowing. Even a superficial observer might see that the simplifications, and elimination of interferences, which are possible and may be desirable in a laboratory experiment, may be by no means possible in an industrial process which the laboratory experiment aims to elucidate. To know the ingredients of a rice pudding and the appearance of a rice pudding when well made does not mean, dear reader, that you are able to make one." He went on to remark: "What a man sees through the microscope is more of less, and his vision has been known to be thereby so limited that he misses what he is looking for, which has been apparent at the first glance to the man whose eye is informed by experience." That view of things has never entirely died out.

At the same time as Brearley was discovering stainless steel and building up scepticism about the usefulness of metallographists, Walter Rosenhain, whom we have already met in Section 3.1.3 and who had quickly become the most influential metallurgist in Britain, was preparing to release a bombshell. In 1906 he had become the Superintendent of the Metallurgy Division of the new National Physical Laboratory at the edge of London and with his team of scientists was using a variety of physical methods to study the equilibria and properties of alloys. In 1913 he was writing his masterpiece, a book entitled *An Introduction to the Study of Physical Metallurgy*, which was published a year later (Rosenhain 1914). This book (which appeared in successive editions until 1934) recorded the transition of scientific metallurgy from being in effect a branch of applied chemistry to becoming an aspect of applied physics. It focused strongly on phase diagrams, a concept which emerged from physical-chemistry principles created by a mechanical engineer turned mathematical physicist. Gibbs single-handedly proved that in the presence of genius, scientific labels matter not at all, but most researchers are not geniuses.

Rosenhain (1917) published a book chapter entitled "The modern science of metals, pure and applied", in which he makes much of the New Metallurgy (which invariably rates capital letters!). In essence, this is an eloquent plea for the importance of basic research on metals; it is the diametric converse of the passage by Brearley which we met earlier.

In the three decades following the publication of Rosenhain's book, the physical science of metals and alloys developed rapidly, so that by 1948 it was possible for Robert Franklin Mehl (1898–1976) (see Smith 1990, Smith and Mullins 2001 and Figure 3.15), a doyen of American physical metallurgy, to bring out a book entitled *A Brief History of the Science of Metals* (Mehl 1948), which he then updated in the

Figure 3.15. Robert Franklin Mehl (courtesy Prof. W.W. Mullins).

historical chapter of the first edition of my multiauthor book, *Physical Metallurgy* (Cahn 1965). The 1948 version already had a bibliography of 364 books and papers. These masterly overviews by Mehl are valuable in revealing the outlook of his time, and for this purpose they can be supplemented by several critical essays he wrote towards the end of his career (Mehl 1960, 1967, 1975). After working with Sauveur at Harvard, Mehl in 1927, aged 29, joined the Naval Research Laboratory in Washington, DC, destined to become one of the world's great laboratories (see Rath and DeYoung 1998), as head of its brandnew Physical Metallurgy Division, which later became just the Metallurgy Division, indicating that 'physical metallurgy' and 'metallurgy' had become synonymous. So the initiative taken by Rosenhain in 1914 had institutional effects just a few years later. In Mehl's 1967 lecture at the Naval Research Laboratory (by this time he had been long established as a professor in Pittsburgh), he seeks to analyse the nature of physical metallurgy through a detailed examination of the history of just one phenomenon, the decomposition (on heat-treatment) of austenite, the high-temperature form of iron and steel. He points out that "physical metallurgy is a very broad field", and goes on later to make a fanciful comparison: "The US is a pluralistic nation, composed of many ethnic strains, and in this lies the strength of the country. Physical metallurgy is comparably pluralistic and has strength in this". He goes on to assert something quite new in the history of metallurgy: "Theorists and experimentalists interplay. Someone has said that 'no one believes experimental data except the man who takes them, but everyone believes the results of a theoretical analysis except the man who makes it'." And at the end, having sucked his particular example dry, he concludes by asking "What is physical metallurgy?", and further, how does it relate to the fundamental physics which in 1967 was well on the way to infiltrating metallurgy? He asks: "Is it not the primary task of metallurgists through research to *try to define a problem*, to do the initial scientific work, nowadays increasingly sophisticated, upon which the solid-state physicist can base his further and relentless probing towards ultimate causes?" That seems to me admirably to define the nature of the discipline which was the direct precursor of modern materials science. I shall rehearse further examples of the subject-matter of physical metallurgy later in this chapter, in the next two and in Chapter 9.

In 1932, Robert Mehl at the age of 34 became professor of metallurgy at Carnegie Institute of Technology in Pittsburgh, and there created the Metals Research Laboratory (Mehl 1975), which was one of *the* defining influences in creating the 'new metallurgy' in America. It is still, today, an outstanding laboratory. In spite of his immense positive influence, after the War Mehl dug in his heels against the materials science concept; it would be fair to say that he led the opposition. He also inveighed against vacancies and dislocations, which he thought tarred with the brush of the physicists whom he regarded as enemies of metallurgy; the consequences of this scepticism for his own distinguished experimental work on diffusion are outlined in Section 4.2.2. Mehl thought that metallurgy incorporated all the variety that was needed. According to a recently completed memoir (Smith and Mullins 2001), Mehl regarded "the move (to MSE) as a hollow gimmick to obtain funds..." Smith and Mullins go on to say "Nevertheless, he undoubtedly played a central and essential role in preparing the ground for the benefits of this broader view of materials". So the foe of materials science inadvertently helped it on its way.

3.2.2 Polymorphism and phase transformations

In Section 3.1.1 we encountered the crystallographer and chemist Eilhardt Mitscherlich who around 1818 discovered the phenomenon of polymorphism in some substances, such as sulphur. This was the first recognition that a solid phase can change its crystal structure as the temperature varies (a phase transformation), or alternatively that the same compound can crystallise (from the melt, the vapour or from a chemical reaction) in more than one crystalline form. This insight was first developed by the mineralogists (metallurgists followed much later). As a recent biography (Schütt 1997) makes clear, Mitscherlich started as an oriental linguist, began to study medicine and was finally sidetracked into chemistry, from where he learned enough mineralogy to study crystal symmetry, which finally led him to isomorphism and polymorphism.

The polymorphism of certain metals, iron the most important, was after centuries of study perceived to be the key to the hardening of steel. In the process of studying iron polymorphism, several decades were devoted to a red herring, as it proved: this was the β-iron controversy. β-iron was for a long time regarded as a phase distinct from α-iron (Smith 1965) but eventually found to be merely the ferromagnetic form of α-iron; thus the supposed transition from β to α-iron was simply the Curie temperature. β-iron has disappeared from the iron–carbon phase diagram and all transformations are between α and γ.

Polymorphism in nonmetals has also received a great deal of study and is particularly clearly discussed in a book by two Indian physicists (Verma and Krishna 1966) which also links to the phenomenon of polytypism, discussed in Section 3.2.3.4.

Of course, freezing of a liquid – or its inverse – are themselves phase transformations, but the scientific study of freezing and melting was not developed until well into the 20th century (Section 9.1.1). Polymorphism also links with metastability: thus aragonite, one polymorphic form of calcium carbonate, is under most circumstances metastable to the more familiar form, calcite.

The really interesting forms of phase transformations, however, are those where a single phase *precipitates* another, as in the age-hardening (= precipitation-hardening) process. Age-hardening is a good example of a *nucleation-and-growth* transformation, a very widespread category. These transformations have several quite distinct aspects which have been separately studied by different specialists – this kind of subdivision in the search for understanding has become a key feature of modern materials science. The aspects are: nucleation mechanism, growth mechanism, microstructural features of the end-state, crystallography of the end-state, and kinetics of the transformation process. Many transformations of this kind in both alloy and ceramic systems lead to a *Widmanstätten structure*, like that in Figure 3.4 but on a much finer scale. A beautiful example can be seen in Figure 3.16, taken from a book mentioned later in this paragraph. An early example of an intense study of one feature, the orientation relationship between parent and daughter phases, is the impressive body of crystallographic research carried out by C.S. Barrett and R.F. Mehl in Pittsburgh in the early 1930s, which led to the recognition that in

Figure 3.16. Widmanstätten precipitation of a hexagonal close-packed phase from a face-centred cubic phase in a Cu–Si alloy. Precipitation occurs on $\{1\ 1\ 1\}$ planes of the matrix, and a simple epitaxial crystallographic correspondence is maintained, $(0\ 0\ 0\ 1)_{hex} \parallel (1\ 1\ 1)_{cub}$ (after Barrett and Massalski 1966).

transformations of this kind, plates are formed in such a way that the atomic fit at the interface is the best possible, and correspondingly the interface energy is minimised. This work, and an enormous amount of other early research, is concisely but very clearly reviewed in one of the classic books of physical metallurgy, *Structure of Metals* (Barrett and Massalski 1966). The underlying mechanisms are more fully examined in an excellent text mentioned earlier in this chapter (Porter and Easterling 1981), while the growth of understanding of age-hardening has been very clearly presented in a historical context by Martin (1968, 1998).

The historical setting of this important series of researches by Barrett and Mehl in the 1930s was analysed by Smith (1963), in the light of the general development of X-ray diffraction and single-crystal research in the 1920s and 1930s. The Barrett/ Mehl work largely did without the use of single crystals and X-ray diffraction, and yet succeeded in obtaining many of the insights which normally required those approaches. The concept of *epitaxy*, orientation relationships between parent and daughter phases involved in phase transformations, had been familiar only to mineralogists when Barrett and Mehl began their work, but by its end, the concept had become familiar to metallurgists also and it soon became a favoured theme of investigation. Mehl's laboratory in Pittsburgh in the 1930s was America's most prolific source of research metallurgists.

The kinetics of nucleation-and-growth phase transformations has proved of the greatest practical importance, because it governs the process of heat-treatment of alloys – steels in particular – in industrial practice. Such kinetics are formulated where possible in terms of the distinct processes of nucleation rates and growth rates, and the former have again to be subdivided according as nuclei form all at once or progressively, and according as they form homogeneously or are restricted to sites such as grain boundaries. The analysis of this problem – as has so often happened in the history of materials science – has been reinvented again and again by investigators who did not know of earlier (or simultaneous) research. Equations of the general form $f = 1 - \exp(-kt^n)$ were developed by Gustav Tammann of Göttingen (Tammann 1898), in America by Melvin Avrami (who confused the record by changing his name soon after) and by Johnson and the above-mentioned Mehl both in 1939, and again by Ulick Evans of Cambridge (Evans 1945), this last under the title "The laws of expanding circles and spheres in relation to the lateral growth of surface films and the grain size of metals". There is a suggestion that Evans was moved to his investigation by an interest in the growth of lichens on rocks. A.N. Kolmogorov, in 1938, was another of the pioneers.

The kinetics of diffusion-controlled phase transformations has long been a focus of research and it is vital information for industrial practice as well as being a fascinating theme in fundamental physical metallurgy. An early overview of the subject is by Aaronson *et al.* (1978).

A quite different type of phase transformation is the *martensitic* type, named by the French metallurgist Floris Osmond after the German 19th-century metallographer Adolf Martens. Whereas the nucleation-and-growth type of transformation involves migration of atoms by diffusive jumps (Section 4.2.2) and is often very slow, martensitic transformations, sometimes termed diffusionless, involve regimented shear of large groups of atoms. The hardening of carbon-steel by quenching from the γ-phase (austenite) stable at high temperatures involves a martensitic transformation. The crystallographic relationships involved in such transformations are much more complex than those in nucleation-and-growth transformations and their elucidation is one of the triumphs of modern transformation theory. Full details can be found in the undisputed bible of phase transformation theory (Christian 1965). Georgi Kurdyumov, the Russian 'father of martensite', appears in Chapter 14.

There are other intermediate kinds of transformations, such as the bainitic and massive transformations, but going into details would take us too far here. However, a word should be said about *order–disorder transformations*, which have played a major role in modern physical metallurgy (Barrett and Massalski 1966). Figure 3.17 shows the most-studied example of this, in the Cu–Au system: the nature of the process shown here was first identified in Sweden in 1925, where there was a flourishing school of "X-ray metallographers" in the 1920s (Johansson and Linde 1925). At high temperatures the two kinds of atom are distributed at random (or nearly at random) over all lattice sites, but on cooling they redistribute themselves on groups of sites which now become crystallographically quite distinct. Many alloys behave in this way, and in the 1930s it was recognised that the explanation was based on the Gibbsian competition between internal energy and entropy: at high temperature entropy wins and disorder prevails, while at low temperatures the stronger bonds between unlike atom pairs win. This picture was quantified by a simple application of statistical mechanics, perhaps the first application to a phase transformation, in a celebrated paper by Bragg and Williams (1934). (Bragg's recollection of this work in old age can be found in Bragg (1975, 1992), p. 212.) The ideas formulated here are equally applicable to the temperature-dependent alignment of magnetic spins in a ferromagnet and to the alignment of long organic molecules in a liquid crystal. Both the experimental study of order–disorder transitions (in some of them, very complex microstructures appear, Tanner and Leamy 1974) and the theoretical convolutions have attracted a great deal of attention, and ordered alloys, nowadays called *intermetallics*, have become important structural materials for use at high temperatures. The complicated way in which order–disorder transformations fit midway between physical metallurgy and solid-state physics has been surveyed by Cahn (1994, 1998).

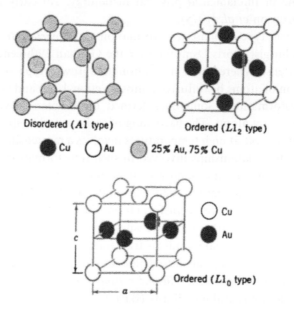

Figure 3.17. Ordering in Cu–Au alloys.

The Bragg–Williams calculation was introduced to metallurgical undergraduates (this was before materials science was taught as such) for the first time in a pioneering textbook by Cottrell (1948), based on his teaching in the Metallurgy Department at Birmingham University, England; Bragg–Williams was combined with the Gibbsian thermodynamics underlying phase diagrams, electron theory of metals and alloys and its applications, and the elements of crystal defects. This book marked a watershed in the way physical metallurgy was taught to undergraduates, and had a long-lasting influence.

The whole field of phase transformations has rapidly become a favourite stamping-ground for solid-state physicists, and has broadened out into the closely related aspects of phase stability and the prediction of crystal structures from first theoretical principles (e.g., de Fontaine 1979, Stocks and Gonis 1989). Even professional mathematicians are moving into the game (Gurtin 1984). The extremely extensive and varied research on phase transformations by mainline materials scientists is recorded in a series of substantial conference proceedings, with a distinct emphasis on microstructural studies (the first in the series: Aaronson *et al.* 1982); a much slimmer volume that gives a good sense of the kind of research done in the broad field of phase transformations is the record of a symposium in honor of John Kirkaldy, a nuclear physicist turned materials scientist (Embury and Purdy 1988); his own wide-ranging contribution to the symposium, on the novel concept of 'thermologistics', is an illustration of the power of the phase-transformation concept! A good example of a treatment of the whole field of phase transformations (including solidification) in a manner which represents the interests of mainline materials scientists while doing full justice to the physicists' extensive input is a multiauthor book edited by Haasen (1991).

While most of the earlier research was done on metals and alloys, more recently a good deal of emphasis has been placed on ceramics and other inorganic compounds, especially 'functional' materials used for their electrical, magnetic or optical properties. A very recent collection of papers on oxides (Boulesteix 1998) illustrates this shift neatly. In the world of polymers, the concepts of phase transformations or phase equilibria do not play such a major role; I return to this in Chapter 8.

The conceptual gap between metallurgists (and nowadays materials scientists) on the one hand and theoretical solid-state physicists and mathematicians on the other, is constantly being bridged (Section 3.3.1) and as constantly being reopened as ever new concepts and treatments come into play in the field of phase transformations; the large domain of critical phenomena, incorporating such recondite concepts as the renormalisation group, is an example. There are academic departments, for instance one of Materials Science at the California Institute of Technology, which are having success in bridging conceptual gaps of this kind.

3.2.2.1 Nucleation and spinodal decomposition. One specific aspect of phase transformations has been so influential among physical metallurgists, and also more recently among polymer physicists, that it deserves a specific summary; this is the study of the nucleation and of the spinodal decomposition of phases. The notion of homogeneous nucleation of one phase in another (e.g., of a solid in a supercooled melt) goes back all the way to Gibbs. Minute embryos of different sizes (that is, transient nuclei) constantly form and vanish; when the product phase has a lower free energy than the original phase, as is the case when the latter is supercooled, then some embryos will survive if they reach a size large enough for the gain in volume free energy to outweigh the energy that has to be found to create the sharp interface between the two phases. Einstein himself (1910) examined the theory of this process with regard to the nucleation of liquid droplets in a vapour phase. Then, after a long period of dormancy, the theory of nucleation kinetics was revived in Germany by Max Volmer and A.Weber (1926) and improved further by two German theoretical physicists of note, Richard Becker and Wolfgang Döring (1935). (We shall meet Volmer again as one of the key influences on Frank's theory of crystal growth in 1953, Section 3.2.3.3.) Reliable experimental measurements became possible much later still in 1950, when David Turnbull, at GE, perfected the technique of dividing a melt up into tiny hermetic compartments so that heterogeneous nucleation catalysts were confined to just a few of these; his measurements (Turnbull and Cech 1950, Turnbull 1952) are still frequently cited.

It took a long time for students of phase transformations to understand clearly that there exists an alternative way for a new phase to emerge by a diffusive process from a parent phase. This process is what the Nobel-prize-winning Dutch physicist Johannes van der Waals (1837–1923), in his doctoral thesis, first christened the "spinodal". He recognised that a liquid beyond its liquid/gas critical point, having a negative compressibility, was unstable towards *continuous changes*. A negative Gibbs free energy has a similar effect, but this took a very long time to become clear. The matter was at last attacked head-on in a famous theoretical paper (based on a 1956 doctoral thesis) by the Swedish metallurgist Mats Hillert (1961): he studied theoretically both atomic segregation and atomic ordering, two alternative diffusional processes, in an unstable metallic solid solution. The issue was taken further by John Cahn and the late John Hilliard in a series of celebrated papers which has caused them to be regarded as the creators of the modern theory of spinodal decomposition; first (Cahn and Hilliard 1958) they revived the concept of a *diffuse* interface which gradually thickens as the unstable parent phase decomposes *continuously* into regions of diverging composition (but, typically, of similar crystal structure); later, John Cahn (1961) generalised the theory to three dimensions. It then emerged that a very clear example of spinodal decomposition in the solid state had been studied in detail as long ago as 1943, at the Cavendish by Daniel and

Lipson (1943, 1944), who had examined a copper–nickel–iron ternary alloy. A few years ago, on an occasion in honour of Mats Hillert, Cahn (1991) mapped out in masterly fashion the history of the spinodal concept and its establishment as a widespread alternative mechanism to classical nucleation in phase transformations, specially of the solid–solid variety. An excellent, up-to-date account of the present status of the theory of spinodal decomposition and its relation to experiment and to other branches of physics is by Binder (1991). The Hillert/Cahn/Hilliard theory has also proved particularly useful to modern polymer physicists concerned with structure control in polymer blends, since that theory was first applied to these materials in 1979 (see outline by Kyu 1993).

3.2.3 Crystal defects

I treat here the principal types of point defects, line defects, and just one of the many kinds of two-dimensional defects. A good, concise overview of all the many types of crystal defects, and their effects on physical and mechanical properties, has been published by Fowler *et al.* (1996).

3.2.3.1 Point defects. Up to now, the emphasis has been mostly on metallurgy and physical metallurgists. That was where many of the modern concepts in the physics of materials started. However, it would be quite wrong to equate *modern* materials science with physical metallurgy. For instance, the gradual clarification of the nature of point defects in crystals (an essential counterpart of dislocations, or line defects, to be discussed later) came entirely from the concentrated study of ionic crystals, and the study of polymeric materials after the Second World War began to broaden from being an exclusively chemical pursuit to becoming one of the most fascinating topics of physics research. And that is leaving entirely to one side the huge field of semiconductor physics, dealt with briefly in Chapter 7. Polymers were introduced in Chapter 2, Section 2.1.3, and are further discussed in Chapter 8; here we focus on ionic crystals.

At the beginning of the century, nobody knew that a small proportion of atoms in a crystal are routinely missing, even less that this was not a matter of accident but of thermodynamic equilibrium. The recognition in the 1920s that such "vacancies" had to exist in equilibrium was due to a school of statistical thermodynamicians such as the Russian Frenkel and the Germans Jost, Wagner and Schottky. That, moreover, as we know now, is only one kind of "point defect"; an atom removed for whatever reason from its lattice site can be inserted into a small gap in the crystal structure, and then it becomes an "interstitial". Moreover, in insulating crystals a point defect is apt to be associated with a local excess or deficiency of electrons, producing what came to be called "colour centres", and this can lead to a strong sensitivity to light: an extreme example of this is the photographic reaction in silver halides. In all kinds of crystal, pairs of vacancies can group into divacancies and they can also become attached to solute atoms; interstitials likewise can be grouped. All this was in the future when research on point defects began in earnest in the 1920s.

At about the same time as the thermodynamicians came to understand why vacancies had to exist in equilibrium, another group of physicists began a systematic experimental assault on colour centres in insulating crystals: this work was mostly done in Germany, and especially in the famous physics laboratory of Robert Pohl (1884–1976) in Göttingen. A splendid, very detailed account of the slow, faltering approach to a systematic knowledge of the behaviour of these centres has recently been published by Teichmann and Szymborski (1992), as part of a magnificent collaborative history of solid-state physics. Pohl was a resolute empiricist, and resisted what he regarded as premature attempts by theorists to make sense of his findings. Essentially, his school examined, patiently and systematically, the wavelengths of the optical absorption peaks in synthetic alkali halides to which controlled "dopants" had been added. (Another approach was to heat crystals in a vapour of, for instance, an alkali metal.) Work with X-ray irradiation was done also, starting with a precocious series of experiments by Wilhelm Röntgen in the early years of the century; he published an overview in 1921. Other physicists in Germany ignored Pohl's work for many years, or ridiculed it as "semiphysics" because of the impurities which they thought were bound to vitiate the findings. Several decades were yet to elapse before minor dopants came to the forefront of applied physics in the world of semiconductor devices. Insofar as Pohl permitted any speculation as to the nature of his 'colour centres', he opined that they were of non-localised character, and the adherents of localised and of diffuse colour centres quarrelled fiercely for some years. Even without a theoretical model, Pohl's cultivation of optical spectroscopy, with its extreme sensitivity to minor impurities, led through collaborations to advances in other fields, for instance, the isolation of vitamin D.

One of the first experimental physicists to work with Pohl on impure ionic crystals was a Hungarian, Zoltan Gyulai (1887–1968). He rediscovered colour centres created by X-ray irradiation while working in Göttingen in 1926, and also studied the effect of plastic deformation on the electrical conductivity. Pohl was much impressed by his Hungarian collaborator's qualities, as reported in a little survey of physics in Budapest (Radnai and Kunfalvi 1988). This book reveals the astonishing flowering of Hungarian physics during the past century, including the physics of materials, but many of the greatest Hungarian physicists (people like Szilard, Wigner, von Neumann, von Karman, Gabor, von Hevesy, Kurti (who has just died at age 90 as I write this), Teller (still alive)) made their names abroad because the unceasing sequence of revolutions and tyrannies made life at home too uncomfortable or even dangerous. However, Gyulai was one of those who returned and he later presided over the influential Roland Eötvös Physical Society in Budapest.

Attempts at a theory of what Pohl's group was discovering started in Russia, whose physicists (notably Yakov Frenkel and Lev Landau) were more interested in Pohl's research than were most of his own compatriots. Frenkel, Landau and Rudolf Peierls, in the early 1930s, favoured the idea of an electron trapped "by an extremely distorted part of the lattice" which developed into the idea of an "exciton", an activated atom. Finally, in 1934, Walter Schottky in Germany first proposed that colour centres involved a pairing between an anion vacancy and an extra (trapped) electron – now sometimes called a "Schottky defect". (Schottky was a rogue academic who did not like teaching and migrated to industry, where he fastened his teeth on copper oxide rectifiers; thus he approached a fundamental problem in alkali halides via an industrial problem, an unusual sequence at that time.)

At this point, German research with its Russian topdressing was further fertilised by sudden and major input from Britain and especially from the US. In 1937, at the instigation of Nevill Mott (1905–1996) (Figure 3.18), a physics conference was held in Bristol University, England, on colour centres (the beginning of a long series of influential physics conferences there, dealing with a variety of topics including also dislocations, crystal growth and polymer physics). Pohl delivered a major experimental lecture while R.W. Gurney and Mott produced a quantum theory of colour centres, leading on soon afterwards to their celebrated model of the photographic effect. (This sequence of events was outlined later by Mitchell 1980.)

The leading spirit in the US was Frederick Seitz (b. 1911) (Figure 3.19). He first made his name with his model, jointly with his thesis adviser, Eugene Wigner, for calculating the electron band structure of a simple metal, sodium. Soon afterwards he spent two years working at the General Electric Company's central research centre (the first and at that time the most impressive of the large industrial laboratories in America), and became involved in research on suitable phosphorescent materials ("phosphors") for use as a coating in cathode-ray tubes; to help him in this quest, he began to study Pohl's papers. (These, and other stages in Seitz's life are covered in some autobiographical notes published by the Royal Society (Seitz 1980) and more recently in an autobiographical book (Seitz 1994).) Conversations with Mott then focused his attention on crystal defects. Many of the people who were to create the theory of colour centres after the War devoted themselves meanwhile to the improvement of phosphors for radar (TV tubes were still in the future), before switching to the related topic of radiation damage in relation to the Manhattan Project. After the War, Seitz returned to the problem of colour centres and in 1946 published the first of two celebrated reviews (Seitz 1946), based on his resolute attempts to unravel the nature of colour centres. Theory was now buttressed by

Figure 3.18. Nevill Francis Mott (courtesy Mrs. Joan Fitch).

purpose-designed experiments. Otto Stern (with two collaborators) was able to show that when ionic crystals had been greatly darkened by irradiation and so were full of colour centres, there was a measurable decrease in density, by only one part in 10^4. (This remarkably sensitive measurement of density was achieved by the use of a flotation column, filled with liquid arranged to have a slight gradient of density from top to bottom, and establishing where the crystal came to rest.) Correspondingly, the concentration of vacancies in metals was measured directly by an equally ingenious experimental approach due to Feder and Nowick (1958), followed up later by Simmons and Balluffi (1960–1963): they compared dilatometry (measurements of changes in length as a function of changing temperature) with precision measurements of lattice parameter, to extract the concentration of vacancies in equilibrium at various temperatures. This approach has proved very fruitful.

Vacancies had at last come of age. Following an intense period of research at the heart of which stood Seitz, he published a second review on colour centres (Seitz 1954). In this review, he distinguished between 12 different types of colour centres, involving single, paired or triple vacancies; many of these later proved to be

Figure 3.19. Frederick Seitz (courtesy Dr. Seitz).

misidentifications, but nevertheless, in the words of Teichmann and Szymborski, "it was to Seitz's credit that, starting in the late 1940s, both experimental and theoretical efforts became more convergent and directed to the solution of clearly defined problems". The symbiosis of quantitative theory and experiment (which will be treated in more detail in Chapter 5) got under way at much the same time for metals and for nonmetals.

Nowick (1996) has outlined the researches done on crystal defects during the period 1949–1959 and called this the "golden age of crystal defects". A recent, very substantial overview (Kraftmakher 1998) admirably surveys the linkage between vacancies in equilibrium and 'thermophysical' properties of metals: this paper includes a historical table of 32 key papers, on a wide range of themes and techniques, 1926–1992.

Point defects are involved in many modern subfields of materials science: we shall encounter them again particularly in connection with diffusion (Chapter 4, Section 4.2.2) and radiation damage (Chapter 5, Section 5.1.3).

3.2.3.2 Line defects: dislocations. The invention of dislocations is perhaps the most striking example in the history of materials science of a concept being recognised as soon as the time is ripe. A dislocation is a *line defect*, in a crystal, which is linked to an elastic stress field within a crystal in such a way that under an external stress, a dislocation is impelled to move through the crystal and thereby causes a permanent change of shape... i.e., plastic deformation. Dislocations were invented – that is the right word, they were *not* initially 'discovered' – mainly because of a huge mismatch between the stress calculated from first principles for the stress needed to deform crystal plastically, and the much smaller stress actually observed to suffice. A subsidiary consideration which led to the same explanatory concept was the observation that any crystalline material subjected to plastic deformation thereby becomes harder – it *work-hardens*. Three scientists reached the same conclusion at almost the same time, and all published their ideas in 1934: Michael Polanyi (1891–1976), Geoffrey Taylor (1886–1975), both of them already encountered, and Egon Orowan (1902–1989): two of these were emigré Hungarians, which shows again the remarkable contributions to science made by those born in this country of brilliant scholars, of whom so many were forced by 20th-century politics into emigration.

The papers which introduced the concept of a dislocation all appeared in 1934 (Polanyi 1934, Taylor 1934, Orowan 1934). Figure 3.20 shows Orowan's original sketch of an edge dislocation and Taylor's schematic picture of a dislocation moving. It was known to all three of the co-inventors that plastic deformation took place by slip on lattice planes subjected to a higher shear stress than any of the other symmetrically equivalent planes (see Chapter 4, Section 4.2.1). Taylor and his collaborator Quinney had also undertaken some quite remarkably precise calorimetric research to determine how much of the work done to deform a piece of metal

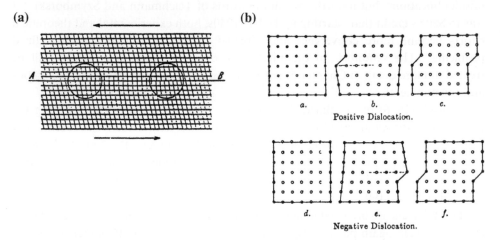

Figure 3.20. An edge dislocation, as delineated by Orowan (a) and Taylor (b).

remained behind as stored energy, and Taylor decided that this stored energy must be localised as elastic distortion at some kind of crystal defect; he also believed that work-hardening must be due to the interaction between these defects, which increased in concentration by some unknown mechanism. Orowan was also intrigued by the fact that some of his zinc crystals when stressed deformed in a discontinuous, jerky fashion (he reflected about this observation all his life, as many great scientists tend to do about their key observations) and decided that each 'jerk' must be due to the operation of one defect. All three were further moved by the recognition that plastic deformation begins at stresses very much lower (by a factor of ≈ 1000) than would be necessary if the whole slip plane operated at once. The defects illustrated in Figure 3.20 can move under quite small stresses, in effect because only a small area of slip plane glides at any one instant. In the 3 papers, this is presented as the result of a local elastic enhancement of stress, but it is in fact more accurate to present the matter as a reduction in the stress needed to move the defect. Taylor, alone, used his theory to interpret the actual process of work-hardening, and he was no doubt driven to this by consideration of his own measurements of the measured retained energy of cold work (Taylor and Quinney 1934).

The above very abbreviated account of the complicated thought processes that led Polanyi, Taylor and Orowan to their simultaneous papers can be expanded by reference to detailed accounts, including autobiographical notes by all three. One interesting fact that emerges from Polanyi's own account (Polanyi 1962) is that his paper was actually ready several months before Orowan's, but he was already in regular contact with Orowan and, learning that Orowan's ideas were also rapidly gelling, Polanyi voluntarily waited and submitted his paper at the same time as Orowan's, and they appeared side by side in the same issue of *Zeitschrift für Physik*. Polanyi was a gentleman of the old school; his concern with ethics was no doubt one of the impulses which drove him later in life to become a professional philosopher; he dropped crystal plasticity after 1934. The movement of Taylor's ideas can be found in a recent biography (Batchelor 1996). This includes a passage contributed by Nevill Mott and another by Taylor himself. At the end of this passage, Taylor points out that when he had finished the work on crystal plasticity, he went back promptly to his beloved fluid mechanics and to the design of novel anchors (he was an enthusiastic yachtsman). Nevertheless, over the years Taylor did a great deal of work on the mechanics of monocrystals and polycrystals, on the calorimetric determination of retained energy of cold work (he took several bites at this hard cherry) and on the nature of work-hardening: his 41 papers in this broad area have been collected in one impressive volume (Batchelor 1958). However, dislocations featured very little in these papers.

Only Orowan remained with the topic and contributed a number of seminal ideas to the theory of the interaction between moving dislocations and other dislocations or other obstacles inside a crystal. In an excellent biographical memoir of Orowan (Nabarro and Argon 1995) we learn Orowan's side of things. He confirms Polanyi's self-denying decision; he is quoted as writing: "...slowly I recognised that dislocations were important enough to warrant a publication, and I wrote to Polanyi, with whom I discussed them several times, suggesting a joint paper. He replied that it was my bird and I should publish it; finally we agreed that we would send separate papers to Professor Scheel, editor of the *Zeitschrift für Physik*, and ask him to print them side by side. This he did." He also expressed, 50 years after the event, his sceptical reaction to Taylor's version; indeed he went so far as to say in a letter to one of the memoirists that "his theory was no theory at all"! In the memoir, among many other fascinating things, we learn how Orowan escaped from the practice of electrical engineering which his father sought to impose upon him (to ensure that his son could earn a living). Orowan was at Göttingen University and, in between designing transformers, he proposed to spend one day a week in an advanced physics laboratory. In late 1928 he visited Professor Richard Becker (a highly influential solid-state physicist whom we shall meet again) to get an enrollment card signed. In Orowan's own words, recorded in the memoir, "my life was changed by the circumstance that the professor's office was a tremendously large room... Becker was a shy and hesitating man; but by the time I approached the door of the huge room he struggled through with his decision making, called me back and asked whether I would be interested in checking experimentally a 'little theory of plasticity' he (had) worked out three years before. Plasticity was a prosaic and even humiliating proposition in the age of de Broglie, Heisenberg and Schrödinger, but it was better than computing my sixtieth transformer, and I accepted with pleasure. I informed my father that I had changed back to physics; he received the news with stoic resignation." In fact, by another trivial accident (a fellow student asked a challenging question) he worked for his doctorate not on plasticity but on cleavage of mica! The work that led to the dislocation came afterwards. On such small accidents can a researcher's lifetime work depend.

After 1934, research on dislocations moved very slowly, and little had been done by the time the War came. After the War, again, research at first moved slowly. In my view, it was not coincidence that theoretical work on dislocations accelerated at about the same time that the first experimental demonstrations of the actual existence of dislocations were published and turned 'invention' into 'discovery'. In accord with my remarks in Section 3.1.3, it was a case of 'seeing is believing'; all the numerous experimental demonstrations involved the use of a microscope. The first demonstration was my own observation, first published in 1947, of the process of polygonization, stimulated and christened by Orowan (my thesis adviser). When a metal crystal is plastically bent, it is geometrically necessary that it contains an excess of positive over negative dislocations; when the crystal is then heated, most of the dislocations of opposite signs 'climb' and demolish one another, but the excess dislocations remain behind and arrange themselves into stable walls of subgrain-boundaries, which can be revealed by suitable etching. Elastic theory quickly proved that such walls would actually be the most stable configuration for an array of dislocations of the same sign. The detailed story of the discovery of polygonization has been told (Cahn 1985). At Bell Laboratories, Vogel *et al.* (1953) took my observation a notch further and proved, using germanium crystals, that the density of etchpits along a small-angle subgrain-boundary exactly matched the density of dislocations needed to produce the measured angular misorientation along the boundary.

Following this, there was a rapid sequence of observations: J.W. Mitchell in Bristol 'decorated' networks of dislocations in silver chloride by irradiating the crystals with ultraviolet light to nucleate minute silver crystals at favoured sites, viz., dislocation lines. He has given a circumstantial account of the sequence of events that led to this indirect method of observing dislocation geometries (Mitchell 1980). We have already seen Dash's method of revealing dislocations in silicon by 'decorating' them with copper (Figure 3.14). Another group (Gilman and Johnston) at General Electric were able to reveal successive positions of dislocations in lithium fluoride by repeated etching; at the place where a dislocation line reaches the surface, etching generates a sharp-bottomed etchpit, a place where it previously surfaced and was etched but where it is no longer located turns into a blunt-bottomed etchpit. This technique played a major part in determining how the speed of moving dislocations related to the magnitude of applied stress. All these microscopic techniques of revealing dislocation lines were surveyed in masterly fashion by an expert microscopist (Amelinckx 1964). A much more recent survey of the direct observation of dislocations has been provided by Braun (1992) as part of his account of the history of the understanding of the mechanical properties of solids.

The 'clincher' was the work of Peter Hirsch and his group at the Cavendish Laboratory in 1956. A transmission electron microscope was acquired by this group in 1954; the next year images were seen in deformed aluminium foils which Michael Whelan suspected to reveal dislocation lines (because the lattice nearby is distorted and so the Bragg reflection of the electron beam is diverted to slightly different angles). Once both imaging and local-area diffraction from the same field of view became possible, in mid-1956, the first convincing images of moving dislocations were obtained – more than 20 years after the original publication of the dislocation hypothesis. The history of this very important series of researches is systematically told by Hirsch (1986) and is outlined here in Section 6.2.2.1. Nevill Mott has told of his delight when "his young men burst into his office" and implored him to come and see a moving dislocation, and Geoffrey Taylor also, working in Cambridge at the time on quite different matters, was highly pleased to see his hypothesis so elegantly vindicated.

One of the big problems initially was to understand how the relatively few dislocations that are grown into crystals can multiply during plastic deformation, increasing their concentration by a factor of more than thousandfold. The accepted answer today is the Frank–Read source, of which Figure 3.14 is a specimen. The segment of dislocation line between two powerful pinning points (constituted by other dislocations skew to the plane of the source) moves rapidly under stress, emits a complete dislocation ring and returns to its initial geometry to start over again. Charles Frank (1911–1998) has recorded in brief and pithy form how this configuration acquired its name (Frank 1980). He and his co-originator, Thornton Read (W.T. Read, Jr.), who worked at Bell Laboratories, in 1950 were introduced to each other in a hotel in Pittsburgh, just after Frank had given a lecture at Cornell University and conceived the source configuration. Frank was told at the hotel that Read had something to tell him; it was exactly the same idea. On checking, they found that they had their brainwaves within an hour of each other two days previously. So their host remarked: "There is only one solution to that, you must write a joint paper", which is what they did (Frank and Read 1950). Coincidence rarely comes more coincident than this!

Mott played a major part, with his collaborator Frank Nabarro (b. 1917) and in consultation with Orowan, in working out the dynamics of dislocations in stressed crystals. A particularly important early paper was by Mott and Nabarro (1941), on the flow stress of a crystal hardened by solid solution or a coherent precipitate, followed by other key papers by Koehler (1941) and by Seitz and Read (1941). Nabarro has published a lively sequential account of their collaboration in the early days (Nabarro 1980). Nabarro originated many of the important concepts in dislocation theory, such as the idea that the contribution of grain boundaries to the flow stress is inversely proportional to the square root of the grain diameter, which was later experimentally confirmed by Norman Petch and Eric Hall.

The early understanding of the geometry and dynamics of dislocations, as well as a detailed discussion of the role of vacancies in diffusion, is to be found in one of the early classics on crystal defects, a hard-to-find book entitled *Imperfections in Nearly Perfect Crystals*, based on a symposium held in the USA in 1950 (Shockley *et al.* 1952).[3] Since in 1950, experimental evidence of dislocations was as yet very sparse, more emphasis was placed on a close study of slip lines (W.T. Read, Jr.,

---

[3] The Shockley involved in this symposium was the same William Shockley who had participated in the invention of the transistor in 1947. Soon after that momentous event, he became very frustrated at Bell Laboratories (and virtually broke with his coinventors, Walter Brattain and John Bardeen), as depicted in detail in a rivetting history of the transistor (Riordan and Hoddeson 1997). For some years, while still working at Bell Laboratories, he became closely involved with dislocation geometry, clearly as a means of escaping from his career frustrations, before eventually turning fulltime to transistor manufacture.

p. 129), following in Ewing and Rosenhain's footsteps. Orowan did not participate in this symposium, but his detailed reflections on dislocation dynamics appeared two years later in another compilation (Koehler *et al.* 1954). The first systematic account of the elastic theory of dislocations, based to a considerable degree on his own work, was published by Cottrell (1953). This book has had a lasting influence and is still frequently cited. In Chapter 5, I shall reexamine his approach to these matters.

Dislocations are involved in various important aspects of materials apart from mechanical behaviour, such as semiconducting behaviour and crystal growth. I turn next to a brief examination of crystal growth.

3.2.3.3 *Crystal growth.* As we saw in the preceding section, before World War II the dislocation pioneers came to the concept through the enormous disparity between calculated and measured elastic limiting stresses that led to plastic deformation. The same kind of disparity again led to another remarkable leap of imagination in post-war materials science.

Charles Frank (1911–1998; Figure 3.21), a physicist born in South Africa, joined the productive physics department at Bristol University, in England, headed by Nevill Mott, soon after the War. According to Braun's interview with Frank (Braun 1992), Mott asked Frank to lecture on crystal growth (a subject of which at first he knew little) and Frank based himself upon a textbook published in Germany just before the War, which a friend had sent him as a 'postwar present' (Frank 1985). This book, by the physical chemist Max Volmer (1939), was about the kinetics of phase transformations, and devoted a good deal of space to discussing the concept of *nucleation*, a topic on which Volmer had contributed one of the key papers of the interwar years (Volmer and Weber 1926). We have already met this crucial topic in Section 3.2.2.1; suffice it to say here that the point at issue is the obstacle to creating the first small 'blob' of a stable phase within a volume of a phase which has been rendered metastable by cooling or by supersaturation (in the case of a solution). I avowedly use the term 'metastable' here rather than 'unstable': random thermal fluctuations generate minute 'embryos' of varying sizes, but unless these exceed a critical size they cannot survive and thus redissolve, and that is the essence of metastability. The physical reason behind this is the energy needed to create the interface between the embryo of the stable phase and the bulk of the metastable phase, and the effect of this looms the larger, the smaller the embryo. The theory of this kind of 'homogeneous' nucleation, also known as the 'classical theory', dates back to Volmer and Weber (see a survey by Kelton 1991).

While Charles Frank was soaking up Volmer's ideas in 1947, Volmer himself was languishing as a slave scientist in Stalin's Russia, as described in a recent book about

Figure 3.21. Charles Frank (courtesy Prof. J.-P. Poirier).

the Soviet race for the atom bomb (Riehl and Seitz 1996); so Frank could not consult him. Instead he argued with his roommates, N. Cabrera and J. Burton. Volmer in his book had described the growth of iodine crystals from the vapour at just 1%

supersaturation, and Burton and Cabrera, stimulated by the argumentative Frank, calculated what supersaturation would be needed for a *perfect* (defect-free) iodine crystal to continue to grow, using methods based on Volmer's work and on another key German paper by Becker and Döring (1935) devoted to two-dimensional nucleation, and they concluded that a supersaturation of 50% would be necessary. The point here is that a deposited iodine atom skittering across the crystal surface would readily attach itself to a ledge, one atom high, of a growing layer (a small supersaturation would suffice for this), but once the layer is complete, an incoming atom then needs to join up with several others to form a stable nucleus, and do so before it re-evaporates. Only at a very high supersaturation would enough iodine atoms hit the surface, close together in space and time, to form a viable nucleus quickly enough.

At the same time as Burton and Cabrera were making their calculation, Frank Nabarro, who was to become a high priest of dislocations in his later career, drew Frank's attention to the (postulated) existence of screw dislocations. These differ from the edge dislocations sketched in Figure 3.20, because the (Burgers) vector that determines the quantum of shear displacement when a dislocation passes a point in a crystal is now not normal to the dislocation line, as in Figure 3.20, but parallel to it, as in Figure 3.22. In a flash of inspiration, Frank realized that this kind of defect provides an answer to the mismatch between theory and experiment pinpointed by Burton, because the growing layer can never be complete: as the layer rotates around the dislocation axis, there is always a step to which arriving iodine atoms can attach themselves.

Burton and Cabrera explained their calculations at the famed 1949 Faraday Discussion on Crystal Growth in Bristol (Faraday Society 1949, 1959a), and Frank

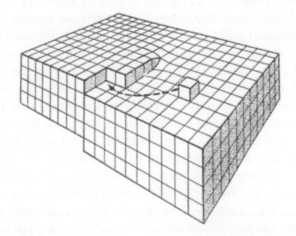

Figure 3.22. Screw dislocation and crystal growth, after W.T. Read.

described his dislocation model; he had only developed it days before the conference opened. The three together set out the whole story briefly in *Nature* in 1949 and in extenso in a famous and much-cited paper (Burton *et al.* 1951). Volmer was of course unable to attend the Faraday Society Discussion, but Richard Becker was there and contributed a theoretical paper. Thus Becker had a double link with dislocations: in 1928 he gave Orowan the opportunity that led to his 1934 paper, and he coauthored a paper that helped lead Burton, Cabrera and Frank to the inspiration that they revealed in Bristol in 1949 and developed fully by 1951.

Frank's model implies as an unavoidable corollary that the growing surface takes the form of a spiral; each rotation of the growing step mounts on the previous rotations which also keep on growing. Nobody had, apparently, reported such spirals, until a young mineralogist working in another physics department, L.J. Griffin, at another Bristol conference later in 1949 tried to attract Frank's attention, at first without success: when at last he succeeded, Griffin showed Frank beautiful growth spirals on a surface of a crystal of the mineral beryl, revealed by phase contrast microscopy (which can detect step heights very much smaller than a wavelength of light). Braun (1992) tells the entire story of the Bristol crystal growth theory, on the basis of an interview with Frank, and remarks that the effect of Griffin's revelation "was shattering...The pictures were shown to all and aroused great excitement". I was there and can confirm the excitement. Once Griffin's pictures had been publicised, all sorts of other microscopists saw growth spirals within months on all kinds of other crystals. It was a fine illustration of the fact that observers often do not see what is staring them in the face until they know exactly what they are looking for.

What is really important about the events of 1934 and 1949 is that on each occasion, theoretical innovation was driven directly by a massive mismatch between measurement and old theory. The implications of this are examined in Chapter 5.

Frank's prediction of spiral growth on crystal surfaces, followed by its successful confirmation, had an indirect but major effect on another aspect of modern science. In his 1968 book, *The Double Helix: A Personal Account of the Discovery of the Structure of DNA*, Watson (1968) describes how, not long before the final confirmation of the helical structure of DNA, he and Crick were arguing whether tobacco mosaic virus (TMV) has a helical structure; Crick was sceptical. Watson wrote: "My morale automatically went down, until I hit upon a foolproof reason why subunits should be helically arranged. In a moment of after-supper boredom I had read a Faraday Society Discussion on 'The Structure of Metals' (he remembered wrong: it was actually devoted to Crystal Growth). It contained an ingenious theory by the theoretician F.C. Frank on how crystals grow. Every time the calculations were properly done, the paradoxical answer emerged that the crystals could not grow at anywhere near the observed rates. Frank saw that the paradox vanished if crystals were not as regular as suspected, but contained dislocations resulting in the perpetual presence of cosy corners into which new molecules could fit. Several days later...the notion came to me that each TMV particle should be thought of as a tiny crystal growing like other crystals through the possession of cosy corners. Most important, the simplest way to generate cosy corners was to have the subunits helically arranged. The idea was so simple that it had to be right." Crick remained sceptical for the time being, but the seed that led to the double helix was firmly sown in Watson's mind.

3.2.3.4 Polytypism. Just after Frank and his colleagues had announced their triumph, in 1950, a young Indian physicist, Ajit Ram Verma, was awarded a fellowship to undertake research in the laboratory of a noted microscopist, S. Tolansky, in London University. Tolansky was experienced in detecting minute steps at surfaces, of the order of single atom height, by two methods: phase-contrast microscopy (as used by Griffin, one of his students) and multiple beam interferometry, a subtle technique which produces very narrow and sharp interference fringes that show small discontinuities where there is a surface step. In the immediate aftermath of the Bristol innovations, Tolansky asked Verma to concentrate on studying crystal surfaces; Verma had brought a variety of crystals with him from India, and some of these were of silicon carbide, SiC, as he explains in an autobiographical essay (Verma 1982). He now set out to look for growth spirals. Using ordinary optical microscopy he was successful in observing his first spirals by simply breathing on the surface; as he later recognised, water drops condensed preferentially at the ledges of the spiral, and rendered the very low steps visible; thus, one form of nucleation was called into service to study another form of nucleation. Then, using phase contrast and multiple-beam interferometry to measure step heights, he published his first growth spirals on silicon carbide in *Nature*, only to find that the adjacent paper on the same page, by Severin Amelinckx in Belgium (Verma and Amelinckx, 1951), showed exactly the same thing (Figure 3.23). Both measured the step height and found that it matched the unit cell height, as it should. (This episode is reminiscent of the adjacent but entirely independent publication of Letters to *Nature* concerning the mechanism of age-hardening, by Guinier and by Preston, in 1938.)

On silicon carbide, it is easier to see and measure step heights than in crystals like beryl, because SiC has *polytypes*, first discovered by the German crystallographer Baumhauer (1912). The crystal structure is built up of a succession of close-packed layers of identical structure, but stacked on top of each other in alternative ways (Figure 3.24). The simplest kind of SiC simply repeats steps ABCABC, etc., and the step height corresponds to three layers only. Many other stacking sequences

Figure 3.23. A growth spiral on a silicon carbide crystal, originating from the point of emergence of a screw dislocation (courtesy Prof. S. Amelinckx).

are found, for instance, ABCACBCABACABCB; for this "15R" structure, the repeat height must be five times larger than for an ABC sequence. Such polytypes can have 33 or even more single layers before the sequence repeats. Verma was eventually able to show that in all polytypes, spiral step height matched the height of the expanded unit cell, and later he did the same for other polytypic crystals such as CdI_2 and PbI_2. The details can be found in an early book (Verma 1953) and in the aforementioned autobiographical memoir. Like all the innovations outlined here, polytypism has been the subject of burgeoning research once growth spirals had been detected; one recent study related to polytypic phase transformations: dislocation mechanisms have been detected that can transform one polytype into another (Pirouz and Yang 1992).

The varying stacking sequences, when they are found irregularly rather than reproducibly, are called *stacking faults*; these are one of several forms of two-dimensional crystal defects, and are commonly found in metals such as cobalt where there are two structures, cubic and hexagonal close-packed, which differ very little in free energy. Such stacking faults are also found as part of the configuration of edge dislocations in such metals; single dislocations can split up into partial dislocations,

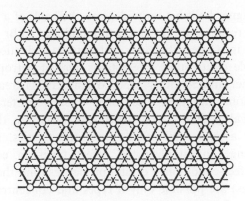

Figure 3.24. Projection of silicon carbide on the (0 0 0 1) plane (after Verma 1953).

separated by stacking faults, and this splitting has substantial effects on mechanical behaviour. William Shockley with his collaborator R.D. Heidenreich was responsible for this discovery, in 1948 just after he had helped to create the first transistor.

Stacking faults and sometimes proper polytypism are found in many inorganic compounds – to pick out just a few, zinc sulphide, zinc oxide, beryllium oxide. Interest in these faults arises from the present-day focus on electron theory of phase stability, and on computer simulation of lattice faults of all kinds; investigators are attempting to relate stacking-fault concentration on various measurable characteristics of the compounds in question, such as "ionicity", and thereby to cast light on the electronic structure and phase stability of the two rival structures that give rise to the faults.

3.2.3.5 Crystal structure, crystal defects and chemical reactions.

Most chemical reactions of interest to materials scientists involve at least one reactant in the solid state: examples include surface oxidation, internal oxidation, the photographic process, electrochemical reactions in the solid state. All of these are critically dependent on crystal defects, point defects in particular, and the thermodynamics of these point defects, especially in ionic compounds, are far more complex than they are in single-component metals. I have space only for a superficial overview.

Two German physical chemists, W. Schottky and C. Wagner, founded this branch of materials science. The story is very clearly set out in a biographical memoir of Carl Wagner (1901–1977) by another pioneer solid-state chemist, Hermann Schmalzried (1991), and also in Wagner's own survey of "point defects and their interaction" (Wagner 1977) – his last publication. Schottky we have already briefly met in connection with the Pohl school's study of colour centres

(Section 3.2.3.1). Wagner built his early ideas on the back of a paper by a Russian, J. Frenkel, who first recognised that in a compound like AgBr some Ag ions might move in equilibrium into interstitial sites, balancing a reduction in internal energy because of favourable electrostatic interactions against entropy increase. Wagner and Schottky (Wagner and Schottky 1930, Wagner 1931) treated point defects in metallic solid solutions and then also ionic crystals in terms of temperature, pressure and chemical potential as independent variables; these were definitive papers. Schmalzried asserts firmly that "since the thirties, it has remained an undiminished challenge to establish the defect types in equilibrated crystals. Predictions about defect-conditioned crystal properties (and that includes inter alia all reaction properties) are possible only if types and concentrations of defects are known as a function of the chemical potentials of the components." Wagner, in a productive life, went on to study chemical reactions in solids, especially those involving electrical currents, diffusion processes (inseparable from reactions in solids). For instance, he did some of the first studies on stabilised zirconia, a crucial component of a number of chemical sensors: he was the first to recognise (Wagner 1943) that in this compound, it is the ions and not the electrons which carry the current, and thus prepared the way for the study of superionic conductors which now play a crucial role in advanced batteries and fuel cells. Wagner pioneered the use of intentionally non-stoichiometric compounds as a way of controlling point-defect concentrations, with all that this implies for the control of compound (oxide) semiconductors. He also performed renowned research on the kinetics and mechanism of surface oxidation and, late in his life, of 'Ostwald ripening' (the preferential growth of large precipitates at the cost of small ones). There was a scattering of other investigations on defects in inorganic crystals; one of the best known is the study of defects in ferrous oxide, FeO, by Foote and Jette, in the first issue of *Journal of Chemical Physics* in 1933, already mentioned in Section 2.1.1. The systematic description of such defects, in ionic crystals mostly, and their interactions formed the subject-matter of a remarkable, massive book (Kröger 1964); much of it is devoted to what the author calles "imperfection chemistry".

The subject-matter outlined in the last paragraph also forms the subject-matter of a recent, outstanding monograph by Schmalzried (1995) under the title *Chemical Kinetics of Solids*. While the role of point defects in governing chemical kinetics received pride of place, the role of dislocations in the heterogeneous nucleation of product phases, a neglected topic, also receives attention; the matter was analysed by Xiao and Haasen (1989). Among many other topics, Wagner's theory of oxidation receives a thorough presentation. It is rare to find different kinds of solid-state scientists brought together to examine such issues jointly; one rare example was yet another Faraday Discussion (1959b) on *Crystal Imperfections and the Chemical Reactivity of Solids*. Another key overview is a book by Rao and Gopalakrishnan

(1986, 1997) which introduces defects and in a systematic way relates them to non-stoichiometry, including the 'shear planes' which are two-dimensional defects in off-stoichiometric compounds such as the niobium oxides. This book also includes a number of case-histories of specific compounds and also has a chapter on the *design* of a great variety of chemicals to fulfil specified functional purposes. Yet another excellent book which covers a great variety of defects, going far beyond simple point defects, is a text entitled *Disorder in Crystals* (Parsonage and Staveley 1978). It touches on such recondite and apparently paradoxical states as 'glassy crystals' (also reviewed by Cahn 1975): these are crystals, often organic, in which one structural component rotates freely while another remains locked immobile in the lattice, and in which the former are then 'frozen' in position by quenching. These in turn are closely related to so-called 'plastic crystals', in which organic constituents are freely rotating: such crystals are so weak that they will usually deform plastically merely under their own weight.

A word is appropriate here about the most remarkable defect-mediated reaction of all – the photographic process in silver bromide. The understanding of this in terms of point defects was pioneered in Bristol by Mott and Gurney (1940, 1948).[4] The essential stages are shown in Figure 3.25: the important thing is that a captured photon indirectly causes a neutral silver atom to sit on the surface of a crystallite. It was subsequently established that a nucleus of only 4 atoms suffices; this is large enough to be developable by subsequent chemical treatment which then turns the whole crystallite into silver, and contributes locally to the darkening of the photographic emulsion. AgBr has an extraordinary range of physical properties, which permit light of long wavelengths to be absorbed and generate electron/hole pairs at very high efficiencies (more than 10% of all photons are thus absorbed). The photoelectrons have an unusually long lifetime, several microseconds. Also, only a few surface sites on crystallites manage to attract all the silver ions so that the 4-atom nuclei form very efficiently. The American physicist Lawrence Slifkin (1972, 1975) has analysed this series of beneficial properties, and others not mentioned here, and estimates the probability of the various separate physical properties that must come together to make high-sensitivity photography possible. The product of all these independent probabilities $\approx 10^{-8}$ and it is thus not surprising that all attempts to find a cheaper, efficient substitute for AgBr have uniformly failed (unless one regards the recently introduced digital (filmless) camera as a substitute). Slifkin asserts baldly: "The photographic process is a miracle – well, perhaps not quite a miracle, but certainly an extraordinary phenomenon".

---

[4] Frederick Seitz has recently remarked (Seitz 1998) that he has long thought that Nevill Mott deserved the Nobel Prize for this work alone, and much earlier in his career than the Prize he eventually received.

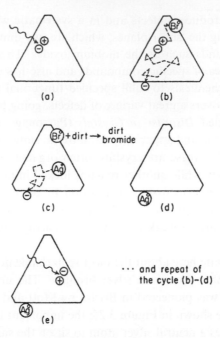

Figure 3.25. The Gurney–Mott model for the formation of a latent image (after Slifkin 1972).

Yet another category of chemical behaviour which is linked to defects, including under that term ultrasmall crystal size and the presence of uniformly sized microchannels which act as filters for molecules of different sizes, is *catalysis*. It is open to discussion whether heterogeneous catalysis, a field of very great current activity, belongs to the domain of materials science, so nothing more will be said here than to point the reader to an outstanding historical overview by one of the main protagonists, Thomas (1994). He starts his account with Humphry Davy's discovery at the Royal Institution in London that a fine platinum wire will glow when in contact with an inflammable mixture (e.g., coal gas and air) and will remain so until the mixture is entirely consumed. This then led a German, Döbereiner, to produce a gas-lighter based upon this observation. It was some considerable time before advances in surface science allowed this observation to be interpreted; today, catalysis is a vast, commercially indispensable and very sophisticated branch of materials design.

3.2.4 Crystal chemistry and physics

The structure of sodium chloride determined by the Braggs in 1913 was deeply disturbing to many chemists. In a letter to *Nature* in 1927, Lawrence Bragg made (not for the first time) the elementary point that "In sodium chloride there appear to be no molecules represented by NaCl. The equality in number of sodium and chlorine atoms is arrived at by a chessboard pattern of these atoms; it is a result of geometry and not of a pairing-off of the atoms." The irrepressible chemist Henry Armstrong, whom we have already met in Chapter 2 pouring ridicule on the pretensions of the 'ionists' (who believed that many compounds on dissolving in water were freely dissociated into ions), again burst into print in the columns of *Nature* (Armstrong 1927) to attack Bragg's statement as "more than repugnant to common sense, as absurd to the *n*th degree, not chemical cricket. Chemistry is neither chess nor geometry, whatever X-ray physics may be. Such unjustified aspersion of the molecular character of our most necessary condiment must not be allowed any longer to pass unchallenged". He went on to urge that "it were time that chemists took charge of chemistry once more and protected neophytes against the worship of false gods..." One is left with the distinct impression that Armstrong did not like ions! Two years earlier, also in *Nature*, he had urged that "dogmatism in science is the negation of science". He never said a truer word.

This little tale reveals the difficulties that the new science of crystal structure analysis posed for the chemists of the day. Lawrence Bragg's own researches in the late 1920s, with W.H. Taylor and others, on the structures of a great variety of silicates and their crucial dependence on the Si/O ratio required completely new principles of what came to be called *crystal chemistry*, as is described in a masterly retrospective overview by Laves (1962). The crucial intellectual contribution came from a Norwegian geochemist of genius, Viktor Moritz Goldschmidt (1888–1947) (Figure 3.26); his greatest work in crystal chemistry, a science which he created, was done between 1923 and 1929, even while Bragg was beginning to elucidate the crystal structures of the silicates.

Goldschmidt was born in Switzerland of Jewish parents, his father a brilliant physical chemist; he was initially schooled in Amsterdam and Heidelberg but moved to Norway at the age of 13 when his father became professor in Oslo. Young Goldschmidt himself joined the university in Christiania (=Oslo) to study chemistry (with his own father), mineralogy and geology, three disciplines which he later married to astonishing effect. He graduated young and at the age of 23 obtained his doctorate, a degree usually obtained in Norway between the ages of 30 and 40. He spent some time roaming Europe and learning from masters of their subjects such as the mineralogist Groth, and his initial researches were in petrography – that is, mainline geology. In 1914, at the age of 26, he applied for a chair in Stockholm, but the usually ultra-sluggish Norwegian academic authorities moved with lightning speed to preempt this application, and before the Swedish king had time to approve the appointment (this kind of formality was and is common in Continental universities), Oslo University got in first and made him an unprecedently young

Figure 3.26. Viktor Goldschmidt (courtesy Royal Society).

professor of mineralogy. 15 years later, he moved to Göttingen, but Nazi persecution
forced him to flee back to Norway in 1935, abandoning extensive research equipment
that he had bought with his own family fortune. Then, during the War, he again had
a very difficult time, especially since he used his geological expertise to mislead the
Nazi occupiers about the location of Norwegian mineral deposits and eventually the
Gestapo caught up with him. Again, all his property was confiscated; he just avoided
being sent to a concentration camp in Poland and escaped via Sweden to Britain.
After the War he returned once more to Norway, but his health was broken and he
died in 1947, in a sad state of paranoia towards his greatest admirers. He is generally
regarded as Norway's finest scientist.

There are a number of grim anecdotes about him in wartime; thus, at that time
he always carried a cyanide capsule for the eventuality of his capture, and when a
fellow professor asked him to find him one too, he responded: "This poison is for
professors of chemistry only. You, as a professor of mechanics, will have to use the
rope".

For our purposes, the best of the various memoirs of Goldschmidt are a lecture
by the British crystallographer and polymath John Desmond Bernal (Bernal 1949),

delivered in the presence of Linus Pauling who was carrying Goldschmidt's work farther still, and the Royal Society obituary by an eminent petrologist (Tilley 1948–1949). For geologists, Goldschmidt's main claim to fame is his systematisation of the distribution of the elements geochemically, using his exceptional skills as an analytical inorganic chemist. His lifetime's geochemical and mineralogical researches appeared in a long series of papers under the title "Geochemical distribution laws of the elements". For materials scientists, however, as Bernal makes very clear, Goldschmidt's claim to immortality rests upon his systematisation of crystal chemistry, which in fact had quite a close linkage with his theories concerning the factors that govern the distribution of elements in different parts of the earth.

In the course of his work, he trained a number of eminent researchers who inhabited the borderlands between mineralogy and materials science, many of them from outside Norway – e.g., Fritz Laves, a German mineralogist and crystal chemist, and William Zachariasen, a Norwegian who married the daughter of one of Goldschmidt's Norwegian teachers and became a professor in Chicago for 44 years; he first, in the 1930s, made fundamental contributions to crystal structure analysis and to the understanding of glass structure (Section 7.5), then (at Los Alamos during the War) made extensive additions to the crystallography of transuranium elements (Penneman 1982). Incidentally, Zachariasen obtained his Oslo doctorate at 22, even younger than his remarkable teacher had done. Goldschmidt's own involvement with many lands perhaps led his pupils to become internationalists themselves, to a greater degree than was normal at the time.

During 1923–1925 Goldschmidt and his collaborators examined (and often synthesized) more than 200 compounds incorporating 75 different elements, analysed the natural minerals among them by X-ray fluorescence (a new technique based on Manne Siegbahn's discoveries in Sweden) and examined them all by X-ray diffraction. His emphasis was on oxides, halides and sulphides. A particularly notable study was of the rare-earth sesquioxides (A_2X_3 compounds), which revealed three crystal structures as he went through the lanthanide series of rare-earth elements, and from the lattice dimensions he discovered the renowned 'lanthanide contraction'. He was able to determine the standard sizes of both cations and anions, which differed according to the charge on the ion. He found that the ratio of ionic radii was the most important single factor governing the crystal structure because the *coordination number* of the ions was governed by this ratio. For Goldschmidt, coordination became *the* governing factor in crystal chemistry. Thus simple binary AX compounds had 3:3 coordination if the radius ratio < 0.22, 4:4 if it was in the range 0.22–0.41, 6:6 up to 0.73 and 8:8 beyond this. This, however, was only the starting-point, and general rules involving (a) numerical proportions of the constituent ions, (b) radius ratios, (partly governed by the charge on each kind of ion) and (c) polarisability of large anions and polarising power of small cations which together determined the shape distortion of ions, governed crystal structures of ionic compounds and also their geochemical distributions. All this early work was published in two classical (German-language) papers in Norway in 1926.

Later in the 1920s he got to work on covalently bonded crystals and on intermetallic compounds and found that they followed different rules. He confirmed that normal valency concepts were inapplicable to intermetallic compounds. He established the 'Goldschmidt radii' of metal atoms, which are a function of the coordination number of the atoms in their crystal structures; for many years, all undergraduate students of metallurgy learnt about these radii at an early stage in their education. Before Goldschmidt, ionic and atomic radii were vague and handwaving concepts; since his work, they have been precise and useful quantities. It is now recognised that such radii are not strictly constant for a particular coordination number but vary somewhat with bond length and counter-ion to which a central ion is bonded (e.g., Gibbs *et al.* 1997), but this does not detract from the great practical utility of the concepts introduced by Goldschmidt.

Together with the structural principles established by the Bragg school concerning the many types of silicates, Goldschmidt's ideas were taken further by Linus Pauling in California to establish the modern science of crystal chemistry. A good early overview of the whole field can be found in a book by Evans (1939, 1964).

In his heyday, Goldschmidt "was a man of amazing energy and fertility of ideas. Not even periods of illness could diminish the ardour of his mind, incessantly directed to the solution of problems he set himself" (Tilley). His knowledge and memory were stupendous; Max Born often asked him for help in Göttingen and more often than not Goldschmidt was able to dictate long (and accurate) tables of figures from memory. This ability went with unconventional habits of organisation. According to Tilley, "he remembered at once where he had buried a paper he wanted, and this was all the more astonishing as he had a system not to tidy up a writing-desk but to start a new one when the old one was piled high with papers. So gradually nearly every room in his house came to have a writing-desk until there was only a kitchen sink in an unused kitchen left and even this was covered with a board and turned to the prescribed use."

Perhaps the most influential of Goldschmidt's collaborators, together with W.H. Zachariasen, was the German Fritz Laves (1906–1978), who (after becoming devoted to mineralogy as a 12-year-old when the famous Prof. Mügge gave him the run of his mineralogical museum) joined Goldschmidt in Göttingen in 1930, having taken his doctorate with Paul Niggli (a noted crystallographer/mineralogist) in Zürich. He divided his most active years between several German universities and Chicago (where Zachariasen also did all his best work). Laves made his name with the study of feldspars, one of the silicate families which W.L. Bragg was studying so successfully at the same time as Laves's move to Göttingen. He continued

Goldschmidt's emphasis on the central role of geometry (radius ratios of ions or atoms) in determining crystal structure. The additional role of electronic factors was identified in England a few years later (see Section 3.3.1, below). A good example of Laves's insights can be found in a concise overview of the crystal structures of intermetallics (Laves 1967). A lengthy obituary notice in English of Laves, which also gives an informative portrait of the development of mineralogical crystallography in the 20th century and provides a complete list of his publications, is by Hellner (1980).

3.2.5 *Physical mineralogy and geophysics*

As we have seen, mineralogy with its inseparable twin sister, crystallography, played a crucial role in the establishment of the atomic hypothesis. For centuries, however, mineralogy was a systematiser's paradise (what Rutherford called 'stamp-collecting') and modern science really only touched it in earnest in the 1920s and 1930s, when Goldschmidt and Laves created crystal chemistry. In a survey article, Laves (1959) explained why X-ray diffraction was so late in being applied to minerals in Germany particularly: traditionally, crystallography belonged to the great domain of the mineralogists, and so the physicists, who were the guardians of X-ray diffraction, preferred to keep clear, and the mineralogists were slow to pick up the necessary skills.

While a few mineralogists, such as Groth himself, did apply physical and mathematical methods to the study of minerals, tensor descriptions of anisotropy in particular – an approach which culminated in a key text by Nye (1957) – 'mineral physics' in the modern sense did not get under way until the 1970s (Poirier 1998), and then it merged with parts of modern geophysics. A geophysicist, typically, is concerned with physical and mechanical properties of rocks and metals under extremely high pressure, to enable him to interpret heat flow, material transport and phase transformations of material deep in the earth (including the partially liquid iron core). The facts that need to be interpreted are mostly derived from sophisticated seismometry. Partly, the needed information has come from experiments, physical or mechanical, in small high-pressure cells, including diamond cells which allow X-ray diffraction under hydrostatic pressure, but lately, first-principles calculations of material behaviour under extreme pressure and, particularly, computer simulation of such behaviour, have joined the geophysicist's/mineralogist's armoury, and many of the scientists who have introduced these methods were trained either as solid-state physicists or as materials scientists. They also brought with them basic materials scientist's skills such as transmission electron microscopy (D. McConnell, formerly in Cambridge and now in Oxford, was probably the first to apply this technique to minerals), and crystal mechanics. M.S. Paterson in Canberra,

Australia, is the doyen of materials scientists who study the elastic and plastic properties of minerals under hydrostatic pressure and also phase stability under large shear stresses (Paterson 1973). J.-P. Poirier, in Paris, a professor of geophysics, was trained as a metallurgist; one of his special skills is the use of analogue materials to help understand the behaviour of inaccessible high-pressure polymorphs, e.g., $CaTiO_3$ perovskite to stand in for $(Mg, Fe)SiO_3$ in the earth's mantle (Poirier 1988, Besson *et al.* 1996).

A group of physicists and chemists at the atomic laboratory at Harwell, led by A.M. Stoneham, were among the first to apply computer simulation techniques (see Chapter 12) to minerals; this approach is being energetically pursued by G.D. Price at University College, London: an example is the computer-calculation of ionic diffusion in MgO at high temperatures and pressures (Vocadlo *et al.* 1995); another impressive advance is a study of the melting behaviour of iron at pressures found at the earth's core, from *ab initio calculations* (Alfè *et al.* 1999). This was essential for getting a good understanding of the behaviour of iron in the core; its melting temperature at the relevant pressure was computed to be 6670 K. In a commentary on this research, in the same issue of *Nature*, Bukowinski remarks that "the earth can be thought of as a high-pressure experiment, a vast arena for the interplay of geophysical observation with experimental and computational materials science. For research, it is a clear win–win situation".

'Computational mineralogy' has now appeared on the scene. First-principles calculations have been used, inter alia, to estimate the transport properties of both solid and molten iron under the extreme pressures characteristic of the earth's core (Vocadlo *et al.* 1997). The current professor of mineralogy, Ekhard Salje, in Cambridge's Department of Earth's Sciences is by origin a mathematical physicist, and he uses statistical mechanics and critical theory to interpret phenomena such as ferroelasticity in minerals; he also applies lessons garnered from the study of minerals to the understanding of high-temperature superconductors. Generally, modern mineralogists and geophysicists interact much more freely with various kinds of materials scientists, physicists, solid-state chemists and engineers than did their predecessors in the previous generation, and new journals such as *Physics and Chemistry of Minerals* have been created.

3.3. EARLY ROLE OF SOLID-STATE PHYSICS

To recapitulate, the legs of the imaginary tripod on which the structure of materials science is assembled are: atoms and crystals; phase equilibria; microstructure. Of course, these are not wholly independent fields of study. Microstructure consists of phases geometrically disposed, phases are controlled by Gibbsian thermodynamics, crystal structures identify phases. Phases and their interrelation can be understood in physical terms; in fact, Gibbsian thermodynamics are a major branch of physics, and one expert in statistical physics has characterised Gibbs as "a great pioneer of modern physics". To round out this long chapter, it is time now to outline the physical underpinning of modern materials science.

3.3.1 Quantum theory and electronic theory of solids

When Max Planck wrote his remarkable paper of 1901, and introduced what Stehle (1994) calls his "time bomb of an equation, $\varepsilon = h v$", it took a number of years before anyone seriously paid attention to the revolutionary concept of the quantisation of energy; the response was as sluggish as that, a few years later, which greeted X-ray diffraction from crystals. It was not until Einstein, in 1905, used Planck's concepts to interpret the photoelectric effect (the work for which Einstein was actually awarded his Nobel Prize) that physicists began to sit up and take notice. Niels Bohr's thesis of 1911 which introduced the concept of the quantisation of electronic energy levels in the free atom, though in a purely empirical manner, did not consider the behaviour of atoms assembled in solids.

It took longer for quantum ideas to infect solid-state physics; indeed, at the beginning of the century, the physics of the solid state had not seriously acquired an identity. A symposium organised in 1980 for the Royal Society by Nevill Mott under the title of *The Beginnings of Solid State Physics* (Mott 1980) makes it clear that there was little going on that deserved the title until the 1920s. My special concern here is the impact that quantum theory had on the theory of the behaviour of electrons in solids. In the first quarter of the century, attention was focused on the Drude–Lorentz theory of free electrons in metals; anomalies concerning the specific heat of solids proved obstinately resistant to interpretation, as did the understanding of why some solids conducted electricity badly or not at all. Such issues were destined to continue to act as irritants until quantum theory was at last applied to the theory of solids, which only happened seriously after the creation of wave mechanics by Erwin Schrödinger and Werner Heisenberg in 1926, the introduction of Pauli's exclusion principle and the related conception of Fermi–Dirac statistics in the same year. This familiar story is beyond my remit here, and the reader must turn to a specialist overview such as that by Rechenberg (1995).

In the above-mentioned 1980 symposium (p. 8), the historians Hoddeson and Baym outline the development of the quantum-mechanical electron theory of metals from 1900 to 1928, most of it in the last two years of that period. The topic took off when Pauli, in 1926, examined the theory of paramagnetism in metals and proved, in a famous paper (Pauli 1926) that the observations of weak paramagnetism in various metals implied that metals obeyed Fermi–Dirac statistics – i.e., that the electrons in metals obeyed his exclusion principle. Soon afterwards, Arnold Sommerfeld applied these statistics to generate a hybrid classical-quantum theory of metals (the story is outlined by Hoddeson and Baym), but real progress was not made until the band theory of solids was created. The two key early players were Felix Bloch, who in 1928 applied wave mechanics to solids, treating 'free' electrons as waves propagating through the lattice, *unscattered by the individual stationary metal ions constituting the lattice*, and Léon Brillouin (1930) who showed that some of these same electron waves must be diffracted by planes of ions when the Bragg Law was satisfied – and this, in turn, limited the velocities at which the electrons can migrate through the lattice. Bloch (in Mott 1980, p. 24) offers his personal *memories of electrons in crystals*, starting with his thesis work under Heisenberg's direction which began in 1927. The best place to read the history of these developments in clear, intelligible terms is in Pippard's treatment of "electrons in solids" (Pippard 1995) – which here largely means electrons in metals; this excellent account starts with Drude–Lorentz and the complexities of the early work on the Hall Effect and thermoelectricity, and goes on to modern concerns such as magnetoresistance... but the heroic era was concentrated in the years 1926–1930.

The other place to read an authoritative history of the development of the quantum-mechanical theory of metals and the associated evolution of the band theory of solids is in Chapters 2 and 3 of the book, *Out of the Crystal Maze*, which is a kind of official history of solid-state physics (Hoddeson *et al.* 1992).

The recognition of the existence of semiconductors and their interpretation in terms of band theory will be treated in Chapter 7, Section 7.2.1. Pippard, in his chapter, includes an outline account of the early researches on semiconductors.

Pippard, in his historical chapter, also deals with some of his own work which proved to have a notable effect on theoretical metallurgy in the 1950s. The "anomalous skin effect", discovered in 1940, is an enhanced electrical resistivity in the surface layers of a (non-superconductive) metal when tested with a high-frequency field; at high frequencies, most of the current is restricted to a surface "skin". Sondheimer (1954) developed the theory of this effect and showed its relation to the form of the Fermi surface, the locus of the maximum possible electron kinetic energies in a solid ion in different crystal directions. This was initially taken to be always spherical, but Pippard himself was stimulated by Sondheimer's work to make experiments on the anomalous skin effect in copper crystals and succeeded, in a virtuoso piece of research, in making the first determination (Pippard 1957) of the true shape of a Fermi surface (Figure 3.27). The figure is drawn in *k*-space... i.e., each vector from the origin represents an electron moving with a momentum (k) defined by the vector.

One other classical pair of papers should be mentioned here. Eugene Wigner, an immigrant physicist of Hungarian birth, and his student Frederick Seitz whom we

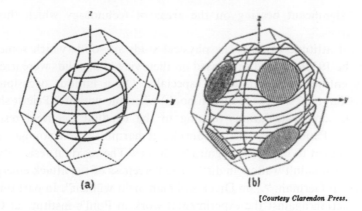

[*Courtesy Clarendon Press.*]

Figure 3.27. The first Brillouin zone of the face-centred cubic structure, after Pippard.

have already met (Figure 3.19) wrote theoretical papers (Wigner and Seitz 1933, 1934) about the origin of the cohesion of solid sodium – i.e., what holds the metal together. They chose this esoteric metal because it was easier to handle with acceptable accuracy than the more familiar metals. The task was to calculate the wave-function of the free (valence) electrons in the neighbourhood of a sodium ion: in very simplified terms, the valence electrons have greater freedom in the metal than in the isolated atom, and the potential energy of an electron in the regions between ions is less than at the same distance from an isolated atom. This circumstance in effect holds the ions together in the lattice. The methods used by Wigner and Seitz to make these calculations are still frequently cited, and in fact these two papers are regarded by many as marking the effective birth of modern solid-state physics. The success of his collaboration with Wigner encouraged Seitz to write the first comprehensive book on solid-state physics, *The Modern Theory of Solids* (Seitz 1940), which must have alerted thousands of students of the solid state to the central importance of quantum theory. About this extremely influential book, Seitz, in a recent autobiography, has remarked with undue modesty: "It has since been reissued by Dover Press and presumably possesses at least archaeological value" (Seitz 1994, p. 83).

24 years later, another standard text, *Physics of Solids*, was brought out by Wert and Thomson (1964). In his foreword to this book, Seitz has this to say: "This fine book, which was inspired by my old book but has outgrown it in almost all respects, is a preparatory text for the young engineer of today. A generation ago *it would have provided sound material for a graduate student of physics with an interest in solid-state science* (my emphasis). The fact that it is written by two members of a modern active metallurgy department (at the University of Illinois) demonstrates that a field of engineering has now reached out to absorb another newly developed field of science

which has a significant bearing on the areas of technology which this field of engineering serves."

The critical attitude towards the physical study of solids which some eminent physicists in the 1930s evinced was based on their view that solids were irremediably dirty, messy entities, semiconductors especially. On a famous occasion in 1933 (recorded in Chapter 2 of the Hoddeson book) when the youthful Peierls showed his adviser, Pauli, some calculations relating to the residual electrical resistivity in (impure) solids, Pauli burst out: "I consider it harmful when younger physicists become accustomed to order-of-magnitude physics. The residual resistivity is a dirt effect, and one shouldn't wallow in dirt". The fierceness of the attack emerges better from the original German: "...im Dreck soll man nicht wühlen". In part this attitude was also a reaction against the experimental work in Pohl's institute at Göttingen where colour centres in intentionally doped ionic crystals were systematically studied. One of those who was infected by this critical attitude was the eminent American physicist Isidore Rabi (1898–1988), who spent some years in Germany in the 1920s. To one of his graduate students at Columbia University, towards the end of the 1940s, he declared: "The physics department at Columbia will never occupy itself with the physics of dirt". Ironically, he said this just as the transistor, which depends on controlled impurities, was being developed at the Bell Laboratories.

3.3.1.1 Understanding alloys in terms of electron theory. The band theory of solids had no impact on the thinking of metallurgists until the early 1930s, and the link which was eventually made was entirely due to two remarkable men – William Hume-Rothery in Oxford and Harry Jones in Bristol, the first a chemist by education and the second a mathematical physicist.

Hume-Rothery (1899–1968; Figure 3.28; for biographical memoirs, see Raynor 1969 and Pettifor 2000) was educated as a chemist in Oxford, where he spent all of his later scientific career, but took his Ph.D. at Imperial College, London, with Harold Carpenter, the professor of metallurgy there (we shall meet him again in Section 4.2.1), on the structure and properties of intermetallic compounds. Such compounds were sure to interest a bright chemist at a time when the nature of valence was a leading concern in chemistry, since they do not follow normal valence rules: the experience converted Hume-Rothery into a dedicated metallurgist who eventually, after sustained struggles, succeeded in introducing metallurgy as a fully fledged undergraduate subject at Oxford University from 1949 – rather later than in Cambridge. For 23 years he performed his notable researches, initially at a single bench in a small room, without longterm security as a Warren Research Fellow of the Royal Society, before eventually his admirers provided the means for creating first a Readership (associate professorship) and soon after, an endowed chair of metallurgy. He was in frequent communication with, and had the support of, many of the notable chemists and physicists of his time, notably the physical chemist Cyril Hinshelwood in Oxford and the theoretical physicist Nevill Mott (1905–1996, Figure 3.18) in Bristol. Mott has already appeared many times in this chapter, especially in connection with dislocation theory, and his role in the evolution of modern materials science was massive.

In a brief note in Mott's historical symposium (Mott 1980, p. 54), written after Hume-Rothery's death, B.R. Coles (a metallurgist turned experimental physicist... it does sometimes happen) remarked that "Hume-Rothery was the first to recognise explicitly that one should regard a random substitutional alloy of two metals as a giant molecule possessing an electron gas to which both components contributed. The essential quantity of interest was therefore the average number of outer electrons per atom...". He and his students determined a number of phase diagrams, especially of alloys based on copper, silver and gold, with great precision and then worked out regularities governing the appearance of successive intermetallic phases in these systems. Starting with a precocious key paper (Hume-Rothery 1926) and culminating in a classic paper on silver- and copper-based phases (Hume-Rothery *et al.* 1934), Hume-Rothery established empirically that the successive phases turned up at specific values (such as 3/2 or 21/13) of the ratio of free (valence) electrons to metallic atoms. Since solvent and solute in general bring different numbers of valence electrons into the alloys, this ratio is bound to change as the solute concentration increases. The phases thus examined by Hume-Rothery became known as *electron phases*. The precision study of phase diagrams and conclusions drawn from them continued for many years thereafter, and he also followed in the footsteps of Moritz Goldschmidt (a near-contemporary) by focusing on the role of atomic size in governing solubilities. This in turn led to a sustained programme of analysing the stability of alloy phases in the light of their lattice parameters.

Harry Jones, as a young researcher in Mott's physics department in Bristol heard about Hume-Rothery's empirical regularities in a lecture by W.L. Bragg in 1932 or 1933 (see Jones 1980), and at once began trying to understand the reasons for the formation of γ-brass, Cu_5Zn_8, the crystal structure of which had been determined by one of Bragg's students, Albert Bradley. The Jones theory, to simplify drastically, was based on the notion that as polyvalent solute (Zn) is added to monovalent face-centred cubic solvent (Cu), the (supposedly) spherical Fermi surface expands and eventually touches the first Brillouin zone (Figure 3.27). When that happens, the density of electronic energy states changes drastically, and that in turn, by simple arguments, can be shown to raise the Gibbsian free energy of the initial phase sufficiently for an alternative crystal structure to become stabilised instead. In that way, first the β-brass and subsequently the γ-brass structure become stabilised. A theory based purely on the quantum theory of electrons in solids had thereby been shown to interpret a set of metallurgical observations on phase stability (Jones 1934). This work became much more widely known after the publication of a key theoretical book by Mott and Jones (1936), still frequently cited today.

Hume-Rothery popularised his findings, and also the theoretical superstructure initiated by Jones, in a series of influential books, beginning with a 1931 volume (*The Metallic State*) and peaking with *The Structure of Metals and Alloys*, first published in 1936 by the Institute of Metals in London and updated through many editions over the years with a number of distinguished coauthors. Another, more elementary book, republished from short articles in an industrial metallurgy journal, consisted of conversations between an older and a younger metallurgist. He encountered much opposition from those older metallurgists (like the steelmaker, Harry Brearley, whom we have already met) who even thought that their professional body, the Institute of Metals, had no business publishing such a cloudy volume as *The Structure of Metals and Alloys*, but Hume-Rothery persisted and succeeded in transforming metallurgical education, starting with the Department of Physical Metallurgy at Birmingham University where Geoffrey Raynor, Hume-Rothery's most distinguished student, from 1948 spread the 'gospel' of the new metallurgy. The

Figure 3.28. William Hume-Rothery as a young man (courtesy Mrs. Jennifer Moss).

reader will recall that in 1917, Rosenhain was proselytising for his own 'new metallurgy'; 20 years later, Hume-Rothery was rewriting the meaning of 'new' in that pregnant phrase. Many books followed Hume-Rothery's in successive attempts to interpret modern electron theory of metals to scientists trained as metallurgists or materials scientists; notable examples are books by Cottrell (1988) and by Pettifor and Cottrell (1992), in addition to Cottrell's classic textbook of 1948 which we have already met.

At one point it seemed that the entire theoretical superstructure advanced to explain Hume-Rothery's electron phases had collapsed, because of Pippard's (1957) discovery that the Fermi surface of pure copper was not after all spherical and already touched the first Brillouin zone *even before any polyvalent solute was added* (Figure 3.27, right). This seemed to remove the essential concept from Jones's theory, and thus the agreement between Hume-Rothery's experimental findings and Jones's theory appeared to be merely fortuitous. But, as the science-historian Gerald Holton once remarked, "The graveyard of failed scientists is littered with those who did not suspend disbelief when their ideas were first shown to be wrong". In due course, the apparent disaster was seen not to be one after all. Cottrell, in a little textbook on electron theory published just half a century after his first book (Cottrell 1998) explains what happened: Because of the absence of computers in the 1930s, Jones had to make a number of simplifying approximations in developing his theory, one being the so-called "rigid-band approximation" – that the form of the density-of-states distribution remains fixed as the electron-to-atom ratio increases, another being that the Fermi surface remains spherical even when it touches a Brillouin zone boundary. Even though Jones modified some of his approximations in 1937, Pippard's study still seemed to undermine the theory, but in fact it became clear later that some of the theoretical errors revealed by this study cancelled each other. (This is a not uncommon experience in the history of theoretical science.) The new theory (Paxton *et al.* 1997) avoids Jones's approximations, takes proper account of the influence of *d* electrons (which Jones could not do), and, in Cottrell's words: "The modern theory, by taking full advantage of present-day computer power, has been able to avoid both approximations and so, because of their mutual cancellation, has achieved the same success – or even better – but on an intrinsically more sound basis".

Hume-Rothery's position as one of the key creators of modern physical metallurgy remains unchallenged.

Hume-Rothery's ideas and their theoretical development by Mott and Jones stimulated much consequential research around the world. The most impressive early 'convert' was a French physicist, Jacques Friedel, who should have been mentioned in connection with dislocations, in the theory of which he played an early part (see the Corrigenda). After a very disturbed war, which ranged from study at the

Ecole Polytechnique to work down a coalmine, he resolved to make himself an expert in quantum mechanics, a theme until then gravely neglected in France, and decided that the place to learn it was as a doctoral student with Nevill Mott in Bristol. The idea of going abroad to study a branch of physics in depth was at that time novel among the French. In his autobiography (Friedel 1994), he describes "le choc de Bristol (1949–1952)" and the difficulties he had in being assigned a research topic that fitted his objective. He finally wrote his thesis on the electron theory of metallic solid solutions, in which he became a renowned innovator. A first account was published soon after (Friedel 1952) and some more sophisticated developments followed later, notably his treatment of the distribution of conduction electrons round an alloy atom of valency different from that of the host. The screening charge was shown (Friedel 1958) to exist as a series of concentric haloes, of higher and lower electron density, around a dissolved solute atom...the 'Friedel oscillations'. A number of other developments followed later, and Friedel created a distinguished school in Paris with offshoots elsewhere in France. An account of his role, from a French perspective, is given in a book chapter devoted to the history of solid-state physics in France (Guinier 1988).

Meanwhile, electron theory was revived effectively in Hume-Rothery's own base of Oxford, and is now led by a distinguished mathematical physicist, David Pettifor.

Nevill Mott, first in Bristol and then in Cambridge, has repeatedly surfaced in this chapter; a few more words about his remarkable personality are in order here. He was a superb theorist who interacted effortlessly with experimentalists and had his own idiosyncratic way of pursuing theory. At a recent unveiling of his magnificent bronze bust in the Cavendish Laboratory (June 2000), Malcolm Longair quoted Mott's own words about himself: "I am neither an experimentalist nor a real mathematician – my theory stops at Schrödinger's equation. What I have done in this subject is to look at the evidence, do calculations on the back of an envelope and say to the theoretician: 'If you apply your techniques to this problem, this is how it will come out.' And to the experimentalist, just the same thing." And, Longair concluded, Mott's work epitomises the very best of the Cavendish tradition. A series of short memoirs of Mott are assembled in a book (Davis 1998).

3.3.2 Statistical mechanics

It is one of the wonders of the history of physics that a rigorous theory of the behaviour of a chaotic assembly of molecules – a gas – preceded by several decades the experimental uncovering of the structure of regular, crystalline solids. Attempts to create a kinetic theory of gases go all the way back to the Swiss mathematician, Daniel Bernouilli, in 1738, followed by John Herapath in 1820 and John James Waterston in 1845. But it fell to the great James Clerk Maxwell in the 1860s to take the first accurate steps – and they were giant steps – in interpreting the pressure–volume–temperature relationship of a gas in terms of a probabilistic (or statistical) analysis of the behaviour of very large populations of mutually colliding molecules – the *kinetic theory of gases*. He was the first to recognise that the molecules would *not* all have the same kinetic energy. The *Maxwell distribution* of kinetic energies of such a population has made his name immortal... even if it had not been immortalised by his electromagnetic equations. The science he created is sometimes called *statistical mechanics*, sometimes *statistical thermodynamics*.

For many years this kind of theory was applied to fluids of various kinds, and it became interestingly applicable to solids much later, in 1925, when W. Lenz in Germany, together with his student Ising, created the theory of *critical phenomena*, which covers phenomena in solids such as ferromagnetism and order–disorder transitions. This important field of theory, which has no proper name even today, has become a major domain of research in its own right and has been recognised with a Nobel Prize awarded to Kenneth Wilson in 1982. The issue was whether an array of spins attached to atoms in a regular array would automatically generate spin alignment and ferromagnetism. Ising only managed a theory in one dimension and wrongly surmised that in higher dimensions there would be no ferromagnetism. The many attempts to generalise the theory to two or three dimensions began with Rudolf Peierls in 1936; he showed that Ising's surmise was wrong.

A population of theorists floating uneasily between physics and materials science (but a number of them working in materials science departments) have become specialists in the statistical thermodynamics of solids, critical phenomena in particular, working in specific fields such as order-disorder transitions; to go into any details of critical phenomena here would take us much too far into the domain of mathematical physics. Two splendid historical accounts of the whole field are by Domb (1995, 1996); another important historical treatment is by Brush (1967). It is intriguing that Ising's name was immortalised in the Ising Model, but in Domb's opinion (private communication), "Ising was a low-grade scientist who by a quirk of fate managed to get his name on thousands of papers, many of them outstandingly good. His own contributions to the field were negligible." Naming of phenomena sometimes rewards the wrong person!

From the historical point of view, an interesting dispute concerns the relative claims of Maxwell in England, Josiah Willard Gibbs in America and Ludwig Boltzmann in Austria to be regarded as the true father of statistical thermodynamics – as distinct from macroscopic chemical thermodynamics, where Gibbs' claims are undisputed. Gibbs' claim rests on a book in 1902 (Gibbs 1902), but this is a good deal later than the various classic papers by Boltzmann. The most important of these were his study of the process by which a gas, initially out of equilibrium, approaches the Maxwell–Boltzmann distribution (as it has since become known), and his profound investigation in 1877 of the probabilistic basis of entropy, culminating in the relation $S = k \log W$, where S is entropy and W is the probability of a microstate; this immortal equation is carved on Boltzmann's tomb. It is Boltzmann's work which has really made possible the modern flowering of statistical thermo-dynamics of solids.

The sequence of events is traced with historical precision in a new biography of Boltzmann (Cercignani 1998). An entire chapter (7) is devoted to the Gibbs/Boltzmann connection, culminating in a section entitled "Why is statistical mechanics usually attributed to Gibbs and not to Boltzmann?". Cercignani attributes this to the unfamiliarity of many physicists early in this century with Boltzmann's papers, partly because of the obscurity of his German style (but Gibbs is not easy to read, either!), and partly because the great opinion-formers of early 20th-century physics, Bohr and Einstein, knew little of Boltzmann's work and were inclined to decry it. The circumstances exemplify how difficult it can be to allocate credit appropriately in the history of science.

3.3.3 *Magnetism*

The study of the multifarious magnetic properties of solids, followed in due course by the sophisticated control of those properties, has for a century been a central concern both of physicists and of materials scientists. The history of magnetism illustrates several features of modern materials science.

That precocious Cambridge engineer, Alfred Ewing, whom we have already met as the adviser of the young Walter Rosenhain, was probably the first to reflect seriously (Ewing 1890) about the origin of ferromagnetism, i.e., the characteristics of strong permanent magnets. He recognised the possibility that the individual magnetic moments presumed to be associated with each constituent atom in a solid somehow kept each other aligned, and he undertook a series of experiments with a lattice of magnetised needles that demonstrated that such an interaction could indeed take place. This must have been one of the first mechanical simulations of a physical process, and these became increasingly popular until eventually they were displaced by computer simulations (Chapter 12). Ewing also did precocious work in the 1880s on the nature of (ferro)magnetic hysteresis, and indeed he invented the term *hysteresis*, deriving from the Greek for 'to be late'.

The central mystery about lodestones and magnetised needles for compasses was where the strong magnetism (what today we call *ferromagnetism*) comes from... what is the basis for all magnetic behaviour? The first written source about the behaviour of (natural) lodestones was written in 1269, and in 1600 William Gilbert (1544–1603) published a notable classic, *De magnete, magnetisque corporibus, et de magno magnete tellure* ...the last phrase referring to 'the great magnet, the earth'. One biographer says of this: "It is a remarkably 'modern' work – rigorously experimental, emphasising observation, and rejecting as unproved many popular beliefs about magnetism, such as the supposed ability of diamond to magnetise iron. He showed that a compass needle was subject to magnetic dip (pointing downward) and, reasoning from experiments with a spherical lodestone, explained this by concluding that the earth acts as a bar magnet. ... The book... was very influential in the creation of the new mechanical view of science" (Daintith *et al.* 1994). Ever since, the study of magnetism has acted as a link between sciences.

Early in the 20th century, attention was focused on diamagnetic and paramagnetic materials (the great majority of elements and compounds); I do not discuss this here for lack of space. The man who ushered in the modern study of magnetism was Pierre Weiss (1865–1940); he in effect returned to the ideas of Ewing and conceived the notion of a 'molecular field' which causes the individual atomic magnets, the existence of which he felt was inescapable, to align with each other and in this way the feeble magnetisation of each atomic magnet is magnified and becomes macroscopically evident (Weiss 1907). The way Weiss's brilliant idea is put in one excellent historical overview of magnetics research (Keith and Quédec 1992) is: "The interactions within a ferromagnetic substance combine to give the same effects as a fictional mean field..."; such fictional mean fields subsequently became very common devices in the theory of solids. However, the purely magnetic interaction between neighbouring atomic minimagnets was clearly not large enough to explain the creation of the fictional field.

The next crucial step was taken by Heisenberg when he showed in 1928 that the cause of ferromagnetism lies in the quantum-mechanical exchange interaction between electrons imposed by the Pauli exclusion principle; this exchange interaction acts between neighbouring atoms in a crystal lattice. This still left the puzzle of where the individual atoms acquired their magnetic moments, bearing in mind that the crucial component of these moments resides in the *unbalanced* spins of populations of individual electrons. It is interesting here to cite the words of Hume-Rothery, taken from another of his influential books of popularization, *Atomic Theory for Students of Metallurgy* (Hume-Rothery 1946): "The electrons at absolute zero occupy the $N/2$ lowest energy states, each state containing two electrons of opposite spins. Since each electron state cannot contain more than one electron of a given spin, it is clear that any preponderance of electrons of a given spin must increase the Fermi energy, and ferromagnetism can only exist if some other factor lowers the energy." He goes on to emphasize the central role of Heisenberg's exchange energy, which has the final effect of stabilising energy bands containing unequal numbers of positive and negative spin vectors. In 1946 it was also a sufficient approximation to say that the *sign* of the exchange energy depended on the separation of neighbouring atoms, and if that separation was too small, ferromagnetism (with parallel atomic moments) was impossible and, instead, neighbouring atomic moments were aligned *anti*parallel, creating *antiferromagnetism*. This phenomenon was predicted for manganese in 1936 by a remarkable physicist, Louis Néel (1904–2000), Pierre Weiss's star pupil, in spite of his self-confessed neglect of quantum mechanics. (His portrait is shown in Chapter 7, Figure 7.8.) There was then no direct way of proving the reality of such antiparallel arrays of atomic moments, but later it became possible to establish the arrangements of atomic spins by neutron diffraction and many antiferromagnets were then discovered. Néel went on to become one of the most influential workers in the broad field of magnetism; he ploughed his own idiosyncratic furrow and it became very fertile (see 'Magnetism as seen by Néel' in Keith and Quédec's book chapter, p. 394). One proof of the importance of interatomic distance in determining whether atomic moments were aligned parallel or antiparallel was the accidental discovery in 1889 of the Heusler alloy, Cu_2MnAl, which was ferromagnetic though none of its constituent elements was thought to be magnetic (the antiferromagnetism of manganese was unknown at the time). This alloy occasioned widespread curiosity long before its behaviour was understood. Thus, the American physicist Robert Wood wrote about it to Lord Rayleigh in 1904: "I secured a small amount in Berlin a few days ago and enclose a sample. Try the filings with a magnet. I suppose the al. and cu. in some way loosen up the manganese molecules so that they can turn around" (Reingold and Reingold 1981); he was not so far out! In 1934 it was found that this phase underwent an order–disorder transition, and that the ordered form was ferromagnetic while the disordered form was apparently non-magnetic (actually, it turned out later, antiferromagnetic). In the ordered form, the distance between nearest-neighbour manganese atoms in the crystal structure was greater than the mean distance was in the disordered form, and this brought about the ferromagnetism. The intriguing story is outlined by Cahn (1998).

The inversion from ferromagnetic to antiferromagnetic interaction between neighbouring atoms is expressed by the "Néel–Slater curve", which plots magnitude and sign of interaction against atomic separation. This curve is itself being subjected to criticism as some experimental observations inconsistent with the curve are beginning to be reported (e.g., Schobinger-Papamantellos *et al.* 1998). In physics and materials science alike, simple concepts tend to be replaced by increasingly complicated ones.

The nature of the exchange energy, and just how unbalanced spin systems become stabilised, was studied more deeply after Hume-Rothery had written, and a very clear non-mathematical exposition of the present position can be found in (Cottrell 1988, p. 101).

The reader interested in this kind of magnetic theory can find some historical memories in an overview by the American physicist, Anderson (1979).

Up to this point, I have treated only the fundamental quantum physics underlying the existence of ferromagnetism. This kind of theory was complemented by the application of statistical mechanics to the understanding of the progressive misalignment of atomic moments as the temperature is raised – a body of theory which led Bragg and Williams to their related mean-field theory of the progressive loss of atomic order in superlattices as they are heated, which we have already met. Indeed, the interconnection between changes in atomic order and magnetic order (i.e., ferromagnetism) is a lively subspeciality in magnetic research; a few permanent magnet materials have superlattices.

Quite separate and distinct from this kind of science was the large body of research, both experimental and theoretical, which can be denoted by the term *technical magnetism*. Indeed, I think it is fair to say that no other major branch of materials science evinces so deep a split between its fundamental and technical branches. Perhaps it would be more accurate to say that the quantum- and statistical-mechanical aspects have become so ethereal that they are of no real concern even to sophisticated materials scientists, while most fundamental physicists (Néel is an exception) have little interest in the many technical issues; their response is like Pauli's.

When Weiss dreamt up his molecular-field model of ferromagnetism, he was at once faced by the need to explain why a piece of iron becomes progressively more strongly magnetised when placed in a gradually increasing energising magnetic field. He realized that this could only be explained by two linked hypotheses: first, that the atomic moments line up along specific crystal directions (a link between the lattice and magnetism), and second, that a crystal must be split into *domains*, each of which is magnetised along a different, crystallographically equivalent, vector... e.g., (1 0 0), (0 1 0) or (0 0 1), each in either a positive or negative direction of magnetisation. In the absence of an energising field, these domains cancel each other out macroscopically and the crystal has no resultant magnetic moment. The stages of Ewing's hysteresis cycle involve the migration of domain boundaries so that some domains (magnetised nearly parallel to the external field) grow larger and 'unfavourable' ones disappear. The alternative mechanism, of the bodily rotation of atomic moments as a group, requires much larger energy input and is hard to achieve.

Domain theory was the beginning of what I call technical magnetism; it had made some progress by the time domains were actually observed in the laboratory. There was then a long period during which the relation between two-phase microstructures in alloys and the 'coercive field' required to destroy macroscopic magnetisation in a material was found to be linked in complex ways to the pinning of domain boundaries by dispersed phases and, more specifically, by local strain fields created by such phases. This was closely linked to the improvement of permanent magnet materials, also known as 'hard' magnets. The terms 'hard' and 'soft' in this context point up the close parallel between the movement of dislocations and of domain boundaries through local strain fields in crystals.

The intimate interplay between the practitioners of microstructural and phase-diagram research on the one hand, and those whose business it was to improve both soft and hard magnetic materials can be illustrated by many case-histories; to pick just one example, some years ago Fe–Cr–Co alloys were being investigated in order to create improved permanent magnet materials which should also be ductile. Thermodynamic computation of the phase diagram uncovered a miscibility gap in the ternary phase diagram and, according to a brief account (Anon. 1982), "Homma *et al.* experimentally confirmed the existence of a ridge region of the miscibility gap and found that thermomagnetic treatment in the ridge region is effective in aligning and elongating the ferromagnetic particles parallel to the applied magnetic field direction, resulting in a remarkable improvement of the magnetic properties of the alloys". This sentence refers to two further themes of research in technical magnetism: the role of the shape and dimensions of a magnetic particle in determining its magnetic properties, and the mastery of heat-treatment of alloys in a magnetic field.

A separate study was the improvement of magnetic permeability in 'soft' alloys such as are used in transformers and motors by lining up the orientations of individual crystal grains, also known as a preferred orientation; this became an important subspeciality in the design of transformer laminations made of dilute Fe–Si alloys, introduced more than 100 years ago and still widely used.

Another recent success story in technical magnetism is the discovery around 1970 that a metallic glass can be ferromagnetic in spite of the absence of a crystal lattice; but that very fact makes a metallic glass a very 'soft' magnetic material, easy to magnetise and thus very suitable for transformer laminations. In recent years this has become a major market. Another success story is the discovery and intense development, during the past decade, of compounds involving rare earth metals, especially samarium and neodymium, to make extraordinarily powerful permanent magnets (Kirchmayr 1996). Going further back in time, the discovery during the last War, in the Philips laboratories in the Netherlands, of magnetic 'ferrites' (complex oxides including iron), a development especially associated with the name of the Dutch physicist Snoek, has had major industrial consequences, not least for the growth of tape-recorders for sound and vision which use powders of such materials. These materials are *ferrimagnetic*, an intriguing halfway house between ferromagnetic and antiferromagnetic materials: here, the total magnetic moments of the two families of atoms magnetised in opposing directions are unequal, leaving a macroscopic balance of magnetisation. The ferrites were the first insulating magnetic materials to find major industrial use (see Section 7.3).

This last episode points to the major role, for a period, of industrial laboratories such as the giant Philips (Netherlands), GE (USA) and Siemens (Germany) laboratories in magnetic research, a role very clearly set out in the book chapter by Keith and Quédec. GE, for instance, in the 1950s developed a family of permanent magnets exploiting the properties of small, elongated magnetic particles. Probably the first laboratory to become involved in research on the fringes of magnetism was the Imphy laboratory in France at the end of the nineteenth century: a Swiss metallurgist named Charles-Edouard Guillaume (1861–1938), working in Paris, had in 1896 discovered an iron–nickel alloy which had effectively zero coefficient of thermal expansion near room temperature, and eventually (with the support of the Imphy organisation) tracked this down to a loss of ferromagnetism near room temperature, which entails a 'magnetostrictive' contraction that just compensates the normal thermal expansion. This led to a remarkable programme of development in what came to be known as 'precision metallurgy' and products, 'Invar' and 'Elinvar', which are still manufactured on a large scale today and are, for instance, essential components of colour television tubes. Guillaume won the Nobel Prize for Physics in 1920, the only such prize ever to be awarded for a metallurgical achievement. The story is told in full detail in a centenary volume (Béranger *et al.* 1996).

Most recently, industrial magnetics research has taken an enormous upswing because of the central importance of magnetic recording in computer memories. Audio-recording on coated tape was perfected well before computer memories came on the scene: the first step (1900) was recording on iron wires, while plastic recording tape coated with iron oxide was developed in Germany during the First World War. Magnetic computer memories, old and new, are treated in Section 7.4. Not all the innovations here have been successful: for instance, the introduction of so-called 'bubble memories' (with isolated domains which could be nudged from one site to a neighbouring one to denote a bit of memory) (Wernick and Chin 1992) failed because they were too expensive. However, a remarkable success story, to balance this, is the magnetoresistant multilayer thin film. This apparently emerged from work done in Néel's Grenoble laboratory in the 1960s: thin films of a ferromagnet and an antiferromagnet in contact acquire a new kind of magnetic anisotropy from exchange coupling (à la Heisenberg) and this in turn was found to cause an unusually large change of electrical resistivity when a magnetic field is applied normal to the film (a phenomenon known as magnetoresistivity). This change in resistivity can be used to embody an electronic signal to be recorded. The matter languished for a number of years and around 1978 was taken up again. Multilayers such as Co–Pt are now used on a huge scale as magnetoresistive memories, as is outlined in a survey by Simonds (1995). (See also Section 7.4.) It could be said that this kind of development has once again brought about a rapprochement between the quantum theorists and the hard-headed practical scientist.

Not only information technology has benefited from research in technical magnetism. Both permanent magnets and electromagnets have acquired manifold uses in industry; thus automotive engines nowadays incorporate ever more numerous permanent magnets. An unexpected application of magnets of both kinds is to magnetic bearings, in which a rotating component is levitated out of contact with an array of magnets under automatic control, so that friction-free operation is achieved. As I write this, the seventh international symposium on magnetic bearings is being planned in Zurich. The ultracentrifuges which played such an important part in determining molecular weights of polymers (see Chapter 8, Section 8.7) rely on such magnetic bearings.

Magnetism intrudes in the most unexpected places. A very recent innovation is the use of 'magnetorheological finishing'. An American company, QED Technologies in Rochester, NY, has developed a polishing agent, a slurry of carbonyl iron, cerium oxide (a hard abrasive) and other materials. A magnetic field converts this slurry from a mobile liquid to a rigid solid. Thus a coating of the slurry can take up the shape of a rough object to be polished and then 'solidified' to accelerate polishing without use of a countershape. This is useful, for instance, in polishing aspheric lenses.

The literature of magnetics research, both in journals and in books, is huge, and a number of important titles help in gaining a historical perspective. A major classic is the large book (Bozorth 1951), simply called *Ferromagnetism*, by Richard Bozorth (1896–1981). An English book, more angled towards fundamental themes, is by Bates (1961). An excellent perspective on the links between metallurgy and magnetism is offered by an expert on permanent magnets, Kurt Hoselitz (1952), also by one of the seminar volumes formerly published by the American Society for Metals (ASM 1959), a volume which goes in depth into such arcane matters as the theory of the effects caused by annealing alloys in a magnetic field. An early, famous book which, precociously, strikes a judicious balance between fundamental physics and technical considerations, is by Becker and Döring (1939), also simply called *Ferromagnetismus*. An excellent perspective on the gradually developing ideas of technological (mostly industrial) research on ferromagnetic materials can be garnered from two survey papers by Jacobs (1969, 1979), the second one being subtitled "a quarter-century overview". An early overview of research in technical magnetism, with a British slant, is by Sucksmith (1949).

REFERENCES

Aaronson, H.I., Lee, J.K. and Russell, K.C. (1978) Diffusional nucleation and growth, in *Precipitation Processes in Solids*, eds. Russell, K.C. and Aaronson, H.I. (Metallurgical Society of AIME, New York) p. 31.
Aaronson, H.I., Laughlin, D.E., Sekerka, R.F. and Wayman, C.M. (1982) *Solid → Solid Phase Transformations* (The Metallurgical Society of AIME, Warrendale, PA).
Adams, B.L., Wright, S.I. and Kunze, K. (1993) *Metall. Trans.* **24A**, 819.

Agricola, G. (1556, 1912) *De re metalica* (1556). Translated into English, 1912, by Herbert Hoover and Lou Henry Hoover, The Mining Magazine, London. Reprinted in facsimile by the AIME, New York, 1950 (see also entry on Agricola in *Encyclopedia Britannica*, 15th edition, 1974, by R.W. Cahn).

Aitchison, L. (1960) *A History of Metals*, 2 volumes (Macdonald and Evans, London).

Alfè, D., Gillan, M.J. and Price, G.D. (1999) *Nature* **401**, 462; see also *News and Views* comment by M.S.T. Bukowinski, p. 432 of the same issue.

Amelinckx, S. (1964) *The Direct Observation of Dislocations* (Academic Press, New York).

Anderson, P.W. (1979) *J. Appl. Phys.* **50**, 7281.

Andrade, E.N. da C. (1923) *The Structure of the Atom* (Bell and Sons, London).

Anon. (editor) (1981) *Gefüge der Metalle: Entstehung, Beeinflussung, Eigenschaften* (Deutsche Gesellschaft für Metallkunde, Oberursel).

Anon. (1982) *Bull. Alloy Phase Diagrams* **2**, 423.

Anon. (1998) *IUCr Newsletter* **6**(2), 15.

Armstrong, H.E. (1927) *Nature* **120**, 478.

ASM (1959) *Magnetic Properties of Metals and Alloys* (multiple authors, no editor) (American Society for Metals, Cleveland, Ohio).

Barr, L.W. (1997) *Def. Diff. Forum* **143–147**, 3.

Barrett, C.S. and Massalski, T.B. (1966) *Structure of Metals: crystallographic methods, principles and data*, 3rd edition, Chapters 11 and 18 (McGraw-Hill, New York). The first and second editions appeared in 1943 and 1952, under Barrett's sole authorship.

Batchelor, G.K. (1958) *The Scientific Papers of Sir Geoffrey Ingram Taylor; Mechanics of Solids*, vol. 1 (The University Press, Cambridge).

Batchelor, G.K. (1996) *The Life and Legacy of G.I. Taylor*, Chapter 11 (Cambridge University Press, Cambridge) esp. pp. 150, 152.

Bates, L.F. (1961) *Modern Magnetism*, 4th edition (Cambridge University Press, London).

Baumhauer, H. (1912) *Z. Krist.* **50**, 33.

Becker, R. and Döring, W. (1935) *Ann. Physik* **24**, 719.

Becker, R. and Döring, W. (1939) *Ferromagnetismus* (Springer, Berlin).

Béranger, G. *et al.* (1996) *A Hundred Years after the Dicovery of Invar... the Iron–Nickel Alloys* (Lavoisier Publishing, Paris).

Bernal, J.D. (1949) *J. Chem. Soc.* p. 2108 (This journal at that time carried no volume numbers).

Besson, P., Poirier, J.-P. and G.D. Price (1996) *Phys. Chem. Minerals* **23**, 337.

Bever, M.B. and Rocca, R. (1951) *Revue de Métallurgie* **48**(5), 3.

Bijvoet, J.M., Burgers, W.G. and Hägg, G. (1972) *Early Papers on Diffraction of X-rays by Crystals* (Int. Union of Crystallography, Utrecht) pp. 5.

Binder, K. (1991) *Phase Transformations in Materials*, ed. Haasen, P., in *Materials Science and Technology: A Comprehensive Treatment*, eds. Cahn, R.W., Haasen, P. and Kramer E.J. (VCH, Weinheim) p. 405.

Biringuccio, V. (1540, 1945) *Pirotechnia* (Venice, 1540). An English translation by C.S. Smith and M.T. Gnudi was published by the AIME (American Institute of Mining and Metallurgical Engineers), New York, 1945.

Boulesteix, C. (editor) (1998) *Oxides: Phase Transitions, Non-Stoichiometry, Supercon-ductors* in *Key Engineering Materials*, vol. 155–156.

Bozorth, R.M. (1951) *Ferromagnetism* (Van Nostrand, New York).

Bradley, A.J. and Thewlis, J. (1926) *Proc. Roy. Soc. (Lond.) A* **112**, 678.

Bragg, W.H. (1917) in *Science and the Nation*, ed. Seward, A.C. (Cambridge University Press, Cambridge) p. 24.

Bragg, W.L. (1975, 1992) *The Development of X-ray Analysis* (G. Bell and Sons, London) 1975 (Dover Editions, Mineola, NY) 1992.

Bragg, W.L. and Williams, E.J. (1934) *Proc. Roy. Soc. (Lond.) A* **145**, 699.

Braun, E. (1992) in *Out of the Crystal Maze: Chapters from the History of Solid-State Physics*, eds. Hoddeson, L., Braun, E., Teichmann, J. and Weart, S. (Oxford University Press, Oxford) p. 317.

Brearley, H. (1995) *Steel-makers and Knotted String* (The Institute of Materials, London). This book is a combination of *Steel-makers* (originally published in 1933) and *Knotted String* (originally published in 1941).

Brillouin, L. (1930) *J. Phys. Radium* **1**, 377.

Brown, L.M., Pais, A. and Pippard, B. (1995) *Twentieth Century Physics*, 3 volumes (Institute of Physics Publishing and American Institute of Physics, Bristol, Philadelphia and New York).

Brush, S. (1967) *Rev. Mod. Phys.* **39**, 883.

Burke, J.G. (1966) *The Origins of the Science of Crystals* (University of California Press, Berkeley and Los Angeles).

Burton, W.K., Cabrera, N. and Frank, F.C. (1951) *Phil. Trans. Roy. Soc. (Lond.) A* **243**, 299.

Cahn, J.W. and Hilliard, J.E. (1958) *J. Chem. Phys.* **28**, 258.

Cahn, J.W. (1961) *Acta Metall.* **9**, 795.

Cahn, J.W. (1991) *Scandinavian J. Metall.* **20**, 9.

Cahn, R.W. (editor) (1965) *Physical Metallurgy* (North-Holland, Amsterdam). Subsequent editions appeared in 1970, 1983 and 1996, the last two edited jointly with P. Haasen.

Cahn, R.W. (1975) *Nature* **253**, 310.

Cahn, R.W. (1985) The discovery of polygonization, in *Dislocations and Properties of Real Materials* (The Institute of Metals, London) p. 12.

Cahn, R.W. (1994) The place of atomic order in the physics of solids and in metallurgy, in *Physics of New Materials*, ed. Fujita, F.E. (Springer, Berlin) p. 179. (1998) Second, updated edition.

Cahn, R.W. (1998) Metals combined – or chaos vanquished, *Proc. Roy. Inst. Great Britain* **69**, 215.

Cahn, R.W., Haasen, P. and Kramer, E.J. (1991–1998), *Materials Science and Technology: A Comprehensive Treatment*, 18 volumes plus a Cumulative Index volume and 2 Supplementary volumes (VCH; now Wiley-VCH, Weinheim).

Caroe, G.M. (1978) *William Henry Bragg* (Cambridge University Press, Cambridge).

Cercignani, C. (1998) *Ludwig Boltzmann: The Man who Trusted Atoms* (Oxford University Press, Oxford).

Chaloner, C. (1997) *British J. Hist. Sci.* **30**, 357.

Christian, J.W. (1965) *The Theory of Transformations in Metals and Alloys* (Pergamon Press, Oxford).

Cottrell, A.H. (1948) *Theoretical Structural Metallurgy*, 2nd edition, 1955 (Edward Arnold, London).

Cottrell, A.H. (1953) *Dislocations and Plastic Flow in Crystals* (Clarendon Press, Oxford).

Cottrell, A. (1988) *Introduction to the Modern Theory of Metals* (The Institute of Metals, London).

Cottrell, A. (1998) *Concepts in the Electron Theory of Metals* (IOM Communications Ltd., London) p. 84.

Dahl, P.F. (1997) *Flash of the Cathode Rays: A History of J.J. Thomson's Electron* (Institute of Physics Publishing, Bristol and Philadelphia).

Daintith, J., *et al.* (1994), in *Biographical Encyclopedia of Scientists*, eds. William, G., Ernst, M. and Bakhuis, R. 2 volumes, 2nd edition (Institute of Physics Publishing, Bristol and Philadelphia).

Dalton, J. (1808) *A New System of Chemical Philosophy*, vol. 1 (Manchester) p. 211.

Daniel, V. and Lipson, H. (1943) *Proc. Roy. Soc. A (Lond.)* **181**, 368; (1944) *ibid* **182**, 378.

Darken, L. and Gurry, R. (1953) *Physical Chemistry of Metals* (McGraw-Hill, New York) p. 290.

Davis, E.A. (ed.) (1998) *Nevill Mott: Reminiscences and Appreciations* (Taylor & Francis, London).

De Fontaine, D. (1979) Configurational thermodynamics of solid solutions, in *Solid State Physics*, eds. Ehrenreich, H., Seitz, F. and Turnbull, D. (Academic Press, New York) p. 74.

DeHoff, R.T. (1986) Stereology in *Encyclopedia of Materials Science and Engineering* vol. 6, ed. Bever, M.B. (Pergamon Press, Oxford) p. 4633.

Desch, C.H. (1919) *J. Inst. Met.* **22**, 241.

Diu, B. (1997) *Les Atomes: Existent-ils Vraiment?* (Editions Odile Jacob, Paris).

Domb, C. (1995) in ed. Hoddeson *et al.*, p. 521.

Eigler, D.M. and Schweizer, E.K. (1990) *Nature* **344**, 524.

Einstein, A. (1910) *Ann. Phys.* **33**, 1275.

Embury, J.D. and Purdy, G.R. (1988) *Advances in Phase Transitions* (Pergamon Press, Oxford).

Emerton, N.E. (1984) *The Scientific Reinterpretation of Form* (Cornell University Press, Ithaca).

Evans, R.C. (1939, 1964) *An Introduction to Crystal Chemistry* (Cambridge University Press, Cambridge).

Evans, U.R. (1945) *Trans. Faraday Soc.* **41**, 365.

Ewald, P.P. (editor) (1962) *Fifty Years of X-ray Diffraction* (Oosthoek, Utrecht).

Ewing, J.A. (1890) *Proc. Roy. Soc. Lond. A* **48**, 342.

Ewing, J.A. and Rosenhain, W. (1900) *Phil. Trans. R. Soc. (Lond.) A* **193**, 353.

Exner, H.E. (1996), in *Physical Metallurgy*, 4th edition, vol. 2, eds. Cahn, R.W. and Haasen, P. p. 996.

Faraday Society, Discussion No. 5 (1949, 1959a) *A General Discussion on Crystal Growth*, Faraday Society, London, 1949 (reprinted, London: Butterworths Scientific Publications, 1959).

Faraday Society, Discussion No. 28 (1959b) *Crystal Imperfections and the Chemical Reactivity of Solids* (Aberdeen University Press).

Feder, R. and Nowick, A.S. (1958) *Phys. Rev.* **109**, 1959.

Fedorov, E. (1895) *Z. Krystallographie* **24**, 209.

Fleischer, R.L., Price, P.B. and Walker, R.M. (1968) *Geochimica et Cosmochimica Acta* **32**, 21.

Fowler, B.W., Phillips, Rob and Carlsson, Anders E. (1996) *Encyclopedia of Applied Physics*, vol. 14 (Wiley-VCH, Weinheim) p. 317.

Frank, F.C. (1980) *Proc. Roy. Soc. A* **371**, 136.

Frank, F.C. (1985) Some personal reminiscences of the early days of crystal dislocations, in *Dislocations and Properties of Real Materials* (The Institute of Metals, London) p. 9.

Frank, F.C. and Read, W.T. (1950) *Phys. Rev.* **79**, 722.

Friedel, J. (1952) *Adv. Phys.* **3**, 446.

Friedel, J. (1958) *Suppl. to Il Nuovo Cimento* **7**, 287.

Friedel, J. (1994) *Graine de Mandarin* (Editions Odile Jacob, Paris) p. 169.

Furley, D.J. (1973) Lucretius, in *Dictionary of Scientific Biography*, ed. Gillispie, C.C. (Ch. Scribner's Sons, New York) p. 536.

Galison, P. (1997) *Image and Logic* (University of Chicago Press, Chicago) p. 65.

Gibbs, J.W. (1875–1878) *Trans. Connecticut Acad. Arts and Sci.* **3**, 108, 343.

Gibbs, J.W. (1878) Abstract of preceding paper, *Amer. J. Sci.*, 3rd series **16**, 441.

Gibbs, J.W. (1902) *Elementary Principles in Statistical Mechanics, Developed with Special Reference to the Rational Foundations of Thermodynamics* (Yale University Press, New Haven).

Gibbs, G.V., Tamada, O. and Boisen Jr., M.B. (1997) *Phys. Chem. Minerals* **24**, 432.

Glazebrook, R.T. (1932–1935) Biographical memoir on Alfred Ewing, in *Obituary Notices of the Royal Society*, vol. 1, p. 475.

Greer, A.L. (1996) in *Sidney Sussex College, Cambridge: Historical Essays*, eds. Beales, D.E.D. and Nisbet, H.B. (The Boydell Press, Woodbridge, UK) p. 195.

Guinier, A. (1938) *Nature* **142**, 569.

Guinier, A. (1988) The development of solid-state physics in France after 1945, in *The Origins of Solid-State Physics in Italy: 1945–1960*, ed. Giuliani, G. (Società Italiana di Fisica, Bologna) p. 249.

Gurtin, M.E. (ed.) (1984) *Phase Transformations and Material Instabilities in Solids* (Academic Press, Orlando) (Proceedings of a conference conducted by the Mathematics Research Center, University of Wisconsin).

Haasen, P. (1991) Phase transformations in materials, in *Materials Science and Technology: A Comprehensive Treatment*, vol. 5, eds. Cahn, R.W., Haasen, P. and Kramer, E.J. (VCH, Weinheim).

Hägg, G. (1950) *Acta Chem. Scandinavica* **4**, 88.

Hellner, E. (1980) *Z. Kristallographie* **151**, 1.

Heycock, C.T. and Neville F.H. (1904) *Phil. Trans. R. Soc. A (Lond.)* **202**, 1.

Hillert, M. (1961) *Acta Metall.* **9**, 525.

Hirsch, P.B. (1986) *Mater. Sci. Eng.* **84**, 1.

Hoddeson, L., Braun, E., Teichmann, J. and Weart, S. (eds.) (1992) *Out of the Crystal Maze: Chapters from the History of Solid-State Physics* (Oxford University Press, Oxford, New York).

Hoffmann, R. and Schmidt, S.L. (1997) *Old Wine in New Flasks: Reflections on Science and the Jewish Tradition* (Freeman, New York).

Hoselitz, K. (1952) *Ferromagnetic Properties of Metals and Alloys* (Clarendon Press, Oxford).

Hume-Rothery, W. (1926) *J. Inst. Metals* **35**, 295.

Hume-Rothery (1936) *The Structure of Metals and Alloys* (The Institute of Metals, London).

Hume-Rothery (1946) *Atomic Theory for Students of Metallurgy* (The Institute of Metals, London).

Hume-Rothery, W., Mabbott, G.W. and Channel-Evans, K.M. (1934) *Phil. Trans. Roy. Soc. Lond. A* **233**, 1.

Humphreys, F.J. and Hatherly, M. (1995) *Recrystallization and Related Annealing Phenomena* (Pergamon Press, Oxford) p. 3.

Jacobs, I.S. (1969) *J. Appl. Phys.* **40**, 917; (1979) *ibid* **50**, 7294.

Jenkin, J. (1995) Lecture to a history of science group at the Royal Institution, London. Prof. Jenkin, of LaTrobe University in Australia, is writing a joint biography of the Braggs, father and son, at the time they were in Adelaide (The episode is confirmed in an unpublished autobiography by W.L. Bragg, in possession of his son Stephen.).

Johansson, C.H. and Linde, J.O. (1925) *Ann. Physik* **78**, 305.

Jones, H. (1934) *Proc. Roy. Soc. Lond. A* **144**, 225.

Jones, H. (1980) in ed. Mott, p. 52.

Keith, S.T. and Quédec, P. (1992) in ed. Hoddeson *et al.*, p. 359.

Kelly, A. and Nicholson, R.B. (1963) *Progr. Mat. Sci.* **10**, 149.

Kelton, K.F. (1991) *Solid State Phys.* **45**, 75.

Kirchmayr, H.R. (1996) *J. Phys. D: Appl. Phys.* **29**, 2763.

Klein, M.J. (1970–1980) in *Dictionary of Scientific Biography*, vol. 5, ed. Gillispie, C.C. (entry on J.W. Gibbs) (Scribner's, New York) p. 386.

Koehler, J.S. (1941) *Phys. Rev.* **60**, 397.

Koehler, J.S., Seitz, F., Read Jr., W.T., Shockley, W. and Orowan, E. (eds.) (1954) *Dislocations in Metals* (The Institute of Metals Division, American Institute of Mining and Metallurgical Engineers, New York).

Köster, W. (1939, 1961) *Z. Metallkde.* **30**/2 (1939) (obituary of G. Tammann). *Ibid*, **52**/6 (1961) (memoir for centenary of Tammann's birth).

Kraftmakher, Y. (1998) *Phys. Rep.* **299**, 80.

Kröger, F.A. (1964) *The Chemistry of Imperfect Crystals* (North-Holland, Amsterdam).

Kyu, T. (1993) in *Supplementary Volume 3 of the Encyclopedia of Materials Science and Engineering*, ed. Cahn, R.W. (Pergamon Press, Oxford) p. 1893.

Laudan, R. (1987) *From Mineralogy to Geology: The Foundations of a Science, 1650–1830* (University of Chicago Press, Chicago and London).

Laves, F. (1959) *Fortschr. Miner.* **37**, 21.

Laves, F. (1962) in ed. Ewald, p. 174.

Laves, F. (1967) in *Intermetallic Compounds*, ed. Westbrook, J.H. (Wiley, New York) p. 129.

Lipson, H. (1990) in *Physicists Look Back: Studies in the History of Physics*, ed. Roche, J. (Adam Hilger, Bristol) p. 226.

Martin, J.W. (1968) *Precipitation Hardening* (Pergamon Press, Oxford); (1998), 2nd edition (Butterworth-Heinemann, Oxford).

Mehl, R.F. (1948) *A Brief History of the Science of Metals* (American Institute of Mining and Metallurgical Engineers, New York).

Mehl, R.F. (1960) A commentary on metallurgy, *Trans. Metall. Soc. AIME* **218**, 386.

Mehl, R.F. (1967) Methodology and mystique in physical metallurgy, report published by the Naval Research Laboratory, Washington, DC (Metallurgy Division Annual Lecture, No. 1).

Mehl, R.F. (1975) A department and a research laboratory in a university, *Annu. Rev. Mater. Sci.* **5**, 1.

Merica, P.D., Waltenberg, R.G. and Scott, H. (1920) *Trans. Amer. Inst. of Mining and Metall. Engrs.* **64**, 41.

Mitchell, J.W. (1980) The beginnings of solid state physics (organised by N.F. Mott), *Proc. Roy. Soc. (Lond.) A* **371**, 126.

Morowitz, H.J. (1993) *Entropy and the Magic Flute* (Oxford University Press, Oxford) p. 12.

Mott, N.F. (1980) The beginnings of solid state physics, *Proc. Roy. Soc. A* **371**, 1–177.

Mott, N.F. and Gurney, R.W. (1940, 1948) *Electronic Processes in Ionic Crystals*, 1st edition 1940 (Oxford University Press, Oxford); 2nd edition 1948.

Mott, N.F. and Jones, H. (1936) *The Theory of the Properties of Metals and Alloys* (Oxford University Press, London).

Mott, N.F. and Nabarro, F.R.N. (1941) *Proc. Phys. Soc.* **52**, 86.

Nabarro, F.R.N. (1980) *Proc. Roy. Soc. A* **371**, 131.

Nabarro, F.R.N. and Argon, A.S. (1995) Memoir of Egon Orowan, in *Biographical Memoirs of Fellows of the Royal Society*, vol. 41 (The Royal Society, London) p. 315.

Niggli, P. (1917) *Geometrische Kristallographie des Diskontinuums* (Bornträger, Leipzig).

Nowick, A.S. (1996) *Annu. Rev. Mater. Sci.* **26**, 1.

Nutting, J. and Baker, R.G. (1965) *The Microstructure of Metals* (The Institute of Metals, London).

Nye, J.F. (1957) *Physical Properties of Crystals* (Clarendon Press, Oxford).

Nye, M.J. (1972) *Molecular Reality: A Perspective on the Scientific Work of Jean Perrin* (Macdonald, London and American Elsevier, New York).

Orowan, E. (1934) *Z. Physik*, **89**, 605, 614, 634 (The last of these is the paper in which dislocations are explicitly introduced).

Ostwald, W. (1893) *Lehrbuch der Allgemeinen Chemie, 2nd edition*, vol. 2, part 1 (Engelmann, Leipzig) p. 516.

Parsonage, N.G. and Staveley, L.A.K. (1978) *Disorder in Crystals* (Clarendon Press, Oxford).

Paterson, M.S. (1973) *Rev. Geophys. and Space Phys.* **11**, 355.

Pauli, W. (1926) *Z. Physik* **41**, 91.

Paxton, A.T., Methfessel, M. and Pettifor, D.G. (1997) *Proc. Roy. Soc. Lond. A* **453**, 1493.

Penneman, R.A. (1982) in *Actinides in Perspective*, ed. Edelstein, N.M. (Pergamon Press, Oxford) p. 57.

Percy, J. (1864) *Metallurgy – Iron and Steel* (London).

Perrin, J. (1909) Annales de Chimie et de Physique *18*, 1. English translation by F. Soddy, *Brownian Movement and Molecular Reality* (Taylor & Francis, London) 1910.

Perrin, J. (1913) *Les Atomes* (Paris: Alcan); 4th edition, 1914. English translation by D.L. Hammick based on a revised edition (Van Nostrand, New York) 1923.

Pettifor, D.G. (2000) William Hume-Rothery: his life and science, in *The Science of Alloys for the 21st Century: A Hume-Rothery Symposium Celebration*, eds. Turchi, P. *et al.* (TMS, Warrendale).

Pettifor, D.G. and Cottrell, A.H. (1992) *Electron Theory in Alloy Design* (The Institute of Materials, London).

Petzow, G. and Henig, E.-T. (1977) *Z. Metallkde.* **68**, 515.

Pippard, A.B. (1957) *Phil. Trans. R. Soc. A* **250**, 325.

Pippard, B. (1995) in ed. Brown *et al.*, p. 1279.

Pirouz, P. and Yang, J.W. (1992) *Ultramicroscopy* **51**, 189.

Poirier, J.-P. (1988) *Bull. Geol. Inst. Univ. Uppsala, N.S.* **14**, 49.

Poirier, J.-P. (1998) Letter dated 1 June 1998.

Polanyi, M. (1934) *Z. Physik* **89**, 660.

Polanyi, M. (1962) Personal Reminscences in *Fifty Years of X-ray Diffraction*, ed. Ewald, P.P. p. 629.

Porter, D.A. and Easterling, K.E. (1981) *Phase Transformations in Metals and Alloys* (Van Nostrand Reinhold, Wokingham) pp. 1, 263.

Preston, G.D. (1938) *Nature* **142**, 570.

Price, G.D. and Vocadlo, L. (1996) *C.R. Acad. Sci. Paris, Ser. IIa* **323**, 357.

Radnai, R. and Kunfalvi, R. (1988) *Physics in Budapest* (North-Holland, Amsterdam) p. 74.

Rath, B.B. and DeYoung, D.J. (1998) *JOM* (formerly *Journal of Metals*) July issue, 14.

Rao, C.N.R. and Gopalakrishnan, J. (1986, 1997) *New Directions in Solid State Chemistry*, 1st edition 1986 (Cambridge University Press, Cambridge) (2nd edition 1997).

Raynor, G.V. (1969) *Biographical Memoirs of Fellows of the Royal Society*, vol. 13, p. 109.

Rechenberg, H. (1995) *Quanta and Quantum Mechanics*, ed. Brown *et al.*, p. 143.

Reingold N. and Reingold, I.H. (1981) *Science in America: A Documentary History* (University of Chicago Press, Chicago) p. 101.

Riehl, N. and Seitz, F. (1996) *Stalin's Captive: Nikolaus Riehl and the Soviet Race for the Bomb* (American Chemical Society and the Chemical Heritage Foundation).

Riordan, M. and Hoddeson, L. (1997) *Crystal Fire: The Birth of the Information Age* (W.W. Norton and Company, New York).

Roozeboom, H.W.B. (1900) *Z. phys. Chem.*, **34**, 437 (English translation): *J. Iron and Steel Inst.* **58**, 311.

Rosenhain, W. (1914) *An Introduction to the Study of Physical Metallurgy* (Constable, London).

Rosenhain, W. (1917) *Science and the Nation*, ed. Seward, A.C. (Cambridge University Press, Cambridge) p. 49.

Ruoff, A.L. (1973) *Materials Science* (Prentice-Hall, Englewood Cliffs) p. 330.

Sauveur, A. (1916, 1918, 1926, 1938) *The Metallography and Heat Treatment of Iron and Steel* (successive editions) (McGraw-Hill, New York and London).

Scherz, G. (translated by A.J. Pollock) (1969) *Steno, Geological Papers* (Oxford University Press, Oxford).

Schmalzried, H. (1991) Memoir on Carl Wagner, in *Ber. Bunsengesellschaft für Phys. Chem.*, vol. 95, p. 936 (In German).

Schmalzried, H. (1995) *Chemical Kinetics of Solids* (VCH, Weinheim).

Schmid, E. and Wassermann, G. (1928) *Metallwirtschaft* **7**, 1329.

Schobinger-Papamantellos, P., Buschow, K.H.J., de Boer, F.R., Ritter, C., Isnard, O. and Fauth, F. (1998) *J. Alloys Compounds* **267**, 59.

Schrödinger, E. (1954, 1996) *Nature and the Greeks* (Cambridge University Press, Cambridge).

Schütt, H.-W. (1997) *Eilhard Mitscherlich, Prince of Prussian Chemistry* (American Chemical Society and the Chemical Heritage Foundation).

Seitz, F. (1940) *The Modern Theory of Solids* (McGraw-Hill, New York).

Seitz, F. (1946) *Rev. Modern Phys.* **18**, 384.

Seitz, F. (1954) *Rev. Modern Phys.* **26**, 7.

Seitz, F. (1980) The beginnings of solid state physics (organised by N.F. Mott); *Proc. Roy. Soc. (Lond.) A* **371**, 84.

Seitz, F. (1994) *On the Frontier: My Life in Science* (American Institute of Physics, New York).

Seitz, F. (1998) in *Nevill Mott: Reminiscences and Appreciations*, ed. Davis, E.A. (Taylor & Francis, London) p. 96.

Seitz, F. and Read, T.A. (1941) *J. Appl. Phys.* **12**, 100, 170, 470.

Shi, W.C. and Greer, A.L. (1997) *Thin Solid Films* **292**, 103.

Shockley, W., Hollomon, J.H., Maurer, R. and Seitz, F. (eds.) (1952) *Imperfections in Nearly Perfect Crystals* (Wiley, New York).

Siilivask, K. (1998) Europe, Science and the Baltic Sea, in *Euroscientia Forum* (European Commission, Brussels) p. 29.

Simmons, R.O. and Balluffi, R.W. (1960–1963) *Phys. Rev.* **117** (1960) 52; *ibid* **119** (1960) 600; *ibid* **125** (1962) 862; *ibid* **129** (1963) 1533.

Simonds, J.L. (1995) *Phys. Today* April, 26.

Slifkin, L.M. (1972) *Sci. Prog. Oxf.* **60**, 151.

Slifkin, L.M. (1975) in *Radiation Damage Processes in Materials*, ed. Dupuy, C.H.S. (Noordhoff, Leiden).

Smith, C.S. (1952, 1981) Grain shapes and other metallurgical applications of topology, in *Metal Interfaces*, ASM Seminar Report (American Society for Metals, Cleveland) p. 65. Grain shapes and other metallurgical applications of topology, in *A Search for Structure*, ed. Smith, C.S. (MIT Press, Cambridge, MA) p. 3.

Smith, C.S. (1960) *A History of Metallography* (The University of Chicago Press, Chicago).

Smith, C.S. (1963) *Four Outstanding Researches in Metallurgical History* (The 1963 Lecture on Outstanding Research) (American Society for Testing and Materials).

Smith, C.S. (ed.) (1965) *The Sorby Centennial Symposium on the History of Metallurgy* (Gordon and Breach, New York).

Smith, C.S. (1977) *Metallurgy as a Human Experience: An Essay on Man's Relationship to his Materials in Science and Practice Throughout History* (Published jointly by American Society for Metals, Metals Park, Ohio, and The Metallurgical Society of AIME, New York).

Smith, C.S. (1981) *A Search for Structure* (MIT Press, Cambridge, MA) p. 33.

Smith, C.S. (1985) *Isis* **76**, 584.

Smith, C.S. (1990) R.F. Mehl, in *Dictionary of Scientific Biography*, ed. Gillispie, C.C. (Ch. Scribner's Sons, New York) Supplement II, p. 611.

Smith, C.S. and Mullins, W.W. (2001) *Biographical memoir of R.F. Mehl for the National Academy of Sciences*, in press.

Sondheimer, E.H. (1954) *Proc. Roy. Soc. Lond. A* **224**, 260.

Stehle, P. (1994) *Order, Chaos, Order: The Transition from Classical to Quantum Physics* (Oxford University Press, Oxford) pp. 55, 123.

Stockdale, D. (1946) *Metal Prog.* p. 1183.

Stocks, G.M. and Gonis, A. (eds.) (1989) *Alloy Phase Stability* (Kluwer Academic Publishers, Dordrecht).

Sucksmith, W. (1949) *J. Iron and Steel Inst.*, September, p. 51.

Tammann, G.A. (1970–1980) in *Dictionary of Scientific Biography*, vol. 13, ed. Gillispie, C.C. (entry on G.H.J.A. Tammann) (Scribner's, New York) p. 242.

Tammann, G. (1898) *Z. Phys. Chem.* **25**, 442.

Tanner, L.E. and Leamy, H.J. (1974) The microstructure of order–disorder transitions, in *Order–Disorder Transformations in Alloys*, ed. Warlimont, H. (Springer, Berlin) p. 180.

Taylor, G.I. (1934) *Proc. Roy. Soc. A* **145**, 362.

Taylor, G.I. and Quinney H. (1934) *Proc. Roy. Soc. A* **143**, 307.

Teichmann, J. and Szymborski, K. (1992) *Out of the Crystal Maze*, ed. Hoddeson, L. *et al.* (Oxford University Press, Oxford) p. 236.

Thomas, J.M. (1994) Angewandte Chemie, *Int. Edition in English* **33**, 913.

Tilley, C.E. (1948–1949) *Obituary Notices of Fellows of the Royal Society* **6**, 51.

Turnbull, D. and Cech, R.E. (1950) *J. Appl. Phys.* **21**, 804.

Turnbull, D. (1952) *J. Chem. Phys.* **20**, 411.

Van den Broek, J.J. and Dirks, A.G. (1987) *Philips Tech. Rev.* **43**, 304.

Verma, A.R. (1953) *Crystal Growth and Dislocations* (Butterworths Scientific Publications, London).

Verma, A.R. (1982) in *Synthesis, Crystal Growth and Characterization*, ed. Lal, K. (North-Holland, Amsterdam) p. 1.

Verma, A.R. and Amelinckx, S. (1951) *Nature* **167**, 939.

Verma, A.R. and Krishna, P. (1966) *Polymorphism and Polytypism in Crystals* (Wiley, New York).

Vocadlo, L., Wall, A., Parker, S.C. and Price, G.D. (1995) *Physics of the Earth and Planetary Interiors* **88**, 193.

Vocadlo, L. *et al.* (1997) *Faraday Disc.* **106**, 205.

Vogel, F.L., Pfann, W.G., Corey, H.E. and Thomas, E.E. (1953) *Phys. Rev.* **90**, 489.
Volmer, M. (1939) *Kinetik der Phasenbildung* (Steinkopff, Dresden).
Volmer, M. and Weber, A. (1926) *Z. Phys. Chem.* **119**, 227.
Von Laue, M. (1962) in ed. Ewald, p. 278.
Wagner, C. (1931) *Z. Phys. Chem., Bodenstein-Festband*, p. 177.
Wagner, C. (1943) *Naturwissenschaften*, **31**, 265.
Wagner, C. (1977) *Annu. Rev. Mater. Sci.* **7**, 1.
Wagner, C. and Schottky, W. (1930) *Z. Phys. Chem. B* **11**, 163.
Watson, J.D. (1968) *The Double Helix: A Personal Account of the Discovery of the Structure of DNA*, Chapter 16 (Weidenfeld and Nicolson, London).
Weiss, P. (1907) *J. Physique* **5**, 70.
Wernick, J.H. and Chin, G.Y. (1992) in *Concise Encyclopedia of Magnetic and Superconducting Materials*, ed. Evetts, J.E. (Pergamon Press, Oxford) p. 55.
Wert, C.A. and Thomson, R.M. (1964) *Physics of Solids* (McGraw-Hill, New York).
Wertime, T.A. (1961) *The Coming of the Age of Steel*, ed. Brill, E.J. (Leiden, Netherlands).
Westgren, A. and Phragmén, G. (1922) *J. Iron and Steel. Inst. (Lond.)* **105**, 241.
Wigner, E. and Seitz, F. (1933) *Phys. Rev.* **43**, 804; (1934) *ibid* **46**, 509.
Xiao, S.Q. and Haasen, P. (1989) *Scripta Metall.* **23**, 365.

Chapter 4
The Virtues of Subsidiarity

Chapter 4
The Virtues of Subsidiarity

4.1. THE ROLE OF PAREPISTEMES IN MATERIALS SCIENCE

Physical metallurgy, like other sciences and technologies, has its mainline topics: examples, heat transfer in mechanical engineering, distillation theory in chemical engineering, statistical mechanics in physics, phase transformations in physical metallurgy. But just as one patriarch after a couple of generations can have scores of offspring, so mainline topics spawn subsidiary ones. The health of any science or technology is directly dependent on the vigour of research on these subsidiary topics. This is so obvious that it hardly warrants saying... except that 200 years ago, hardly anyone recognised this truth. The ridiculous doctrine of yesteryear has become the truism of today.

What word should we use to denote such subsidiary topics? All sorts of dry descriptors are to hand, such as 'subfield', 'subdiscipline', 'speciality', 'subsidiary topic', but they do not really underline the importance of the concept in analysing the progress of materials science. So, I propose to introduce a neologism, suggested by a classicist colleague in Cambridge: *parepisteme*. This term derives from the ancient Greek 'episteme' (a domain of knowledge, a science... hence 'epistemology'), plus 'par(a)-', a prefix which among many other meanings signifies 'subsidiary'. The term *parepisteme* can be smoothly rendered into other Western languages, just as Greek- or Latin-derived words like entropy, energy, ion, scientist have been; and another requirement of a new scientific term, that it can be turned into an adjective (like 'energetic', 'ionic', etc.) is also satisfied by my proposed word... 'parepistemic'.

A striking example of the importance of narrowing the focus in research, which is what the concept of the parepisteme really implies, is the episode (retailed in Chapter 3, Section 3.1.1) of Eilhard Mitscherlich's research, in 1818, on the crystal forms of potassium phosphate and potassium arsenate, which led him, quite unexpectedly, to the discovery of isomorphism in crystal species and that, in turn, provided heavyweight evidence in favour of the then disputed atomic hypothesis. As so often happens, the general insight comes from the highly specific observation.

Some parepistemes are pursued by small worldwide groups whose members all know each other, others involve vast communities which, to preserve their sanity, need to sub-classify themselves into numerous subsets. They all seem to share the feature, however, that they are not disciplines in the sense that I have analysed these

159

in Chapter 2: although they all form components of degree courses, none of the parepistemes in materials science that I exemplify below are degree subjects at universities – not even crystallography, huge field though it is.

The essence of the concept of a parepisteme, to me, is that parepistemic research is *not* directly aimed at solving a practical problem. Ambivalent views about the justifiability of devoting effort to such research can be found in all sciences. Thus a recent overview of a research programme on the genome of a small worm, *C. elegans* (the first animal genome to be completely sequenced) which was successfully concluded after an intense 8-year effort (Pennisi 1998), discusses some reactions to this epoch-making project. Many did not think it would be useful to spend millions of dollars "on something which didn't solve biological problems right off", according to one participant. Another, commenting on the genetic spinoffs, remarked that "suddenly you have not just your gene, but context revealed. You're looking at the forest, not just the tree." Looking at the forest, not just the tree – that is the value of parepistemic research in any field.

A good way of demonstrating the importance of parepistemes, or in other terms, the virtues of subsidiarity, is to pick and analyse just a few examples, out of the many hundreds which could be chosen in the broad field of materials science and engineering.

4.2. SOME PAREPISTEMES

4.2.1 Metallic single crystals

As we saw in Section 3.1.3, Walter Rosenhain in 1900 published convincing micrographic evidence that metals are assemblies of individual crystal grains, and that plastic deformation of a metal proceeds by slip along defined planes in each grain. It took another two decades before anyone thought seriously of converting a piece of metal into a *single crystal*, so that the crystallography of this slip process could be studied as a phenomenon in its own right. There would, in fact, have been little point in doing so until it had become possible to determine the crystallographic orientation of such a crystal, and to do that with certainty required the use of X-ray diffraction. That was discovered only in 1912, and the new technique was quite slow in spreading across the world of science. So it is not surprising that the idea of growing metallic single crystals was only taken seriously around the end of World War I.

Stephen Keith, a historian of science, has examined the development of this parepisteme (Keith 1998), complete with the stops and starts caused by fierce competition between individuals and the discouragement of some of them, while a shorter account of the evolution of crystal-growing skill can be found in the first chapter of a book by one of the early participants (Elam 1935). There are two approaches to the problem: one is the 'critical strain-anneal' approach, the other, crystal growth from the melt.

The strain-anneal approach came first chronologically, apparently because it emerged from the chance observation, late in the 19th century, of a few large grains in steel objects. This was recognised as being deleterious to properties, and so some research was done, particularly by the great American metallurgist Albert Sauveur, on ways of *avoiding* the formation of large grains, especially in iron and steel. In 1912, Sauveur published the finding that large grains are formed when initially strain-free iron is given a *small* (critical) strain and subsequently annealed: the deformed metal recrystallises, forming just a few large new grains. If the strain is smaller than the critical amount, there is no recrystallisation at all; if it is larger, then many grains are formed and so they are small. This can be seen in Figure 4.1, taken from a classic 'metallographic atlas' (Hanemann and Schrader 1927) and following on an observation recorded by Henri Le Chatelier in France in 1911: A hardened steel ball was impressed into the surface of a piece of mild steel, which was then annealed; the further from the impression, the smaller the local strain and the larger the resultant grains, and the existence of a critical strain value is also manifest. This critical-strain method, using tensile strain, was used in due course for making large iron crystals (Edwards and Pfeil 1924) – in fact, because of the allotropic transformations during cooling of iron from its melting-point, no other method would have worked for iron – but first came the production of large aluminium crystals.

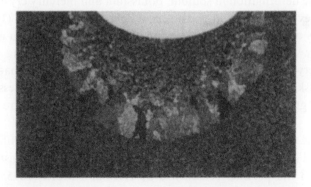

Figure 4.1. Wrought low-carbon mild steel, annealed and impressed by a Brinell ball (12 mm diameter), then annealed 30 min at 750°C and sectioned. The grain size is largest just inside the zone beyond which the critical strain for recrystallisation has not quite been attained (after Hanemann and Schrader 1927, courtesy M. Hillert).

The history of the researches that led to large aluminium crystals is somewhat confused, and Keith has gone into the sequence of events in some detail. Probably the first relevant publication was by an American, Robert Anderson, in 1918; he reported the effects of strain preceding annealing (Anderson 1918). My late father-in-law, Daniel Hanson (1892–1953), was working with Rosenhain in the National Physical Laboratory near London during World War I, and told me that he had made the first aluminium crystals at that time; but the circumstances precluded immediate publication. I inherited two of the crystals (over 100 cm^3 in size) and presented them to the Science Museum in London; Jane Bowen of that Museum (Bowen 1981) undertook some archival research and concluded that Hanson may indeed have made the first crystals around the end of the War. Another early 'player' was Richard Seligman, then working in H.C.H. Carpenter's department of metallurgy at Imperial College. Seligman became discouraged for some reason, though not until he had stated in print that he was working on making single crystals of aluminium, in consultation with Rosenhain. (Clearly he loved the metal, for later he founded a famous enterprise, the Aluminium Plant and Vessel Company.) It appears that when Carpenter heard of Hanson's unpublished success, he revived Seligman's research programme, and jointly with Miss Constance Elam, he published in 1921 the first paper on the preparation of large metal crystals by the strain-anneal method, and their tensile properties (Carpenter and Elam 1921). Soon, aluminium crystals made in this way were used to study the changes brought about by fatigue testing (Gough *et al.* 1928), and a little later, Hanson used similar crystals to study creep mechanisms.

The other method of growing large metal crystals is controlled freezing from the melt. Two physicists, B.B. Baker and E.N. da C. Andrade, in 1913–1914 published studies of plastic deformation in sodium, potassium and mercury crystals made from the melt. The key paper however was one by a Pole, Jan Czochralski (1917), who dipped a cold glass tube or cylinder into a pan of molten Pb, Sn or Zn and slowly and steadily withdrew the crystal which initially formed at the dipping point, making a long single-crystal cylinder when the kinetics of the process had been judged right. Czochralski's name is enshrined in the complex current process, based on his discovery, for growing huge silicon crystals for the manufacture of integrated circuits.

Probably the first to take up this technique for purposes of scientific research was Michael Polanyi (1891–1976) who in 1922–1923, with the metallurgist Erich Schmid (1896–1983) and the polymer scientist-to-be Hermann Mark (1895–1992), studied the plastic deformation of metal crystals, at the Institute of Fibre Chemistry in Berlin-Dahlem; in those days, good scientists often earned striking freedom to follow their instincts where they led, irrespective of their nominal specialisms or the stated objective of their place of work. In a splendid autobiographical account of those days, Polanyi (1962) explains how Mark made the Czochralski method work well for tin by covering the melt surface with a mica sheet provided with a small hole. In 1921, Polanyi had used natural rocksalt crystals and fine tungsten crystals extracted from electric lamp filaments to show that metal crystals, on plastic stretching, became work-hardened. The grand old man of German metallurgy, Gustav Tammann, was highly sceptical (he was inclined to be sceptical of everything not done in Göttingen), and this reaction of course spurred the young Polanyi on, and he studied zinc and tin next (Mark *et al.* 1922). Work-hardening was confirmed and accurately measured, and for good measure, Schmid about this time established the law of critical shear stresses for plastic deformation. In Polanyi's own words: "We were lucky in hitting on a problem ripe for solution, big enough to engage our combined faculties, and the solution of which was worth the effort". Just before their paper was out, Carpenter and Robertson published their own paper on aluminium; indeed, the time was ripe. By the end of 1923, Polanyi had moved on to other things (he underwent many intellectual transitions, eventually finishing up as a professor of philosophy in Manchester University), but Erich Schmid never lost his active interest in the plastic deformation of metal crystals, and in 1935, jointly with Walter Boas, he published *Kristallplastizität*, a deeply influential book which assembled the enormous amount of insight into plastic deformation attained since 1921, insight which was entirely conditional on the availability of single metal crystals. "Ripeness" was demonstrated by the fact that *Kristallplastizität* appeared simultaneously with Dr. Elam's book on the same subject. Figure 4.2 shows a medal struck in 1974 to mark the 50th anniversary of Schmid's discovery, as a corollary of the 1922 paper by

Figure 4.2. Medal struck in Austria to commemorate the 50th anniversary of the discovery of the critical shear stress law by Erich Schmid. The image represents a stereographic triangle with 'isobars' showing crystal orientations of constant resolved shear stress (courtesy H.P. Stüwe).

Mark, Polanyi and Schmid, of the constant resolved shear-stress law, which specifies that a crystal begins to deform plastically when the shear stress on the most favoured potential slip plane reaches a critical value.

Aside from Czochralski, the other name always associated with growth of metal crystals from the melt is that of Percy Bridgman (1882–1961), an American physicist who won the Nobel Prize for his extensive researches on high-pressure phenomena (see below). For many of his experiments on physical properties of metals (whether at normal or high pressure) – for instance, on the orientation dependence of thermoelectric properties – he needed single crystals, and in 1925 he published a classic paper on his own method of doing this (Bridgman 1925). He used a metal melt in a glass or quartz ampoule with a constriction, which was slowly lowered through a thermal gradient; the constriction ensured that only one crystal, nucleated at the end of the tube, made its way through into the main chamber. In a later paper (Bridgman 1928) he showed how, by careful positioning of a glass vessel with many bends, he could make crystals of varied orientations. In the 1925 paper he recorded that growing a single crystal from the melt 'sweeps' dissolved impurities into the residual melt, so that most of the crystal is purer than the initial melt. He thus foreshadowed by more than 20 years the later discovery of zone-refining.

Metallic monocrystals were not used only to study plastic deformation. One of the more spectacular episodes in single-crystal research was F.W. Young's celebrated use of spherical copper crystals, at Oak Ridge National Laboratory in America, to examine the anisotropy of oxidation rates on different crystal planes (Young *et al.* 1956). For this purpose, spheres were machined from cylindrical copper crystals, carefully polished by mechanical means and then made highly smooth by anodic electrolytic polishing, thereby removing all the surface damage that was unavoidably caused by mechanical polishing. Figure 4.3 shows the optical interference patterns on such a crystal after oxidation in air, clearly showing the cubic symmetry of the crystal. Such patterns were used to study the oxidation kinetics on different crystal faces, for comparison with the then current theory of oxidation kinetics. Most of Young's extensive researches on copper crystals (1951–1968) concerned the etching of dislocations, but the oxidation study showed how important such crystals could be for other forms of fundamental metallurgical research.

Detailed, critical surveys of the variants and complexities of crystal growth from the melt were published for low-melting metals by Goss (1963) and for high-melting metals (which present much greater difficulties) by Schadler (1963).

It is worth while, now, to analyse the motivation for making metallic single crystals and how, in turn, their production affected physical metallurgy. Initially, metallurgists were concerned to prevent the accidental generation of coarse grains in parts of objects for load-bearing service, and studied recrystallisation with this objective in view. To quote Keith, "Iron crystals... were achieved subsequently by

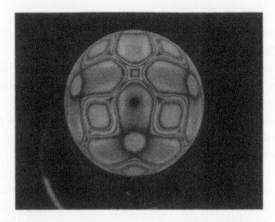

Figure 4.3. Polished spherical copper monocrystal, oxidised to show anisotropy of oxidation rates (after Young *et al.* 1956).

Edwards and Pfeil on the back of investigations... motivated initially by the commercial importance of avoiding coarse recrystallisation in metals during manufacturing processes". Then, a few foreseeing metallurgists like Hanson (1924) and Honda (1924) saw the latent possibilities for fundamental research; thus Hanson remarked: "It (the production of metal crystals) opened up the possibility of the study of behaviour of metals, and particularly of iron and steel, such as had not presented itself before". During the 10 years following, this possibility was energetically pursued all over the world. That precocious physicist, Bridgman, saw the same possibilities from a physicist's perspective. So a parepisteme developed, initially almost accidentally, by turning on its head a targeted practical objective, and many novel insights followed.

Growth of nonmetallic crystals developed partly as a purely academic study that led to major insights, such as Charles Frank's prediction of spiral growth at dislocation sites (Chapter 3, Section 3.2.3.3), and partly as a targeted objective because items such as quartz and ruby crystals were needed for frequency standards, quartz watches, lasers and watch bearings. Some extraordinary single crystals have been grown, including crystals of solid helium grown at 0.1 μm per second at about 1 K (Schuster *et al.* 1996). Crystal growth has become a very major field with numerous books and several journals (e.g., the massive *Journal of Crystal Growth*), but only for metals did single-crystal growth emerge from an initial desire to avoid large grains.

While for many years, metal single crystals were used only as tools for fundamental research, at the beginning of the 1970s single-crystal gas-turbine blades began to be made in the hope of improving creep performance, and today all such blades are routinely manufactured in this form (Duhl 1989).

4.2.2 Diffusion

The migration of one atomic species in another, in the solid state, is the archetype of a materials-science parepisteme. From small beginnings, just over a century ago, the topic has become central to many aspects of solid-state science, with a huge dedicated literature of its own and specialised conferences attended by several hundred participants.

A recent historian of diffusion, Barr (1997), has rediscovered a precociously early study of solid-state diffusion, by the 17th-century natural philosopher, Robert Boyle, (1684); Boyle was one of those men who, in Restoration England, were described as 'the curious'. He describes several experiments involving copper and several other elements and goes on to say: "...there is a way, by which, without the help of salts sulphur or arsenic, one may make a solid and heavy body soak into the pores of that metal and give it a durable colour. I shall not mention the way, because of the bad use that may be made of it..." Barr concludes, from internal evidence, that Boyle had diffused zinc into copper and preceded by fifty years the discovery, in 1732, by Christopher Pinchbeck of the Cu–Zn alloy later called 'pinchbeck' and used as a cheap substitute for gold. Boyle was clearly worried that his experiment, if fully described, might clear the way for forgery of apparent gold coins. Boyle verified that the zinc really had penetrated deeply into the copper (without the copper having been melted), by filing a cross-section and examining it. Boyle's findings were promptly forgotten for over 300 years... the time was not ripe for them. It is ironic, however, that this first attempt to examine solid-state diffusion was partly suppressed precisely because it was too practical.

The next historical waystop is the research of Thomas Graham, also in England, whom we have already encountered (Section 2.1.4) as the originator of colloid science, and again in Section 3.1.1, described as "the leading atomist of his age". In the 1830s (Graham 1833) he studied the diffusion of various gases into air through a porous plug that slowed down the motion of gas molecules, and found that the rate of motion of a gas is linked to its molecular weight. This was the first attempt at a quantitative study of diffusion, albeit not in a solid. Graham's researches were perhaps the first to indicate that the then standard static lattice model of a gas (according to which the gas molecules are arranged on a very dilute lattice subject to mutual repulsion of the molecules... see Mendoza 1990) needed to be replaced by a dynamic model in which all the molecules are in ceaseless motion. Later on, Thomas studied diffusion of solutes in liquids.

Next, the German Adolph Fick (1829–1901), stimulated by Graham's researches, sought to turn diffusion into a properly quantitative concept and formulated the law named after him, relating the rate of diffusion to the steepness of the concentration gradient (Fick 1855), and confirmed his law by measurements of diffusion in liquids. In a critical examination of the influence of this celebrated piece of theory, Tyrrell

(1964) opined that the great merit of Fick's work lay in the stimulus it has given for over a century to *accurate* experimental work in the field, and goes on to remark: "A glance at Graham's extensive, and almost unreadable, descriptions of quantitative studies on diffusion, will show how great a contribution it (Fick's work) was".

All the foregoing were precursors to the first accurate research on diffusion in *solids*, which was performed by William Roberts-Austen (1843–1902), who spent his working life in London (Figure 4.4). It has been said that Graham's greatest contribution to science was to employ Roberts-Austen as his personal assistant at the London Mint (a factory for producing coinage), where he became a skilled assayer, learning to analyse metal concentrations quantitatively. Roberts-Austen, an immensely hard worker, not only became immortal for his researches on diffusion but also played a major role in the introduction of binary metallic phase diagrams; thus in 1897 he presented the first *T-concentration* diagram for Fe–C, which the Dutchman Roozeboom (Section 3.1.2) soon after turned into a proper phase diagram. The face-centred cubic form of iron, *austenite*, was in due course named after Roberts-Austen (there is no phase with a double-barrelled name!). This aspect of his distinguished career, as also features of his life, are outlined in a recent review (Kayser and Patterson 1998). His work on diffusion is discussed by Barr (1997) and

W. CHANDLER ROBERTS-AUSTEN.

Figure 4.4. W. Roberts-Austen (courtesy of M. McLean, Imperial College, London).

also in a lively manner by Koiwa (1998), who further discusses Fick's career in some detail.

In his classic paper on solid-state diffusion (Roberts-Austen 1896a), he remarks that "my long connection with Graham's researches made it almost a duty to attempt to extend his work on liquid diffusion to metals". He goes on to say that initially he abandoned this work because he had no means of measuring high temperatures accurately. This same problem was solved at about the same time by Heycock and Neville (Section 3.1.2) by adopting the then novel platinum-resistance thermometer; Roberts-Austen in due course made use of Le Chatelier's platinum/platinum–rhodium thermocouple, combined with his own instrument for recording temperature as a function of time. His researches on solid-state diffusion became feasible for three reasons: the concept was instilled in his mind by his mentor, Graham; the theoretical basis for analysing his findings had been provided by Fick; and the needful accuracy in temperature came from instrumental improvements. All three... stimulus, theory, instruments... are needed for a major advance in experimental research.

Roberts-Austen's research was focused primarily on the diffusion of gold in solid lead, a fortunate choice, since this is a fast-diffusing couple and this made his sectioning measurements easier than they would have been for many other couples. He chose a low-melting solvent because he surmised, correctly, that the melting-temperature played a dominant role in determining diffusivity. About the same time he also published the first penetration profile for carbon diffusing in iron (Roberts-Austen 1896b); indeed, this was the very first paper in the new *Journal of the Iron and Steel Institute*. It is not clear, according to Barr, whether Roberts-Austen recognised that the diffusion kinetics were related exponentially to temperature, in accordance with Arrhenius's concept of activation energy (Section 2.1.1), but by 1922 that linkage had certainly been recognised by Dushman and Langmuir (1922).

Slight experimental departures from the Arrhenius relation in turn led to recognition of anomalous diffusion mechanisms. Indeed, after a gap in activity of a quarter century, in the 1920s, interest veered to the *mechanism(s)* involved in solid-state diffusion. The history of these tortuous discussions, still in progress today, has been told by Tuijn (1997) and also discussed in Koiwa's papers mentioned above. In 1684, Boyle had in passing referred to his solute 'soaking into the pores of copper', and in a way this was the centre of all the debates in the 1920s and 1930s: the issue was whether atoms simply switched lattice sites without the aid of crystal defects, or whether diffusion depends on the presence, and migration, of vacant lattice sites (vacancies) or, alternatively, on the ability of solute atoms to jump off the lattice and into interstitial sites. The history of the point-defect concept has already been outlined (Chapter 3, Section 3.2.3.1), but one important player was only briefly mentioned. This was a Russian, Yakov Frenkel, who in 1924, while visiting Germany, published a crucial paper (Frenkel 1924). In this he argued that since atoms in a crystal can sublime (evaporate) from the surface, so they should be able to do *inside* the crystal, that is, an atom should be able to wander from its proper site into an interstitial site, creating what has since been termed a 'Frenkel defect' (a vacant lattice site *plus* an interstitial atom nearby). He followed this up by a further paper (Frenkel 1926) which Schmalzried, in his important textbook on chemical kinetics of solids, describes as a "most seminal theoretical paper" (Schmalzried 1995). Here he points out that in an 'ionic' crystal such as silver bromide, some of the silver ions will 'evaporate' into interstitial sites, leaving silver vacancies behind; the two kinds of ion will behave differently, the size being an important variable. Frenkel recognised that point defects are an *equilibrium feature* of a crystal, the concentration being determined by, in Schmalzried's words, "a compromise between the ordering interaction energy and the entropy contribution of disorder (point defects, in this case)". In its own way, this was as revolutionary an idea as Willard Gibbs's original notion of chemical equilibrium in thermodynamic terms.

There is no space here to map the complicated series of researches and sustained debates that eventually led to the firm recognition of the crucial role of crystal vacancies in diffusion, and Tuijn's brief overview should be consulted for the key events. A key constituent in these debates was the observation in 1947 of the *Kirkendall effect* – the motion of an inert marker, inserted between two metals welded together before a diffusion anneal, relative to the location of the (now diffuse) interface after the anneal. This motion is due to the fact that vacancies in the two metals move at different speeds. The effect was reported by Smigelskas and Kirkendall (1947). It then met the unrelenting scepticism of Kirkendall's mentor, Robert Mehl (a highly influential metallurgist whom we met in Section 3.2.1), and so took some time to make its full impact. In due course, in 1951, one of Mehl's later students, Carrea da Silva, himself put the phenomenon beyond doubt, and on his deathbed in 1976, Mehl was reconciled with Kirkendall (who had by then long since left research to become a scientific administrator – the fate of so many fine researchers). This affecting tale is told in detail in a historical note on the Kirkendall effect by Nakajima (1997); it is well worth reading.

In some materials, semiconductors in particular, interstitial atoms play a crucial role in diffusion. Thus, Frank and Turnbull (1956) proposed that copper atoms dissolved in germanium are present both substitutionally (together with vacancies) *and* interstitially, and that the vacancies and interstitial copper atoms diffuse independently. Such diffusion can be very rapid, and this was exploited in preparing the famous micrograph of Figure 3.14 in the preceding chapter. Similarly, it is now recognised that transition metal atoms dissolved in silicon diffuse by a very fast, predominantly interstitial, mechanism (Weber 1988).

Turnbull was also responsible for another insight of great practical importance. In the late 1950s, while working at the General Electric research laboratories during their period of devotion to fundamental research, he and his collaborators (Desorbo *et al.* 1958) were able to explain the fact that Al–Cu alloys quenched to room temperature initially age-harden (a diffusion-linked process) several orders of magnitude faster than extrapolation of measured diffusion rates at high temperatures would have predicted. By ingenious electrical resistivity measurements, leading to clearly defined activation energies, they were able to prove that this disparity was due to excess vacancies 'frozen' into the alloy by the high-speed quench from a high temperature. Such quenched-in vacancies are now known to play a role in many metallurgical processes.

Another subsidiary field of study was the effect of high concentrations of a diffusing solute, such as interstitial carbon in iron, in slowing diffusivity (in the case of carbon in fcc austenite) because of mutual repulsion of neighbouring dissolved carbon atoms. By extension, high carbon concentrations can affect the mobility of substitutional solutes (Babu and Bhadeshia 1995). These last two phenomena, quenched-in vacancies and concentration effects, show how a parepisteme can carry smaller parepistemes on its back.

From diffusion of one element in another it is a substantial intellectual step to the study of the diffusion of an element in itself... self-diffusion. At first sight, this concept makes no sense; what can it matter that identical atoms change places in a crystalline solid? In fact, self-diffusion plays a key role in numerous processes of practical consequence, for instance: creep, radiation damage, pore growth, the evolution of microstructure during annealing; the attempts to understand how self-diffusion operates has led to a wider understanding of diffusion generally. To study self-diffusion, some way has to be found to distinguish some atoms of an element from others, and this is done either by using radioactive atoms and measuring radioactivity, or by using stable isotopes and employing mass-spectrometry. The use of radio-isotopes was pioneered by a Hungarian chemist, György von Hevesy (1885–1966): he began in 1921 with natural radio-isotopes which were the end-product of a radioactive decay chain (^{210}Pb and ^{212}Pb), and later moved on to artificial radio-isotopes. As Koiwa (1998) recounts, he was moved to his experiments with lead by his total failure to separate radium D (in fact, as it proved, a lead isotope) from a mass of lead in which the sample had been intentionally embedded. Here, as in the attempts to prevent excessive grain growth in iron, a useful but unexpected concept emerged from a frustrating set of experiments. Later, von Hevesy moved on to other exploits, such as the discovery of the element hafnium.

There is no space here to go into the enormous body of experiment and theory that has emerged from von Hevesy's initiative. The reader is referred to an excellent critical overview by Seeger (1997). Important concepts such as the random-walk model for the migration of vacancies, modified by non-random aspects expressed by the 'correlation coefficient', emerged from this work; the mathematics of the random walk find applications in far-distant fields, such as the curling-up of long polymer chains and the elastic behaviour of rubber. (Indeed, the random walk concept has recently been made the basis of an 'interdisciplinary' section in a textbook of materials science (Allen and Thomas 1999).) When it was discovered that some plots of the logarithm of diffusion coefficients against reciprocal temperature were curved, the recognition was forced that divacancies as well as monovacancies can be involved in self-diffusion; all this is set out by Seeger.

The transport of charged ions in alkali halides and, later on, in (insulating) ceramics is a distinct parepisteme, because electric fields play a key role. This large field is discussed in Schmalzried's 1995 book, already mentioned, and also in a review by one of the pioneers (Nowick 1984). This kind of study in turn led on to the developments of superionic conductors, in which ions and not electrons carry substantial currents (touched on again in Chapter 11, Section 11.3.1.1).

Diffusion now has its own specialised journal, *Defect and Diffusion Forum*, which published the successive comprehensive international conferences devoted to the parepisteme.

Some of the many fields of MSE in which an understanding of, and quantitative knowledge of, diffusion, self-diffusion in particular, plays a major role will be discussed in the next chapter.

4.2.3 High-pressure research

In Section 3.2.5 something was said about the central role of measurements of physical and mechanical properties at high pressures as a means of understanding processes in the interior of the earth. This kind of measurement began early in the 20th century, but in a tentative way because the experimental techniques were unsatisfactory. Pressures were usually generated by hydraulic means but joints were not properly pressure-tight, and there were also difficulties in calibration of pressures. All this was changed through the work of one remarkable man, Percy (known as Peter) Bridgman (1882–1961). He spent his entire career, student, junior researcher and full professor (from 1919) at Harvard University, and although all his life (except during the Wars) he was fiercely devoted to the pursuit of basic research, as an unexpected byproduct he had enormous influence on industrial practice. Good accounts of his career can be found in a biographical memoir prepared for the National Academy of Sciences (Kemble and Birch 1970) and in an intellectual biography (Walter 1990). Figure 4.5 is a portrait. His numerous papers (some 230 on high-pressure research alone) were published in collected form by Harvard University Press in 1964. Two books by Bridgman himself give accounts of his

Figure 4.5. P.W. Bridgman (courtesy of G. Holton, Harvard University).

researches from 1906 onwards. One (Bridgman 1931, 1949) includes a useful historical chapter: here we learn that in the nineteenth century, attention focused largely on the liquefaction of gases and on supercritical behaviour that removed the discontinuity between gaseous and liquid states, whereas early in the twentieth century, attention began to be focused on condensed matter, both liquids and solids, with geological laboratories well to the fore. Bridgman's other relevant book (Bridgman 1952) was devoted entirely to plasticity and fracture in pressurised solids. A very recent book (Hazen 1999) on diamond synthesis includes an excellent chapter on *The legacy of Percy Bridgman*.

Bridgman came to high-pressure research through a project to check the predicted relationship between the density of a glass of specified composition and its refractive index. He quickly became so fascinated by the technical problems of creating high pressures while allowing measurements of properties to be made that he focused on this and forgot about refractive indices. (This sort of transfer of attention is how a variety of parepistemes were born.) Bridgman was an excellent mechanic who did not allow professional craftsmen into his own home – his memoirists refer to his "fertile mechanical imagination and exceptional manipulative dexterity" – and he quickly designed a pressure seal which became the tighter, the greater the pressure on it; the Bridgman seal solved the greatest problem in high-pressure research. He also learned enough metallurgy to select appropriate high-strength steels for the components of his apparatus. He had few research students and did most of his research with his own hands. (It is said of him that when news of his Nobel Prize came through in 1946, a student sought to interrupt him during a period of taking experimental readings to tell him the news but was told to go away and tell him the details later.) Once his apparatus worked well, he focused on electrical properties for preference, especially of single crystals (see Section 4.2.1) but became so interested by the occasional distortions and fractures of his equipment that he undertook extensive research on enhanced plastic deformability of metals and minerals, some of them normally completely brittle, under superimposed hydrostatic pressure; he undertook research for the US armed forces on this theme that led to several important military applications, and eventually he wrote the aforementioned book dedicated to this (Bridgman 1952). These researches cleared the path for much subsequent research in geological laboratories.

Bridgman had strong views on the importance of empirical research, influenced as little as possible by theory, and this helped him test the influence of numerous variables that lesser mortals failed to heed. He kept clear of quantum mechanics and dislocation theory, for instance. He became deeply ensconced in the philosophy of physics research; for instance, he published a famous book on dimensional analysis, and another on 'the logic of modern physics'. When he sought to extrapolate his ideas into the domain of social science, he found himself embroiled in harsh disputes; this has happened to a number of eminent scientists, for instance, J.D. Bernal. Walter's book goes into this aspect of Bridgman's life in detail.

It is noteworthy that though Bridgman set out to undertake strictly fundamental research, in fact his work led to a number of important industrial advances. Thus his researches on mechanical properties led directly to the development of high-pressure metal forming in industry: the story of this is told by Frey and Goldman (of the Ford Motor Company) (1967). Thus, copper at the relatively low hydrostatic pressure of 100 000 psi (0.7 GPa) can be deformed to enormous strains without fracture or reannealing, and connectors of complex shape can be cold-formed in a single operation. Frey and Goldman claim that their development programme proved "exceedingly profitable", and they directly credit Bridgman for its genesis.

In the same volume, two former research directors of the GE Corporate Research Center (Suits and Bueche 1967) record the case-history of GE's 'diamond factory'. The prolonged research effort began in 1941 with a contract awarded to Bridgman; the War intervened and prevented Bridgman from working on the theme; in any case, Bridgman was insufficiently versed in chemistry to recognise the need for metallic catalysts. After the War was over GE acquired high-pressure equipment from Bridgman and did in-house research which eventually, in late 1954, when a method of reaching the very high temperatures and pressures required had been perfected and after the crucial role of catalysts had been established, led to the large-scale synthesis of industrial diamond grit at high temperatures and pressures. According to Hazen's book, roughly 100 tons of synthetic diamond (mostly grit for grinding and cutting tools) are now manufactured every year, "providing almost nine out of every ten carats used in the world". In recent years, methods have been perfected of making synthetic diamond in the form of thins sheets and coatings, by a vapour-based method operating at low pressure. This approach also has increasing applications, though they do not overlap with the pressure-based approach. This latest advance is an instance of *challenge and response*, rather like the great improvements made in crystalline transformer steel sheets to respond to the challenge posed by the advent of metallic glass ribbons.

The GE research program, although it was the most successful effort, was far from being the only attempt to make synthetic diamond. There was much research in Russia, beginning in the 1930s; language barriers and secrecy meant that this valuable work was not widely recognised for many years, until DeVries *et al.* (1996) published a detailed account. Another determined attempt to synthesise diamond which led to a success in 1953 but was not followed through was by the ASEA company in Sweden. This episode is racily told in Chapter 4 of Hazen's book under the title *Baltzar von Platen and the Incredible Diamond Machine*.

Hot isostatic pressing (HIP), a technique which was introduced in 1955 and became widespread in advanced materials processing from 1970 onwards, was developed by ASEA and derived directly from the Swedish diamond research in the early 1950s. In this apparatus, material is heated in a furnace which is held within a large (cold) pressure vessel filled with highly pressurised argon. Elaborate techniques, including reinforcement of the pressure vessel by pretensioned wire windings, had to be developed for this technique to work reliably. By HIP, microporosity within a material, whether caused during manufacture or during service, can be closed up completely. HIP has been used for such purposes as the containment of radioactive waste in ceramic cylinders, strength improvement of cemented carbides (Engel and Hübner 1978), the homogenisation of high-speed tool steels, the 'healing' of porous investment castings (by simply pressing the pores into extinction), and the 'rejuvenation' of used jet-engine blades again by getting rid of the porous damage brought about by creep in service. Lately, HIP has been widely used to permit complete densification of 'difficult' powder compacts. Apparently, HIP was even used at GE to repair damaged carborundum pressure-transmitting blocks needed for their production process. HIP is an excellent example of a process useful to materials engineers developed as spin-off from what was initially a piece of parepistemic research.

It appears that HIP was independently invented, also in 1955, at the Battelle Memorial Institute in Columbus, Ohio, under contract to the Atomic Energy Commission and with the immediate objective of bonding nucelar fuel elements with precise dimensional control.

The various densification mechanisms at different temperatures can be modelled and displayed in HIP diagrams, in which relative temperature is plotted against temperature normalised with respect to the melting-point (Arzt *et al.* 1983). This procedure relates closely to the deformation-mechanism maps discussed in Section 5.1.2.2.

Bridgman's personal researches, as detailed in his 1931 book, covered such themes as electrical resistivity, electrical and thermal conductivity, thermoelectricity and compressibility of solids, and viscosity of liquids. The ability to measure all these quantities in small pressure chambers is a mark of remarkable experimental skill. There is also a chapter on pressure-induced phase transformations, including what seem to have been the first studies of the pressure-induced polymorphs of ice (and 'heavy ice'). In recent decades research emphasis has shifted more and more towards polymorphism under pressure. Pressures now readily attainable in ordinary pressure chambers exceed 20 GPa, while minute diamond anvils have also been developed that permit X-ray diffraction under pressures well over 200 GPa. Nowadays, pressure effects are often created transiently, by means of shock waves, and studied by techniques such as X-ray flash radiography. Recent researches are reviewed by Ruoff (1991), and a lively popular account of these methods makes up the end of Hazen's (1999) book. A good example of a research programme that falls between several specialities (it is often classified as chemical physics) is the analysis of crystal structures of ice at different temperatures and pressures, pioneered by Bridgman in 1935. A few years ago, nine different ice polymorphs, all with known crystal structures, had been recorded (Savage 1988); by now, probably even more polymorphs are known. Indeed, many of the elements have been found to have pressure-induced polymorphs, which often form very sluggishly (Young 1991).

The impact of high pressures on crystal structure research generally is considerable, to the extent that the International Union of Crystallography has set up a high-pressure commission; a recent (1998) "workshop" organised by this commission at the Argonne National Laboratory in Illinois (home to a synchrotron radiation source) attracted 117 researchers. At the big Glasgow Congress of the International Union of Crystallography in 1999, the high-pressure commission held several meetings that attracted very varied contributions, summarised in IUCr (2000). One finding was that carbon dioxide forms a polymer under extreme pressure!

Robert Hazen's excellent 1999 book on the diamond-makers has been repeatedly cited. Earlier, he had brought out a popular account of high-pressure research generally, under the title *The New Alchemists: Breaking Through the Barriers of High Pressure* (Hazen 1993).

The high-pressure community is now drawn from many fields of interest and many branches of expertise. A recent symposium report (Wentzcovich *et al.* 1998) gives a flavour of this extraordinary variety, drawing in not only earth science but microelectronics, supercritical phase transformations in fluids studied by chemical engineers (the wheel coming full circle), powder processing under extreme conditions, etc. One paper focuses on one characterisation tool, the Advanced Photon Source (a synchrotron radiation facility), which has been used in 11 different ways to characterise materials at 'ultrahigh pressures and temperatures', including time-resolved X-ray diffraction. Perhaps because the high-pressure parepisteme is so very diffuse, it has taken a long time for a journal exclusively devoted to the field to emerge: *High Pressure Research*. Much research on high pressures is still divided between materials-science and earth-science journals.

This summary shows how research undertaken by one brilliant scientist for his own interest has led to steadily enhanced experimental techniques, unexpected applications and a merging of many skills and interests.

4.2.4 Crystallography

In Chapter 3, from Section 3.1.1.1 onwards, I discuss a range of aspects of crystals – X-ray diffraction, polymorphism and phase transformations, crystal defects, crystal growth, polytypism, the relation of crystal structure to chemical reactivity, crystal chemistry and physics. All these topics belong, more or less closely, to the vast parepisteme of *crystallography*. In that Chapter, I treated the study of crystals as one of the central *precursors* of materials science, and so indeed it is, but all the above-mentioned component topics, and others too, were parts of a huge parepisteme because none of them was directly aimed, originally, at the solution of specific practical problems.

Crystallography is an exceptional parepisteme because of the size of its community and because it has an 'aristocracy' – the people who use X-ray diffraction to determine the structures of crystals. This probably came about because, alone among the parepistemes I have discussed, crystallographers have had their own scientific union, the International Union of Crystallography (IUCr), affiliated to the International Council of Scientific Unions (ICSU), since 1948. Its origin is discussed by a historian of ICSU (Greenaway 1996), who remarks that the IUCr "was brought into existence because of the development, not of crystallography, which had its origin in the 17th century, but of X-ray crystallography which originated in about 1913. By 1946 there were enough X-ray crystallographers in the world and in touch with each other for them to want to combine. Moreover, though publication was important, a mere learned society would not quite meet their needs. *The reason for this was that their subject was already useful in throwing light on problems in other fields of science, pure and applied* (my italics). A Union, with its ICSU-guided links with other Unions, was a better form". We have seen in Chapter 3 that the old crystallographic journal, *Zeitschrift für Kristallographie*, was very tardy in recognising the importance of X-ray diffraction after 1912. The new Union, founded in 1948, created its own giant journal, *Acta Crystallographica*; as with some other journals founded in this period, the title resorts to Latin to symbolise the journal's international outlook. Incidentally, while the IUCr flourishes mightily, materials science and engineering has no scientific union. A social historian is needed to attempt an analysis of the reasons for this omission.

In addition to the overarching role of the IUCr, there are numerous national crystallographic associations in various countries, some of them under the umbrella of bodies like the Institute of Physics in Britain. I doubt whether there is any other parepisteme so generously provided with professional assemblies all over the world.

Metallurgists originally, and now materials scientists (as well as solid-state chemists) have used crystallographic methods, certainly, for the determination of the structures of intermetallic compounds, but also for such subsidiary parepistemes as the study of the orientation relationships involved in phase transformations, and the study of preferred orientations, alias 'texture' (statistically preferential alignment of the crystal axes of the individual grains in a polycrystalline assembly); however, those who pursue such concerns are not members of the aristocracy! The study of texture both by X-ray diffraction and by computer simulation has become a huge sub-subsidiary field, very recently marked by the publication of a major book (Kocks *et al.* 1998).

Physics also is intimately linked with crystallography in many ways. One mode of connection is through the detailed study of crystal perfection, which substantially influences the diffraction behaviour: the most recent review of this involved topic, which has been studied since the earliest days of X-ray diffraction, is by Lal (1998) (his paper is titled 'Real structure of real crystals'). A famous systematic presentation of the mathematical theory of crystal anisotropy, still much cited, is a book by Nye (1957); this study goes back in its approach to the great German mineralogists of the 19th century. Nevertheless, physicists feel increasingly uneasy about the proper nature of their linkage with crystallography; thus in 1999, the Physical Crystallography Group of the (British) Institute of Physics decided to change its name to 'Structural Condensed Matter Physics Group'; the word 'crystallography' has vanished from the name.

Perhaps the last general overview of crystallography in all its many aspects, including crystal chemistry and crystal physics and the history of crystallographic concepts, as well as the basics of crystal structure determination, was a famous book by the Braggs, father and son (Bragg and Bragg 1939), both of them famous physicists as well as being the progenitors of X-ray diffraction.

Chemical crystallographers are also beginning to reconsider their tasks. Thus, in a prologue to a new book (Rogers and Zaworotko 1999), G.R. Desiraju comments: "...the determination of most small-molecule structures became a straightforward operation and crystallographic databases began to be established..." (see Section 13.2.2). The interest of the chemical crystallographer, now more properly called a structural chemist, has changed from crystal structure determination to crystal structure synthesis. The question now becomes 'How does one go about designing a particular crystal structure that is associated with a particular architecture, geometry, form or function?'.

The broad appeal of crystallography across a wide gamut of sciences is demonstrated by a book brought out to mark the 50th anniversary of *Acta Crystallographica* and the International Union of Crystallography (Schenk 1998). Physics, chemistry, biochemistry, superconductivity, neutron diffraction and the teaching of crystallography all find their champions here; several of these essays are written with a historical emphasis.

The 'aristocrats' who determine crystal structures have garnered a remarkable number of Nobel Prizes; no fewer than 26 have gone to scientists best described as crystallographers, some of them in physics, some in chemistry, latterly some in biochemistry. Crystallography is one of those fields where physics and chemistry have become intimately commingled. It has also evinced more than its fair share of quarrelsomeness, since many physicists regard it as a mere technique rather than a respectable science, while crystal structure analysts, as we have seen, were for years inclined to regard anyone who studied the many other aspects of crystals as second-class citizens.

It is striking that, in spite of the huge importance of crystallography in physics, chemistry, biochemistry, pharmacology and materials science, few degree courses leading to bachelor's degrees in crystallography are on record. The famous Institute of Crystallography in Moscow in its heyday gave degrees in crystallography (it certainly trained students at the research level), as did some other Russian institutes; Birkbeck College in London University has a famous Department of Crystallography, based on the early fame of J.D. Bernal, which awards degrees, and there is a degree course in crystallography in the Netherlands. There was a brief attempt to award a degree in crystallography in the physics department of Cambridge University, but it did not last. Now students in Cambridge who wish to specialise early in this parepisteme need to take a degree in earth sciences. So, the small parepisteme of colloid science and the large parepisteme of crystallography are in this respect on a par – one cannot easily get degrees in colloid science or in crystallography.

4.2.5 Superplasticity

To conclude this selection of examples from the wide range of parepistemes in MSE, I have chosen a highly specialised one which has developed into a major industrial technique. Superplasticity has recently been defined, in a formulation agreed at a major international conference devoted to the subject, as follows: "Superplasticity is the ability of a polycrystalline material to exhibit, in a generally isotropic manner, very high tensile elongations prior to failure". In this connection, 'high' means thousands of percent; the world record is currently held by a Japanese, Higashi, at 8000% elongation.

The first recorded description of the phenomenon was in 1912 by an English metallurgist (Bengough 1912). He studied a two-phase brass, pulling it at a modest strain rate at a range of temperatures up to 800°C, and securing a maximum strain of ≈160% at 700°C. His thumbnail description is still very apposite: "A certain special brass... pulled out to a fine point, just like glass would do, having an enormous elongation". In the following 35 years occasional studies of various two-phase alloys confirmed this type of behaviour, which mimics the behaviour of glasses like pyrex and amorphous silica while the alloys remain crystalline throughout. Thus, Pearson (1934) stretched a Bi–Sn alloy to nearly 2000% (see Figure 4.6). The stress σ required to maintain a strain rate $d\varepsilon/dt$ is approximately given by $\sigma = (d\varepsilon/dt)^m$; for a glass, $m = 1$, for metals it is usually much lower. When m is high, the formation of a neck in tension is impeded because in a neck, the local strain rate becomes enhanced and

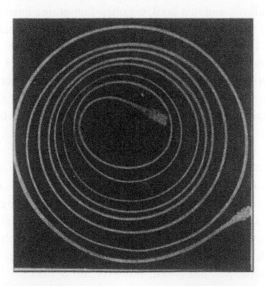

Figure 4.6. Pearson's famous photograph in 1934 of a Bi–Sn alloy that has undergone 1950% elongation.

so a much higher stress would be needed to sustain the formation of the neck. In a glass, very rapid drawing out is feasible; for instance, many years ago it was found that a blob of amorphous silica can be drawn into a very fine fibre (for instrument suspensions) by shooting the hot blob out like an arrow from a stretched bow. In alloys, the mechanism of deformation is quite different: it involves Nabarro–Herring creep, in which dislocations are not involved; under tension, strain results from the stress-biased diffusion of vacancies from grain boundaries transverse to the stress to other boundaries lying parallel to the stress. The operation of this important mechanism, which is the key to superplasticity, can be deduced from the mathematical form of the grain-size dependence of the process (Nabarro 1948, Herring 1950); it plays a major part in the deformation-mechanism maps outlined in the next chapter (Section 5.1.2.2). For large superplastic strains to be feasible, very fine grains (a few micrometres in diameter) and relatively slow strain rates (typically, 0.01/second) are requisite, so that the diffusion of vacancies can keep pace with the imposed strain rate. Sliding at grain boundaries is also involved. Practical superplastic alloys are always two-phase in nature, because a second phase is needed to impede the growth of grains when the sample is held at high temperature, and a high temperature is essential to accelerate vacancy diffusion.

The feasibility of superplastic forming for industrial purposes was first demonstrated, half a century after the first observation, by a team led by Backofen at MIT in 1964; until then, the phenomenon was treated as a scientific curiosity... a parepisteme, in fact. In 1970, the first patent was issued, with reference to superplastic nickel alloys, and in a book on ultra-fine-grained metals published in the same year, Headley et al. (1970) gave an account of 'the current status of applied superplasticity'. In 1976, the first major industrial advance was patented and then published in Britain (Grimes et al. 1976), following a study 7 years earlier on a simple Al–Cu eutectic alloy. The 1976 alloy (Al–6 wt% Cu–0.5 wt% Zr), trade name SUPRAL, could be superplastically formed at a reasonably fast strain rate and held its fine grains because of a fine dispersion of second-phase particles. It was found that such forming could be undertaken at modest stresses, using dies (to define the end-shape) made of inexpensive materials; it is therefore suitable for small production runs, without incurring the extravagant costs of tool-steel dies like those used in pressing automobile bodies of steel. A wide variety of superplastically formable aluminium alloys was developed during the following years. There was then a worldwide explosion of interest in superplasticity, fuelled by the first major review of the topic (Edington et al. 1976), which surveyed the various detailed mechanistic models that had recently been proposed. The first international conference on the topic was not called, however, until 1982.

In 1986, Wakai et al. (1986) in Japan discovered that ultra-fine-grained ceramics can also be superplastically deformed; they may be brittle with respect to dislocation behaviour, but can readily deform by the Nabarro–Herring mechanism. This recognition was soon extended to intermetallic compounds, which are also apt to be brittle in respect of dislocation motion. Rapid developments followed after 1986 which are clearly set out in the most recent overview of superplasticity (Nieh *et al.* 1997). Very recently – after Nieh's book appeared – research in Russia by R. Valiev showed that it is possible to deform an alloy very heavily, in a novel way, so as to form a population of minute subgrains within larger grains and thereby to foster superplastic capability in the deformed alloy.

This outline case-history is an excellent example of a parepisteme which began as a metallurgical curiosity and developed, at a leisurely pace, into a well-understood phenomenon, from which it became, at a much accelerated pace, an important industrial process.

4.3. GENESIS AND INTEGRATION OF PAREPISTEMES

Parepistemes grow from an individual's curiosity, which in turn ignites curiosity in others; if a piece of research is directly aimed at solving a specific practical problem, then it is part of mainline research and not a parepisteme at all. However, the improvement of a technique used for solving practical problems constitutes a parepisteme.

Curiosity-driven research, a term I prefer to 'fundamental' or 'basic', involves following the trail wherever it may lead and, in Isaac Newton's words (when he was asked how he made his discoveries): "by always thinking unto them. I keep the subject constantly before me and wait until the first dawnings open little by little into full light". The central motive, curiosity, has been rendered cynically into verse by no less a master than A.E. Housman:

> Amelia mixed some mustard,
> She mixed it strong and thick:
> She put it in the custard
> And made her mother sick.
> And showing satisfaction
> By many a loud "huzza!",
> "Observe" she said "the action
> Of mustard on mamma".

A further motive is the passion for clarity, which was nicely illustrated many years ago during a conversation between Dirac and Oppenheimer (Pais 1995). Dirac was astonished by Oppenheimer's passion for Dante, and for poetry generally, side by side with his obsession with theoretical physics. "Why poetry?" Dirac wanted to know. Oppenheimer replied: "In physics we strive to explain in *simple terms* what no one understood before. With poetry, it is just the opposite". Perhaps, to modify this bon mot for materials science, we could say: "In materials science, we strive to achieve by reproducible means what no one could do before...".

"Simple terms" can be a trap and a delusion. In the study of materials, we must be prepared to face complexity and we must distrust elaborate theoretical systems advanced too early, as Bridgman did. As White (1970) remarked with regard to Descartes: "Regarding the celebrated 'vorticist physics' which took the 1600s by storm... it had all the qualities of a perfect work of art. Everything was accounted for. It left no loose ends. It answered all the questions. Its only defect was that it was not true".

The approach to research which leads to new and productive parepistemes, curiosity-driven research, is having a rather difficult time at present. Max Perutz, the crystallographer who determined the structure of haemoglobin and for years led the Laboratory for Molecular Biology in Cambridge, on numerous occasions in recent years bewailed the passion for *directing* research, even in academic environments, and pointed to the many astonishing advances in his old laboratory resulting from free curiosity-driven research. That is often regarded as a largely lost battle; but when one contemplates the numerous, extensive and apparently self-directing parepistemic 'communities', for instance, in the domains of diffusion and high pressures, one is led to think that perhaps things are not as desperate as they sometimes seem.

My last point in this chapter is the value of integrating a range of parepistemes in the pursuit of a practical objective: in materials science terms, such integration of curiosity-driven pursuits for practical reasons pays a debt that parepistemes owe to mainline science. A good example is the research being done by Gregory Olson at Northwestern University (e.g., Olson 1993) on what he calls 'system design of materials'. One task he and his students performed was to design a new, ultrastrong martensitic bearing steel for use in space applications. He begins by formulating the objectives and restrictions as precisely as he can, then decides on the broad category of alloy to be designed, then homes in on a desirable microstructure type, going on to exploit a raft of *distinct* parepistemes relating to: (1) the strengthening effect of dispersions as a function of scale and density, (2) stability against coarsening, (3) grain-refining additives, (4) solid-solution hardening, (5) grain-boundary chemistry, including segregation principles. He then goes on to invoke other parepistemes relating microstructures to processing strategies, and to use CALPHAD (phase-diagram calculation from thermochemical inputs). After all this has been put through successive cycles of theoretical optimisation, a range of prospective compositions emerges. At this point, theory stops and the empirical stage, never to be bypassed entirely, begins. What the pursuit and integration of parepistemes makes possible is to narrow drastically the range of options that need to be tested experimentally.

REFERENCES

Allen, S.M. and Thomas, E.L. (1999) *The Structure of Materials*, Chapter 2 (Wiley, New York).
Anderson, R.P. (1918) *Trans. Faraday Soc.* **14**, 150.
Arzt, E., Ashby, M.F. and Easterling, K.E. (1983) *Metall. Trans.* **14A**, 211.
Babu, S.S. and Bhadeshia, H.K.D.H. (1995) *J. Mater. Sci. Lett.* **14**, 314.
Barr, L.W. (1997) *Defect Diffusion Forum* **143–147**, 3.
Bengough, G.D. (1912) *J. Inst. Metals* **7**, 123.
Bowen, J.S.M. (1981) Letter to RWC dated 24 August.
Boyle, R. (1684) *Experiments and Considerations about the Porosity of Bodies in Two Essays*.
Bragg, W.H. and Bragg, W.L. (1939) *The Crystalline State: A General Survey* (Bell and Sons, London).
Bridgman, P.W. (1925) *Proc. Amer. Acad. Arts Sci.* **60**, 305; also (1928) *ibid* **63**, 351; Most easily accessible, in *Collected Experimental Papers*, 1964, ed. Bridgman, P.W. (Harvard University Press, Cambridge, MA).
Bridgman, P.W. (1931, 1949) *The Physics of High Pressure*, 1st and 2nd editions (Bell and Sons, London).
Bridgman, P.W. (1952) *Studies in Large Plastic Flow and Fracture, with Special Emphasis on the Effects of Hydrostatic Pressure* (McGraw-Hill, New York).
Carpenter, H.C.H. and Elam, C.F. (1921) *Proc. Roy. Soc. Lond.* **A100**, 329.
Czochralski, J. (1917) *Z. Phys. Chem.* **92**, 219.
Desorbo, W., Treaftis, H.N. and Turnbull, D. (1958) *Acta Metall.* **6**, 401.
DeVries, R.C., Badzian, A. and Roy, R. (1996) *MRS Bull.* **21**(2), 65.
Duhl, D.N. (1989) Single crystal superalloys, in *Superalloys, Supercomposites and Superceramics*, ed. Tien, J.K. and Caulfield, T. (Academic press, Boston) p. 149.
Dushman, S. and Langmuir I. (1922) *Phys. Rev.* **20**, 113.
Edington, J.W., Melton K.W. and Cutler, C.P. (1976) *Progr. Mater. Sci.* **21**, 61.
Edwards, C.A. and Pfeil, L.B. (1924) *J. Iron Steel Inst.* **109**, 129.
Elam, C.F. (1935) *Distortion of Metal Crystals* (Clarendon Press, Oxford).
Engel, U. and Hübner, H. (1978) *J. Mater. Sci.* **13**, 2003.
Fick, A. (1855) *Poggendorf Ann.* **94**, 59; *Phil. Mag.* **10**, 30.
Frank, F.C. and Turnbull, D. (1956) *Phys. Rev.* **104**, 617.
Frenkel, Y. (1924) *Z. f. Physik* **26**, 117.
Frenkel, Y. (1926) *Z. f. Physik* **35**, 652.
Frey, D.N. and Goldman, J.E. (1967) in *Applied Science and Technological Progress* (US Government Printing Office, Washington, DC) p. 273.
Goss, A.J. (1963) in *The Art and Science of Growing Crystals*, ed. Gilman, J.J. (Wiley, New York) p. 314.

Gough, H.J., Hanson, D. and Wright, S.J. (1928) *Phil. Trans. Roy. Soc. Lond.* **A226**, 1.

Graham, T. (1833) *Phil. Mag.* **2**, 175.

Greenaway, F. (1996) *Science International: A History of the International Council of Scientific Unions* (Cambridge University Press, Cambridge).

Grimes, R., Stowell, M.J. and Watts, B.M. (1976) *Metals Technol.* **3**, 154.

Hanemann, H. and Schrader, A. (1927) *Atlas Metallographicus* (Bornträger, Berlin) Table 101.

Hanson, D. (1924) *J. Inst. Metals* **109**, 149.

Hazen, R. (1993) *The New Alchemists: Breaking Through the Barriers of High Pressure* (Times Books, Random House, New York).

Hazen, R. (1999) *The Diamond Makers* (Cambridge University Press, Cambridge).

Headley, T.J., Kalish, D. and Underwood, E.E. (1970) The current status of applied superplasticity, in *Ultrafine-Grain Metals*, ed. Burke, J.J. and Weiss, V. (Syracuse University Press, Syracuse, NY) p. 325.

Herring, C. (1950) *J. Appl. Phys.* **21**, 437.

Honda, K. (1924) *J. Inst. Metals* **109**, 156.

IUCr (2000) The high-pressure program from Glasgow, *IUCr Newsletter* **8**(2), 11.

Kayser, F.X. and Patterson, J.W. (1998) *J. Phase Equili.* **19**, 11.

Keith, S. (1998) unpublished paper, private communication.

Kemble, E.C. and Birch, F. (1970) *Biographical Memoirs of Members of the National Academy of Sciences*, Vol. 41, Washington (Columbia University Press, New York and London) (Memoir of P.W. Bridgman) p. 23.

Kocks, U.F., Tomé, C.N. and Wenk, H.-R (1998) *Texture and Anisotropy: Preferred Orientations in Polycrystals and their Effects on Materials Properties* (Cambridge University Press, Cambridge).

Koiwa, M. (1998) *Mater. Trans. Jpn. Inst. Metals* **39**, 1169; *Metals Mater.* **4**, 1207.

Lal, K. (1998) *Curr. Sci. (India)* **64A**, 609.

Mark, H., Polanyi, M. and Schmid, E. (1922) *Z. Physik* **12**, 58.

Mendoza, E. (1990) *J. Chem. Educat.* **67**, 1040.

Nabarro, F.R.N (1948) *Report, Conference on the Strength of Solids* (Physical Society, London) p. 75.

Nakajima, H. (1997) *JOM* **49** (June), 15.

Nieh, T.G., Wadsworth, J. and Sherby, O.D. (1997) *Superplasticity in Metals and Ceramics* (Cambridge University Press, Cambridge).

Nowick, A.S. (1984) in *Proceedings of the International Conference on Defects in Insulating Crystals*, ed. Lüty, F. (Plenum Press, New York).

Nye, J.F. (1957) *Physical Properties of Crystals: Their Representation by Tensors and Matrices* (Oxford University Press, Oxford).

Olson, G.B. (1993) in *Third Supplementary Volume of the Encyclopedia of Materials Science and Engineering*, ed. Cahn, R.W. (Pergamon Press, Oxford) p. 2041.

Pais, A. (1995) From a memorial address for P.A.M. Dirac at the Royal Society, London.

Pearson, C.E. (1934) *J. Inst. Metals.* **54**, 111.

Pennisi, E. (1998) *Science* **282**, 1972.

Polanyi, M. (1962) My time with X-rays and crystals, in *Fifty Years of X-ray Diffraction*, ed. Ewald, P.P. (The International Union of Crystallography, Utrecht) p. 629.

Roberts-Austen, W. (1896a) *Phil. Trans. R. Soc. Lond.* **187**, 383.
Roberts-Austen, W. (1896b) *J. Iron Steel Inst.* **1**, 1.
Rogers, R.D. and Zaworotko, M.J. (1999) *Proc. Symp. Crystal Engineering, ACA Trans.* **33**, 1.
Ruoff, A.L. (1991) in *Phase Transformations in Materials*, ed. Haasen, P.; *Materials Science and Technology*, Vol. 5, ed. Cahn, R.W., Haasen, P. and Kramer, E.J. (VCH, Weinheim) p. 473.
Savage, H. (1988) in *First Supplementary Volume of the Encyclopedia of Materials Science and Engineering*, ed. Cahn, R.W. (Pergamon Press, Oxford) p. 553.
Schadler, H.W. (1963) in *The Art and Science of Growing Crystals*, ed. J.J. Gilman (Wiley, New York) p. 343.
Schenk, H. (ed.) (1998) *Crystallography Across the Sciences: A Celebration of 50 Years of Acta Crystallographica and the IUCr* (Munksgaard, Copenhagen). Originally published in *Acta Cryst. A* **54**(6), 1.
Schmalzried, H. (1995) *Chemical Kinetics of Solids* (VCH, Weinheim).
Schmid, E. and Boas, W. (1935) *Kristallplastizität* (Springer, Berlin).
Schuster, I., Swirsky, Y., Schmidt, E.J., Polturak, E. and Lipson, S.G. (1996) *Europhys. Lett.* **33**, 623.
Seeger, A. (1997) *Defect Diffusion Forum* **143–147**, 21.
Smigelskas, A.D. and Kirkendall, E.O. (1947) *Trans. AIME* **171**, 130.
Suits, C.G. and Bueche, A.M. (1967) in *Applied Science and Technological Progress* (US Government Printing Office, Washington, DC) p. 299.
Tuijn, C. (1997) *Defect Diffusion Forum* **143–147**, 11.
Tyrrell, H.J.V. (1964) *J. Chem. Educ.* **41**, 397.
Wakai, F. Sakaguchi, S. and Matsuno, Y. (1986) *Adv. Ceram. Mater.* **1**, 259.
Walter, M.L. (1990) *Science and Cultural Crisis: An Intellectual Biography of Percy William Bridgman* (Stanford University Press, Stanford, CA).
Weber, E.R. (1988) *Properties of Silicon* (INSPEC, London) p. 236.
Wentzcovich, R.M., Hemley, R.J., Nellis, W.J. and Yu, P.Y. (1998) *High-Pressure Materials Research* (Materials Research Society, Warrendale, PA) Symp. Proc. vol. **499**.
White, R.J. (1970) *The Antiphilosophers* (Macmillan, London).
Young, D.A. (1991) *Phase Diagrams of the Elements* (University of California Press, Berkeley).
Young, Jr., F.W., Cathcart, J.V. and Gwathmey, A.T. (1956) *Acta Metall.* **4**, 145.

Chapter 5
The Escape from Handwaving

Chapter 5
The Escape from Handwaving

5.1. THE BIRTH OF QUANTITATIVE THEORY IN PHYSICAL METALLURGY

In astrophysics, reality cannot be changed by anything the observer can do. The classical principle of 'changing one thing at a time' in a scientific experiment, to see what happens to the outcome, has no application to the stars! Therefore, the acceptability of a hypothesis intended to interpret some facet of what is 'out there' depends entirely on rigorous *quantitative self-consistency* – a rule that metallurgists were inclined to ignore in the early decades of physical metallurgy.

The matter was memorably expressed recently in a book, *GENIUS – The Life of Richard Feynman*, by James Gleick: "So many of his witnesses observed the utter freedom of his flights of thought, yet when Feynman talked about his own methods he emphasised not freedom but constraint... For Feynman the essence of scientific imagination was a powerful and almost painful rule. What scientists create must match reality. It must match what is already known. Scientific imagination, he said, is imagination in a straitjacket... The rules of harmonic progression made (for Mozart) a cage as unyielding as the sonnet did for Shakespeare. As unyielding and as liberating – for later critics found the creators' genius in the counterpoint of structure and freedom, rigour and inventiveness."

This also expresses accurately what was new in the breakthroughs of the early 1950s in metallurgy.

Rosenhain (Section 3.2.1), the originator of the concept of physical metallurgy, was much concerned with the fundamental physics of metals. In his day, ≈1914, that meant issues such as these: What is the structure of the boundaries between the distinct crystal grains in polycrystalline metals (most commercial metals are in fact polycrystalline)? Why does metal harden as it is progressively deformed plastically... i.e., why does it work-harden? Rosenhain formulated a generic model, which became known as the amorphous metal hypothesis, according to which grains are held together by "amorphous cement" at the grain boundaries, and work-hardening is due to the deposition of layers of amorphous material within the slip bands which he had been the first to observe. These erroneous ideas he defended with great skill and greater eloquence over many years, against many forceful counterattacks. Metallurgists at last had begun to argue about basics in the way that physicists had long done. Concerning this period and the amorphous grain-boundary cement theory in particular, Rosenhain's biographer has this to say (Kelly 1976): "The theory was wrong in scientific detail but it was of great utility. It enabled the metallurgist to reason and recognise that at high temperatures grain boundaries are fragile, that heat-treatment involving hot or cold work coupled with annealing can lead to benefits in some instances and to catastrophes such as 'hot shortness' in others (this term means brittleness at high temperatures)... Advances in technology and practice do not always require exact theory. This must always be striven for, it is true, but a 'hand-waving' argument which calls salient facts to attention, if readily grasped in apparently simple terms, can be of great practical utility." This controversial claim goes to the heart of the relation between metallurgy as it was, and as it was fated to become under the influence of physical ideas and, more important, of the physicist's approach. We turn to this issue next.

As we have seen, Rosenhain fought hard to defend his preferred model of the structure of grain boundaries, based on the notion that layers of amorphous, or glassy, material occupied these discontinuities. The trouble with the battles he fought was twofold: there was no theoretical treatment to predict what properties such a layer would have, for an assumed thickness and composition, and there were insufficient experimental data on the properties of grain boundaries, such as specific energies. This lack, in turn, was to some degree due to the absence of appropriate experimental techniques of characterisation, but not to this alone: no one measured the energy of a grain boundary as a function of the angle of misorientation between the adjacent crystal lattices, not because it was difficult to do, even then, but because metallurgists could not see the point of doing it. Studying a grain boundary *in its own right* – a parepisteme if ever there was one – was deemed a waste of time; only grain boundaries as they directly affected useful properties such as ductility deserved attention. In other words, the cultivation of parepistemes was not yet thought justifiable by most metallurgists.

Rosenhain's righthand collaborator was an English metallurgist, Daniel Hanson, and Rosenhain infected him with his passion for understanding the plastic deformation of metals (and metallurgy generally) in atomistic terms. In 1926, Hanson became professor of metallurgy at the University of Birmingham. He struggled through the Depression years when his university department nearly died, but after the War, when circumstances improved somewhat, he resolved to realise his ambition. In the words of Braun (1992): "When the War was over and people could begin to think about free research again, Hanson set up two research groups, funded with money from the Department of Scientific and Industrial Research. One, headed by Geoffrey Raynor from Oxford (he had worked with Hume-Rothery, Section 3.3.1.1) was to look into the constitution of alloys; the other, headed by Hanson's former student Alan Cottrell, was to look into strength and plasticity. Cottrell had been introduced to dislocations as an undergraduate in metallurgy, when Taylor's 1934 paper was required reading for all of Hanson's final-year students." Cottrell's odyssey towards a proper understanding of dislocations during his years at

Birmingham is set out in a historical memoir (Cottrell 1980). Daniel Hanson, to whose memory this book is dedicated, by his resolve and organisational skill reformed the understanding and teaching of physical metallurgy, introducing interpretations of properties in atomistic terms and giving proper emphasis to theory, in a way that cleared the path to the emergence of materials science a few years after his untimely death.

5.1.1 Dislocation theory

In Section 3.2.3.2, the reader was introduced to dislocations (and to that 1934 paper by Geoffrey Taylor) and an account was also presented of how the sceptical response to these entities was gradually overcome by visual proofs of various kinds. However, by the time, in the late 1950s, that metallurgists and physicists alike had been won over by the principle 'seeing is believing', another sea-change had already taken place.

After World War II, dislocations had been taken up by some adventurous metallurgists, who held them responsible, in a purely handwaving (qualitative) manner and even though there was as yet no evidence for their very existence, for a variety of phenomena such as brittle fracture. They were claimed by some to explain everything imaginable, and therefore 'respectable' scientists reckoned that they explained nothing.

What was needed was to escape from handwaving. That milestone was passed in 1947 when Cottrell formulated a rigorously quantitative theory of the discontinuous yield-stress in mild steel. When a specimen of such a steel is stretched, it behaves elastically until, at a particular stress, it *suddenly* gives way and then continues to deform at a lower stress. If the test is interrupted, then after many minutes holding at ambient temperature the former yield stress is restored... i.e., the steel strengthens or *strain-ages*. This phenomenon was of practical importance; it was much debated but not understood at all. Cottrell, influenced by the dislocation theorists Egon Orowan and Frank Nabarro (as set out by Braun 1992) came up with a novel model. The essence of Cottrell's idea was given in the abstract of his paper to a conference on dislocations held in Bristol in 1947, as cited by Braun:

> "It is shown that solute atoms differing in size from those of the solvent (carbon, in fact) can relieve hydrostatic stresses in a crystal and will thus migrate to the regions where they can relieve the most stress. As a result they will cluster round dislocations forming 'atmospheres' similar to the ionic atmospheres of the Debye–Hückel theory of electrolytes. The conditions of formation and properties of these atmospheres are examined and the theory is applied to problems of precipitation, creep and the yield point."

The importance of this advance is hidden in the simple words "It is shown...", and furthermore in the parallel drawn with the D–H theory of electrolytes. This was one of the first occasions when a quantitative lesson for a metallurgical problem was derived from a neighbouring but quite distinct science.

Cottrell (later joined by Bruce Bilby in formulating the definitive version of his theory), by precise application of elasticity theory to the problem, was able to work out the concentration gradient across the carbon atmospheres, what determines whether the atmosphere 'condenses' at the dislocation line and thus ensures a well-defined yield-stress, the integrated force holding a dislocation to an atmosphere (which determines the drop in stress after yield has taken place) and, most impressively, he was able to predict the time law governing the reassembly of the atmosphere after the dislocation had been torn away from it by exceeding the yield stress – that is, the strain-ageing kinetics. Thus it was possible to compare accurate measurement with precise theory. The decider was the strain-ageing kinetics, because the theory came up with the prediction that the fraction of carbon atoms which have rejoined the atmosphere is strictly proportional to $t^{2/3}$, where t is the time of strain-ageing after a steel specimen has been taken past its yield-stress.

In 1951, this strain-ageing law was checked by Harper (1951) by a method which perfectly encapsulates the changes which were transforming physical metallurgy around the middle of the century. It was necessary to measure the change with time of *free* carbon dissolved in the iron, and to do this in spite of the fact that the solubility of carbon in iron at ambient temperature is only a minute fraction of one per cent. Harper performed this apparently impossible task and obtained the plots shown in Figure 5.1, by using a torsional pendulum, invented just as the War began by a Dutch physicist, Snoek (1940, 1941), though his work did not become known outside the Netherlands until after the War. Harper's/Snoek's apparatus is shown in Figure 5.2(a). The specimen is in the form of a wire held under slight tension in the elastic regime, and the inertia arm is sent into free torsional oscillation. The amplitude of oscillation gradually decays because of internal friction, or damping: this damping had been shown to be caused by dissolved carbon (and nitrogen, when that was present also). Roughly speaking, the dissolved carbon atoms, being small, sit in interstitial lattice sites close to an edge of the cubic unit cell of iron, and when that edge is elastically compressed and one perpendicular to it is stretched by an applied stress, then the equilibrium concentrations of carbon in sites along the two cube edges become slightly different: the carbon atoms "prefer" to sit in sites where the space available is slightly enhanced. After half a cycle of oscillation, the compressed edge becomes stretched and vice versa. When the frequency of oscillation matches the most probable jump frequency of carbon atoms between adjacent sites, then the damping is a maximum. By finding how the temperature of peak damping varies with the (adjustable) pendulum frequency (Figure 5.2(b)), the jump frequency and hence the diffusion coefficient can be determined, even below

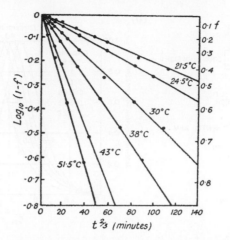

Figure 5.1. Fraction, f, of carbon atoms restored to the 'atmosphere' surrounding a dislocation, as determined by means of a Snoek pendulum.

room temperature where it is very small (Figure 5.2(c)). The subtleties of this "anelastic" technique, and other related ones, were first recognised by Clarence Zener and explained in a precocious text (Zener 1948); the theory was fully set out later in a classic text by two other Americans, Nowick and Berry (1972). The magnitude of the peak damping is proportional to the amount of carbon in solution. A carbon atom situated in an 'atmosphere' around a dislocation is locked to the stress-field of the dislocation and thus cannot oscillate between sites; it therefore does not contribute to the peak damping.

By the simple expedient of stretching a steel wire beyond its yield-stress, clamping it into the Snoek pendulum and measuring the decay of the damping coefficient with the passage of time at temperatures near ambient, Harper obtained the experimental plots of Figure 5.1: here f is the fraction of dissolved carbon which had migrated to the dislocation atmospheres. The $t^{2/3}$ law is perfectly confirmed, and by comparing the slopes of the lines for various temperatures, it was possible to show that the activation energy for strain-ageing was identical with that for diffusion of carbon in iron, as determined from Figure 5.2(a). After this, Cottrell and Bilby's model for the yield-stress and for strain-ageing was universally accepted and so was the existence of dislocations, even though nobody had seen one as yet at that time. Cottrell's book on dislocation theory (1953) marked the coming of age of the subject; it was the first rigorous, quantitative treatment of how the postulated dislocations must react to stress and obstacles. It is still cited regularly. Cottrell's research was aided by the theoretical work of Frank Nabarro in Bristol, who worked out the response of stressed dislocations to obstacles in a crystal: he has devoted his whole

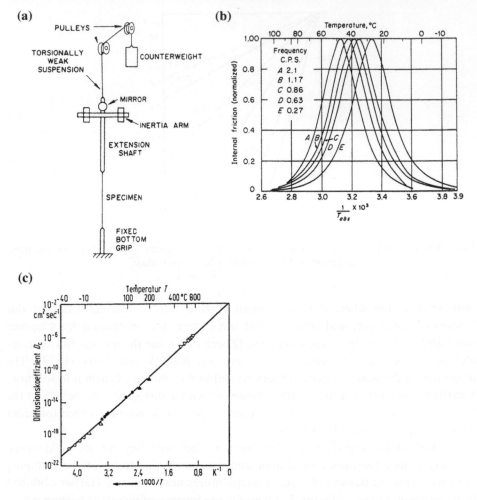

Figure 5.2. (a) Arrangement of a Snoek pendulum. (b) Internal friction as a function of temperature, at different pendulum frequencies, for a solution of carbon in iron. (c) Diffusion of carbon in iron over 14 decades, using the Snoek effect (−30–200°C) and conventional radioisotope method (400–700°C).

scientific life to the theory of dislocations and has written or edited many major texts on the subject.

Just recently (Wilde *et al.* 2000), half a century after the indirect demonstration, it has at last become possible to see carbon atmospheres around dislocations in steel directly, by means of atom-probe imaging (see Section 6.2.4). The maximum carbon concentration in such atmospheres was estimated at 8 ± 2 at.% of carbon.

It is worthwhile to present this episode in considerable detail, because it encapsulates very clearly what was new in physical metallurgy in the middle of the century. The elements are: an accurate theory of the effects in question, preferably without disposable parameters; and, to check the theory, the use of a technique of measurement (the Snoek pendulum) which is simple in the extreme in construction and use but subtle in its quantitative interpretation, so that theory ineluctably comes into the measurement itself. It is impossible that any handwaver could ever have conceived the use of a pendulum to measure dissolved carbon concentrations!

The Snoek pendulum, which in the most general sense is a device to measure relaxations, has also been used to measure relaxation caused by tangential displacements at grain boundaries. This application has been the central concern of a distinguished Chinese physicist, Tingsui Kê, for all of the past 55 years. He was stimulated to this study by Clarence Zener, in 1945, and pursued the approach, first in Chicago and then in China. This exceptional fidelity to a powerful quantitative technique was recognised by a medal and an invitation to deliver an overview lecture in America, recently published shortly before his death (Kê 1999).

This sidelong glance at a grain-boundary technique is the signal to return to Rosenhain and *his* grain boundaries. The structure of grain boundaries was critically discussed in Cottrell's book, page 89 *et seq*. Around 1949, Chalmers proposed that a grain boundary has a 'transition lattice', a halfway house between the two bounding lattices. At the same time, Shockley and Read (1949, 1950) worked out how the specific energy of a simple grain boundary must vary with the degree of misorientation, for a specified axis of rotation, on the hypothesis that the transition lattice consists in fact of an array of dislocations. (The Shockley in this team was the same man who had just taken part in the invention of the transistor; his working relations with his co-inventors had become so bad that for a while he turned his interests in quite different directions.) Once this theory was available, it was very quickly checked by experiment (Aust and Chalmers 1950); the technique depended on measurement of the dihedral angle where three boundaries meet, or where one grain boundary meets a free surface. As can be seen from Figure 5.3, theory (with one adjustable parameter only) fits experiment very neatly. The Shockley/Read theory provided the motive for an experiment which had long been feasible but which no one had previously seen a reason for undertaking.

A new parepisteme was under way: its early stages were mapped in a classic text by McLean (1957), who worked in Rosenhain's old laboratory. Today, the atomic structure of interfaces, grain boundaries in particular, has become a virtual scientific industry: a recent multiauthor book of 715 pages (Wolf and Yip 1992) surveys the present state, while an even more recent equally substantial book by two well-known authors provides a thorough account of all kinds of interfaces (Sutton and Balluffi 1995). In a paper published at about the same time, Balluffi

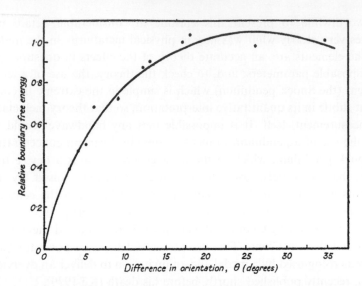

Figure 5.3. Variation of grain-boundary specific energy with difference of orientation. Theoretical curve and experimental values (•) (1950).

and Sutton (1996) discuss "why we should be interested in the atomic structure of interfaces".

One of the most elegant experiments in materials science, directed towards a particularly detailed understanding of the energetics of grain boundaries, is expounded in Section 9.4.

5.1.2 Other quantitative triumphs

The developments described in the preceding section took place during a few years before and after the exact middle of the 20th century. This was the time when the *quantitative revolution* took place in physical metallurgy, leading the way towards modern materials science. A similar revolution in the same period, as we have seen in Section 3.2.3.1, affected the study of point defects, marked especially by Seitz's classic papers of 1946 and 1954 on the nature of colour centres in ionic crystals; this was a revolution in solid-state physics as distinct from metallurgy, and was a reaction to the experimental researches of an investigator, Pohl, who believed only in empirical observation. At that time these two fields, physics and physical metallurgy, did not have much contact, and yet a quantitative revolution affected the two fields at the same time.

The means and habit of making highly precise measurements, with careful attention to the identification of sources of random and systematic error, were well established by the period I am discussing. According to a recent historical essay by

Dyson (1999), the "inventor of modern science" was James Bradley, an English astronomer, who in 1729 found out how to determine the positions of stars to an accuracy of ≈1 part in a million, a hundred times more accurately than the contemporaries of Isaac Newton could manage, and thus discovered stellar aberration. Not long afterwards, still in England, John Harrison constructed the first usable marine chronometer, a model of precision that was designed to circumvent a range of sources of systematic error. After these events, the best physicists and chemists knew how to make ultraprecise measurements, and recognised the vital importance of such precision as a path to understanding. William Thomson, Lord Kelvin, the famous Scottish physicist, expressed this recognition in a much-quoted utterance in a lecture to civil engineers in London, in 1883: "I often say that when you can measure what you are speaking about, and express it in numbers, you know something about it; but when you cannot measure it, when you cannot express it in numbers, your knowledge is of a meagre and unsatisfactory kind; it may be the beginning of knowledge, but you have scarcely, in your own thoughts, advanced to the state of science". Habits of precision are not enough in themselves; the invention of entirely new kinds of instrument is just as important, and to this we shall be turning in the next chapter.

Bradley may have been the inventor of modern *experimental* science, but the equally important habit of interpreting exact measurements in terms of equally exact theory came later. Maxwell, then Boltzmann in statistical mechanics and Gibbs in chemical thermodynamics, were among the pioneers in this kind of theory, and this came more than a century after Bradley. In the more applied field of metallurgy, as we have seen, it required a further century before the same habits of *theoretical* rigour were established, although in some other fields such rigour came somewhat earlier.: Heyman (1998) has recently surveyed the history of 'structural analysis' applied to load-bearing assemblies, where accurate quantitative theory was under way by the early 19th century.

Rapid advances in understanding the nature and behaviour of materials required both kinds of skill, in measurement and in theory, acting in synergy; among metallurgists, this only came to be recognised fully around the middle of the twentieth century, at about the same time as materials science became established as a new discipline.

Many other parepistemes were stimulated by the new habits of precision in theory. Two important ones are the entropic theory of rubberlike elasticity in polymers, which again reached a degree of maturity in the middle of the century (Treloar 1951), and the calculation of phase diagrams (CALPHAD) on the basis of measurements of thermochemical quantities (heats of reaction, activity coefficients, etc.); here the first serious attempt, for the Ni–Cr–Cu system, was done in the Netherlands by Meijering (1957). The early history of CALPHAD has recently been set out (Saunders and Miodownik 1998) and is further discussed in chapter 12 (Section 12.3), while rubberlike elasticity is treated in Chapter 8 (Section 8.5.1).

Some examples of the synergy between theory and experiment will be outlined next, followed by two other examples of quantitative developments.

5.1.2.1 Pasteur's principle. As MSE became ever more quantitative and less handwaving in its approach, one feature became steadily more central – the power of surprise. Scientists learned when something they had observed was mystifying... in a word, surprising... or, what often came to the same thing, when an observation was wildly at variance with the relevant theory. The importance of this *surprise factor* goes back to Pasteur, who defined the origin of scientific creativity as being "savoir s'étonner à propos" (to know when to be astonished with a purpose in view). He applied this principle first as a young man, in 1848, to his precocious observations on optical rotation of the plane of polarisation by certain transparent crystals: he concluded later, in 1860, that the molecules in the crystals concerned must be of unsymmetrical form, and this novel idea was worked out systematically soon afterwards by van 't Hoff, who thereby created stereochemistry. A contemporary corollary of Pasteur's principle was, and remains, "accident favours the prepared mind". *Because* the feature that occasions surprise is so unexpected, the scientist who has drawn the unavoidable conclusion often has a sustained fight on his hands. Here are a few exemplifications, in outline form and in chronological sequence, of Pasteur's principle in action:

(1) Pierre Weiss and his recognition in 1907 that the only way to interpret the phenomena associated with ferromagnetism, which were inconsistent with the notions of paramagnetism, was to postulate the existence of ferromagnetic domains, which were only demonstrated visually many years later.

(2) Ernest Rutherford and the structure of the atom: his collaborators, Geiger and Marsden, found in 1909 that a very few (one in 8000) of the alpha particles used to bombard a thin metal foil were deflected through 90° or even more. Rutherford commented later, "it was about as credible as if you had fired a 15 inch. shell at a piece of tissue paper and it came back and hit you". The point was that, in the light of Rutherford's carefully constucted theory of scattering, the observation was wholly incompatible with the then current 'currant-bun' model of the atom, and his observations forced him to conceive the planetary model, with most of the mass concentrated in a very small volume; it was this concentrated mass which accounted for the unexpected backwards scatter (see Stehle 1994). Rutherford's astonished words have always seemed to me *the* perfect illustration of Pasteur's principle.

(3) We have already seen how Orowan, Polanyi and Taylor in 1934 were independently driven by the enormous mismatch between measured and calculated

yield stresses of metallic single crystals to postulate the existence of dislocations to bridge the gap.

(4) Alan Arnold Griffith, a British engineer (1893–1963, Figure 5.4), who just after the first World War (Griffith 1920) grappled with the enormous mismatch between the fracture strength of brittle materials such as glass fibres and an approximate theoretical estimate of what the fracture strength should be. He postulated the presence of a population of minute surface cracks and worked out how such cracks would amplify an applied stress: the amplification factor would increase with the depth of the crack. Since fracture would be determined by the size of the deepest crack, his hypothesis was also able to explain why thicker fibres are on average weaker (the larger surface area makes the presence of at least one deep crack statistically more likely). Griffith's paper is one of the most frequently cited papers in the entire history of MSE. In an illuminating commentary on Griffith's great paper, J.J. Gilman has remarked: "One of the lessons that can be learned from the history of the Griffith theory is how exceedingly influential a good fundamental idea can be. Langmuir called such an idea 'divergent', that is, one that starts from a small base and spreads in depth and scope."

(5) Charles Frank and his recognition, in 1949, that the observation of ready crystal growth at small supersaturations required the participation of screw dislocations emerging from the crystal surface (Section 3.2.3.3); in this way the severe mismatch with theoretical estimates of the required supersaturation could be resolved.

Figure 5.4. Portrait of A.A. Griffith on a silver medal sponsored by Rolls-Royce, his erstwhile employer.

(6) Andrew Keller (1925–1999) who in 1957 found that the polymer polyethylene, in unbranched form, could be crystallised from solution, and at once recognised that the length of the average polymer molecule was much greater than the observed crystal thickness. He concluded that the polymer chains must fold back upon themselves, and because others refused to accept this plain necessity, Keller unwittingly launched one of the most bitter battles in the history of materials science. This is further treated in Chapter 8, Section 8.4.2.

In all these examples of Pasteur's principle in action, surprise was occasioned by the mismatch between initial quantitative theory and the results of accurate measurement, and the surprise led to the resolution of the paradox. The principle remains one of the powerful motivating influences in the development of materials science.

5.1.2.2 Deformation-mechanism and materials selection maps.

Once the elastic theory of dislocations was properly established, in mid-century, quantitative theories of various kinds of plastic deformation were established. Issues such as the following were clarified theoretically as well as experimentally: What is the relation between stress and strain, for a particular material, specified imposed strain rate, temperature and grain size? What is the creep rate for a given material, stress, grain size and temperature? Rate equations were well established for such processes by the 1970s. An essential point is that the *mechanism* of plastic flow varies according to the combination of stress, temperature and grain size. For instance, a very fine-grained metal at a low stress and moderate temperature will flow predominantly by 'diffusion-creep', in which dislocations are not involved at all but deformation takes place by diffusion of vacancies through or around a grain, implying a counterflow of matter and therefore a strain.

In the light of this growing understanding, a distinguished materials engineer, Ashby (1972), and his colleague Harold Frost invented the concept of the *deformation-mechanism map*. Figure 5.5(a) and (b) are examples, referring to a nickel-based jet-engine superalloy, MAR-M200, of two very different grain sizes. The axes are shear stress (normalised with respect to the elastic shear modulus) and temperature, normalised with respect to the melting-point. The field is divided into combinations of stress and temperature for which a particular deformation mechanism predominates; the graphs also show a box which corresponds to the service conditions for a typical jet-engine turbine blade. It can be seen that the predicted flow rate (by diffusion-creep involving grain boundaries) for a blade is lowered by a factor of well over 100 by increasing the grain size from 100 μm to 10 mm.

The construction, meaning and uses of such maps has been explained with great clarity in a monograph by Frost and Ashby (1982). The various mechanisms and rate-limiting factors (such as 'lattice friction' or dislocation climb combined with glide, or Nabarro-Herring creep – see Section 4.2.5) are reviewed, and the corresponding constitutive equations (alternatively, rate equations) critically examined. The iterative stages of constructing a map such as that shown in Figure 5.5 are then explained; a simple computer program is used. The boundaries shown by thick lines correspond to conditions under which two neighbouring mechanisms are predicted to contribute the same strain rate. Certain assumptions have to be made about the superposition of parallel deformation mechanisms. Critical judgment has to be exercised by the mapmaker concerning the reliability of different, incompatible measurements of the same plastic mechanism for the same material. Maps are included in the book for a variety of metals, alloys and ceramic materials. Finally, a range of uses for such maps is rehearsed, and illustrated by a number of case-histories: (1) the flow mechanism under specific conditions can be identified, so that, for a particular use, the law which should be used for design purposes is known. (2) The total strain in service can be approximately estimated. (3) A map can offer guidance for purposes of alloy selection. (4) A map can help in designing experiments to obtain further insight into a particular flow mechanism. (5) Such maps have considerable pedagogical value in university teaching.

Ten years later, the deformation-mechanism map concept led Ashby to a further, crucial development – *materials selection charts*. Here, Young's modulus is plotted against density, often for room temperature, and domains are mapped for a range of quite different materials... polymers, woods, alloys, foams. The use of the diagrams is combined with a criterion for a minimum-weight design, depending on whether the important thing is resistance to fracture, resistance to strain, resistance to buckling, etc. Such maps can be used by design engineers who are not materials experts. There is no space here to go into details, and the reader is referred to a book (Ashby 1992) and a later paper which covers the principles of material selection maps for high-temperature service (Ashby and Abel 1995). This approach has a partner in what Sigmund (2000) has termed "topology optimization: a tool for the tailoring of structures and materials"; this is a systematic way of designing complex load-bearing structures, for instance for airplanes, in such a way as to minimise their weight. Sigmund remarks in passing that "any material is a structure if you look at it through a microscope with sufficient magnification".

Ashby has taken his approach a stage further with the introduction of physically based estimates of material properties where these have not been measured (Ashby 1998, Bassett *et al.* 1998), where an independent check on values is thought desirable or where property ranges of categories of materials would be useful. Figure 5.5(c) is one example of the kind of estimates which his approach makes possible. A still more recent development of Ashby's approach to materials selection is an analysis in depth of the total financial cost of using alternative materials (for different number of identical items manufactured). Thus, an expanded metallic foam beam offers the

(a)

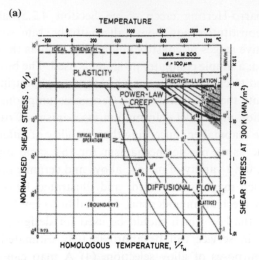

(b)

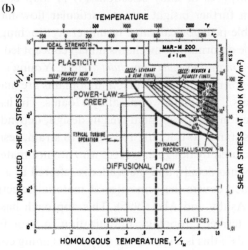

(c)

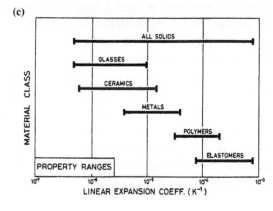

Figure 5.5. Deformation-mechanism maps for MAR-M200 superalloy with (a) 100 μm and (b) 10 mm grain size. The rectangular 'box' shows typical conditions of operation of a turbine blade. (after Frost and Ashby 1982). (c) A bar chart showing the range of values of expansion coefficient for generic materials classes. The range for all materials spans a factor of almost 3000; that for a class spans, typically, a factor of 20 (after Ashby 1998).

same stiffness as a solid metallic beam but at a lower mass. However, in view of the high manufacturing cost of such a foam, a detailed analysis casts doubt on the viability of such a usage (Maine and Ashby 2000).

These kinds of maps and optimisation approaches represent impressive applications of the quantitative revolution to purposes in materials engineering.

5.1.2.3 Stereology. In Section 3.1.3, the central role of *microstructure* in materials science was underlined. Two-phase and multiphase microstructures were treated and so was the morphology of grains in polycrystalline single phase microstructures. What was not discussed there in any detail was the relationship between properties, mechanical properties in particular, and such quantities as average grain size, volume fraction and shapes of precipitates, mean free path in two-phase structures: such correlations are meat and drink to some practitioners of MSE. To establish such correlations, it is necessary to establish reliable ways of measuring such quantities. This is the subject-matter of the parepisteme of *stereology*, alternatively known as *quantitative metallography*. The essence of stereological practice is to derive statistical information about a microstructure in three dimensions from measurements on two-dimensional sections. This task has two distinct components: first, *image analysis*, which nowadays involves computer-aided measurement of such variables as the area fraction of a disperse phase in a two-phase mixture or the measurement of mean free paths from a micrograph; second, a theoretical framework is required that can convert such two-dimensional numbers into three-dimensional information, with an associated estimate of probable error in each quantity. All this is much less obvious than appears at first sight: thus, crystal grains in a single phase polycrystal have a range of sizes, may be elongated in one or more directions, and it must also be remembered that a section will not cut most grains through their maximum diameter; all such factors must be allowed for in deriving a valid average grain size from micrographic measurements.

Stereology took off in the 1960s, under pressure not only from materials scientists but also from anatomists and mineralogists. Figure 3.13 (Chapter 3) shows two examples of property-microstructure relationships, taken from writings by one of the leading current experts, Exner and Hougardy (1988) and Exner (1996). Figure 3.13(a) is a way of plotting a mechanical indicator (here, indentation hardness) against grain geometry: here, the amount of grain-boundary surface is plotted instead of the reciprocal square root of grain size. Determining interfacial area like this is one of the harder tasks in stereology. Figure 3.13(b) is a curious correlation: the ferromagnetic coercivity of the cobalt phase in a Co/WC 'hard metal' is measured as a function of the amount of interface between the two phases per unit volume. Figure 5.6 shows yield strength in relation to grain size or particle spacing for unspecified alloys: the linear relation between yield strength and the reciprocal square root of (average) grain size is known as the *Hall-Petch law* which is one of the early exemplars of the quantitative revolution in metallurgy.

The first detailed book to describe the practice and theory of stereology was assembled by two Americans, DeHoff and Rhines (1968); both these men were famous practitioners in their day. There has been a steady stream of books since then; a fine, concise and very clear overview is that by Exner (1996). In the last few years, a specialised form of microstructural analysis, entirely dependent on computerised image analysis, has emerged – *fractal analysis*, a form of measurement of roughness in two or three dimensions. Most of the voluminous literature of fractals, initiated by a mathematician, Benoit Mandelbrot at IBM, is irrelevant to materials science, but there is a sub-parepisteme of fractal analysis which relates the fractal dimension to fracture toughness: one example of this has been analysed, together with an explanation of the meaning of 'fractal dimension', by Cahn (1989).

This whole field is an excellent illustration of the deep change in metallurgy and its inheritor, materials science, wrought by the quantitative revolution of mid-century.

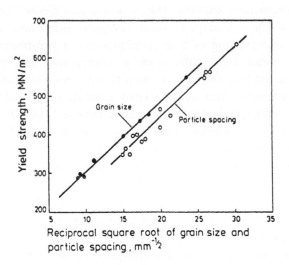

Figure 5.6. Yield strength in relation to grain size or particle spacing (courtesy of H.E. Exner).

5.1.3 Radiation damage

The first nuclear reactors were built during the Second World War in America, as an adjunct to the construction of atomic bombs. Immediately after the War, the development of civil atomic power began in several nations, and it became clear at once that the effects of neutron and gamma irradiation, and of neutron-induced fission and its products, on fuels, moderators and structural materials, had to be taken into account in the design and operation of nuclear reactors, and enormous programmes of research began in a number of large national laboratories in several countries. The resultant body of knowledge constitutes a striking example of the impact of basic research on industrial practice and also one of the best exemplars of a highly quantitative approach to research in materials science. A lively historical survey of the sequence of events that led to the development of civil atomic power, with a certain emphasis on events in Britain and a discussion of radiation damage, can be found in a recent book (West and Harris 1999).

In the early days of nuclear power (and during the War), thermal reactors all used metallic uranium as fuel. The room-temperature allotrope of uranium (the α form) is highly anisotropic in structure and properties, and it was early discovered that polycrystalline uranium normally has a preferred orientation of the population of grains (i.e., a 'texture') and then suffers gross dimensional instability on (a) thermal cycling or (b) neutron irradiation. A fuel rod could easily elongate to several times its original length. Clearly, this would cause fuel elements to burst their containers and release radioactivity into the cooling medium, and so something had to be done quickly to solve this early problem. The solution found was to use appropriate heat treatment in the high-temperature (β phase) domain followed by cooling: this way, the α grains adopted a virtually random distribution of orientations and the problem was much diminished. Later, the addition of some minor alloying constituents made the orientations even more completely random, and also generated reasonably fine grains (to obviate the 'orange peel effect'). These early researches are described in the standard text on uranium (Holden 1958). In spite of this partial success in eliminating the problems arising from anisotropy, "the metallurgy of uranium proved so intractable that in the mid-1950s the element was abandoned as a fuel worldwide" (Lander *et al.* 1994). These recent authors, in their superb review of the physical metallurgy of metallic uranium, also remark: "Once basic research (mostly, extensive research with single crystals) had shown that the anisotropic thermal expansion and consequent dimensional instability during irradiation by neutrons was an *intrinsic* property of the metal, it was abandoned in favour of oxide. *This surely represents one of the most rapid changes of technology driven by basic research!*" (my italics).

A close study of the chronology of this episode teaches another lesson. The early observation on irradiation-induced growth of uranium was purely phenomenological[1]. As Holden tells it, many contradictory theories were put forward for irradiation-induced growth in particular (and also for thermal cycling-induced growth), and it was not until the late 1950s, following some single crystal research, that there was broad agreement that (Seigle and Opinsky 1957) anisotropic diffusion was the aetiological cause; the theory specified that interstitial atoms will migrate preferentially in [0 1 0] directions and vacancies in [1 0 0] directions of a crystal, and this leads to shape distortion as observed. But the phenomenological facts alone sufficed to lead to practical policy shifts... there was no need to wait for full understanding of the underlying processes. However, the researches of Seigle and Opinsky opened the way for the understanding of many other radiation-induced phenomena in solids, in later years. It is also to be noted that the production of single crystals (which, as my own experience in 1949 confirmed, was very difficult to achieve for uranium) and the detailed understanding of diffusion, two of the parepistemes discussed in Chapter 4, were involved in the understanding of irradiation growth of uranium.

The role of interstitial point defects (atoms wedged between the normal lattice sites) came to be central in the study of irradiation effects. At the same time as Seigle and Opinsky's researches, in 1957, one of the British reactors at Windscale suffered a serious accident caused by sudden overheating of the graphite moderator (the material used to slow down neutrons in the reactor to speeds at which they are most efficient in provoking fission of uranium nuclei). This was traced to the 'Wigner effect', the elastic strain energy due to carbon atoms wedged in interstitial positions. When this energy becomes too substantial, the strain energy 'heals' when the graphite warms up, and as it warms the release becomes self-catalytic and ultimately catastrophic. This insight led to an urgent programme of research on how Wigner energy in graphite could be safely released: the key experiments are retrospectively described by Cottrell (1981), who was in charge of the programme.

In Britain, a population of thermal reactors fuelled by metallic uranium have remained in use, side by side with more modern ones (to that extent, Lander *et al.* were not quite correct about the universal abandonment of metallic uranium). In 1956, Cottrell (who was then working for the Atomic Energy Authority) identified from first principles a mechanism which would cause metallic (α) uranium to creep rapidly under small applied stress: this was linked with the differential expansion of

---

[1] This adjective is freely used in the scientific literature but it refers to a complex concept. According to the Oxford English Dictionary, the scientific philosopher William Whewell, in 1840, remarked that "each... science, when complete, must possess three members: the phenomenology, the aetiology, and the theory." The OED also tells us that "aetiology" means "the assignment of a cause, the rendering of a reason." So the phenomenological stage of a science refers to the mere observation of visible phenomena, while the hidden causes and the detailed origins of these causes come later.

individual anisotropic grains, a process which generated local stresses between grains, in effect mechanically 'sensitising' the polycrystal. A simple experiment with a uranium spring inside a nuclear reactor proved that the effect indeed exists (Cottrell and Roberts 1956), and this experiment led to the immediate redesign of a reactor then under construction. Here was another instance of a rapid change in technology driven by basic research.

By 1969, when a major survey (Thompson 1969) was published, the behaviour of point defects and also of dislocations in crystals subject to collisions with neutrons and to the consequential 'collision cascades' had become a major field of research. Another decade later, the subject had developed a good deal further and a highly quantitative body of theory, as well as of phenomenological knowledge, had been assembled. Gittus (1978) published an all-embracing text that covered a number of new topics: chapter headings include "Bubbles", "Voids" and "Irradiation(-enhanced) Creep".

The success of theory in interpreting the behaviour of irradiated solids and in making useful predictions of behaviour as yet unknown was largely due to the creation of the parepisteme of 'atomistic modelling'; this was lauded in a Festschrift for the 60th birthday of Ronald Bullough, a Harwell scientist who played a major part in establishing such models. In this book, Cottrell (1992) remarks: "Although atomistic theory has remained partly phenomenological, for example taking in the measured energies of point defects in its calculations of the processes of radiation damage, two features have enabled it to take giant strides into our understanding of materials. The first is that the lattice distortion produced by extended defects, such as dislocations and cracks, is long range and so can be described accurately by linear elasticity. Second, crystallographic constraints largely dictate the forms which lattice defects can take; for a regular structure can go wrong only in regular ways. As a result, atomistic theories have now been able to achieve incredibly detailed and accurate representations of complex structures and properties."

It is worth spending a moment on the subject of bubbles in irradiated nuclear fuels. Much of the related research was done in the 1960s. It was preceded by a study of the fragmentation, on heating, of a natural mineral, thorianite; this contains helium from radioactive decay of thorium and when that helium precipitates at flaws in the mineral, it "fragments explosively" (words from Barnes and Mazey 1957). The fission of uranium generates a range of fission products, including the noble gases, helium particularly. These are insoluble in equilibrium and can nucleate at lattice defects and on precipitates, causing swelling of the fuel. The study of these bubbles (some studies were based on model systems, such as helium injected into copper) led to a number of important advances in materials science, of value beyond nuclear engineering, including a detailed understanding of the migration of bubbles *in solids* in a thermal gradient. To minimise the effect of bubbles, it was important to nucleate as many of them as possible, because many small bubbles led to much less swelling than the same amount of gas precipitated in a few large bubbles. To achieve a high bubble density, in turn, required a detailed understanding of the process of bubble nucleation. Gittus describes this important research very clearly. This subject has recently even been treated in a textbook of materials physics for undergraduates (Quéré 1998).

The subject of voids marks the coming of age of research on radiation damage. Voids are like bubbles, but do not depend on the availability of fission gases. They are produced when the equal numbers of interstitials and vacancies generated by collision cascades behave differently, so that the end result is an unbalanced population of vacancies coagulating to form *empty* bubbles, that is, 'voids'. This process is particularly important in connection with so-called fast reactors, in which no moderator is present to slow down the fission neutrons. Understanding the formation and properties of voids required an extremely sophisticated understanding of point defect behaviour, as explained in Gittus's book, and also in a very substantial American publication (Corbett and Ianniello 1972). A fine example of the very subtle quantitative theory of the differential absorption of distinct kinds of point defects at various 'sinks' – grain boundaries, dislocations, stacking faults and voids themselves – is a substantial paper by Brailsford and Bullough (1981), which might be said to represent the apotheosis of theory applied to radiation damage research. There was also a spinoff from void research in the way of novel phenomena which played no role in radiation damage but have kept solid-state scientists happily engaged for many years: a prime example is the formation of *void lattices*, first observed by Evans (1971). Populations of minute voids in metals such as molybdenum form lattices that mimic that of the underlying metal but with periodicities much greater than the lattice spacing of the metal. The interest lies in trying to understand what kinds of interactions lead to the alignment of voids into a lattice. This is still under debate, nearly 30 years later.

It is worth considering what role the study of radiation damage has played in furthering the broad domain of materials science as a whole. The question is briefly addressed by Mansur (1993) in the preface of Volume 200 of the *Journal of Nuclear Materials*. He points out that everything that is known about the behaviour of self-interstitial point defects and much of what is known about vacancies is derived from studies of atomic displacements in solids by radiation. The development of radiation-resistant structural materials for fission and fusion reactors is based on this knowledge and on studies of consequential diffusion-induced structural changes. Finally, Mansur emphasises that some technologies for improving surface properties, ion-implantation in particular, stem directly from research on radiation damage. The mutual help between MSE and research on nuclear materials is a perfect example of two-way benefits in action.

REFERENCES

Ashby, M.F. (1972) *Acta Metall.* **20**, 887.
Ashby, M.F. (1992) *Materials Selection in Mechanical Design* (Pergamon Press, Oxford).
Ashby, M.F. and Abel, C.A. (1995) in *High-Temperature Structural Materials*, eds. Cahn, R.W., Evans, A.G. and McLean, M. (Chapman & Hall, London) p. 33.
Ashby, M.F. (1998) *Proc. Roy. Soc. Lond. A* **454**, 1301.
Aust, K.T. and Chalmers, B. (1950) *Proc. Roy. Soc. Lond. A* **204**, 359.
Balluffi, R.W. and Sutton, A.P. (1996) *Mater. Sci. Forum* **207–209**, 1.
Barnes, R.S. and Mazey, D.J. (1957) *J. Nucl. Energy* **5**, 1.
Bassett, D., Brechet, Y. and Ashby, M.F. (1998) *Proc. Roy. Soc. Lond. A* **454**, 1323.
Brailsford, A.D. and Bullough, R. (1981) *Phil. Trans. Roy. Soc. Lond.* **302**, 87.
Braun, E. (1992) in *Out of the Crystal Maze*, ed. Hoddeson *et al.*, L. (Oxford University Press, Oxford) p. 317.
Cahn, R.W. (1989) *Nature* **338**, 201.
Corbett, J.W. and Ianniello, L.C. (1972) *Radiation-Induced Voids in Metals* (US Atomic Energy Commission).
Cottrell, A.H. (1953) *Dislocations and Plastic Flow in Crystals* (Clarendon Press, Oxford).
Cottrell, A.H. (1980) Dislocations in metals: the Birmingham school, 1945–1955, *Proc. Roy. Soc. A* **371**, 144.
Cottrell, A. (1981) *J. Nucl. Mater.* **100**, 64.
Cottrell, A. (1992) Theoretical models in materials science, in *Materials Modelling: From Theory to Technology*, ed. English, C.A. *et al.* (Institute of Physics Publishing, Bristol and Phildelphia) p. 3.
Cottrell, A.H. and Roberts, A.C. (1956) *Phil. Mag.* **1**, 711.
DeHoff, R.T. and Rhines, F.N. (eds.) (1968) *Quantitative Microscopy* (McGraw-Hill, New York).
Dyson, F.J. (1999) *Nature* **400**, 27.
Evans, J.H. (1971) *Nature* **229**, 403.
Exner, H.E. (1996) in *Physical Metallurgy*, vol. 2, ed. Cahn, R.W. and Haasen, P., p. 996.
Exner, H.E. and Hougardy, H.P. (1988) *Quantitative Image Analysis of Microstructures* (DGM Informationsgesellschaft Verlag, Oberursel).
Frost, H.J. and Ashby, M.F. (1982) *Deformation-Mechanism Maps* (Pergamon Press, Oxford).
Gittus, J. (1978) *Irradiation Effects in Crystalline Solids* (Applied Science Publishers, London).
Griffith, A.A. (1920) *Phil. Trans. Roy. Soc. Lond. A* **221**, 163.
Harper, S. (1951) *Phys. Rev.* **83**, 709.
Heyman, J. (1998) *Structural Analysis: A Historical Approach* (Cambridge University Press, Cambridge).
Holden, A.N. (1958) *Physical Metallurgy of Uranium*, Chapter 11 (Addison-Wesley, Reading).
Kê, T.S. (1999) *Metall. Mater. Trans.* **30A**, 2267.
Kelly, A. (1976) *Phil. Trans. Roy. Soc. Lond. A* **282**, 5.
Lander, G.H., Fisher, E.S. and Bader, S.D. (1994) *Adv. Phys.* **43**, 1.

Maine, E. and Ashby, M.F. (2000) *Adv. Eng. Mat.* **2**, 205.

McLean, D. (1957) *Grain Boundaries in Metals* (Oxford University Press, Oxford).

Mansur, L.K. (1993) *J. Nucl. Mater.* **200**, v.

Meijering, J.L. (1957) *Acta Metall.* **5**, 257.

Nowick, A.S. and Berry, B.S. (1972) *Anelastic Relaxations in Crystalline Solids* (Academic Press, New York).

Quéré, Y. (1998) *Physics of Materials* (Gordon and Breach Science Publishers, Amsterdam) p. 427.

Saunders, N. and Midownik, A.P. (1998) *Calphad: Calculation of Phase Diagrams* (Pergamon Press, Oxford).

Seigle, L.L. and Opinsky, A.J. (1957) *Nucl. Sci. Eng.* **2**, 38.

Shockley, W. and Read, W.T. (1949) *Phys. Rev.* **75**, 692; (1950) *ibid* **78**, 275.

Sigmund, O. (2000) *Phil. Trans. Roy. Soc. Lond. A* **358**, 211.

Snoek, J.L. (1940) *Ned. Tijd. v. Nat.*, **7**, 133; (1941) *ibid* **8**, 177; (1941) *Physica*, **8**, 711.

Stehle, P. (1994) *Order, Chaos, Order: The Transition from Classical to Quantum Physics* (Oxford University Press, New York) p. 218.

Sutton, A.P. and Balluffi, R.W. (1995) *Interfaces in Crystalline Materials* (Clarendon Press, Oxford).

Thompson, M.W. (1969) *Defects and Radiation Damage in Metals* (Cambridge University Press, Cambridge).

Treloar, L.R.G. (1951) *The Physics of Rubber Elasticity* (Clarendon Press, Oxford).

West, D.R.F. and Harris, J.E. (1999) *Metals and the Royal Society* (IOM Communications Ltd., London), Chapter 18.

Wilde, J., Cerezo, A. and Smith, G.D.W. (2000) *Scripta Mater.* **43**, 39.

Wolf, D. and Yip, S. (eds.) (1992) *Materials Interfaces: Atomic-level Structure and Properties* (Chapman & Hall, London).

Zener, C. (1948) *Elasticity and Anelasticity of Metals* (The University of Chicago Press, Chicago).

Chapter 6
Characterisation

Chapter 6
Characterisation

6.1. INTRODUCTION

The characterisation of materials is a central necessity of modern materials science. Effectively, it signifies making precise distinctions between different specimens of what is nominally the same material. The concept covers qualitative and quantitative analysis of chemical composition and its variation between phases; the examination of the spatial distribution of grains, phases and of minor constituents; the crystal structures present; and the extent, nature and distribution of structural imperfections (including the stereological analysis outlined in Chapter 5).

The word *characterisation* is hard to pin down precisely. The Oxford English Dictionary, Second Edition, defines it as "the marking out of the precise form of anything; the form, mould or stamp thus impressed"; this rather implies that the form of the 'anything' is being modified by the person doing the characterizing. An alternative definition is "description of characteristics or essential features; portrayal in words". That brings us closer to the nature of characterisation as applied to materials science, although "portrayal in numbers or in images" would be closer to the mark than "portrayal in words". A third definition offered by the OED, "creation of fictitious characters", is better avoided by anyone who wishes to be taken seriously in the materials science profession.

An eminent specialist in characterisation, Eric Lifshin, has informed me that two groups of American specialists in General Electric's research centre, one group devoted to metallography, the other to analytical chemistry, came together some time in the 1960s, soon after the concept of materials science had been inaugurated in America, and formed a joint group which they named the "Materials Characterisation Operation." This is probably where the use of the word in relation to materials originated and I believe that it originally related to people working in industrial laboratories, like Lifshin himself who had long been in charge of the large characterisation group in General Electric's Corporate R&D Centre. I have checked through a selection of university textbooks of materials science from the 1960s and 1970s, and the term does not feature in any of them, so its entry into general use must have been delayed until the 1980s. In 1986, the *Encyclopedia of Materials Science and Engineering* published by Pergamon Press in Oxford included a large group of articles on 'techniques for investigation and characterisation of materials', edited by Lifshin; he also wrote a substantial overview of the whole field as it was at that time (Lifshin 1986). Another encyclopedia covering an extremely wide range of techniques more cursorily is by Brundle *et al.* (1992).

Much earlier than these encyclopedias is a book first published in 1941 (Chalmers and Quarrell 1941, 1960) and devoted to the 'physical examination of metals'. This multiauthor book includes some recondite methods, such as the study of the damping capacity of solids (Section 5.1). In the second edition, the authors remark: "Not the least of the many changes that have taken place since the first edition appeared has been in the attitude of the metallurgist to pure science and to modern techniques involving scientific principles." The two editions span the period to which I have attributed the 'quantitative revolution', in Chapter 5.

I have described Lifshin as a 'specialist in characterisation'. This is almost a contradiction in terms, because the techniques that are sheltered under the *characterisation* umbrella are so numerous, varied and sophisticated that nobody can be truly expert in them all, even if his entire working time is devoted to the pursuit of characterisation. The problem is more serious for other materials scientists whose primary interest lies elsewhere. As Lifshin has expressed it in the preface to an encyclopedia of materials characterisation (Cahn and Lifshin 1993), "scientists and engineers have enough difficulty in keeping up with advances in their own fields without having to be materials characterisation experts. *However, it is essential to have enough basic understanding of currently used analytical methods to be able to interact effectively with such experts* (my italics)."

In this chapter, I propose to take a strongly historical approach to the field, and focus on just a few of the numerous techniques 'of investigation and characterisation'. What is not in doubt is that these techniques, and the specialised research devoted to improving them in detail, are at the heart of modern materials science.

6.2. EXAMINATION OF MICROSTRUCTURE

We have met, in Section 3.1.3, microstructure as one of the 'legs of the tripod', as a crucial precursor-concept of modern materials science. The experimental study of microstructure, by means of microscopes, is called *metallography*; in recent years, neologisms such as ceramography and even materialography have been proposed but have not taken hold. The original term is still enshrined in the names of journals and in the titles of competitions for the best and most striking micrographs, and is now taken to cover all materials, not only metals. Almost 100 years ago, in German-speaking countries, 'metallography' meant, more or less, what we now call physical metallurgy; this is exemplified by Gustav Tammann's famous *Lehrbuch der Metallographie* of 1912; at that time there was also a British journal entitled *The Metallographist*. In England, practitioners were then indeed called 'metallographists', and as we saw in Section 3.2.1, practical men had a reserved attitude towards them. To repeat here the remarks of Harry Brearly, a notable practical man of that time: "What a man sees through the microscope is more of less, and his vision has been known to be thereby so limited that he misses what he is looking for, which has been apparent at the first glance to the man whose eye is informed by experience". A more light-hearted version of a related sentiment was expressed earlier, in 1837, by the cockney Sam Weller in Dickens' novel, *Pickwick Papers*: " 'Yes, I have a pair of eyes', replied Sam, 'and that's just it. If they wos a pair o' patent double million magnifyin' gas microscopes of hextra power, p'raps I might be able to see through a flight of stairs and a deal door; but bein' only eyes, you see my wision's limited'."

That view of things has never entirely died out, and indeed there is something in it: the higher the magnification at which an object is examined, the harder it is to be sure that what is being looked at is typical of the object as a whole. The old habit of looking at specimens at a range of very different magnifications is on the way out, and that is a pity.

6.2.1 The optical microscope

The optical microscope was first used systematically in England by Robert Hooke, as exemplified by his celebrated book of 1665, *Micrographia*, and soon after by the Dutchman Antoni van Leeuwenhoek, from 1702 onwards. The observations, some of them on metal artefacts, are analysed by Smith (1960) in his key book, *A History of Metallography*. Leeuwenhoek was the first to depict metallic dendrites, but these early microscopists did not see metallic microstructures with any clarity. That decisive step was taken by Henry Sorby (Section 3.1.2) in 1864–1865, as set out on p. 175 by Smith (1960) in his book chapter devoted to that precocious scientist. Sorby invented 'normal illumination' in a microscope (Figure 6.1). This allowed him to examine polished sections of opaque materials, notably metals and alloys, and by chemical etching he was able to reveal distinct phases in a microstructure. It took Sorby a long time to overcome widespread scepticism about his mode of using the microscope. It was not until Sorby's micrographic method was used by the first investigators of phase diagrams (Section 3.1.2) at the turn of the century that its true value was at length accepted.

Sorby's method of 'vertical-illumination' microscopy has become standard in all metallographic laboratories. Meanwhile, mineralogists developed the microscope for use by transmitted light through thin transparent rock sections. The fact that (non-cubic) crystals of low symmetry have two distinct refractive indexes, according to the plane in which the incident light is polarised, gave mineralogical microscopy a complete new aspect, especially once the use of convergent incident light was introduced late in the nineteenth century and the so-called *indicatrix* was used as a

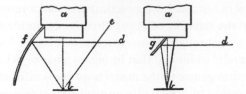

Figure 6.1. Beck's parabolic reflector and Sorby's flat mirror, both invented by Sorby (from *Quarterly Journal of the Microscopical Society*, 1865).

means of identifying minerals in polyphase rock sections. Polarised light is also used by metallurgists, but only as a means of rendering microstructure visible in non-cubic metals. This can be a vital aid in achieving clear contrast in alloys which cannot readily be etched, as it was in the examination of trigonal CuPt polycrystals (Figure 6.2) (Irani and Cahn 1973).

Most treatments of polarised light in transmission are to be found in the mineralogical literature, but a fine book presenting the subject in relation to crystal identification and structure analysis is by Bunn (1945).

In the second half of the 20th century, a number of advanced variants of optical microscopy were invented. They include phase-contrast microscopy (invented in France) and multiple-beam interference microscopy (invented in England), methods

Figure 6.2. Stoichiometric CuPt, ordered at 550°C for 157 hours. Viewed under polarised light in reflection. Shows growth of ordered domains, heterogeneously nucleated at grain boundaries and surface scratches (after Irani and Cahn 1973).

that allow depth differences very much less than one wavelength of light to be clearly distinguished; these techniques were particularly useful in investigating polytypism (Section 3.2.3.4). More recently, the radically novel technique of confocal optical microscopy (Turner and Szarowski 1993) has been introduced; this allows the sharp resolution of successive layers in depth of a specimen, whether transparent or opaque, by focusing on one layer at a time and rejecting the out-of-focus scattered light. There are also variants of microscopy that use either ultraviolet or infrared light. The former improves the resolving power of a microscope, which is diffraction-limited by the light wavelength; the latter has been used for special investigations, like that illustrated by Figure 3.14.

The techniques, instrumentation and underlying theory of optical microscopy for materials scientists have been well surveyed by Telle and Petzow (1992). One of the last published surveys including metallographic techniques of *all* kinds, optical and electronic microscopy and also techniques such as microhardness testing, was a fine book by Phillips (1971).

The impact of electron-optical instruments in materials science has been so extreme in recent years that optical microscopy is seen by many young research workers as faintly fuddy-duddy and is used less and less in advanced research; this has the unfortunate consequence, adumbrated above, that the beneficial habit of using a wide range of magnifications in examining a material is less and less followed.

6.2.2 Electron microscopy

It has long been known that the resolving power of optical microscopes is limited by the wavelength of light, according to the Abbe theory, to around 1 micrometer. If one attempts to use a higher magnification with light, all one gets is fuzzy diffraction fringes instead of sharp edges. Once it had been made clear by de Broglie, in 1924, that a beam of monoenergetic electrons has the characteristics of a wave, with a wavelength inversely proportional to the kinetic energy – and for accelerating voltages of a few thousand volts, a wavelength very much smaller than that of visible light – it became quite clear that if such a beam could be used to generate an image, then the Abbe resolution limit would be very much finer than for any light microscope. De Broglie's theory was triumphantly confirmed in 1927 by G.P. Thomson (J.J. Thomson's son) and A. Reid who showed that a beam of electrons is diffracted from a crystal, just like X-rays; moreover, electrical engineers developed the cathode-ray tube, and oscilloscope, as engineering tools soon after. The way was then clear for attempts to construct an electron microscope.

The difficulty, of course, was that electrons cannot be focused by a glass lens, and it was necessary to use either magnetic or electrostatic 'lenses'. A German, Hans Busch, in 1926/27 published some seminal papers on the analogy between the effect of magnetic or electric fields on electrons, and formal geometric optics, and thereafter a group of German engineers, aided by two young doctoral students, a German – Ruska – and a Hungarian – Gabor, by stages developed electron-optical columns with (mostly) magnetic lenses, until in 1931 the first primitive 2-lens microscope was constructed. The leading spirit was E. Ruska, who many years later (Ruska 1980) published an account of the early work; he also eventually, very belatedly, received a Nobel Prize for his achievement. (So did Dennis Gabor, but that was for another piece of work.) By 1933, Ruska had obtained a resolution considerably greater than attainable with the best optical microscope; it soon became clear that other instrumental factors limited the resolution of electron microscopes, and that the extreme, sub-Ångström resolution deduced from the Abbé theory was not even remotely attainable. In 1937, Siemens in Germany announced the first 'serially produced' transmission electron microscope, with a remarkable resolution of $\approx$7 nm, and many Siemens microscopes were again manufactured soon after the War, beginning about 1955, and used all over the world.

In an excellent historical overview of these stages and the intellectual and practical problems which had to be overcome, Mulvey (1995) remarks that the first production microscopes pursued exactly the same electron-optical design as Ruska's first experimental microscope. The stages of subsequent improvement are outlined by Mulvey, to whom the reader is referred for further details.

6.2.2.1 Transmission electron microscopy. The penetrating power of electrons, even if accelerated by a potential difference of 100–200 kV, through dense materials such as metals is of the order of a micrometre, and therefore in the early days of transmission electron microscopy, the standard way of examining samples was by polishing, etching and then making a thin (low-density) plastic replica of the surface contour. Such features as slip lines on the surface of deformed metals were examined in this way. Meanwhile, biologists were able to send electron beams through thin sections of low-density materials prepared by microtomy, but materials scientists never thought of attempting this.

All this was changed by Peter Hirsch, then a physics research student at the Cavendish Laboratory in Cambridge (Figure 6.3). His story (told in detail in two papers, Hirsch 1980, 1986) shows how innovations of great consequence can stem from small beginnings. In brief, Hirsch (together with his colleague J.N. Kellar) in 1946 was set by his doctoral supervisor, W.H. Taylor, what in retrospect seems a trivial problem in metal physics: is the broadening of X-ray diffraction lines in plastically deformed metals due to internal strains or to the breakdown of the lattice into minute particles? The former idea was espoused by a Cambridge physicist, Henry Lipson, the latter, by a somewhat eccentric Australian, W.A. Wood, who

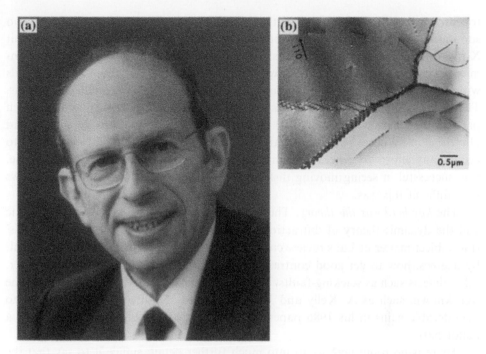

Figure 6.3. (a) Portrait of Peter Hirsch (courtesy Sir Peter Hirsch). (b) Transmission electron micrograph of dislocations in a sub-boundary in a (Ni, Fe)Al intermetallic compound (Kong and Munroe 1994) (courtesy editor of *Intermetallics*).

incidentally was an impassioned foe of the dislocation concept. Hirsch's professor, Lawrence Bragg, was happy for him to use microbeams of X-rays to settle this issue: if Wood was right, then the usually smooth diffraction rings should become spotty, because each 'particle' would generate its own diffraction spot. Because the postulated particles were so small, for this strategy to work it was necessary to use a very fine beam. Some of the experiments supported Wood's idea. It turned out that there were difficulties in generating a strong enough microbeam and Hirsch conceived the idea of replacing the X-ray beam by a fine electron beam in an electron microscope. While the main idea was to use the instrument purely as a diffraction device, after a while he was inspired by an American, R.D. Heidenreich, who in 1949 looked at deformed aluminium crystals in an electron microscope and found that they were broken down into 'particles', to exploit the image-forming function as well. With beaten gold foils, Hirsch saw some features along {1 1 1} planes which *might* have something to do with dislocations.

Hirsch obtained his Ph.D. and went off into industry; in 1953 he was awarded a fellowship which allowed him to return to Cambridge, and here he set out, together with M.J. Whelan and with the encouragement of Sir Nevill Mott who soon after succeeded Bragg as Cavendish professor, to apply what he knew about X-ray diffraction theory to the task of making dislocations visible in electron-microscopic images. The first step was to perfect methods of thinning metal foils without damaging them; W. Bollmann in Switzerland played a vital part in this. Then the hunt for dislocations began. The important thing was to control which part of the diffracted 'signal' was used to generate the microscope image, and Hirsch and Whelan decided that 'selected-area diffraction' always had to accompany efforts to generate an image. Their group, in the person of R. Horne, was successful in seeing moving dislocation lines in 1956; the 3-year delay shows how difficult this was.

The *key here was the theory*. The pioneers' familiarity with both the kinematic and the dynamic theory of diffraction and with the 'real structure of real crystals' (the subject-matter of Lal's review cited in Section 4.2.4) enabled them to work out, by degrees, how to get good contrast for dislocations of various kinds and, later, other defects such as stacking-faults. Several other physicists who have since become well known, such as A. Kelly and J. Menter, were also involved; Hirsch goes to considerable pains in his 1986 paper to attribute credit to all those who played a major part.

There is no room here to go into much further detail; suffice it to say that the diffraction theory underlying image formation in an electron microscope plays a much more vital part in the intelligent use of an electron microscope in transmission mode than it does in the use of an optical microscope. In the words of one recent reviewer of a textbook on electron microscopy, "The world of TEM is quite different (from optical microscopy). Almost no image can be intuitively understood." For instance, to determine the Burgers vector of a dislocation from the disappearance of its image under particular illumination conditions requires an exact knowledge of the mechanism of image formation, and moreover the introduction of technical improvements such as the weak-beam method (Cockayne *et al.* 1969) depends upon a detailed understanding of image formation. As the performance of microscopes improved over the years, with the introduction of better lenses, computer control of functions and improved electron guns allowing finer beams to be used, the challenge of interpreting image formation became ever greater. Eventually, the resolving power crept towards 1–2 Å (0.1–0.2 nm) and, in high-resolution microscopes, atom columns became visible.

Figure 6.3(b) is a good example of the beautifully sharp and clear images of dislocations in assemblies which are constantly being published nowadays. It is printed next to the portrait of Peter Hirsch to symbolise his crucial contribution to modern metallography. It was made in Australia, a country which has achieved an enviable record in electron microscopy.

To form an idea of the highly sophisticated nature of the analysis of image formation, it suffices to refer to some of the classics of this field – notably the early book by Hirsch *et al.* (1965), a recent study in depth by Amelinckx (1992) and a book from Australia devoted to the theory of image formation and its simulation in the study of interfaces (Forwood and Clarebrough 1991).

Transmission electron microscopes (TEM) with their variants (scanning transmission microscopes, analytical microscopes, high-resolution microscopes, high-voltage microscopes) are now crucial tools in the study of materials: crystal defects of all kinds, radiation damage, off-stoichiometric compounds, features of atomic order, polyphase microstructures, stages in phase transformations, orientation relationships between phases, recrystallisation, local textures, compositions of phases... there is no end to the features that are today studied by TEM. Newbury and Williams (2000) have surveyed the place of the electron microscope as "the materials characterisation tool of the millennium".

A special mention is in order of high-resolution electron microscopy (HREM), a variant that permits columns of atoms normal to the specimen surface to be imaged; the resolution is better than an atomic diameter, but the nature of the image is not safely interpretable without the use of computer simulation of images to check whether the assumed interpretation matches what is actually seen. Solid-state chemists studying complex, non-stoichiometric oxides found this image simulation approach essential for their work. The technique has proved immensely powerful, especially with respect to the many types of defect that are found in microstructures.

One of the highly skilled experts working on this technique has recently (Spence 1999) assessed its impact as follows: "What has materials science learnt from HREM? In most general terms, since about 1970, HREM has taught materials scientists that real materials – from minerals to magnetic ceramics and quasicrystals – are far less perfect on the atomic scale than was previously believed. A host of microphases has been discovered by HREM, and the identification of polytypes (cf. Section 3.2.3.4) and microphases has filled a large portion of the HREM literature. The net effect of all these HREM developments has been to give theoreticians confidence in their atomic models for defects." One of the superb high-resolution micrographs shown in Spence's review is reproduced here (Figure 6.4); the separate atomic columns are particularly clear in the central area.

The improvement of transmission electron microscopes, aiming at ever higher resolutions and a variety of new and improved functions, together with the development of image-formation theory, jointly constitute one of the broadest and most important parepistemes in the whole of materials science, and enormous sums of money are involved in the industry, some 40 years after Siemens took a courageous gamble in undertaking the series manufacture of a very few microscopes at the end of the 1950s.

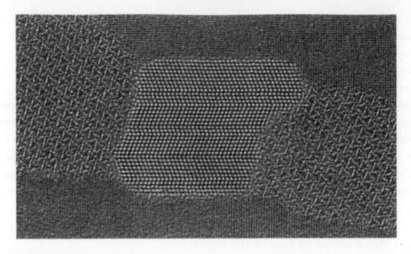

Figure 6.4. Piston alloy, showing strengthening precipitates, imaged by high-resolution electron microscopy. The matrix (top and bottom) is aluminium, while the central region is silicon. The outer precipitates were identified as $Al_5Cu_2Mg_8Si_5$. (First published by Spence 1999, reproduced here by courtesy of the originator, V. Radmilovic).

An important variant of transmission electron microscopy is the use of a particularly fine beam that is scanned across an area of the specimen and generates an image on a cathode ray screen – scanning transmission electron microscopy, or STEM. This approach has considerable advantages for composition analysis (using the approach described in the next section) and current developments in counteracting various forms of aberration in image formation hold promise of a resolution better than 1 Å (0.1 nm). This kind of microscopy is much younger than the technique described next.

6.2.2.2 Scanning electron microscopy. Some materials (e.g., fiber-reinforced composites) cannot usefully be examined by electron beams in transmission; some need to be studied by imaging a surface, and at much higher resolution than is possible by optical microscopy. This is achieved by means of the scanning electron microscope. The underlying idea is that a very finely focused 'sensing' beam is scanned systematically over the specimen surface (typically, the scan will cover rather less than a square millimetre), and secondary (or back-scattered) electrons emitted where the beam strikes the surface will be collected, counted and the varying signal used to modulate a synchronous scanning beam in a cathode-ray oscilloscope to form an enlarged image on a screen, just as a television image is formed. These instruments are today as important in materials laboratories as the transmission instruments, but they had a more difficult birth. The first commercial instruments were delivered in 1963.

The genesis of the modern scanning microscope is described in fascinating detail by its principal begetter, Oatley (1904–1996) (Oatley 1982). Two attempts were made before he came upon the scene, both in industry, one by Manfred von Ardenne in Germany in 1939, and another by Vladimir Zworykin and coworkers in America in 1942. Neither instrument worked well enough to be acceptable; one difficulty was that the signal was so weak that to scan one frame completely took minutes. Oatley was trained as a physicist, was exposed to engineering issues when he worked on radar during the War, and after the War settled in the Engineering Department of Cambridge University, where he introduced light electrical engineering into the curriculum (until then, the Department had been focused almost exclusively on mechanical and civil engineering). In 1948 Oatley decided to attempt the creation of an effective scanning electron microscope with the help of research students for whom this would be an educative experience: as he says in his article, prior to joining the engineering department in Cambridge he had lectured for a while in physics, and so he was bound to look favourably on potential research projects which "could be broadly classified as applied physics."

Oatley then goes on to say: "A project for a Ph.D. student must provide him with good training and, if he is doing experimental work, there is much to be said for choosing a problem which involves the construction or modification of some fairly complicated apparatus. Again, I have always felt that university research in engineering should be adventurous and should not mind tackling speculative projects. This is partly to avoid direct competition with industry which, with a 'safe' project, is likely to reach a solution much more quickly, but also for two other reasons which are rarely mentioned. In the first place, university research is relatively cheap. The senior staff are already paid for their teaching duties (remember, this refers to 1948) and the juniors are Ph.D. students financed by grants which are normally very low compared with industrial salaries. Thus the feasibility or otherwise of a speculative project can often be established in a university at a small fraction of the cost that would be incurred in industry. So long as the project provides good training and leads to a Ph.D., failure to achieve the desired result need not be a disaster. (The Ph.D. candidate must, of course, be judged on the excellence of his work, not on the end result.)" He goes on to point out that at the end of the normal 3-year stay of a doctoral student in the university (this refers to British practice) the project can then be discontinued, if that seems wise, without hard feelings.

Oatley and a succession of brilliant students, collaborating with others at the Cavendish Laboratory, by degrees developed an effective instrument: a key component was an efficient plastic scintillation counter for the image-forming electrons which is used in much the same form today. The last of Oatley's students was A.N. Broers, who later became head of engineering in Cambridge and is now the university's vice-chancellor (=president).

Oatley had the utmost difficulty in persuading industrial firms to manufacture the instrument, and in his own words, "the deadlock was broken in a rather roundabout way." In 1949, Castaing and Guinier in France reported on an electron microprobe analyser to analyse local compositions in a specimen (see next section), and a new research student, Peter Duncumb, in the Cavendish was set by V.E. Cosslett, in 1953, to add a scanning function to this concept; he succeeded in this. Because of this new feature, Oatley at last succeeded in interesting the Cambridge Instrument Company in manufacturing a small batch of scanning electron microscopes, with an analysing attachment, under the tradename of 'Stereoscan'. That name was well justified because of the remarkable depth of focus and consequent stereoscopic impression achieved by the instrument's images. Figure 6.5 shows an image of 'metal whiskers', made on the first production instrument sold by the Company in 1963 (Gardner and Cahn 1966), while Figure 6.6 shows a remarkable surface configuration produced by the differential 'sputtering' of a metal surface due to bombardment with high-energy unidirectional argon ions (Stewart

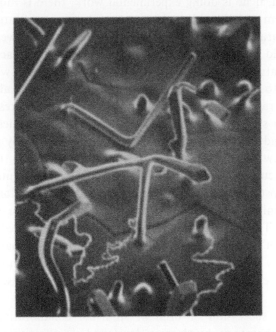

Figure 6.5. Whiskers grown at 1150°C on surface of an iron–aluminium alloy, imaged in an early scanning electron microscope ×250 (Gardner and Cahn 1966).

Figure 6.6. The surface of a tin crystal following bombardment with 5 keV argon ions, imaged in a scanning electron microscope (Stewart and Thompson 1969).

and Thompson 1969). Stewart had been one of Oatley's students who played a major part in developing the instruments.

A book chapter by Unwin (1990) focuses on the demanding *mechanical* components of the Stereoscan instrument, and its later version for geologists and mineralogists, the 'Geoscan', and also provides some background about the Cambridge Instrument Company and its mode of operation in building the scanning microscopes.

Run-of-the-mill instruments can achieve a resolution of 5–10 nm, while the best reach ≈ 1 nm. The remarkable depth of focus derives from the fact that a very small numerical aperture is used, and yet this feature does not spoil the resolution, which is not limited by diffraction as it is in an optical microscope but rather by various forms of aberration. Scanning electron microscopes can undertake compositional analysis (but with much less accuracy than the instruments treated in the next section) and there is also a way of arranging image formation that allows 'atomic-number contrast', so that elements of different atomic number show up in various degrees of brightness on the image of a polished surface.

Another new and much used variant is a procedure called 'orientation imaging microscopy' (Adams *et al.* 1993): patterns created by electrons back-scattered from a grain are automatically interpreted by a computer program, then the grain examined is automatically changed, and finally the orientations so determined are used to create an image of the polycrystal with the grain boundaries colour- or thickness-coded to represent the magnitude of misorientation across each boundary. Very recently, this form of microscopy has been used to assess the efficacy of new methods of making a polycrystalline ceramic superconductor designed to have no large misorientations anywhere in the microstructure, since the superconducting behaviour is degraded at substantially misoriented grain boundaries.

The Stereoscan instruments were a triumphant success and their descendants, mostly made in Britain, France, Japan and the United States, have been sold in thousands over the years. They are indispensable components of modern materials science laboratories. Not only that, but they have uses which were not dreamt of when Oatley developed his first instruments: thus, they are used today to image integrated microcircuits and to search for minute defects in them.

6.2.2.3 *Electron microprobe analysis.*

The instrument which I shall introduce here is, in my view, the most important development in characterisation since the 1939–1945 War. It has completely transformed the study of microstructure in its compositional perspective.

Henry Moseley (1887–1915) in 1913 studied the X-rays emitted by different pure metals when bombarded with high-energy electrons, using an analysing crystal to classify the wavelengths present by diffraction. He found strongly emitted 'characteristic wavelengths', different for each element, superimposed on a weak background radiation with a continuous range of wavelengths, and he identified the mathematical regularity linking the characteristic wavelengths to atomic numbers. His research cleared the way for Niels Bohr's model of the atom. It also cleared the way for compositional analysis by purely physical means. He would certainly have achieved further great things had he not been killed young as a soldier in the 'Great' War. His work is yet another example of a project undertaken to help solve a fundamental issue, the nature of atoms, which led to magnificent practical consequences.

Characteristic wavelengths can be used in two different ways for compositional analysis: it can be done by Moseley's approach, letting energetic electrons fall on the surface to be analysed and analysing the X-ray output, or else very energetic (shortwave) X-rays can be used to bombard the surface to generate secondary, 'fluorescent' X-rays. The latter technique is in fact used for compositional analysis, but until recently only by averaging over many square millimetres. In 1999, a group of French physicists were reported to have checked the genuineness of a disputed van Gogh painting by 'microfluorescence', letting an X-ray beam of the order of 1mm across impinge on a particular piece of paint to assess its local composition nondestructively; but even that does not approach the resolving power of the microprobe, to be presented here; however, it has to be accepted that a van Gogh painting could not be non-destructively stuffed into a microprobe's vacuum chamber.

In practice, it is only the electron-bombardment approach which can be used to study the distribution of elements in a sample on a microscopic scale. The instrument was invented in its essentials by a French physicist, Raimond Castaing (1921–1998) (Figure 6.7). In 1947 he joined ONERA, the French state aeronautics laboratory on the outskirts of Paris, and there he built the first microprobe analyser as a doctoral project. (It is quite common in France for a doctoral project to be undertaken in a state laboratory away from the university world.) The suggestion came from the great French crystallographer André Guinier, who wished to determine the concentration of the pre-precipitation zones in age-hardened alloys, less than a micrometre in thickness. Castaing's preliminary results were presented at a conference in Delft in 1949, but the full flowering of his research was reserved for his doctoral thesis (Castaing 1951). This must be the most cited thesis in the history of materials science, and has been described as "a document of great interest as well

Figure 6.7. Portrait of Raimond Castaing (courtesy Dr. P.W. Hawkes and Mme Castaing).

as a moving testimony to the brilliance of his theoretical and experimental investigations".

The essence of Castaing's instrument was a finely focused electron beam and a rotatable analysing crystal plus a detector which together allowed the wavelengths and intensities of X-rays emitted from the impact site of the electron beam; there was also an optical microscope to check the site of impact in relation to the specimen's microstructure. According to an obituary of Castaing (Heinrich 1999): "Castaing initially intended to achieve this goal in a few weeks. He was doubly disappointed: the experimental difficulties exceeded his expectations by far, and when, after many months of painstaking work, he achieved the construction of the first electron probe microanalyser, he discovered that... the region of the specimen excited by the entering electrons exceeded the micron size because of diffusion of the electrons within the specimen." He was reassured by colleagues that even what he had achieved so far would be a tremendous boon to materials science, and so continued his research. He showed that for accurate quantitative analysis, the (characteristic) line intensity of each emitting element in the sample needed to be compared with the output of a standard specimen of known composition. He also identified the corrections to be applied to the measured intensity ratio, especially for X-ray absorption and fluorescence within the sample, also taking into account the mean atomic number of the sample. Heinrich remarks: "Astonishingly, this strategy remains valid today".

We saw in the previous Section that Peter Duncumb in Cambridge was persuaded in 1953 to add a scanning function to the Castaing instrument (and this in fact was the key factor in persuading industry to manufacture the scanning electron microscope, the *Stereoscan*... and later also the microprobe, the *Microscan*). The result was the generation of compositional maps for each element contained in the sample, as in the early example shown in Figure 6.8. In a symposium dedicated to Castaing, Duncumb has recently discussed the many successive mechanical and electron-optical design versions of the microprobe, some for metallurgists, some for geologists, and also the considerations which went into the decision to go for scanning (Duncumb 2000) as well as giving an account of '50 years of evolution'. At the same symposium, Newbury (2000) discusses the great impact of the microprobe on materials science. A detailed modern account of the instrument and its use is by Lifshin (1994).

The scanning electron microscope (SEM) and the electron microprobe analyser (EMA) began as distinct instruments with distinct functions, and although they have slowly converged, they are still distinct. The SEM is nowadays fitted with an 'energy-dispersive' analyser which uses a scintillation detector with an electronic circuit to determine the quantum energy of the signal, which is a fingerprint of the atomic number of the exciting element; this is convenient but less accurate than a crystal

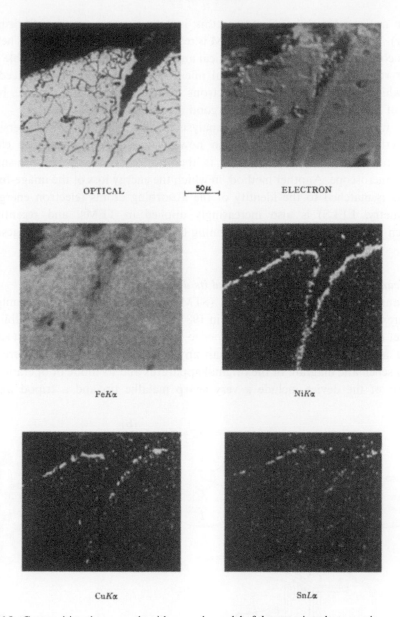

OPTICAL |—50μ—| ELECTRON

FeKα   NiKα

CuKα   SnLα

Figure 6.8. Compositional map made with an early model of the scanning electron microprobe. The pictures show the surface segregation of Ni, Cu and Sn dissolved in steel as minor constituents; the two latter constituents enriched at the surface cause 'hot shortness' (embrittlement at high temperatures), and this study was the first to demonstrate clearly the cause (Melford 1960).

detector as introduced by Castaing (this is known as a wavelength-dispersive analyser). The main objective of the SEM is resolution and depth of focus. The EMA remains concentrated on accurate chemical analysis, with the highest possible point-to-point resolution: the original optical microscope has long been replaced by a device which allows back-scattered electrons to form a topographic image, but the quality of this image is nothing like as good as that in an SEM.

The methods of compositional analysis, using either energy-dispersive or wavelength-dispersive analysis are also now available on transmission electron microscopes (TEMs); the instrument is then called an analytical transmission electron microscope. Another method, in which the energy loss of the image-forming electrons is matched to the identity of the absorbing atoms (electron energy loss spectrometry, EELS) is also increasingly applied in TEMs, and recently this approach has been combined with scanning to form EELS-generated images.

6.2.3 Scanning tunneling microscopy and its derivatives

The scanning tunnelling microscope (STM) was invented by G. Binnig and H. Rohrer at IBM's Zürich laboratory in 1981 and the first account was published a year later (Binnig *et al.* 1982). It is a device to image atomic arrangements at surfaces and has achieved higher resolution than any other imaging device. Figure 6.9(a) shows a schematic diagram of the original apparatus and its mode of operation. The essentials of the device include a very sharp metallic tip and a tripod made of

(a)   **(b)**

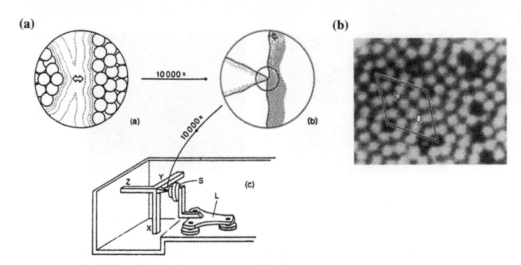

Figure 6.9. (a) Schematic of Binnig and Rohrer's original STM. (b) An image of the "7 × 7" surface rearrangement on a (1 1 1) plane of silicon, obtained by a variant of STM by Hamers *et al.* (1986).

piezoelectric material in which a minute length change can be induced by purely electrical means. In the original mode of use, the tunneling current between tip and sample was held constant by movements of the legs of the tripod; the movements, which can be at the Ångström level (0.1 nm) are recorded and modulate a scanning image on a cathode-ray monitor, and in this way an atomic image is displayed in terms of height variations. Initially, the IBM pioneers used this to display the changed crystallography (Figure 6.9(b)) in the surface layer of a silicon crystal – a key feature of modern surface science (Section 10.4). Only three years later, Binnig and Rohrer received a Nobel Prize.

According to a valuable 'historical perspective' which forms part of an excellent survey of the whole field (DiNardo 1994) to which the reader is referred, "the invention of the STM was preceded by experiments to develop a surface imaging technique whereby a non-contacting tip would scan a surface under feedback control of a tunnelling current between tip and sample." This led to the invention, in the late 1960s, of a device at the National Bureau of Standards near Washington, DC working on rather similar principles to the STM; this failed because no way was found of filtering out disturbing laboratory vibrations, a problem which Binnig and Rohrer initially solved in Zürich by means of a magnetic levitation approach.

DiNardo's 1994 survey includes about 350 citations to a burgeoning literature, only 11 years after the original papers – and that can only have been a fraction of the total literature. A comparison with the discovery of X-ray diffraction is instructive: the Braggs made their breakthrough in 1912, and they also received a Nobel Prize three years later. In 1923, however, X-ray diffraction had made little impact as yet on the crystallographic community (as outlined in Section 3.1.1.1); the mineralogists in particular paid no attention. Modern telecommunications and the conference culture have made all the difference, added to which a much wider range of issues were quickly thrown up, to which the STM could make a contribution.

In spite of the extraordinarily minute movements involved in STM operation, the modern version of the instrument is not difficult to use, and moreover there are a large number of derivative versions, such as the Atomic Force Microscope, in which the tip touches the surface with a measurable though minute force; this version can be applied to non-conducting samples. As DiNardo points out, "the most general use of the STM is for topographic imaging, not necessarily at the atomic level but on length scales from < 10 nm to ≥1 μm." For instance, so-called quantum dots and quantum wells, typically 100 nm in height, are often pictured in this way. Many other uses are specified in DiNardo's review.

The most arresting development is the use of an STM tip, manipulated to move both laterally and vertically, to 'shepherd' individual atoms across a crystal surface to generate features of predeterminate shapes: an atom can be contacted, lifted, transported and redeposited under visual control. This was first demonstrated at

IBM in California by Eigler and Schweizer (1990), who manipulated individual xenon atoms across a nickel (1 1 0) crystal surface. In the immediate aftermath of this achievement, many other variants of atom manipulation by STM have been published, and DiNardo surveys these.

Such an extraordinary range of uses for the STM and its variants have been found that this remarkable instrument can reasonably be placed side by side with the electron microprobe analyser as one of the key developments in modern characterisation.

6.2.4 Field-ion microscopy and the atom probe

If the tip of a fine metal wire is sharpened by making it the anode in an electrolytic circuit so that the tip becomes a hemisphere 100–500 nm across and a high negative voltage is then applied to the wire when held in a vacuum tube, a highly magnified image can be formed. This was first discovered by a German physicist, E.W. Müller, in 1937, and improved by slow stages, especially when he settled in America after the War.

Initially the instrument was called a field-emission microscope and depended on the field-induced emission of electrons from the highly curved tip. Because of the sharp curvature, the electric field close to the tip can be huge; a voltage of 20–50 V/nm can be generated adjacent to the curved surface with an applied voltage of 10 kV. The emission of electrons under such circumstances was interpreted in 1928 in wave-mechanical terms by Fowler and Nordheim. Electrons spreading radially from the tip in a highly evacuated glass vessel and impinging on a phosphor layer some distance from the tip produce an image of the tip which may be magnified as much as a million times. Müller's own account of his early instrument in an encyclopedia (Müller 1962) cites no publication earlier than 1956. By 1962, field-emission patterns based on electron emission had been studied for many high-melting metals such as W, Ta, Mo, Pt, Ni; the metal has to be high-melting so that at room temperature it is strong enough to withstand the stress imposed by the huge electric field. Müller pointed out that if the field is raised sufficiently (and its sign reversed), the metal ions themselves can be forced out of the tip and form an image.

In the 1960s, the instrument was developed further by Müller and others by letting a small pressure of inert gas into the vessel; then, under the right conditions, gas atoms become ionised on colliding with metal atoms at the tip surface and it is now these gas ions which form the image – hence the new name of *field-ion microscopy*. The resolution of 2–3 nm quoted by Müller in his 1962 article was gradually raised, in particular by cooling the tip to liquid-nitrogen temperature, until individual atoms could be clearly distinguished in the image. Grain boundaries, vacant lattice sites, antiphase domains in ordered compounds, and especially details of phase transformations, are examples of features that were studied by the few groups who used the technique from the 1960s till the 1980s (e.g., Haasen 1985). A book about the method was published by Müller and Tsong (1969). The highly decorative tip images obtainable with the instrument by the early 1970s were in great demand to illustrate books on metallography and physical metallurgy.

From the 1970s on, and accelerating in the 1980s, the field-ion microscope was metamorphosed into something of much more extensive use and converted into the *atom probe*. Here, as with the electron microprobe analyser, imaging and analysis are combined in one instrument. All atom probes are run under conditions which extract metal ions from the tip surface, instead of using inert gas ions as in the field-ion microscope. In the original form of the atom probe, a small hole was made in the imaging screen and brief bursts of metal ions are extracted by applying a nanosecond voltage pulse to the tip. These ions then are led by the applied electric field along a path of 1–2 m in length; the heavier the ion, the more slowly it moves, and thus mass spectrometry can be applied to distinguish different metal species. In effect, only a small part of the specimen tip is analysed in such an instrument, but by progressive field-evaporation from the tip, composition profiles in depth can be obtained.

Various ion-optical tricks have to be used to compensate for the spread of energies of the extracted ions, which limit mass resolution unless corrected for. In the latest version of the atom probe (Cerezo *et al.* 1988), spatial as well as compositional information is gathered. The hole in the imaging screen is dispensed with and it is replaced by a position-sensitive screen that measures at each point on the screen the time of flight, and thus a compositional map with extremely high (virtually atomic) resolution is attained. Extremely sophisticated computer control is needed to obtain valid results.

The evolutionary story, from field-ion microscopy to spatially imaging time-of-flight atom probes is set out in detail by Cerezo and Smith (1994); these two investigators at Oxford University have become world leaders in atom-probe development and exploitation. Uses have focused mainly on age-hardening and other phase transformations in which extremely fine resolution is needed. Very recently, the Oxford team have succeeded in imaging a carbon 'atmosphere' formed around a dislocation line, fully half a century after such atmospheres were first identified by highly indirect methods (Section 5.1.1). Another timely application of the imaging atom probe is a study of Cu–Co metallic multilayers used for magnetoresistive probes (Sections 7.4, 10.5.1.2); the investigators (Larson *et al.* 1999) were able to relate the magnetoresistive properties to variables such as curvature of the deposited layers, short-circuiting of layers and fuzziness of the compositional discontinuity between successive layers. This study could not have been done with any other technique.

Several techniques which combine imaging with spectrometric (compositional) analysis have now been explained. It is time to move on to straight spectrometry.

6.3. SPECTROMETRIC TECHNIQUES

Until the last War, variants of optical emission spectroscopy ('spectrometry' when the technique became quantitative) were the principal supplement to wet chemical analysis. In fact, university metallurgy departments routinely employed resident analytical chemists who were primarily experts in wet methods, qualitative and quantitative, and undergraduates received an elementary grounding in these techniques. This has completely vanished now.

The history of optical spectroscopy and spectrometry, detailed separately for the 19th and 20th centuries, is retailed by Skelly and Keliher (1992), who then go on to describe present usages. In addition to emission spectrometry, which in essentials involves an arc or a flame 'contaminated' by the material to be analysed, there are the methods of fluorescence spectrometry (in which a specimen is excited by incoming light to emit characteristic light of lower quantum energy) and, in particular, the technique of atomic absorption spectrometry, invented in 1955 by Alan Walsh (1916–1997). Here a solution that is to be analysed is vaporized and suitable light is passed through the vapor reservoir: the composition is deduced from the absorption lines in the spectrum. The absorptive approach is now very widespread.

Raman spectrometry is another variant which has become important. To quote one expert (Purcell 1993), "In 1928, the Indian physicist C.V. Raman (later the first Indian Nobel prizewinner) reported the discovery of frequency-shifted lines in the scattered light of transparent substances. The shifted lines, Raman announced, were independent of the exciting radiation and characteristic of the sample itself." It appears that Raman was motivated by a passion to understand the deep blue colour of the Mediterranean. The many uses of this technique include examination of polymers and of silicon for microcircuits (using an exciting wavelength to which silicon is transparent).

In addition to the wet and optical spectrometric methods, which are often used to analyse elements present in very small proportions, there are also other techniques which can only be mentioned here. One is the method of mass spectrometry, in which the proportions of separate isotopes can be measured; this can be linked to an instrument called a field-ion microscope, in which as we have seen individual atoms can be observed on a very sharp hemispherical needle tip through the mechanical action of a very intense electric field. Atoms which have been ionised and detached can then be analysed for isotopic mass. This has become a powerful device for both curiosity-driven and applied research.

Another family of techniques is chromatography (Carnahan 1993), which can be applied to gases, liquids or gels: this postwar technique depends typically upon the separation of components, most commonly volatile ones, in a moving gas stream, according to the strength of their interaction with a 'partitioning liquid' which acts like a semipermeable barrier. In gas chromatography, for instance, a sensitive electronic thermometer can record the arrival of different volatile components. One version of chromatography is used to determine molecular weight distributions in polymers (see Chapter 8, Section 8.7).

Yet another group of techniques might be called non-optical spectrometries: these include the use of Auger electrons which are in effect secondary electrons excited by electron irradiation, and photoelectrons, the latter being electrons excited by incident high-energy electromagnetic radiation – X-rays. (Photoelectron spectrometry used to be called ESCA, electron spectrometry for chemical analysis.) These techniques are often combined with the use of magnifying procedures, and their use involves large and expensive instruments working in ultrahigh vacuum. In fact, radical improvements in vacuum capabilities in recent decades have brought several new characterisation techniques into the realm of practicality; ultrahigh vacuum has allowed a surface to be studied at leisure without its contamination within seconds by molecules adsorbed from an insufficient vacuum environment (see Section 10.4).

Quite generally, each sensitive spectrometric approach today requires instruments of rapidly escalating cost, and these have to be centralised for numerous users, with resident experts on tap. The experts, however, often prefer to devote themselves to improving the instruments and the methods of interpretation: so there is a permanent tension between those who want answers from the instruments and those who have it in their power to deliver those answers.

6.3.1 Trace element analysis

A common requirement in MSE is to identify and quantify elements present in very small quantities, parts per million or even parts per billion – trace elements. The difficulty of this task is compounded when the amount of material to be analysed is small: there may only be milligrams available, for instance in forensic research. A further requirement which is often important is to establish whereabouts in a solid material the trace element is concentrated; more often than not, trace elements segregate to grain boundaries, surfaces (including internal surfaces in pores) and interphase boundaries. Trace elements have frequent roles in such phenomena as embrittlement at grain boundaries (Hondros *et al.* 1996), neutron absorption in nuclear fuels and moderators, electrical properties in electroceramics (Section 7.2.2), age-hardening kinetics in aluminium alloys (and kinetics of other phase transformations, such as ordering reactions), and notably in optical glass fibres used for communication (Section 7.5.1).

Sibilia (1988), in his guide to materials characterisation and chemical analysis, offers a concise discussion of the sensitivity of different analytical techniques for trace elements. Thus for optical emission spectrometry, the detection limits for various elements are stated to range from 0.002 μg for beryllium to as much as 0.2 μg for lead or silicon. For atomic absorption spectrometry, detection limits are expressed in mg/litre of solution and typically range from 0.00005 to 0.001 mg/l; since only a small fraction of a litre is needed to make an analysis, this means that absolute detection limits are considerably smaller than for the emission method. A technique widely used for trace element analysis is neutron activation analysis (Hossain 1992): a sample, which can be as small as 1 mg, is exposed to neutrons in a nuclear reactor, which leads to nuclear transmutation, generating a range of radioactive species; these can be recognised and measured by examining the nature, energy and intensity of the radiation emitted by the samples after activation and the half-lives of the underlying isotopes. Thus, oxygen, nitrogen and fluorine can be analysed in polymers, and trace elements in optical fibres.

Trace element analysis has become sufficiently important, especially to industrial users, that commercial laboratories specialising in "trace and ultratrace elemental analysis" are springing up. One such company specialises in "high-resolution glow-discharge mass spectromety", which can often go, it is claimed, to better than parts per billion. This company's advertisements also offer a service, domiciled in India, to provide various forms of wet chemical analysis which, it is claimed, is now "nearly impossible to find in the United States".

Very careful analysis of trace elements can have a major effect on human life. A notable example can be seen in the career of Clair Patterson (1922–1995) (memoir by Flagel 1996), who made it his life's work to assess the origins and concentrations of lead in the atmosphere and in human bodies; minute quantities had to be measured and contaminant lead from unexpected sources had to be identified in his analyses, leading to techniques of 'clean analysis'. A direct consequence of Patterson's scrupulous work was a worldwide policy shift banning lead in gasoline and manufactured products.

6.3.2 Nuclear methods

The neutron activation technique mentioned in the preceding paragraph is only one of a range of 'nuclear methods' used in the study of solids – methods which depend on the response of atomic nuclei to radiation or to the emission of radiation by the nuclei. Radioactive isotopes ('tracers') of course have been used in research ever since von Hevesy's pioneering measurements of diffusion (Section 4.2.2). These techniques have become a field of study in their own right and a number of physics laboratories, as for instance the Second Physical Institute at the University of Göttingen, focus on the development of such techniques. This family of techniques, as applied to the study of condensed matter, is well surveyed in a specialised text

(Schatz and Weidinger 1996). ('Condensed matter' is a term mostly used by physicists to denote solid materials of all kinds, both crystalline and glassy, and also liquids.)

One important approach is Mössbauer spectrometry. This Nobel-prize-winning innovation named after its discoverer, Rudolf Mössbauer, who discovered the phenomenon when he was a physics undergraduate in Germany, in 1958; what he found was so surprising that when (after considerable difficulties with editors) he published his findings in the same year, "surprisingly no one seemed to notice, care about or believe them. When the greatness of the discovery was finally appreciated, fascination gripped the scientific community and many scientists immediately started researching the phenomenon," in the words of two commentators (Gonser and Aubertin 1993). Another commentator, Abragam (1987), remarks: "His immense merit was not so much in having observed the phenomenon as in having found the explanation, which in fact had been known for a long time and only the incredible blindness of everybody had obscured". The Nobel prize was awarded to Mössbauer in 1961, *de facto* for his first publication.

The Mössbauer effect can be explained only superficially in a few words, since it is a subtle quantum effect. Normally, when an excited nucleus emits a quantum of radiation (a gamma ray) to return to its 'ground state', the emitting nucleus recoils and this can be shown to cause the emitted radiation to have a substantial 'line width', or range of frequency – a direct consequence of the Heisenberg Uncertainty Principle. Mössbauer showed that certain isotopes only can undergo recoil-free emissions, where no energy is exchanged with the crystal and the gamma-ray carries the entire energy. This leads to a phenomenally narrow linewidth. If the emitted gamma ray is then allowed to pass through a stationary absorber containing the same isotope, the sharp gamma ray is resonantly absorbed. However, it was soon discovered that the quantum properties of a nucleus can be affected by the 'hyperfine field' caused by the electrons in the neighbourhood of the absorbing nucleus; then the absorber had to be moved, by a few millimetres per second at the most, so that the Doppler effect shifted the effective frequency of the gamma ray by a minute fraction, and resonant absorption was then restored. By measuring a spectrum of absorption versus motional speed, the hyperfine field can be mapped. Today, Mössbauer spectrometry is a technique very widely used in studying condensed matter, magnetic materials in particular.

Nuclear magnetic resonance is another characterisation technique of great practical importance, and yet another that became associated with a Nobel Prize for Physics, in 1952, jointly awarded to the American pioneers, Edward Purcell and Felix Bloch (see Purcell *et al.* 1946, Bloch 1946). In crude outline, when a sample is placed in a strong, homogeneous and constant magnetic field and a small radio-frequency magnetic field is superimposed, under appropriate circumstances the sample can resonantly absorb the radio-frequency energy; again, only some isotopes are suitable for this technique. Once more, much depends on the sharpness of the resonance; in the early researches of Purcell and Bloch, just after the Second World War, it turned out that liquids were particularly suitable; solids came a little later (see survey by Early 2001). Anatole Abragam, a Russian immigrant in France (Abragam 1987), was one of the early physicists to learn from the pioneers and to add his own developments; in his very enjoyable book of memoirs, he vividly describes the activities of the pioneers and his interaction with them. Early on, the 'Knight shift', a change in the resonant frequency due to the chemical environment of the resonating nucleus – distinctly analogous to Mössbauer's Doppler shift – gave chemists an interest in the technique, which has grown steadily. At an early stage, an overview addressed by physicists to metallurgists (Bloembergen and Rowland 1953) showed some of the applications of nuclear magnetic resonance and the Knight shift to metallurgical issues. One use which interested materials scientists a little later was 'motional narrowing': this is a sharpening of the resonance 'line' when atoms around the resonating nucleus jump with high frequency, because this motion smears out the structure in the atomic environment which would have broadened the line. For aluminium, which has no radioisotope suitable for diffusion measurements, this proved the only way to measure self-diffusion (Rowland and Fradin 1969); the ^{27}Al isotope, the only one present in natural aluminium, is very suitable for nuclear magnetic resonance measurements. In fact, this technique applied to ^{27}Al has proved to be a powerful method of studying structural features in such crystals as the feldspar minerals (Smith 1983). This last development indicates that some advanced techniques like nuclear magnetic resonance begin as characterisation techniques for measuring features like diffusion rates but by degrees come to be applied to structural features as supplements to diffraction methods.

A further important branch of 'nuclear methods' in studying solids is the use of high-energy projectiles to study compositional variations in depth, or 'profiling' (over a range of a few micrometres only): this is named Rutherford back-scattering, after the great atomic pioneer. Typically, high-energy protons or helium nuclei (alpha particles), speeded up in a particle accelerator, are used in this way. Such ions, metallic this time, are also used in one approach to making integrated circuits, by the technique of 'ion implantation'. The complex theory of such scattering and implantation is fully treated in a recent book (Nastasi *et al.* 1996).

Another relatively recent technique, in its own way as strange as Mössbauer spectrometry, is positron annihilation spectrometry. Positrons are positive electrons (antimatter), spectacularly predicted by the theoretical physicist Dirac in the 1920s and discovered in cloud chambers some years later. Some currently available radioisotopes emit positrons, so these particles are now routine tools. High-energy positrons are injected into a crystal and very quickly become 'thermalised' by interaction with lattice vibrations. Then they diffuse through the lattice and eventually perish by annihilation with an electron. The whole process requires a few picoseconds. Positron lifetimes can be estimated because the birth and death of a positron are marked by the emission of gamma-ray quanta. When a large number of vacancies are present, many positrons are captured by a vacancy site and stay there for a while, reducing their chance of annihilation: the mean lifetime is thus increased. Vacancy concentrations can thus be measured and, by a variant of the technique which is too complex to outline here, vacancy mobility can be estimated also. The first overview of this technique was by Seeger (1973).

Finally, it is appropriate here to mention neutron scattering and diffraction. It is appropriate because, first, neutron beams are generated in nuclear reactors, and second, because the main scattering of neutrons is by atomic nuclei and not, as with X-rays, by extranuclear electrons. Neutrons are also sensitive to magnetic moments in solids and so the arrangements of atomic magnetic spins can be assessed. Further, the scattering intensity is determined by nuclear characteristics and does not rise monotonically with atomic number: light elements, deuterium (a hydrogen isotope) particularly, scatter neutrons vigorously, and so neutrons allow hydrogen positions in crystal structures to be identified. A chapter in Schatz and Weidinger's book (1996) outlines the production, scattering and measurement of neutrons, and exemplifies some of the many crystallographic uses of this approach; structural studies of liquids and glasses also make much use of neutrons, which can give information about a number of features, including thermal vibration amplitudes. In inelastic scattering, neutrons lose or gain energy as they rebound from lattice excitations, and information is gained about lattice vibrations (phonons), and also about 'spin waves'. Such information is helpful in understanding phase transformations, and superconducting and magnetic properties.

One of the principal places where the diffraction and inelastic scattering of neutrons was developed was Brookhaven National Laboratory on Long Island, NY. A recent book (Crease 1999), a 'biography' of that Laboratory, describes the circumstances of the construction and use of the high-flux (neutron) beam reactor there, which operated from 1965. (After a period of inactivity, it has just – 1999 – been permanently shut down.) Brookhaven had been set up for research in nuclear physics but this reactor after a while became focused on solid-state physics; for years there was a battle for mutual esteem between the two fields. In 1968, a Japanese immigrant, Gen Shirane (b. 1924), became head of the solid-state neutron group and worked with the famous physicist George Dienes in developing world-class solid-state research in the midst of a nest of nuclear physicists. The fascinating details of this uneasy cohabitation are described in the book. Shirane was not however the originator of neutron diffraction; that distinction belongs to Clifford Shull and Ernest Wollan, who began to use this technique in 1951 at Oak Ridge National

Laboratory, particularly to study ferrimagnetic materials. In 1994, a Nobel Prize in physics was (belatedly) awarded for this work, which is mentioned again in the next chapter, in Section 7.3. A range of achievements in neutron crystallography are reviewed by Willis (1998).

6.4. THERMOANALYTICAL METHODS

The procedures of measuring changes in some physical or mechanical property as a sample is heated, or alternatively as it is held at constant temperature, constitute the family of thermoanalytical methods of characterisation. A partial list of these procedures is: differential thermal analysis, differential scanning calorimetry, dilatometry, thermogravimetry. A detailed overview of these and several related techniques is by Gallagher (1992).

Dilatometry is the oldest of these techniques. In essence, it could not be simpler. The length of a specimen is measured as it is steadily heated and the length is plotted as a function of temperature. The steady slope of thermal expansion is disturbed in the vicinity of temperatures where a phase change or a change in magnetic character takes place. Figure 6.10 shows an example; here the state of atomic long-range order in an alloy progressively disappears on heating (Cahn *et al.* 1987). The method has fallen out of widespread use of late, perhaps because it seems too simple and

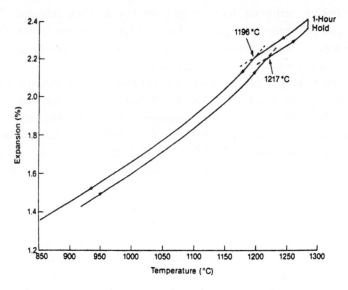

Figure 6.10. Dilatometric record of a sample of a Ni–Al–Fe alloy in the neighbourhood of an order–disorder transition temperature (Cahn *et al.* 1987).

unsophisticated; that is a pity, because the method can be very powerful. Very recently, Li *et al.* (2000) have demonstrated how, by taking into account known lattice parameters, a dilatometer can be used for quantitative analysis of the isothermal decomposition of iron-carbon austenite.

The first really accurate dilatometer was a purely mechanical instrument, using mirrors and lightbeams to record changes in length (length changes of a standard rod were also used to measure temperature). This instrument, one among several, was the brainchild of Pierre Chevenard, a French engineer who was employed by a French metallurgical company, Imphy, early in the 20th century, to set up a laboratory to foster 'la métallurgie de précision'. He collaborated with Charles–Edouard Guillaume, son of a Swiss clockmaker, who in 1883 had joined the International Bureau of Weights and Measures near Paris. There one of his tasks was to find a suitable alloy, with a small thermal expansion coefficient, from which to fabricate subsidiary length standards (the primary standard was made of precious metals, far too expensive to use widely). He chanced upon an alloy of iron with about 30 at.% of nickel with an unusually low (almost zero) thermal expansion coefficient. He worked on this and its variants for many years, in collaboration with the Imphy company, and in 1896 announced INVAR, a Fe–36%Ni alloy with virtually zero expansion coefficient near ambient temperature. Guillaume and Chevenard, two precision enthusiasts, studied the effects of ternary alloying, of many processing variables, preferred crystallographic orientation, etc., on the thermal characteristics, which eventually were tracked down to the disappearance of ferromagnetism and of its associated magnetostriction, compensating normal thermal expansion. In 1920 Guillaume gained the Nobel Prize in physics, the only occasion that a metallurgical innovation gained this honour. The story of the discovery, perfection and wide-ranging use of Invar is well told in a book to mark the centenary of its announcement (Béranger *et al.* 1996). Incidentally, after more than 100 years, the precise mechanism of the 'invar effect' is still under debate; just recently, a computer simulation of the relevant alignment of magnetic spins claims to have settled the issue once and for all (van Schilfgaarde *et al.* 1999).

Thermogravimetry is a technique for measuring changes in weight as a function of temperature and time. It is much used to study the kinetics of oxidation and corrosion processes. The samples are usually small and the microbalance used, operating by electromagnetic self-compensation of displacement, is extraordinarily sensitive (to microgram level) and stable against vibration.

Differential thermal analysis (DTA) and differential scanning calorimetry (DSC) are the other mainline thermal techniques. These are methods to identify temperatures at which specific heat changes suddenly or a latent heat is evolved or absorbed by the specimen. DTA is an early technique, invented by Le Chatelier in France in 1887 and improved at the turn of the century by Roberts-Austen (Section 4.2.2). A sample is allowed to cool freely and anomalies in cooling rate are identified at particular temperatures. The method, simple in essence, is widely used to help in the construction of phase diagrams, because the beginning of solidification or other phase change is easily identified.

Differential scanning calorimetry (DSC) has a more tangled history. In its modern form, developed by the Perkin–Elmer Company in 1964 (Watson *et al.* 1964) two samples, the one under investigation and a reference, are in the same thermal enclosure. Platinum resistance thermometers at each specimen are arranged in a bridge circuit, and any imbalance is made to drive a heater next to one or other of the samples. The end-result is a plot of heat flow versus temperature, quantitatively accurate so that specific heats and latent heats can be determined. The modern form of the instrument normally only reaches about 700°C, which indicates the difficulty of correcting for all the sources of error. DSC is now widely used, for instance to determine glass transition temperatures in polymers and also in metallic glasses, and generally for the study of all kinds of phase transformation. It is also possible, with great care, to use a DSC in isothermal mode, to study the kinetics of phase transformations.

The antecedents of the modern DSC apparatus are many and varied, and go back all the way to the 19th century and attempts to determine the mechanical equivalent of heat accurately. A good way to examine these antecedents is to read two excellent critical reviews (Titchener and Bever 1958, Bever *et al.* 1973) of successive attempts to determine the 'stored energy of cold work', i.e., the enthalpy retained in a metal or alloy when it is heavily plastically deformed. That issue was of great concern in the 1950s and 1960s because it was linked with the multiplication of dislocations and vacancies that accompanies plastic deformation. (Almost all the retained energy is associated with these defects.) Bever and his colleagues examine the extraordinary variety of calorimetric devices used over the years in this pursuit. Perhaps the most significant are the paper by Quinney and Taylor (1937) (this is the same Taylor who had co-invented dislocations a few years earlier) and that by an Australian group, by Clarebrough *et al.* (1952), whose instrument was a close precursor of Perkin–Elmer's first commercial apparatus. The circumstances surrounding the researches of the Australian group are further discussed in Section 14.4.3. The Australian calorimeter was used not only for studying deformed metals but also for studying phase transformations, especially slow ordering transitions. Perhaps the first instrument used specifically to study order–disorder transitions in alloys was a calorimeter designed by Sykes (1935).

Figure 6.11 shows a famous example of the application of isothermal calorimetry. Gordon (1955) deformed high-purity copper and annealed samples in his precision calorimeter and measured heat output as a function of time. In this metal, the heat output is strictly proportional to the fraction of metal recrystallised.

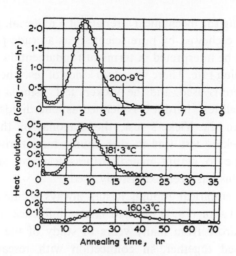

Figure 6.11. Isothermal energy release from cold-worked copper, measured calorimetrically (Gordon 1955).

This approach is an alternative to quantitative metallography and in the hands of a master gives even more accurate results than the rival method. A more recent development (Chen and Spaepen 1991) is the analysis of the isothermal curve when a material which may be properly amorphous or else nanocrystalline (e.g., a bismuth film vapour-deposited at low temperature) is annealed. The form of the isotherm allows one to distinguish nucleation and growth of a crystalline phase, from the growth of a preexisting nanocrystalline structure.

6.5. HARDNESS

The measurement of mechanical properties is a major part of the domain of characterisation. The tensile test is the key procedure, and this in turn is linked with the various tests to measure fracture toughness... crudely speaking, the capacity to withstand the weakening effects of defects. Elaborate test procedures have been developed to examine resistance to high-speed impact of projectiles, a property of civil (birdstrike on aircraft) as well as military importance. Another kind of test is needed to measure the elastic moduli in different directions of an anisotropic crystal; this is, for instance, vital for the proper exploitation of quartz crystal slices in quartz watches.

There is space here for a brief account of only one technique, that is, hardness measurement. The idea of pressing a hard object, of steel or diamond, into a smooth surface under a known load and measuring the size of the indent, as a simple and quick way of classifying the mechanical strength of a material, goes back to the 19th century. It was often eschewed by pure scientists as a crude procedure which gave results that could not be interpreted in terms of fundamental concepts such as yield stress or work-hardening rate. There were two kinds of test: the Brinell test, in which a hardened steel sphere is used, and the Vickers test, using a pyramidally polished diamond. In the Brinell test, hardness is defined as load divided by the curved area of the indentation; in the Vickers test, the diagonal of the square impression is measured. The Vickers test was in due course miniaturised and mounted on an optical microscope to permit microhardness tests on specific features of a microstructure, and this has been extensively used.

The Brinell test, empirical though it has always seemed, did yield to close analysis. A book by Tabor (1951) has had lasting influence in this connection: his interest in hardness arose from many years of study of the true area of contact between solids pressed together, in connection with research on friction and lubrication. The Brinell test suffers from the defect that different loads will give geometrically non-similar indentations and non-comparable hardness values. In 1908, a German engineer, E. Meyer, proposed defining hardness in terms of the area of the indentation projected in the plane of the tested surface. Meyer's empirical law then stated that if W is the load and d the chordal diameter of the indentation, $W = kd^n$, where k and n are material constants. n turned out to be linked to the work-hardening capacity of the test material, and consequently, Meyer analysis was widely used for a time as an economical way of assessing this capacity. Much of Tabor's intriguing book is devoted to a fundamental examination of Meyer analysis and its implications, but this form of analysis is no longer much used today.

A quite different use of a Brinell-type test relates to highly brittle materials, and goes back to an elastic analysis by H.H. Hertz, another German engineer, in 1896. Figure 6.12 shows the Hertzian test in outline. If a hard steel ball is pressed into the polished surface of window glass, at a certain load a sudden conically shaped ring crack will spring into existence. The load required depends on the size of the largest microcrack preexisting in the glass surface (the kind of microcrack postulated in 1922 by A.A. Griffith, see Section 5.1.2.1), and a large number of identical tests performed on the same sample will allow the statistical distribution of preexisting crack depths to be assessed. The value of the Hertzian test in studying brittle materials is explained in detail by Lawn (1993).

The wide use of microhardness testing recently prompted Oliver (1993) to design a 'mechanical properties microprobe' ('nanoprobe' would have been a better name), which generates indentations considerably less than a micrometre in depth. Loads up to 120 mN (one mN ≈ 0.1 g weight) can be applied, but a tenth of that amount is commonly used and hardness is estimated by electronically measuring the depth of impression while the indentor is still in contact. This allows, inter alia, measurement

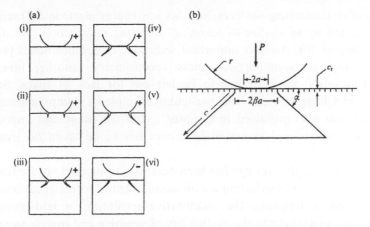

Figure 6.12. Hertzian cone crack evolution and geometry (Lawn 1993).

of a local elastic modulus. The whole process of making a nanoindentation and then moving the indentor to the next site is performed automatically, under computer control. This microprobe is now commercially available. Oliver remarks: "The mechanical properties microprobe will advance our understanding of macroscopic properties in the same way that the chemical microprobe (EMA) has improved our understanding of the chemistry of materials and the transmission electron microscope has improved our understanding of structures." Examples of applications: characterisation of materials surface-hardened by ion-implantation, a process which typically affects only the top micrometre or two of the surface region; rapidly solidified materials, which are often in the form of very thin foils which cannot be reliably examined by ordinary microhardness measurement (because the indentation has a depth of the same order as the specimen thickness).

This progression of techniques reveals how a purely empirical unsophisticated characterisation procedure can mature into a range of advanced methods capable of solving difficult problems.

6.6. CONCLUDING CONSIDERATIONS

Numerous techniques of characterisation have had to be excluded from this historical overview, for simple lack of space. A few are treated elsewhere in the book. Diffraction methods, apart from a few words about neutron diffraction, including the powerful small-angle scattering methods, were left out, partly because X-ray diffraction has also received due attention elsewhere in the book; methods of measuring electrical and magnetic properties have not been discussed; the important techniques of characterising surfaces, such as low-energy electron diffraction, have only been skirted by an outline of scanning tunneling microscopy (but are briefly treated in Chapter 10). Another important technique of compositional profiling in near-surface regions, secondary-ion mass spectrometry (another invention of Raimond Castaing's) has also had to be left out for lack of space. Stereology was treated in Chapter 5, and one particular example of damping measurement among many was also explained in Chapter 5, but the range and importance of damping (internal friction) in materials science has to be taken on trust by the reader.

However, I believe that enough has been described to support my contention that modern methods of characterisation are absolutely central to materials science in its modern incarnation following the quantitative revolution of mid-century. That revolution owed everything to the availability of sensitive and precise techniques of measurement and characterisation.

REFERENCES

Abragam, A. (1987) *De la Physique Avant Toute Chose*, Editions Odile Jacob, Paris.

Adams, B.L., Wright, S.J. and Kunze, K. (1993) *Metall. Trans.* **24A**, 819.

Amelinckx, S. (1992) Electron diffraction and transmission electron microscopy, in *Characterisation of Materials*, ed. Lifshin, E.; *Materials Science and Technology*, vol. 2A, ed. Cahn *et al.* R.W. (VCH, Weinheim) p. 1.

Béranger, G., Duffaut, F., Morlet, J. and Tiers, J.F. (1996) *A Hundred Years after the Discovery of Invar®...the Iron–Nickel Alloys* (Lavoisier Publishing, Paris).

Bever, M.B., Holt, D.L. and Titchener, A.L. (1973) *Prog. Mater Sci.* **17**, 5.

Binnig, G., Rohrer, H., Gerber, C. and Weibel, H. (1982) *Phys. Rev. Lett.* **49**, 57; *Appl. Phys. Lett.* **40**, 178.

Bloch, F. (1946) *Phys. Rev.* **70**, 460.

Bloembergen, N. and Rowland, T.J. (1953) *Acta Metall.* **1**, 731.

Brundle, C.R., Evans, C.A. and Wilson, S. (eds.) (1992) *Encyclopedia of Materials Characterisation* (Butterworth-Heinemann *and* Greenwich: Manning, Boston).

Bunn, C.W. (1945) *Chemical Crystallography: An Introduction to Optical and X-ray Methods* (Clarenson Press, Oxford).

Cahn, R.W. and Lifshin, E. (eds.) (1993) *Concise Encyclopedia of Materials Characterisation* (Pergamon Press, Oxford) p. xxii.

Cahn, R.W., Siemers, P.A., Geiger, J.E. and Bardhan, P. (1987) *Acta Metall.* **35**, 2737.

Carnahan, Jr., C. (1993) Gas and liquid chromatography, in *Concise Encyclopedia of Materials Characterisation*, eds. Cahn, R.W. and Lifshin, E. (Pergamon Press, Oxford) p. 169.

Castaing, R. (1951) Thesis, University of Paris, Application des sondes électroniques à une méthode d'analyse ponctuelle chimique et cristallographique.

Cerezo, A., Hethrington, M.G. and Petford-Long, A.K. (1988) *Rev. Sci. Instr.* **59**, 862.

Cerezo, A. and Smith, G. (1994) Field-ion microscopy and atom probe analysis, in *Characterisation of Materials*, ed. Lifshin, E.; *Materials Science and Technology*, vol. 2B, ed. Cahn, R.W. *et al.* (VCH, Weinheim) p. 513.

Chalmers, B. and Quarrell, A.G. (1941, 1960) *The Physical Examination of Materials*, 1st and 2nd editions (Edward Arnold, London).

Chen, L.C. and Spaepen, F. (1991) *J. Appl. Phys.* **69**, 679.

Clarebrough, L.M., Hargreaves, M.E., Michell, D. and West, G.W. (1952) *Proc. Roy. Soc. (Lond.) A* **215**, 507.

Cockayne, D.J.H., Ray, I.L.E. and Whelan, M.J. (1969) *Phil. Mag.* **20**, 1265.

Crease, R.C. (1999) *Making Physics: A Biography of Brookhaven National Laboratory, 1946–1972*, Chapter 12 (University of Chicago Press, Chicago) p. 316.

DiNardo, N.J. (1994) in *Characterisation of Materials*, ed. Lifshin, E.; *Materials Science and Technology*, vol. 2B, ed. Cahn, R.W. *et al.* (VCH, Weinheim) p. 1.

Duncumb, P. (2000) *Proceedings of Symposium 'Fifty Years of Electron Microprobe Analysis'*, August 1999, *Microscopy and Microanalysis*.

Early, T.A. (2001) Article on Nuclear magnetic resonance in solids, in *Encyclopedia of Materials*, ed. Buschow, K.H.J. *et al.* (Elsevier, Amsterdam).

Eigler, D.M. and Schweizer, E.K. (1990) *Nature* **344**, 524.

Flagel, A.R. (1996) Memoir of Clair C. Patterson, *Nature* **379**, 487.

Forwood, C.T. and Clarebrough, L.M. (1991) *Electron Microscopy of Interfaces in Metals and Alloys* (Adam Hilger, Bristol).

Gallagher, P.K. (1992) in *Characterisation of Materials*, ed. Lifshin, E.; *Materials Science and Technology*, vol. 2A, ed. Cahn, R.W. *et al.* (VCH, Weinheim) p. 491.

Gardner, G.A. and Cahn, R.W. (1966) *J. Mater. Sci.* **1**, 211.

Gonser, U. and Aubertin, F. (eds.) (1993) in *Concise Encyclopedia of Materials Characterisation*, eds. Cahn, R.W. and Lifshin, E. (Pergamon Press, Oxford) p. 259.

Gordon, P. (1955) *Trans. Amer. Inst. Min. (Metall.) Engrs.* **203**, 1043.

Haasen, P. (1985) The early stages of the decomposition of alloys, *Metall. Trans. A* **16**, 1173.

Hamers, R.J., Tromp, R.M. and Demuth, J.E. (1986) *Phys. Rev. Lett.* **56**, 1972.

Heinrich, K.F.J. (1999) *Microscopy and Microanalysis*, p. 517.

Hirsch, P.B. (1980) The beginnings of solid state physics, *Proc. Roy. Soc. (Lond.) A* **371**, 160.

Hirsch, P.B. (1986) *Mater. Sci. Eng.* **84**, 1.

Hirsch, P.B., Nicholson, R.B., Howie, A., Pashley, D.W. and Whelan, M.J. (1965) *Electron Microscopy of Thin Crystals* (Butterworth, London).

Hondros, E.D., Seah, M.P., Hofmann, S. and Lejcek, P. (1996) in *Physical Metallurgy*, vol. 2, eds. Cahn, R.W. and Haasen, P. (North-Holland, Amsterdam) p. 1201.

Hossain, T.Z. (1992) in *Encyclopedia of Materials Characterisation*, eds. Brundle, C.R., Evans, C.A. and Wilson, S. (Butterworth-Heinemann *and* Greenwich: Manning, Boston) p. 671.

Irani, R.S. and Cahn, R.W. (1973) *J. Mater. Sci.* **8**, 1453.

Kong, C.H. and Munroe, P.R. (1994) *Intermetallics*, **2**, 333.

Larson, D.J., Petford-Long, A.K., Cerezo, A. and Smith, G.D.W. (1999) *Acta Mater.* **47**, 4019.

Lawn, B. (1993) *Fracture of Brittle Solids*, 2nd edition (Cambridge University Press, Cambridge) p. 253.

Li, C.-M., Sommer, F. and Mittemeijer, E.J. (2000) *Z. Metallk.* **91**, 5.

Lifshin, E . (1986) Investigation and characterisation of materials, in: *Encyclopedia of Materials Science and Engineering*, vol. 3, ed. Bever, M.B. (Pergamon Press, Oxford) p. 2389.

Lifshin, E. (1994) in *Characterisation of Materials*, ed. Lifshin, E.; *Materials Science and Technology*, vol. 2B, ed. Cahn, R.W. *et al.* (VCH, Weinheim) p. 351.

Melford, D.A. (1960) *Proceedings of Second International Symposium on X-ray Microscopy and X-ray Microanalysis, Stockholm* (Elsevier, Amsterdam) p. 407.

Müller, E.W. (1962) Article on field emission, in *Encyclopaedic Dictionary of Physics*, vol. 3, ed. Thewlis, J. (Pergamon press, Oxford) p. 120.

Müller, E.W. and Tsong, T.T. (1969) *Field-Ion Microscopy: Principles and Applications* (Elsevier, Amsterdam).

Mulvey, T. (1995) Electron-beam instruments, in *20th Century Physics*, vol. 3, ed. Pais, A. *et al.* (Institute of Physics Publishing, and New York: American Institute of Physics Press, Bristol and Philadelphia) p. 1565.

Nastasi, M., Mayer, J.W. and Hirvonen, J.K. (1996) *Ion–Solid Interactions: Fundamentals and Applications* (Cambridge University Press, Cambridge).

Newbury, D.E. (2000) *Proceedings of Symposium 'Fifty Years of Electron Microprobe Analysis', August 1999, Microscopy and Microanalysis* (in press).

Newbury, D.E. and Williams, D.B. (2000) *Acta Mater.* **48**, 323.

Oatley, C.W. (1982) *J. Appl. Phys.* **53**(2), R1.

Oliver, W.C. (1993) in *Concise Encyclopedia of Materials Characterisation*, eds. Cahn, R.W. and Lifshin, E. (Pergamon Press, Oxford) p. 232.

Phillips, V.A. (1971) *Modern Metallographic Techniques and Their Applications* (Wiley-Interscience, New York).

Purcell, E.M., Torrey, H.G. and Pound, R.V. (1946) *Phys. Rev.* **69**, 37.

Purcell, F. (1993) in *Concise Encyclopedia of Materials Characterisation*, ed. Cahn, R.W. and Lifshin, E. (Pergamon Press, Oxford) p. 403.

Quinney, H. and Taylor, G.I. (1937) *Proc. Roy. Soc.* (*Lond.*) *A* **163**, 157.

Rowland, T.J. and Fradin, F.Y. (1969) *Phys. Rev.* **182**, 760.

Ruska, E. (1980) *The Early Development of Electron Lenses and Electron Microscopy* (Hirzel, Stuttgart).

Schatz, G. and Weidinger, A. (1996) *Nuclear Condensed Matter Physics: Nuclear Methods and Applications* (Wiley, Chichester).

Seeger, A. (1973) *J. Phys. F* **3**, 248.

Sibilia, J.P. (1988) *A Guide to Materials Characterisation and Chemical Analysis* (VCH Publishers, New York).

Skelly F.E.M. and Keliher, P.N. (1992) in *Characterisation of Materials*, ed. Lifshin, E.; *Materials Science and Technology*, vol. 2A, ed. Cahn, R.W. *et al.* (VCH, Weinheim) p. 423.

Smith, C.S. (1960) *A History of Metallography* (Chicago University Press, Chicago) pp. 91, 167.

Smith, M.E. (1983) *Appl. Mag. Reson.* **4**, 1.

Spence, J.C.H. (1999) The future of atomic-resolution electron microscopy, *Mat. Sci. Eng.* **R26**, 1.

Stewart, A.D.G. and Thompson, M.W. (1969) *J. Mater. Sci.* **4**, 56.

Sykes, C. (1935) *Proc. Roy. Soc.* (*Lond.*) **145**, 422.

Tabor, D. (1951) *The Hardness of Metals* (Clarendon Press, Oxford) (Recently reissued).

Telle, R. and Petzow, G. (1992) in *Characterisation of Materials*, ed. Lifshin, E.; *Materials Science and Technology*, vol. 2A, ed. Cahn, R.W. *et al.* (VCH, Weinheim) p. 358.

Titchener, A.L. and Bever, M.B. (1958) *Prog. Metal Phys.* **7**, 247.

Turner, J.N. and Szarowski, D.H. (1993) in *Concise Encyclopedia of Materials Characterisation*, eds. Cahn, R.W. and Lifshin, E. (Pergamon Press, Oxford) p. 68.

Unwin, D.J. (1990) in *Physicists Look Back: Studies in the History of Physics*, ed. Roche, J. (Adam Hilger, Bristol) p. 237.

Van Schilfgaarde, M., Abrikosov, I.A. and Johansson, B. (1999) *Nature* **400**, 46.

Watson, E.S., O'Neill, M.J., Justin, J. and Brenner, N. (1964) *Anal. Chem.* **326**, 1233.

Willis, B.T.M. (1998) *Acta Cryst.* **A54**, 914.

Chapter 7
Functional Materials

Chapter 7
Functional Materials

7.1. INTRODUCTION

A major distinction has progressively emerged in materials science and engineering, between *structural materials* and *functional materials*. Structural materials are selected for their load-bearing capacity, functional materials for the nature of their response to electrical, magnetic, optical or chemical stimuli; sometimes a functional material is even chosen for aesthetic reasons. It is much harder to define a functional material accurately than it is to distinguish a structural material. For present purposes, I have decided to include functional ceramics in this chapter. Those ceramics developed for their resistance to mechanical stress or simply their beauty are discussed in Chapter 9. In what follows, I make no attempt to present a comprehensive account of the huge field of functional materials, but instead aim to pick a few historically important aspects and focus on these.

7.2. ELECTRICAL MATERIALS

7.2.1 Semiconductors

Silicon is today the most studied of all materials, with probably a larger accumulated number of scientific papers devoted to its properties than for any other substance. It is *the* archetype of a semiconductor and everybody knows about its transcendent importance in modern technology.

Things looked very different, as little as 70 years ago. The term 'semiconductor' appears to have been used first by Alexander Volta, after whom the volt is named, in a paper to the Royal Society of London in 1782. According to a fine historical overview of early work on semiconductors (Busch 1993), Volta examined the rate at which a charged electrometer would discharge when its contact knob was touched by different substances connected to earth. In this way, Volta was able to distinguish between metals, insulators and (in the middle) semiconductors. In 1840, Humphry Davy was the first to establish clearly that metals become poorer conductors as the temperature is raised, and he was soon followed by Michael Faraday who examined a great range of compounds, many of which changed with temperature in a way opposite to metals. Faraday's researches culminated with silver sulphide which showed an abrupt change in conductivity at about 175°C. This finding moved a German, Johann Hittorf, in 1851 to take up the study of silver sulphide, Ag_2S, as well as Cu_2S. These were the first true semiconductors to be carefully examined. Copper sulphide came to be used for rectification of alternating current in the 1920s and 1930s, in the form of thin films created on copper sheets by an appropriate gas-phase reaction; this application began in empirical mode at a time when the mode of conduction in semiconductors was not yet understood. Another empirical success was the use of selenium photocells in photographic exposure meters; I used one of these as a schoolboy. Many hundreds of research papers were devoted to silver sulphide during the century after Hittorf: it gradually became clear that, in Busch's words, "the results are very sensitive to the preparation of the specimens, essentially purity and deviations from stoichiometry". Eventually, samples were zone-refined under a controlled pressure of sulphur vapour, and Junod in Switzerland (Junod 1959) obtained the results shown in Figure 7.1 for carefully characterised samples. The discontinuity at 170°C represents a semiconductor/metal transition, a rare

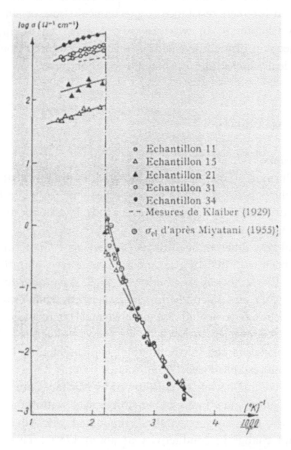

Figure 7.1. Electrical conductivity of Ag_2S as a function of temperature (after Junod, 1959).

means of comparing the characteristics of metals and semiconductors in the same substance.

The great difficulty in obtaining reproducible results with Ag_2S reinforced the prejudices of the many physicists who, in the 1930s particularly, saw no point in spending time with semiconductors. The great quantum physicist Wolfgang Pauli, whose dismissive views on the merits of solid-state physics generally have already been quoted (Section 3.3.1), was specifically contemptuous of semiconductors in a 1931 letter to his student, Rudolf Peierls: "One shouldn't work with semiconductors, that is just a mess (eine Schweinerei); who knows whether semiconductors exist at all". In 1930, B. Gudden, a physicist at Göttingen, reviewed what was known experimentally about semiconductors at that time, and concluded that only impure substances could be semiconductors... i.e., he denied the possibility of intrinsic semiconduction. Busch, in his cited historical article on semiconductors, tells how in 1938, in Switzerland, when he began an investigation of semiconduction in silicon carbide, his friends warned him that working on semiconductors meant scientific suicide. Soon after, when he gave a seminar on his researches, a colleague told him: "What are semiconductors good for? They are good for nothing. They are erratic and not reproducible".

It is no exaggeration to claim that it was the extensive worldwide body of research on semiconductors from the late 1930s onwards that converted physicists to the recognition that scrupulous control of purity, stoichiometry and crystal perfection, together with characterisation methods that could check on these features, are a precondition of understanding the nature of semiconductors and thus also a precondition of exploiting them successfully – indeed, not only semiconductors but, by extension, many kinds of materials.

At about the same time as Gudden and Pauli expressed their sceptical views, the theoretical physicist Alan Wilson of Cambridge (visiting Heisenberg at the time) wrote two classic papers on the band theory of semiconductors (Wilson 1931) and for the first time distinguished between extrinsic and intrinsic semiconductors: he postulated the presence of donors and acceptors. His theory explained clearly why the carrier density goes up sharply with temperature, so that semiconductors, unlike metals, conduct better as they heat up. Heisenberg at about the same time showed that 'holes' are equivalent to positively charged carriers. Wilson's papers marked the true beginning of the modern approach to semiconductors. Much later (Wilson 1980), after he had left science and become a highly successful captain of industry, Wilson reminisced about his glory days in solid-state theory, under title "Opportunities missed and opportunities seized"; he remarked that when he returned to Cambridge from his visit to Germany, he suggested that germanium might be an interesting substance to study in detail, but that in response the "silence was deafening".

Wilson (1939) wrote the first textbook on semiconductors. Sondheimer (1999) in his obituary of Wilson (1906–1995), remarks that "Wilson's aim in this book was to give a clear and simplified, though non-superficial, account of the subject, and Wilson's books were influential in making available to a wide range of physicists, metallurgists and engineers the advances made in solid-state physics during the 1930s."

7.2.1.1 Silicon and germanium. The study of silicon, patchy though it was, began well before the crucial events of 1948 that led to the invention of the transistor. Recently, Frederick Seitz and Norman Einspruch in America have undertaken an extensive programme of historical research on the "tangled prelude to the age of silicon electronics" and published their findings (Seitz 1996, Seitz and Einspruch 1998).

Silicon, very impure, was used by steel metallurgists as well as aluminium-alloy metallurgists early in the 20th century: in particular, iron–silicon alloy sheet was and still is used for transformer laminations, because of its excellent magnetic properties, absence of phase transformation and low cost. As a semiconductor, however, silicon first became visible in 1906, with an American inventor called G.W. Pickard who was concerned to find a good detector for messages sent by the new wireless telegraphy. He experimented extensively with cat's whisker/crystal combinations, in which a fine metallic wire (the cat's whisker) was pressed against a crystal to achieve rectification of radio-frequency signals. According to Seitz, Pickard tried more than 30,000 combinations: crystals included galena, silicon carbide and metallurgical-grade silicon (this last obtained commercially from Westinghouse, presumably in coarse polycrystal form). Pickard was granted a patent for the use of silicon in this way, in 1906. Many amateurs in the early days played with such detectors; it was common knowledge that particular sites on a crystal surface ('hot spots') performed much better than others. In the same year as Pickard's patent, Lee DeForest invented the vacuum triode tube, and when this had been sufficiently improved, it (and its successor, the pentode tube) displaced the short-lived semiconducting detector.

Nothing much happened on the silicon front until early in the Second World War, when intensive research on radar got under way. The detection problem here was that ultra-high-frequency waves had to be used, and vacuum tubes were not appropriate for rectifying these. A German, Hans Hollmann, had published a precocious book on high-frequency techniques just before the War, in which he concluded that cat's whisker/crystal devices were essential for rectification in this frequency range. Another German, Jürgen Rottgard, in 1938 followed these findings up and concluded that a silicon/tungsten combination was best. Fortunately for the Allies, the German high command was not interested, and the German work was read, by chance, by early British radar researchers, Denis Robertson in particular.

He persuaded another physicist, Herbert Skinner, to try this out: the result, in July 1940, was a silicon/tungsten rectifying device, sealed in glass and covered by a vibration-damping liquid, the silicon (still metallurgical grade) brazed on to a tungsten rod which could be sealed into the glass tube. A little earlier (see below) American physicists independently developed a very similar device. Soon after, the Dupont Company in America found a process for making much purer silicon, and thus began the Anglo-American cooperation which quickly led to effective radar measures.

This wartime work persuaded many scientists to take silicon seriously as a *useful* electrical material, and this cleared the way, intellectually, for the researches at Bell Laboratories in New Jersey which began in earnest around the end of the War. The motive energy came from Mervin Kelly, a Ph.D. physicist who became the director of research at Bell Laboratories in 1936. For his doctorate, he worked with Millikan on his renowned oil-drop technique for measuring the charge on the electron and became convinced of the importance of basic research; he joined Bell Labs in the 1920s. When the hiring freeze occasioned by the Great Depression which had begun in 1929 at length eased in 1936, Kelly engaged as his first appointment a physicist, William Shockley, who 11 years later formed part of the triumvirate who invented the transistor. Shockley recalled Kelly's remarks about his long-range goals, soon after Shockley had taken up his post: "Instead of using mechanical devices such as relays, which caused annoying maintenance problems, telephone switching should be done electronically. Kelly stressed the importance of this goal so vividly that it made an indelible impression on me." So Kelly had already formulated his central strategy more than a decade before its eventual resolution. In 1938, he reorganised physical research at Bell Labs, and moved Shockley together with a metallurgist and another physicist into a new, independent group focusing on the physics of the solid state. Kelly insisted that his scientists focused firmly on fundamentals and left the development of their findings to others. As an industrial strategy, this was truly revolutionary.

This information comes from a quite remarkable book, *Crystal Fire: The Birth of the Information Age* (Riordan and Hoddeson 1997), which maps out systematically but very accessibly the events that led to the discovery of the transistor and the aftermath of that episode. I know of no better account of the interlocking sequence of events that led to eventual success, and of the personal characteristics of the principal participants that played such a great part in the story.

Two Bell Labs employees, Russell Ohl and George Southworth, were trying in the late 1930s to detect ultrahigh-frequency radio waves with vacuum tubes, and like Skinner on the other side of the Atlantic, had no success. So, Southworth, a radio ham since childhood, remembered his early silicon-and-cat's-whisker devices and managed to retrieve some old ones from a secondhand radio store. Just as they did for Skinner a little later, the devices worked. Thereupon Ohl and Southworth did systematic comparisons (not knowing about Pickard's similar work in 1906!) and once more silicon came up as the preferred choice; again, preferred hot spots had to be located. The objective now became to understand the nature of these hot spots. At this point, metallurgists began to be involved: Jack Scaff and Henry Theuerer in 1939 grew silicon crystals and, collaborating with Ohl and Southworth, they eventually found boundaries in an ingot on either side of which behaviour was different; the sense of rectification was opposite. The metallurgists named these regions *n-type* and *p-type*. Next, Ohl found that a piece of silicon containing *both* types of region showed an enormous electrical reaction to irradiation with light. The 'p–n barrier' allowed silicon to act as a self-rectifier. The metallurgists undertook painstaking experiments to identify the impurity elements that determined whether a locality was n-type and p-type; they mostly turned out to be in groups 5 and 3 of the periodic table. This research was indispensable to the discovery of transistor action somewhat later.

Ohl demonstrated his results to Kelly early in 1940; Kelly felt that his instincts had been proved justified. Thereupon, Bell Labs had to focus single-mindedly on radar and on silicon rectifiers for this purpose. It was not till 1945 that basic research restarted. This was the year that the theorist John Bardeen was recruited, and he in due course became inseparable from Walter Brattain, an older man and a fine experimenter who had been with Bell since the late 1920s. William Shockley formed the third member of the triumvirate, though from an early stage he and Bardeen found themselves so mutually antagonistic that Bardeen was sometimes close to resignation. But tension can be productive as well as depressing.

Some Bell employees had been trying to focus attention on germanium, the sister element to silicon, but had been discouraged, just as Wilson had been discouraged across the Atlantic a decade earlier. It was Bardeen who eventually focused on the merits of germanium and the research which led to the transistor in 1947 was centred on this element; the fact that on germanium, surface oxide could be washed off with water played a major role.

By 1947, Bardeen, building on some experiments by Shockley, had been driven to the concept of 'surface states' in semiconductors (a notable theoretical breakthrough) – an inversion from n- to p-type close to a semiconductor surface created by a strong external field – and this concept, together with the understanding of p/n junctions, came together in the systematic researches that later in that year led to the transistor, an amplifying device that mimicked the function of a vacuum valve but on a small scale and which worked at the high frequencies at which the valves were useless. (A transistor incorporates several p/n junctions and electrodes.) The original transistor, depending on a point contact rather like a cat's whisker device, was essentially the work of Bardeen and Brattain; Shockley then developed its descendant, the junction transistor which avoided the difficult-to-fabricate point contacts and paved the way for mass production. Figure 7.2 shows the essentials of (any) transistor in purely schematic form.

John Bardeen was a truly remarkable scientist, and also a very private, taciturn man. Pippard (1995) in his obituary of Bardeen, recounts how "John returned home from the Bell Labs and walked into the kitchen to say 'We discovered something today; to (his wife) Jane's regret, all she could find in reply to what proved a momentous statement was 'That's interesting, but I have to get dinner on the table'."

A detailed account of the steps that led to the first transistor, and the steps soon afterwards to improve and miniaturise the device, and to shift from germanium to silicon, would take too much space in this chapter, and the reader must be referred to Riordan and Hoddeson's systematic and rivetting account, though space will be found for a brief account of the subsequent birth of the integrated circuit, the vector of the information age. But before this, some remarks are in order about the crucial interplay of physics and metallurgy in the run-up to the transistor.

7.2.1.2 Physicists, chemists and metallurgists cooperate.

The original invention of the transistor was undoubtedly a physicists' triumph; in particular, John Bardeen's profound capacity for physical insight took the team through their many setbacks and mystifications. The later stages of improvement of semiconducting devices (not only the many kinds of transistors, but light-emitting diodes, photocells and in due course computer memories) remained the province of a kind of physicists' elite. One of the urgent tasks was to find the details of the structure of the electronic energy bands in semiconductors, a task involving theoretical as well as experimental skills. A good impression of the early days of this quest can be garnered from a discursive

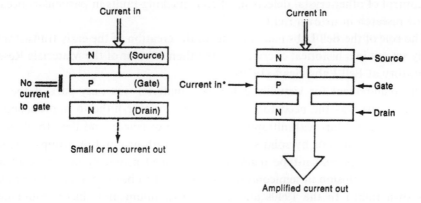

Figure 7.2. Essentials of transistor action.

retrospect by Herman (1984). Bell Labs also had some 'gate-keepers', physicists with encyclopedic solid-state knowledge who could direct researchers in promising new directions: the prince among these was Conyers Herring, characterised by Herman as a "virtual encyclopedia of solid-state knowledge". Herring, not long ago (Herring 1991) wrote an encyclopedia entry on 'Solid State Physics'... an almost but not quite impossible task.

However, physicists alone could never have produced a reliable, mass-produc-able transistor. We have seen that in the run-up to the events of 1947, Scaff and Theuerer had identified p- and n-regions and performed the delicate chemical analyses that enabled their nature to be identified. There was much more to come. The original transistor was successfully made with a slice of germanium cut out of a polycrystal, and early pressure to try single crystals was rebuffed by management. One Bell Labs chemist, Gordon Teal, a natural loner, pursued his obsession with single crystals in secret until at last he was given modest backing by his manager; eventually the preferred method of crystal growth came to be that based on Czochralski's method (Section 4.2.1). It soon became clear that for both germanium and silicon, this was the essential way forward, especially because intercrystalline boundaries proved to be 'electrically active'. It also became clear that dislocations were likewise electrically active and interfered with transistor action, and after a while it transpired that the best way of removing dislocations was by carefully controlled single crystal growth; to simplify, the geometry of the crystal was so arranged that dislocations initially present 'grew out' laterally, leaving a crystal with fewer than 100 dislocation lines per square centimetre, contrasted with a million times that number in ordinary material. This was the achievement of Dash (1958, 1959), whom we have already met in relation to Figure 3.14, an early confirmation of the reality of dislocations. Indeed, the work done at Bell Labs led to some of the earliest demonstrations of the existence of these disputed defects. Later, the study and control of other crystal defects in silicon, stacking-faults in particular, became a field of research in its own right.

The role of the Bell Labs metallurgists in the creation of the early transistors was clearly set out in a historical overview by the then director of the Materials Research Laboratory at Bell Labs, Scaff (1970).

The requirement for virtually defect-free material was only part of the story. The other part was the need for levels of purity never hitherto approached. The procedure was to start with ultrapure germanium or silicon and then to 'dope' that material, by solution or by solid-state diffusion, with group-3 or group-5 elements, to generate p-type and n-type regions of controlled geometry and concentration. (The study of diffusion in semiconductors was fated to become a major parepisteme in its own right.) In the 1940s and 1950s, germanium and silicon could not be extracted and refined with the requisite degree of purity from their ores. The solution was zone-refining, the invention of a remarkable Bell Labs employee, William Pfann.

Pfann has verbally described what led up to his invention, and his account is preserved in the Bell Laboratory archives. As a youth, he was engaged by Bell Laboratories as a humble laboratory assistant, beginning with duties such as polishing samples and developing films. He attended evening classes and finally earned a bachelor's degree (in chemical engineering). He records attending a talk by a famous physical metallurgist of the day, Champion Mathewson, who spoke about plastic flow and crystal glide. Like Rosenhain before him, the youthful Pfann was captivated. Then, while still an assistant, he was invited by his manager, E.E. Schumacher, in the best Bell Labs tradition, to "take half your time and do whatever you want". Astonished, he remembered Mathewson and chose to study the deformation of lead crystals doped with antimony (as used by the Bell System for cable sheaths). He wanted to make crystals of uniform composition, and promptly invented zone-levelling. (He "took it for granted that this idea was obvious to everyone, but was wrong".) Pfann apparently impressed the Bell Director of Research by another piece of technical originality, and was made a full-fledged member of technical staff, though innocent of a doctorate. When William Shockley complained that the available germanium was nothing like pure enough, Pfann, in his own words, "put my feet up on my desk and tilted my chair back to the window sill for a short nap, a habit then well established. I had scarcely dozed off when I suddenly awoke, brought the chair down with a clack I still remember, and realised that a series of molten zones, passed through the ingot of germanium, would achieve the aim of repeated fractional crystallisation." Each zone swept some impurity along with it, until dissolved impurities near one end of the rod are reduced to a level of one in hundreds of millions of atoms. Pfann described his technique, and its mathematical theory, in a paper (Pfann 1954) and later in a book (Pfann 1958, 1966). Incidentally, the invention and perfection of zone-refining was one of the factors that turned solidification and casting from a descriptive craft into a quantitative science.

Today, methods of refining silicon via a gaseous intermediary compound have improved so much that zone-refining is no longer needed, and indeed crystal diameters are now so large that zone-refining would probably be impossible. Present-day chemical methods of preparation of silicon allow impurity levels of one part in 10^{12} to be reproducibly attained. Modern textbooks on semiconductors no longer mention zone-refining; but for more than a decade, zone-refining was an essential factor in the manufacture of transistors.

In the early years, physicists, metallurgists and chemists each formed their own community at Bell Labs, but the experience of collaboration in creating semiconductor devices progressively merged them and nowadays many of the laboratory's employees would rate themselves simply as materials scientists.

7.2.1.3 (Monolithic) integrated circuits. Mervin Kelly had told William Shockley, when he joined Bell Labs in 1936, that his objective was to replace metallic reed relays by electronic switches, because of the unreliability of the former. History repeats itself: by the late 1950s, electronic circuits incorporating discrete transistors (which had swept vacuum tubes away) had become so complex that a few of the large numbers of soldered joints were apt to be defective and eventually break down. Unreliability had arrived all over again. Computers had the most complex circuits: the earliest ones had used tubes and these were apt to burn out. Not only that, but these early computers also used metal relays which sometimes broke down; the term 'bug' still used today by computer programmers originates, some say but others deny, in a moth which had got caught in a relay and impeded its operation. (The distinguished moth is still rumored to be preserved in a glass case.) Now that transistors were used instead, unreliability centred on faulty connections.

In 1958–1959, two American inventors, Jack Kilby and Robert Noyce, men cast in the mould of Edison, independently found a way around this problem. Kilby had joined the new firm of Texas Instruments, Noyce was an employee of another young company, Fairchild Electronics, which William Shockley had founded when he resigned from Bell but mismanaged so badly that his staff grew mutinous: Noyce set up a new company to exploit his ideas. The idea was to create a complete circuit on a single small slice of silicon crystal (a 'chip'), with tiny transistors and condensers fabricated in situ and with metallic interconnects formed on the surface of the chip. The idea worked at once, and triumphantly. Greatly improved reliability was the initial objective, but it soon became clear that further benefits flowed from miniaturisation: (1) low power requirements and very small output of waste heat (which needs to be removed); (2) the ability to accommodate complex circuitry, for instance, for microprocessors or computer memories, in tiny volumes, which was vital for the computers in the Apollo moonlanding project (Figure 7.3); and, most important of all, (3) low circuit costs. Ever since Kilby's and Noyce's original chips, the density of devices in integrated circuits has steadily increased, year by year, and the process has still not reached its limit. The story of the invention and early development of integrated circuits has been well told in a book by Reid (1984). Some of the relatively primitive techniques used in the early days of integrated circuits are described in a fascinating review which covers many materials aspects of electronics and communications, by Baker (1967) who at the time was vice-president for research of Bell Laboratories. Kilby has at last (2000) been awarded a Nobel Prize.

The production of integrated circuits has, in the 40 years since their invention, become the most complex and expensive manufacturing procedure ever; it even leaves the production of airliners in the shade. One circuit requires a sequence of several dozen manufacturing steps, with positioning of successive optically defined layers accurate to a fraction of a micrometer, all interconnected electrically, and

Figure 7.3. The evolution of electronics: a vacuum tube, a discrete transistor in its protective package, and a 150 mm (diameter) silicon wafer patterned with hundreds of integrated circuit chips. *Each* chip, about 1 cm^2 in area, contains over one million transistors, 0.35 μm in size (courtesy M.L. Green, Bell Laboratories/Lucent Technologies).

involving a range of sophisticated chemical procedures and automated inspection at each stage, under conditions of unprecedented cleanliness to keep the smallest dust particles at bay. Epitaxial deposition (ensuring that the crystal lattice of a deposited film continues that of the substrate), etching, oxidation, photoresist deposition to form a mask to shape the distribution of the ensuing layer, localised and differential diffusion of dopants or ion implantation as an alternative, all form major parepistemes in this technology and all involve materials scientists' skills. The costs of setting up a factory for making microcircuits, a 'foundry' as it is called today, are in billions of dollars and steadily rising, and yet the cost of integrated circuits *per transistor* is steadily coming down. According to Paul (2000), current microprocessors (the name of a functional integrated circuit) contain around 11 million transistors, at a cost of 0.003 (US) cents each. The low costs of complex circuits have made the information age possible – it is as simple as that.

The advent of the integrated circuit and its foundry has now firmly integrated materials scientists into modern electronics, their function both to optimise production processes and to resolve problems. To cite just one example, many materials scientists have worked on the problem of *electromigration* in the thin metallic conductors built into integrated circuits, a process which eventually leads to short circuits and circuit breakdown. At high current densities, migrating electrons in a potential gradient exert a mechanical force on metal ions and propel them towards the anode. The solution of the problem involves, in part, appropriate alloying of the aluminium leads, and control of microstructure – this is a matter of controlling the size and shape of crystal grains and their preferred orientation, or texture. Some early papers show the scope of this use of materials science (Attardi and Rosenberg 1970, Ames *et al.* 1970). The research on electromigration in aluminium may soon be outdated, because recently, the introduction of really effective diffusion barriers between silicon and metallisation, such as tungsten nitride, have made possible the replacement of aluminum by copper conductors (Anon. 1998). Since copper is the better conductor, that means less heat output and that in turn permits higher 'clock speeds'... i.e., a faster computer. I am typing this passage on a Macintosh computer of the kind that has a novel chip based on copper conductors.

All kinds of materials science research has to go into avoiding disastrous degradation in microcircuits. Thus in multilayer metallisation structures, polymer films, temperature-resistant polyimides in particular, are increasingly replacing ceramics. One worry here is the diffusion of copper through a polymer film into silicon. Accordingly, the diffusion of metals through polymers has become a substantial field of research (Faupel *et al.* 1998), and it has been established that noble metals (including copper) diffuse very slowly, apparently because of metal-atom-induced crosslinking of polymer chains. MSE fields which were totally distinct are coming to be connected, under the impetus of microcircuit technology.

Recent texts have assembled impressive information about the production, characterisation and properties of semiconductor devices, including integrated circuits, using not only silicon but also the various compound semiconductors such as GaAs which there is no room to detail here. The reader is referred to excellent treatments by Bachmann (1995), Jackson (1996) and particularly by Mahajan and Sree Harsha (1999). In particular, the considerable complexities of epitaxial growth techniques – a major parepisteme in modern materials science – are set out in Chapter 6 of Bachmann's book and in Chapter 6 of that by Mahajan and Sree Harsha.

An attempt to forecast the further shrinkage of integrated circuits has been made by Gleason (2000). He starts out with some up-to-date statistics: during the past 25 years, the number of transistors per unit area of silicon has increased by a factor of 250, and the density of circuits is now such that 20,000 cells (each with a transistor and capacitor) would fit within the cross-section of a human hair. This kind of relentless shrinkage of circuits, following an exponential time law, is known as Moore's law (Moore was one of the early captains of this industry). The question is whether the operation of Moore's Law will continue for some years yet: Gleason says that "attempts to forecast an end to the validity of Moore's Law have failed dismally; it has continued to hold well beyond expectations". The problems at present are largely optical: the resolving power of the projection optics used to transfer a mask to a circuit-to-be (currently costing about a million dollars per instrument) is the current limit. Enormous amounts of research effort are going into the use of novel small-wavelength lasers such as argon fluoride lasers (which need calcium fluoride lenses) and, beyond that, the use of electrons instead of photons. The engineers in latter-day foundries balk at no challenge.

7.2.1.4 Band gap engineering: confined heterostructures.

7.2.1.4 Band gap engineering: confined heterostructures. When the thickness of a crystalline film is comparable with the de Broglie wavelength, the conduction and valence bands will break into subbands and as the thickness increases, the Fermi energy of the electrons oscillates. This leads to the so-called quantum size effects, which had been precociously predicted in Russia by Lifshitz and Kosevich (1953). A piece of semiconductor which is very small in one, two or three dimensions – a *confined structure* – is called a quantum well, quantum wire or quantum dot, respectively, and much fundamental physics research has been devoted to these in the last two decades. However, the world of MSE only became involved when several quantum wells were combined into what is now termed a heterostructure.

A new chapter in the uses of semiconductors arrived with a theoretical paper by two physicists working at IBM's research laboratory in New York State, L. Esaki (a Japanese immigrant who has since returned to Japan) and R. Tsu (Esaki and Tsu 1970). They predicted that in a fine multilayer structure of two distinct semiconductors (or of a semiconductor and an insulator) tunnelling between quantum wells becomes important and a 'superlattice' with minibands and mini (energy) gaps is formed. Three years later, Esaki and Tsu proved their concept experimentally. Another name used for such a superlattice is 'confined heterostructure'. This concept was to prove so fruitful in the emerging field of optoelectronics (the merging of optics with electronics) that a Nobel Prize followed in due course. The central application of these superlattices eventually turned out to be a *tunable laser*.

The optical laser, a device for the generation of coherent, virtually single-wavelength and highly directional light, was first created by Charles Townes in 1960, and then consisted essentially of a rod of doped synthetic ruby with highly parallel mirrors at each end, together with a light source used to 'pump up' the rod till it discharges in a rapid flash of light. At roughly the same time, the light-emitting semiconductor diode was invented and that, in turn, was metamorphosed in 1963 into a semiconductor laser (the Russian Zhores Alferov was the first to patent such a device), using a p–n junction in GaAs and fitted with mirrors: one of its more familiar applications is as the light source for playing compact discs. Its limitation was that the emitted wavelength was defined by the semiconductor used and some colours, especially in the green–blue region, were not accessible. Also, the early semiconductor lasers were unstable, and quickly lost their luminosity. This is where confined heterostructures came in, and with them, the concept of *band gap engineering*. Alferov received a Nobel Prize in Physics in 2000.

To make a confined heterostructure it is necessary to deposit very thin and uniform layers, each required to be in epitaxy with its predecessor, to a precise specification as to successive thicknesses. This is best done with the technique of molecular beam epitaxy (MBE), in which beams from evaporating sources are allowed to deposit on a substrate held in ultrahigh vacuum, using computer-controlled shutters in conjunction with in situ glancing-angle electron diffraction to monitor the layers as they are deposited. MBE is an archetypal example of the kinds of high-technology processing techniques required for modern electronics and optoelectronics. MBE was introduced soon after Esaki and Tsu's pathbreaking proposal, and taken to a high pitch of perfection by A.Y. Cho and F. Capasso at Bell Laboratories and elsewhere (it is used to manufacture most of the semiconductor lasers that go into compact-disc players). R. Kazarinov in Russia in 1971 had built on Esaki and Tsu's theory by suggesting that superlattices could be used to make tunable lasers: in effect, electrons would tunnel from quantum well to quantum well, emitting photons of a wavelength that corresponded to the energy loss in each jump. In 1994, J. Faist, a young physicist, worked out a theoretical 'prescription' for a quantum cascade laser consisting of some 500 layers of varying thickness, consisting of a range of compound semiconductors like GaInAs and AlInAs. Figure 7.4 shows what such a succession of precision-deposited layers looks like, some only 3

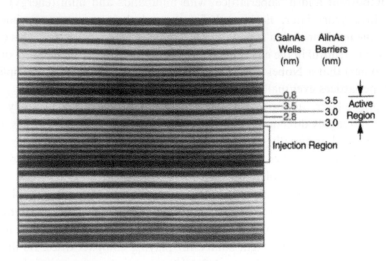

Figure 7.4. Electron micrograph of the cross-section of a quantum cascade semiconductor laser (after Cho 1995).

atoms across. The device produced light of a wavelength not hitherto accessible and of very high brightness. At about the same time, the Bell Labs team produced, by MBE, an avalanche photodiode made with compound semiconductors, required as a sensitive light detector associated with an optical amplifier for 'repeaters' in optical glass-fibre communications. The materials engineering of the glass fibres themselves is outlined later in this chapter. Yet another line of development in band gap engineering is the production of silicon–germanium heterostructures (Whall and Parker 1995) which promise to achieve with the two elementary semiconductors properties hitherto associated only with the more expensive compound semiconductors.

The apotheosis of the line of research just outlined was the development of very bright, blue or green, semiconductor lasers based on heterostructures made of compounds of the group III/nitride type (GaN, InN, AlN or ternary compounds). These have provided wavelengths not previously accessible with other semiconductors, and lasers so bright and long lived that their use as traffic lights is now well under way. Not only are they bright and long lived but the cost of operation per unit of light emitted is only about a tenth that of filament lamps; their lifetime is in fact about 100 times greater (typically, 100,000 h). In conjunction with a suitable phosphor, these devices can produce such bright *white* light that its use for domestic lighting is on the horizon. The opinion is widely shared that gallium nitride, GaN and its "alloys" are the most important semiconductors since silicon, and that light from such sources is about to generate a profound technological revolution. The pioneering work was done by Shuji Nakamura, an inspired Japanese researcher (Nakamura 1996) and by the following year, progress had been so rapid that a review paper was already required (Ponce and Bour 1997). This is characteristic of the speed of advance in this field.

Another line of advance is in the design of semiconductor lasers that emit light at right angle to the heterostructure layers. A remarkable example of such a device, also developed in Japan in 1996, is shown schematically in Figure 7.5. The active region consists of quantum dots (constrained regions small in all three dimensions), spontaneously arranged in a lattice when thin layers break up under the influence of strain. The regions labelled 'DBR' are AlAs/GaAs multilayers so arranged as to act as Bragg reflectors, effectively mirrors, of the laser light. A paper describing this device (Fasor 1997) is headed "Fast, Cheap and Very Bright".

Lasers are not only made *of* semiconductors; old-fashioned pulsed ruby lasers have also been used for some years as production tools to 'heal' lattice damage caused in crystalline semiconductors by the injection ('implantation' is the preferred term) of dopant ions accelerated to a high kinetic energy. This process of pulsed laser annealing has given rise to a fierce controversy as to the mechanism of this healing (which can be achieved without significantly displacing the implanted dopant

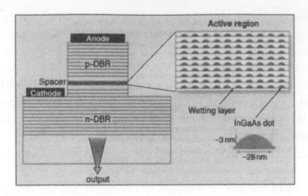

Figure 7.5. Quantum-dot vertical-cavity surface-emitting semiconductor laser, with an active layer consisting of self-assembled $In_{0.5}GaAs_{0.5}$ quantum dots (Fasor 1997).

atoms). The details of the controversy are too complex to go into here, but for many years the Materials Research Society organised annual symposia in an attempt to settle the dispute, which has died down now. For an outline of the points at issue, see Boyd (1985) and a later, comprehensive survey of the issues (Fair 1993).

These brief examples of developments in semiconductor technology and optoelectronics are offered to give the flavour of recent semiconductor research. An accessible technical account of MBE and its triumphs can be found in an overview by Cho (1995), while a more impressionistic but very vivid account of Capasso and his researches at Bell Labs is in a popular book by Amato (1997). A very extensive historical survey of the enormous advances in "optical and optoelectronic physics", with attention to the materials involved, is in a book chapter by Brown and Pike (1995).

The foregoing has only hinted at the great variety of semiconductor devices developed over the past century. A good way to find out more is to look at a selection of 141 of the most important research papers on semiconductor devices, some dating right back to the early years of this century (Sze 1991). A good deal of semiconductor research, even today, is still of the parepistemic variety, aimed at a deeper understanding of the complex physics of this whole group of substances. A good example is the recent research on "isotopically engineered" semiconductors, reviewed by Haller (1995). This began with the study of isotopically enriched diamond, in which the small proportion ($\approx 1.1\%$) of C^{13} is removed to leave almost pure C^{12}, and this results in a $\approx 150\%$ increase of thermal conductivity, because of the reduction in phonon scattering; this was at once applied in the production of synthetically grown isotopically enriched diamond for heat sinks attached to electronic devices. Isotopic engineering was next applied to germanium, and methods were developed to use Ge heterostructures with two distinct stable isotopes as a specially reliable means of measuring self-diffusivity. Haller is of the opinion that a range of isotopically engineered devices will follow. A related claim is that using gaseous deuterium (heavy hydrogen) instead of normal hydrogen to neutralise dangerous dangling bonds at the interface between silicon and silicon oxide greatly reduces the likelihood of circuit failure, because deuterium is held more firmly (Glanz 1996).

A word is in order, finally, about the position of silicon relative to the compound semiconductors. Silicon still, in 2000, accounts for some 98% of the global semiconductor market: low manufacturing cost is the chief reason, added to which the properties of silicon dioxide and silicon nitride, in situ insulating layers, are likewise important (Paul 2000). According to Paul, in the continuing rivalry between silicon and the compound semiconductors, alloying of silicon with germanium is tilting the odds further in favour of silicon. Kasper *et al.* (1975) were the first to make high-quality Si–Ge films, by molecular-beam epitaxy, in the form of a strained-layer superlattice. This approach allows modification of the band gap energy of silicon and allows the engineer to "design many exotic structures". One feature of this kind of material is that faster-acting transistors have been made for use at extreme frequencies.

7.2.1.5 Photovoltaic cells. The selenium photographic exposure meter has already been mentioned; it goes back to Adams and Day's (1877) study of selenium, was further developed by Charles Fritt in 1885 and finally became a commercial product in the 1930s, in competition with a device based on cuprous oxide. This meter was efficient enough for photographic purposes but would not have been acceptable as an electric generator.

The idea of using a thin silicon cell containing a p/n junction parallel to the surface as a means of converting sunlight into DC electricity goes back to a team at Bell Labs, Chaplin *et al.* (1954), who were the first to design a cell of acceptable efficiency. Four years later, the first array of such cells was installed in a satellite, and since then all satellites, many of them incorporating a receiver/transmitter for communications, have been provided with a solar cell array. By degrees procedures were invented to use a progressively wider wavelength range of the incident radiation, and eventually cells with efficiencies approaching 20% could be manufactured. Other materials have been studied as well, but most paths seem eventually to return to silicon. The problem has always been expense; the efficient cells have mostly been made of single crystal slices which cannot be made cheaply, and in general there have to be several layers with slightly different chemistry to absorb different parts of the solar spectrum. Originally, costs of over $20 per watt were quoted. This was down to $10 ten years ago, and today has come down to $5. Until recently, price has restricted solar cells to communications use in remote locations (outer space being a very remote location). The economics of solar cells, and many technical aspects also, were accessibly analysed in a book by Zweibel (1990). A more recent overview is by Loferski (1995). In 1997, the solar cell industry expanded by a massive 38% worldwide, and in Germany, Japan and the USA there is now a rapidly expanding program of fitting arrays of solar cells (≈ 30 m^2), connected to the electric grid, to domestic roofs. Both monocrystalline cells and amorphous cells (discussed below) are being used; it looks as though the long-awaited breakthrough has at last arrived.

One of the old proposals which is beginning to be reassessed today is the notion of using electricity generated by solar cell arrays to electrolyse water to generate hydrogen for use in fuel cells (Section 11.3.2) which are approaching practical use for automotive engines. In several countries, research units are combining activities in photovoltaics with fuel cell research.

An alternative to single crystal solar cells is the use of amorphous silicon. For many years this was found to be too full of electron-trapping defects for p/n junctions to be feasible, but researches beginning in 1969 established that if amorphous silicon was made from a gaseous (silane) precursor in such a way as to trap some of the hydrogen permanently, good rectifying junctions became possible and a group in Scotland (Spear 1974) found that solar cells made from such material were effective. This quickly became a mature technology, with solar-cell efficiencies of $\approx 14\%$, and a large book is devoted to the extensive science and procedures of 'hydrogenated amorphous silicon' (Street 1991). Since then, research on this technology has continued to intensify (Schropp and Zeeman 1998). The material can be deposited *inexpensively* over large areas while yet retaining good semiconducting properties: photovoltaic roof shingles have been developed for the domestic market and are finding a warm response.

It may occasion surprise that an amorphous material has well-defined energy bands when it has no lattice planes, but as Street's book points out, "the silicon atoms have the same tetrahedral local order as crystalline silicon, with a bond angle variation of (only) about 10% and a much smaller bond length disorder". Recent research indicates that if enough hydrogen is incorporated in a-silicon, it transforms from amorphous to microcrystalline, and that the best properties are achieved just as the material teeters on the edge of this transition. It quite often happens in MSE that materials are at their best when they are close to a state of instability.

Yet another alternative is the thin-film solar cell. This cannot use silicon, because the transmission of solar radiation through silicon is high enough to require relatively thick silicon layers. One current favourite is the $Cu(Ga, In)Se_2$ thin-film solar cell, with an efficiency up to 17% in small experimental cells. This material has a very high light absorption and the total thickness of the active layer (on a glass substrate) is only 2 μm.

The latest enthusiasm is for an approach which takes its inspiration from color photography, where special dyes sensitise a photographic emulsion to specific light wavelengths. Photoelectrolysis has a long history but has not been able to compete with silicon photocells. Cahn (1983) surveyed an approach exploiting n-type titanium dioxide, TiO_2. Two Swiss researchers (Regain and Grätzel 1991) used TiO_2 in a new way: colloidal TiO_2 was associated with dye monolayers and immersed in a liquid electrolyte, and they found they could use this system as a photocell with an efficiency of $\approx 12\%$. This work set off a stampede of consequential research, because of the prospect of an inexpensive, impurity-tolerant cell which might be much cheaper than any silicon-based cell. Liquid electrolyte makes manufacture more complex, but up to now, solid polymeric electrolytes depress the efficiency. The long-term stakes are high (Hodgson and Wilkie 2000).

7.2.2 Electrical ceramics

The work on colour centres outlined in Section 3.2.3.1, much of it in the 1930s, and its consequences for understanding electrically charged defects in insulating and semiconducting crystalline materials, helped to stimulate ceramic researches in the electrical/electronic industry. The subject is enormous and here there is space only for a cursory outline of what has happened, most of it in the last 80 years.

The main categories of "electrical/optical ceramics" are as follows: phosphors for TV, radar and oscilloscope screens; voltage-dependent and thermally sensitive resistors; dielectrics, including ferroelectrics; piezoelectric materials, again including ferroelectrics; pyroelectric ceramics; electro-optic ceramics; and magnetic ceramics.

In Section 3.2.3.1 we saw that Frederick Seitz became motivated to study colour centres during his pre-War sojourn at the General Electric Research Laboratory, where he was exposed to studies of phosphors which could convert the energy in an electron beam into visible radiation, as required for oscilloscopes and television receivers. The term 'phosphor' is used generally for materials which fluoresce and those which phosphoresce (i.e., show persistent light output after the stimulus is switched off). Such materials were studied, especially in Germany, early in this century and these early results were assembled by Lenard *et al.* (1928). Phosphors were also a matter of acute concern to Vladimir Zworykin (a charismatic Russian immigrant to America); he wanted to inaugurate a television industry in the late 1920s, but failed to persuade his employers, Westinghouse, that this was a realistic objective. According to an intriguing piece of historical research by Notis (1986), Zworykin then transferred to another company, RCA, which he was able to persuade to commercialise both television and electron microscopes. For the first of these objectives, he needed a reliable and plentiful material to use as phosphors, with a persistence time of less than 1/30 of a second (at that time, he believed that 30 refreshments of the tube image per second would be essential). Zworykin was fortunate to fall in with a ceramic technologist of genius, Hobart Kraner. He had studied crystalline glazes on decorative ceramics (this was an innovation, since most glazes had been glassy), and among these, a zinc silicate glaze (Kraner 1924). He and others later found that when manganese was added as a nucleation catalyst to encourage crystallisation of the glassy precursor, the resulting crystalline glaze was fluorescent. In the meantime, natural zinc silicate, the mineral willemite, was being used as a phosphor, but it was erratic and non-reproducible and anyway in very short supply. Kraner showed Zworykin that synthetic zinc silicate, Zn_2SiO_4, would serve even better as a phosphor when 'activated' by a 1% manganese addition. This serendipitous development came just when Zworykin needed it, and it enabled him to persuade RCA to proceed with the large-scale manufacture of TV tubes. Kraner, a modest man who published little, did present a lecture on creativity and the interactions between people needed to stimulate it (Kraner 1971). The history of materials is full of episodes when the right concatenation of individuals elicited the vitally needed innovation at the right time.

Phosphors to convert X-ray energy into visible light go back to a time soon after X-rays were discovered. Calcium tungstate, $CaWO_4$, was found to be more sensitive to X-rays than the photographic film of that time. Many more efficient phosphors have since been discovered, all doped with rare earth ions, as recently outlined by an Indian physicist (Moharil 1994). The early history of all these phosphors, whether for impinging electrons or X-rays, has been surveyed by Harvey (1957). (The generic term for this field of research is 'luminescence', and this is in the title of Harvey's book.) The subfield of electroluminescence, the emission of light by some crystals when a current flows through them, a theoretically distinctly untidy subject, was reviewed by Henisch (1964).

The relatively simple study of fluorescence and phosphorescence (based on the action of colour centres) has nowadays extended to nonlinear optical crystals, in which the refractive index is sensitive to the light intensity or (in the photorefractive variety (Agullo-López 1994) also to its spatial variation); a range of crystals, the stereotype of which is lithium niobate, is now used.

Ceramic conductors also cover a great range of variety, and a large input of fundamental research has been needed to drive them to their present state of subtlety. A good example is the zinc oxide *varistor* (i.e., voltage-dependent resistors). This consists of semiconducting ZnO grains separated by a thin intergranular layer rich in bismuth, with a higher resistance than the grains; as voltage increases, increasing areas of intergranular film can participate in the passage of current. These important materials have been described in Japan (a country which has achieved an unchallenged lead in this kind of ceramics, which they call 'functional' or 'fine' ceramics) (Miyayama and Yanagida 1988) and in England (Moulson and Herbert

1990). This kind of microstructure also influences other kinds of conductors, especially those with positive (PTC) or negative (NTC) temperature coefficients of resistivity. For instance, PTC materials (Kulwicki 1981) have to be impurity-doped polycrystalline ferroelectrics, usually barium titanate (single crystals do not work) and depend on a ferroelectric-to-paraelectric transition in the dopant-rich grain boundaries, which lead to enormous increases in resistivity. Such a ceramic can be used to prevent temperature excursions (surges) in electronic devices.

Levinson (1985), a varistor specialist, has told the author of the early history of these ceramics. The varistor effect was first found accidentally in a Russian study of the $ZnO-B_2O_3$ system, but was not pursued. In the mid-1960s, it was again stumbled on, in Japan this time, by an industrial scientist, M. Matsuoka and thoroughly studied; this led to manufacture from 1968 and the research was first published in 1969. Matsuoka's company, Matsushita, had long made resistors, fired in hydrogen; the company wished to save money by firing in air, and ZnO was one of the materials they tested in pursuit of this aim. Electrodes were put on the resistors via firable silver-containing paints. One day the temperature control failed, and the ZnO resistor now proved to behave in a non-linear way; it no longer obeyed Ohm's law. It turned out later that the silver paint contained bismuth as an impurity, and this had diffused into the ZnO at high temperature. Matsushita recognised that this was interesting, and the company sought to improve the material systematically by "throwing the periodic table at it", in Levinson's words, with 50–100 staff members working at it, Edison-fashion. Hundreds of patents resulted. Now the bismuth, and indeed other additives, were no longer impurities (undesired) but had become dopants (desired). Parts per million of dopant made a great difference, as had earlier been found with semiconductor devices. Henceforth, minute dopant levels were to be crucial in the development of electroceramics.

A book edited by Levinson (1981) treated grain-boundary phenomena in electroceramics in depth, including the band theory required to explain the effects. It includes a splendid overview of such phenomena in general by W.D. Kingery, whom we have already met in Chapter 1, as well as an overview of varistor developments by the originator, Matsuoka. The book marks a major shift in concern by the community of ceramic researchers, away from topics like porcelain (which is discussed in Chapter 9); Kingery played a major role in bringing this about.

The episode which led to the recognition of varistor action, a laboratory accident, is typical of many such episodes in MSE. The key, of course, is that someone with the necessary background knowledge, and with a habit of observing the unexpected, should be on hand, and it is remarkable how often that happens. The other feature of this story which is characteristic of MSE is the major role of minute dopant concentrations. This was first recognised by metallurgists, then it was the turn of the physicists who had so long ignored imperfect purity when they turned to semiconductors in earnest, and finally the baton was taken over by ceramists. The metallurgical role of impurities, mostly deleterious but sometimes (e.g., in the manufacture of tungsten filaments for electric light bulbs) beneficial, indeed essential, has recently been covered in textbooks (Briant 1999, Bartha *et al.* 1995). The concept of 'science and the drive towards impurity' was outlined in Section 3.2.1, in connection with the role of impurities in 'old-fashioned metallurgy'.

7.2.2.1 Ferroelectrics. In the preceding section, positive-temperature-coefficient (PTC) ceramics were mentioned and it was remarked that they are made of a ferroelectric material.

'Ferroelectric' is a linguistic curiosity, adapted from 'ferromagnetic'. ('Ferro-' here is taken to imply a spontaneous magnetisation, or electrification, and those who invented the name chose to forget that 'ferro' actually refers to iron! The corresponding term 'ferroelastic' for non-metallic crystals which display a sponta-neous strain is an even weirder linguistic concoction!). Ferroelectric crystals are a large family, the modern archetype of which is barium titanate, $BaTiO_3$, although for two centuries an awkward and unstable organic crystal, Rochelle salt (originally discovered by a pharmacist in La Rochelle to be a mild purgative) held sway. Rochelle salt is a form of sodium tartrate, made as a byproduct of Bordeaux wine – a natural source for someone in La Rochelle. It turned out that it is easy to grow large crystals of this compound, and a succession of physicists, attracted by this feature, examined the crystals from 1824 onwards and discovered, first pyroelectric behaviour, and then piezoelectric behaviour – *pyroelectricity* implies an electric polarisation change when a crystal is heated, *piezoelectricity*, a polarisation brought about by strain (or inversely, strain brought about by an applied electric field). After that, a succession of investigators, seduced by the handsome large crystals, measured the *dielectric constant* and studied its relation to the refractive index. Still the ferroelectric character of Rochelle salt eluded numerous investigators in America and Russia, and it was not till Georg Busch, a graduate student in Peter Debye's laboratory in Zürich, began work on particularly perfect crystals which he had grown himself that various anomalies in dielectric constant, and the existence of a Curie temperature, became manifest, and ferroelectric behaviour was at last identified. Busch has recently, in old age, reviewed this intriguing pre-history of ferroelectricity (Busch 1991).

By the 1930s, Rochelle salt had built up an unenviable reputation as a material with irreproducible properties... rather as semiconductors were regarded during those same years. Rochelle salt was abandoned when ferroelectricity was recognised and studied in KH_2PO_4, and then the key compound, barium titanate, $BaTiO_3$, was found to be a strong ferroelectric in a British industrial laboratory during the War;

they kept the material secret. Megaw (1945), in Cambridge, performed a tour de force of crystal structure determination by demonstrating the spontaneous strain associated with the electric moment, and then, in the physics department of Bristol University, leaning partly on Soviet work, Devonshire (1949) finally set out the full phenomenological theory of ferroelectricity. The phenomenon is linked to a symmetry change in the crystal at a critical temperature which breaks it up into minute twinned domains with opposing electric vectors, as was first shown by Kay (1948) in Bristol. Helen Megaw also wrote the first book about ferroelectric crystals (Megaw 1957).

This scientifically fascinating crystal, $BaTiO_3$, is used for its very high dielectric constants in capacitors and also for its powerful piezoelectric properties, for instance for sonar. The essential feature of a ferroelectric is that it has an intrinsic electric moment, disguised in the absence of an exciting field by the presence of domains which leave the material macroscopically neutral... just as magnetic domains do in a ferromagnet. Their very complicated scientific history after 1932, with many vigorous, even acrimonious controversies, has been excellently mapped out by Cross and Newnham (1986) and by Känzig (1991); Känzig had been one of Debye's bright young men in Zürich in the 1930s. One of the intriguing pieces of information in Cross and Newnham's history is that in the 1950s, Bernd Matthias at Bell Laboratories competed with Ray Pepinsky at Pennsylvania State University to see who could discover more novel ferroelectric crystals, just as later he competed again with others to drive up the best superconducting transition temperature in primitive (i.e., metallic) superconductors. Every scientist has his own secret spring of action, if only he has the good fortune to discover it! – Matthias's quite remarkable personality, and his influence on many contemporaries, are portrayed in a Festschrift prepared on the occasion of his 60th birthday (Clogston *et al.* 1978); this issue also included details of his doctoral students and his publications. His own principles of research, and how he succeeded in achieving his "phenomenal record for finding materials with unusual properties" emerge in an instructive interview (Colborn *et al.* 1966).

Other strongly ferroelectric crystals have been discovered and today, PZT – $Pb(Ti, Zr)O_3$ – is the most widely exploited of all piezoelectric (ferroelectric) ceramics.

The PTC materials already mentioned depend directly on the ferroelectric phase transition in solid solutions based on $BaTiO_3$, suitably doped to render them semiconducting. This is a typical example of the interrelations between different electrical phenomena in ceramics.

Due to their high piezoelectric response, 'electrostriction' in ferroelectrics, induced by an applied electric field, can be used as strain-inducing components (just as ferromagnetic materials can be exploited for their magnetostriction). Thus barium titanate is used for the specimen cradle in tunnelling electron microscopes (Section 6.2.3) to allow the minute displacements needed for the operation of these instruments. An intriguing, up-to-date account of uses of electrostriction and magnetostriction in "smart materials" is given by Newnham (1997).

Another important function which ferroelectrics have infiltrated is that of electro-optic activity. In one form of such activity, an electric field applied to a transparent crystal induces birefringence, which can be exploited to modulate a light signal; thus electro-optic crystals (among other uses) can be used in integrated electro-optic devices, in which light takes the place of an electronic current. Very recently (Li *et al.* 2000) a way has been found of using a 'combinatorial materials strategy' to test, in this regard, a series of $Ba_{1-x}Sr_xTiO_3$ crystals. This approach, which is further discussed in Sect. 11.2.7, makes use of a 'continuous phase diagram', in which thin-film deposition techniques are used to prepare a film of continuously varying composition which can then be optically tested at many points.

7.2.2.2 Superionic conductors. A further large family of functional ceramics is that of the *superionic conductors*. This term was introduced by Roth (1972) (working at the GE Central Laboratory); though his work was published in the *Journal of Solid-State Chemistry*, it could with equal justification have appeared in *Physical Review*, but it is usual with crystallographers that people working in this field are polarised between those who think of themselves as chemists and those who think of themselves as physicists. Superionic conductors are electronically insulating ionic crystals in which either cations or anions move with such ease under the influence of an electric field that the crystals function as efficient conductors in spite of the immobility of electrons. The prototype is a sodium-doped aluminium oxide of formula $Na_2O \cdot 11Al_2O_3$, called beta-alumina. Roth substituted silver for some of the sodium, for the sake of easier X-ray analysis, and found that the silver occupied a minority of certain sites on a particular plane in the crystal structure, leaving many other sites vacant. This configuration is responsible for the extraordinarily high mobility of the silver atoms (or the sodium, some of which they replaced); the vacancy-loaded planes have been described as liquid-like. There are now many other superionic conductors and they have important and rapidly increasing uses as electrolytes in all-solid storage batteries and fuel cells (see Chapter 11). They have their own journal, *Solid State Ionics*.

To put the above in perspective, it is necessary to point out that more humdrum ionic conductors (without the 'super' cachet) have been known since the late 19th century, when Nernst developed a lamp based on the use of zirconia which is an ionic conductor (see Section 9.3.2). The use of zirconia for gas sensors is treated in Chapter 11.

7.2.2.3 Thermoelectric materials. Every materials scientist is accustomed to using thermocouples to measure temperature. A thermocouple consists of two dissimilar metals (or, more usually, alloys or semiconductors) welded together; the junction is put in the location where the temperature is to be determined, while the other end of each of the joined wires is welded to a copper wire, these two junctions being kept at a known reference temperature. Each junction generates a Seebeck voltage, called after the German discoverer of this phenomenon, the physician Thomas Seebeck (1770–1831); his discovery was reported in 1822. Not long afterwards, in 1834, the French watchmaker Jean Peltier (1785–1845) discovered the counterpart of the Seebeck effect, a heating or cooling effect when a current is passed through a junction. Thereafter, many years passed before the linked phenomena were either understood or applied.

Pippard (1995), in an overview of 'electrons in solids', sets out the tangled history of the interpretation of these effects, basing himself on an earlier survey (Ziman 1960). He steps back to "a scene of some confusion, some of it the legacy of Maxwell and his followers, in so far as they sought to avoid introducing the concept of charged particles, and looked to the ether as the medium for all electromagnetic processes; the transport of energy along with charge was foreign to their thought". A beginning of understanding had to await the twentieth century and a generation of physicists familiar with electrons; Lorentz and Sommerfeld in the 1920s set out an interpretation of the behaviour of electrons at a junction between two metals. Mott and Jones (1936) expressed the Seebeck coefficient in a form proportional to absolute temperature and also to $(d\sigma/dE_F)$, where σ is the density of electronic states and E_F is the Fermi energy. From this it follows that when the electron state concentration, σ, and E_F are low, as in semimetals such as bismuth and in semiconductors, then a given change in E_F makes a large difference in σ and so the Seebeck coefficient and the electrical output for a given temperature difference will be large.

The man who recognised the importance of this insight and developed thermoelectric devices based on semimetal compounds and on semiconductors was A.F. Ioffe (sometimes transliterated as Joffe) in Leningrad (St. Petersburg), head of a notable applied physics research laboratory – the same laboratory at which, a few years later, Alferov invented the semiconductor laser. In a major review (Joffe and Stil'bans 1959) he set out an analysis of the 'physical problems of thermoelectricity' and went in great detail into the criteria for selecting thermoelectric materials. Ioffe particularly espoused the cause of thermoelectric refrigeration, exploiting the Peltier effect, and set it out in a book (Ioffe 1957). In the West, thermoelectric cooling was popularised by another influential book (Goldsmid 1964). The attainable efficiency however in the end proved to be too small, even with promising materials such as Bi_2Te_3, to make such cooling a practical proposition.

After this, there was a long period of quiescence, broken by a new bout of innovation in the 1990s. Thermoelectric efficiency depends on physical parameters through a dimensionless *figure of merit*, ZT, where $Z = S^2/\kappa\rho$. Here S is the Seebeck coefficient, κ the thermal conductivity and ρ is the electrical resistivity. A high thermal conductivity tends to flatten the temperature gradient and a high resistivity reduces the current for a given value of S. (Such figures of merit are now widely used in selecting materials or engineering structures for well-defined functions; this one may well have been the first such figure to be conceived). Efforts have lately been made to reduce κ, in the hope of raising ZT beyond the maximum value of ≈ 1 hitherto attainable at reasonable temperatures. Slack (1995) sets out some rules for maximising ZT, including the notion that "the ultimate thermoelectric material should conduct electricity like a crystal but heat like a glass." These words are taken from an excellent overview of recent efforts to achieve just this objective (Sales 1997). Among several initiatives described by Sales, he includes his own research on the 'filled skutterudite antimonides', a group of crystals derived from a naturally occurring Norwegian mineral. The derivatives which proved most successful are compositions like $CeFe_3CoSb_{12}$. Rare-earth atoms (here Ce) sitting in capacious 'cages' (Figure 7.6) rattle around and in so doing, confer glass-like characteristic on the phonons in the material and thus on the thermal conductivity; this consequence of 'rattling caged atoms' was predicted by Slack. ZT values matching those for Bi_2Te_3 have already

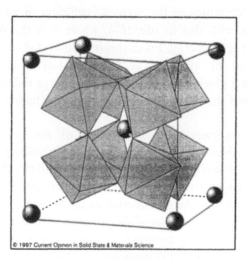

© 1997 Current Opinion in Solid State & Materials Science

Figure 7.6. A filled skutterudite antimonide crystal structure. A transition metal atom (Fe or Co) at the centre of each octahedron is bonded to antimony atoms at each corner. The rare earth atoms (small spheres) are located in cages made by eight octahedra. The large thermal motion of 'rattling' of the rare earth atoms in their cages is believed be responsible for the strikingly low thermal conductivity of these materials (Sales 1997).

been achieved. This episode demonstrates how effective arguments based on crystal chemistry can be nowadays in the conception of completely new materials.

Another strategy reported by Sales links back to the 'superlattices' discussed in Section 7.2.1.4. It was suggested by Mildred Dresselhaus's group at MIT (Hicks *et al.* 1993) that semiconductor quantum wells would have enhanced figures of merit compared with the same semiconductor in bulk form. PbTe quantum wells were confined by suitable intervening barrier layers. From the results, ZT values of ≈ 2 were estimated from single quantum wells. This piece of research shows the intimate links often found nowadays between apparently quite distinct functional features in materials.

Several branches of physics come together in a recent suggestion of a possible way to 'improve' pure bismuth to make it an outstanding candidate for thermo-electric devices, with a target ZT value of at least 2. Shick *et al.* (1999) applied first-principles theoretical methods to assess the electron band structure of bismuth as a function of the interaxial angle (bismuth is rhombohedral, with a unit cell which can be regarded as a squashed cube), and the conclusion was that a modest change in that angle should greatly improve bismuth as a thermoelectric component, by promoting a semimetal-semiconductor phase transition. The authors suggest that depositing Bi epitaxially on a substrate designed to constrain the interaxial angle might do the trick. Being theoreticians, they left the possible implementation to materials scientists.

7.2.2.4 Superconducting ceramics. In 1908, Heike Kamerlingh Onnes in Leiden, The Netherlands, exploiting the first liquefaction of helium in that year in his laboratory, made the measurements that within a few years were to establish the phenomenon of superconductivity – electrical conduction at zero resistivity – in metals. In 1911 he showed that mercury loses all resistivity below 4.2 K. The historical implications of that and what followed in the subsequent decades are set out in a chapter of a history of solid-state physics (Hoddeson *et al.* 1992). Then, Bednorz and Müller (1986) discovered the first of the extensive family of perovskite-related ceramics all containing copper oxide which have critical temperatures up to and even above the boiling point of liquid nitrogen, much higher than any of the metals and alloys, and thereby initiated a fierce avalanche of research. A concise overview of both classes of superconductor is by Geballe and Hulm (1992). Meanwhile, the complex effects of strong magnetic fields in quenching supercon-ductivity had been studied in depth, and intermetallic compounds had been developed that were highly resistant to such quenching and are widely used for windings of superconducting electromagnets, for instance as components of medical computerised tomography scanners.

The electronic theory of metallic superconduction was established by Bardeen, Cooper and Schrieffer in 1957, but the basis of superconduction in the oxides remains a battleground for rival interpretations. The technology of the oxide ("high-temperature") superconductors is currently receiving a great deal of attention; the central problem is to make windable wires or tapes from an intensely brittle material. It is in no way a negative judgment on the importance and interest of these materials that they do not receive a detailed discussion here: it is simply that they do not lend themselves to a superficial account, and there is no space here for a discussion in the detail that they intrinsically deserve.

The intimate mix of basic and technological approaches to the study of high-temperature superconductors, indeed their inseparable nature, was analysed recently by a group of historians of science, led by the renowned scholar Gerald Holton (Holton *et al.* 1996). Holton *et al.* conclude that "historical study of cases of successful modern research has repeatedly shown that the interplay between initially unrelated basic knowledge, technology and products is so intense that, far from being separate and distinct, they are all portions of a single, tightly woven fabric." This paper belongs to a growing literature of analysis of the backgrounds to major technical advances (e.g., Suits and Bueche 1967, TRACES 1968). Holton's analysis is timely in view of the extreme difficulties of applying high-temperature superconductivity to practical tasks (see a group of papers introduced by Goyal 1995).

Just one specific technological aspect of high-temperature superconductors will be explained here. The superconduction in the copper oxide-based ceramics essentially takes place in one crystal plane, and if adjacent crystal grains in a polycrystal (and these materials are always used as polycrystals) are mutually misoriented by more than about 10° then superconduction is impeded to the extent that quite modest magnetic fields can quench superconductivity. It is thus necessary to find a way of constructing thin films in epitaxial orientation so that neighbouring grains are only very slightly misoriented. So, once again, grain boundaries are a key to behaviour. One approach which generates a fairly strong alignment of the crystal grains is 'paramagnetic annealing', solidification of the compound in the presence of a magnetic field (de Rango *et al.* 1991); but the misorientations do not seem to be sufficiently small for practical purposes.

Recent research by a large group of materials scientists at Oak Ridge National Laboratory in America (Goyal *et al.* 1999) has established a means of depositing a superconducting ceramic film with a strong preferred orientation, on a highly oriented alloy sheet made by heavy rolling followed by annealing, using an intermediate epitaxial oxide layer. This is typical of the sophisticated methods in materials processing that are coming to the fore today.

Superconductivity research has reached out to other branches of physics and materials science; perhaps the strangest example of this is a study by Keusin-Elbaum *et al.* (1993) in which the current-carrying capacity of a mercury-bearing ceramic, $HgBa_2CaCu_2O_{6+\delta}$ is greatly enhanced by using a beam of high-energy protons to provoke nuclear fission in some mercury atoms; the consequent radiation damage is responsible for the changes in superconducting behaviour. The authors imply that this might become a production process!

Ceramic superconduction is no longer limited to materials containing copper oxide. Some striking research in India (Nagarajan *et al.* 1994, Gupta 1999) has demonstrated superconduction in a family of alloys of the type RE–Ni–B–C, the quaternary borocarbides, where 'RE' denotes a rare-earth metal. They exhibit interplay of superconductivity and long-range magnetic order. The transition temperatures are not yet exciting but it is reassuring to know that a range of quite distinct ceramics can display superconduction.

7.3. MAGNETIC CERAMICS

The research laboratory of Philips Gloeilampenfabrieken (incandescent lamp factories) in Eindhoven, Netherlands, is one of the glories of industrial science and engineering. A notable Dutch physicist, Hendrik Casimir (1909–2000), joined the company in 1942 after working with Bohr and Pauli, and has a chapter about the history of the company and laboratory in his book of memoirs (Casimir 1983). The company was founded by two Philips brothers, Gerard and Anton, and their father Frederick, in 1891. Shortly before World War I, Gerard, who had been deeply impressed by the research on lamps and tungsten filaments that he had witnessed at the new GE Research Laboratory in Schenectady, NY, resolved to open such a laboratory in Eindhoven. In early 1914, a young Dutch physicist, Gilles Holst, started work as the first research director, and remained until 1946, when Casimir and two others succeeded him as a triumvirate. Casimir's account of Holst's methods and principles is fascinating and is bound to intrigue anyone with a concern for industrial research. Holst "rarely gave his staff accurately defined tasks. He tried to make people enthusiastic about the things he was enthusiastic about – and usually succeeded". Neither did he "believe in strict hierarchic structure". Casimir goes further: he claims that Holst "steered a middle course between individualism and strict regimentation, based authority on real competence, but in case of doubt, preferred anarchy". Also, he did not subdivide the laboratory on disciplinary lines, but created multi-disciplinary teams. All this seems very similar to the principles applied to the GE Laboratory in its heyday.

This is by way of preliminary to an outline account of the genesis of the magnetic ferrites in the Philips Laboratory, before, during and just after World War II. The presiding spirit was Jacobus Louis Snoek (1902–1950), a Dutch physicist whom we have already met in Section 5.1.1, in connection with the torsion pendulum for measuring internal friction which he invented in the late 1930s at Philips. Snoek was just the kind of scientist who would appeal to Holst – a highly original, self-motivated researcher. By 1934, Philips had already begun to diversify away from incandescent lamps, and Holst came to recognise that electromagnets and transformers with iron cores suffered from substantial losses from eddy currents. He reckoned that if it were possible to find an *electrically insulating* magnetic material to replace iron, it might become an extremely valuable industrial property. So Snoek was persuaded to have a look at magnetite (lodestone), the long-familiar oxide magnet, of composition Fe_3O_4. This mineral is better described as $Fe_2O_3 \cdot FeO$; 1/3 of the iron atoms are doubly ionised, 2/3 trebly. The initial plan was to look for other magnetic oxides of the form $Fe_2O_3 \cdot MeO$, where Me is another divalent metal. Cu, Zn, Co, Ni are a few of the Me's that were tried out. This was pursued energetically by Snoek from a physicist's standpoint and by his equally distinguished colleague E.J.W. Verwey from a chemist's perspective. The first of several papers by Snoek appeared soon after he began work (Snoek 1936). Snoek's life ended in a sad way. In 1950, he left Philips and the Netherlands, in search perhaps of a more prosperous lifestyle, or perhaps because he failed to secure the promotion he wished for at Philips (Verwey had become joint director of research). He joined an American consulting firm, but before the year was out, he died at the age of 48 in a car crash.

All these materials, which soon came to be named *ferrites* (no connection with the same word applied as a name for the body-centred allotrope of pure iron) share the spinel structure. Spinel is the type-name for $MgAl_2O_4$, and the crystal structure of that compound was first determined by Lawrence Bragg in 1915, a very early example of crystal structure determination. Figure 7.7 shows the structure. In the type structure, the divalent cations, Mg^{2+}, occupy the tetrahedral (A) sites, while the trivalent cations, Al^{3+}, occupy the octahedral (B) sites: this is the normal spinel structure. The oxygen atoms are slightly displaced from the body diagonals, to an extent depending upon the cation radii. Eventually, studying the crystal structures of the compounds made by Snoek, Verwey and Heilmann (1947) found that some of them had an 'inverse spinel structures; here, instead, Fe^{3+} occupies all the A sites while the B sites are occupied half by Me^{2+} and half by Fe^{3+}. There are also intermediate structures, and polymorphic changes are observed too: magnetite itself is distorted from a cubic form below 120 K. By years of painstaking study, Verwey established the subtle energetic rules that determined the way a particular ferrite crystallises; a clear, concise account of this can be found in Chapter 2 of a recent text (Valenzuela 1994). Snoek and Verwey also found a family of hexagonal ferrites, such as barium ferrite, $BaFe_{12}O_{19}$ ($MeO \cdot 6Fe_2O_3$, generically).

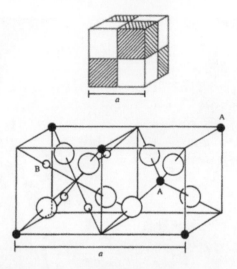

Figure 7.7. The spinel structure. The unit cell can be divided into octants – tetrahedrally coordinated cations A, octahedrally coordinated cations B, and oxygen atoms (large circles) are shown in two octants only (adapted from Smit and Wijn 1959).

The inverse ferrites were found in general to have the most valuable soft magnetic properties (i.e., high permeability); as a family, Snoek called these *ferroxcube*. The hexagonal ferrites, barium ferrite in particular, (*hexaferrites*) were permanent magnets. The gradual development of these two families of ferrites owed everything to the intimate interplay of physical understanding and crystal chemistry. Snoek and Verwey were the joint progenitors of this extremely valuable family of materials.

Many of the ferrites (those which contain two distinct cations with magnetic moments) are *ferrimagnetic* – i.e., there are two populations of cations with oppositely directed but unequal magnetic moments, so that there is a macroscopically resultant magnetic moment. The understanding of this form of magnetism came after Snoek and Verwey had embarked on their study of ferrites, and was a byproduct of Louis Néel's extraordinary prediction, in 1936, of the existence of *antiferromagnetism*, where the two populations of opposed spins both involve the same numbers of the same species of ion so that there is *no* macroscopic resultant magnetisation (Néel 1936). (See also the background outlined in Section 3.3.3.) Néel (1904–2000), a major figure in the history of magnetism (Figure 7.8), recognised that under certain geometrical circumstances, neighboring ions could be so disposed that their magnetic spins line up antiparallel; his paper specifically mentioned manganese, which has no macroscopic magnetism but would have been expected to have shown this. Two years later, another French physicist duly discovered that MnO indeed has all the predicted characteristics of an antiferromagnet, and later the same thing was established for manganese itself.

Figure 7.8. L.E.F. Néel (photograph courtesy of Prof. Néel).

Soon after Verwey had shown that the magnetic spinels studied at Philips were in fact inverse spinels, Néel (1948) applied the ideas he had developed before the War for antiferromagnetism to these structures and demonstrated that the two kinds of cations should have antiparallel spins; he invented the term 'ferrimagnetism', and also recognised that at a certain temperature, analogous to a Curie temperature, both antiferromagnetism and ferrimagnetism would disappear. That temperature is now known as the Néel temperature. A little later, Shull *et al.* (1951) at Oak Ridge used the new technique of neutron diffraction, which is sensitive to magnetic spins, to confirm the presence of antiparallel spins in manganese. The recognition of ferrimagnetism had been achieved, and after that there was no holding back the extensive further development of this family of magnetic materials, both 'soft' and 'hard'. Holst's initial objective had been triumphantly achieved.

The intellectual stages baldly summarised here, and especially Néel's seminal role, are set out fully in an excellent historical treatment (Keith and Quédec 1992).

The many scientific and technological aspects of preparing, treating and understanding ferrites that are essential to their applications are treated in an early work by Smit and Wijn (1959) and the more recent one by Valenzuela (1994). An unusual book (Newnham 1975) treats a very wide range of functional materials (including magnetic materials) from the perspective of crystal chemistry, an approach which paid great dividends for the ferrites; it comes as close as any compilation to a complete overview of functional materials.

An early, striking use of a ferrite began in 1948, when a giant synchrotron was being designed at Brookhaven National Laboratory in America. The designers decided that the use of a metallic core for the electromagnets was impracticable because of the expected energy drain through hysteresis and eddy currents. One of the design team had heard of the brand new ferrites and invited Philips to tender; the shocking response was that Philips did not know how to make any pieces bigger than a matchbox. However, they determined to find out how to satisfy the Brookhaven team and began to supply large pieces in 1949. A recently published history of Brookhaven (Crease 1999) asserts with respect to ferrites that "the effort to make them in sufficiently large sizes at low cost opened the door for their use in many other kinds of devices", and that, when the synchrotron was finally completed in 1952, "it was the first important application of ferrite in the US".

Casimir, in his memoirs, mentions that the first Philips representative to visit Brookhaven gave the design team a set of guaranteed magnetic specifications which were too conservative, and Brookhaven was pleased to find that the material was better than they had been led to expect. Casimir reckons that Brookhaven must have concluded that Philips "was dumb but honest".

7.4. COMPUTER MEMORIES

In his memoirs, Casimir gives an impressionistic account of the way random-access (RAM) memories were manufactured in the 1960s and early 1970s: "A factory of memory stores, for instance one run by North American Philips, always struck me as a remarkable outfit. On one side of the factory a better kind of rust and some equally mundane substances were ground, mixed, pressed into shape, and fired in large ovens... On the other side, ladies with manicured hands were threading coloured wires through little rings and weaving them into mats to the accompaniment of soft music – a kind of glorified kindergarten activity." He is referring here to magnetic core memories (Figure 7.9), an important early application of 'soft' ferrites. The composition chosen was one with a square hysteresis curve, so that the direction of

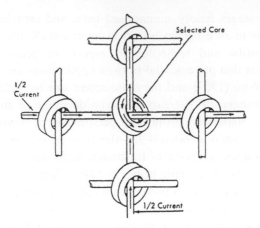

Figure 7.9. A small portion of a magnetic core memory (after *IBM News*, 1967).

magnetization could be reversed by applying a sharply defined critical magnetic field. A current flowing along a wire gives rise to a circular field around it; things were so arranged that a standard current flowing along one wire was insufficient to reverse magnetization in a ring, but if a similar current flows in the orthogonal wire as well, the joint effect of the two currents is enough to reverse the magnetization. Readout of the stored memory is destructive, requiring a rewrite operation for continued storage – which also applies to the semiconducting DRAM memories mentioned below.

According to a recent critical overview of memory design (Yeack-Scranton 1994), such core memories were cost-competitive in the 1960s, with access times of about 1 μs and a cost of about $10,000 per megabyte. (Nowadays, a 48-megabyte semiconducting memory can be bought for less than $100, more than 5000 times cheaper.) The complexity of the manual weaving operation was such that core memories were only made up to a few hundred bytes. It was at this time that "19" was left off the year in recording dates, to save a few bytes of expensive memory, and thus presaged the scare stories 30 years later about the expected 'millennium bug' resulting from the 1900–2000 transition.

Today, dynamic random-access memories (DRAMs) are transistor/capacitor-based semiconductor devices, with access times measured in nanoseconds and very low costs. Core memories were made of magnetic rings not less than a millimetre in diameter, so that a megabyte of memory would have occupied square metres, while a corresponding DRAM would occupy a few square millimetres. Another version of a DRAM is the read-only memory (ROM), essential for the operation of any computer, and unalterable from the day it is manufactured. We see that developments in magnetic memories involved dramatic reductions in cost and volume as well as in cost – very much like the replacement of vacuum tubes, first by discrete transistors and eventually by integrated circuits.

Every computer needs a long-term memory store as well as an evanescent (RAM) store. Things have come a long way since punched paper tapes were used, before World War II. The history of magnetic recording during the past century is surveyed by Livingston (1998). The workhorse of longterm memory storage is the hard disc, in which a minute horseshoe electromagnet is made (by exploiting aerodynamic lift) to float just micrometres (now, a fraction of a micrometre) above a polished ferromagnetic thin film vacuum-deposited on a polished hard non-magnetic substrate such as an Al–Mg alloy, with a thin carbon surface layer to provide wear resistance in case the read/write head accidentally touches the surface. According to the above-mentioned overview, densities of 30 megabytes per cm^2 were current in 1994, but increasing rapidly year by year, with (at that time) a cost of \$1/ megabyte and data-writing rates of 9 megabytes per second. The first hard-disc drive was introduced by IBM in 1956... the RAMAC, or Random Access Method of Accounting and Control. It incorporated 50 aluminium discs 60 cm in diameter, and the device, capable of storing 5 million 'characters' (which today would occupy less than a square centimetre), weighed nearly a ton and occupied the same floor space as two modern refrigerators. This information comes from a survey of future trends in hard-disc design, (Toigo 2000); this article also claims that the magnetic coating of RAMAC was derived from the primer used to paint the Golden Gate Bridge in San Francisco.

One that 'got away' was the magnetic bubble memory, also described by Yeack-Scranton in 1994. A magnetic bubble is a self-contained, cylindrical magnetic domain polarised in direction opposite to the magnetization of the surrounding thin magnetic film, typically a rare-earth iron garnet deposited on a nonmagnetic substrate. Information is stored by creating and moving strings of such bubbles, each about a micrometre across, accessed by means of a magnetic sensor. This very ingenious and unconventional approach received a great deal of research attention for several years in the 1970s; it was popular for a time for such devices as handheld calculators, but in the end the achievable information density and speed of operation were insufficiently attractive. In the field of magnetic memories, competition is red in tooth and claw.

The most recent development in purely magnetic memories exploits so-called giant magnetoresistance. This phenomenon is one of a range of properties of magnetic ultrathin films. Néel, with his usual uncanny perception, foretold in 1954 that in very thin films (1–2 nm thick) the magnetization may be normal to the film, attributed in a way still not wholly understood to the reduced symmetry of surface atoms. The next stage was the study of multilayer thin films, with magnetic and nonmagnetic materials alternating (e.g., Co/Cu or Dy/Y), each layer of the order of a very few nanometers in thickness. It turned out that if the thicknesses were just right, such multilayers could be antiferromagnetic. This was first observed as recently as 1986 (Grunberg *et al.*) and the origin of this curious anomalous magnetic coupling is still under debate. These researches are linked to others in which ultrathin multilayers such as Co/Cu or Fe/Cr were shown to have very large changes of electrical resistance when subjected to a magnetic field normal to the film (e.g., Baibich *et al.* 1988). This phenomenon is being increasingly exploited as an alternative means of storing information in computers; the key is that a readily measurable resistance change happens in small fields of an oerstedt or so, and some multilayer systems are approaching 3% change of resistance for 1 oerstedt of applied field. In such systems, information storage density can be significantly enhanced. These various developments based on ultrathin films and multilayers are explained with great clarity and economy in a review by Howson (1994). Another form altogether of magnetoresistance (sometimes called 'colossal' to distinguish it from the merely 'giant') is found in bulk orthomanganite compounds; these are discussed in Chapter 11. A recent overview (Simonds 1995) analyses computer hard-disc memories, including notably those based on magnetoresistance, in terms of storage density and the linkage of that to the ever-reducing spacing between disc and reading head.

Yet further approaches to computer memories are optical and magneto-optical recording. There is space here only to summarise the great range of developments here. Pure optical recording is used in the production of the various kinds of compact discs; these are not used for computer memories but for recording text, music and videos, mostly on a read-only basis. Strings of grooves and dips are 'burned' into a plastic surface, using semiconductor lasers to achieve the ultrafine focusing of light required to cram information into closely adjacent channels (Burke 1995). The materials used and the elaborate error-correcting codes that have been developed are described by Burke.

Magneto-optical recording depends on the Kerr effect, the rotation of the plane of polarisation of plane-polarised light when it is reflected from a magnetised surface, in a sense which depends on the type of magnetization. By use of highly selective analysers, the presence or absence of a bit of information can be detected. The kind of 'rewritable' system that is now in extensive use depends on phase changes: in outline, one laser beam heats the material (typically, a rare earth metal/ transition metal alloy) above a critical temperature and remanent magnetization is achieved by an applied field as the spot cools; that magnetization is then read by a much weaker laser beam. If the memory is to be erased, a laser beam of suitable strength is used to restore the film to its pristine state. So, light is used both to 'print' the information and then to detect it and eventually to remove it, and magnetism is the form the memory takes (Buschow 1989, Burke 1995).

The foregoing paragraphs make it clear why the making and characterizing of films with thickness of the order of a nanometre (only 4–5 atoms thick) has become a major research field in its own right in recent decades. Molecular beam epitaxy, which we met earlier in this chapter, is one common technique.

7.5. OPTICAL GLASS

One of my most cherished possessions is a Leica camera, and what I like best about it is the superb quality of the lenses and of their combination into objectives. Such quality depends on several factors: the choice of glass compositions; rigorous uniformity of composition and freedom from internal stress; design of the objectives – the province of geometrical optics; precision of lens grinding, polishing and positioning.

Camera lenses, and a huge range of other optical components, are made of oxide glasses, which are also the basis of laboratory glassware, windows and containers of many kinds. Almost all practical glasses are mixes of different oxides, notably SiO_2, Na_2O, B_2O_3, Al_2O_3, CaO, BaO, of which silica is crucial. When a suitable mixture of oxides is cooled in a crucible, even quite slowly, it fails to crystallise and on reheating, the solid gradually softens, without a well-defined melting temperature such as is characteristic of crystals. Curiosity about the internal structure of glasses as a category goes back a long way, but it was not until W.H. Zachariasen (an eminent Norwegian/American crystallographer whom we met in Section 3.2.4) addressed the problem that light was cast on it. In 1932 he published a renowned paper, still frequently cited today (Zachariasen 1932), about the structure of simple oxide glasses Figure 7.10(a) shows (in two dimensions) his proposed model for a hypothetical pure A_2O_3 glass, a glass-former, while Figure 7.10(b), based on Zachariasen's further work (much of it joint with B.E. Warren), shows a model for a sodium silicate glass. SiO_2 and B_2O_3, for example, are glass-formers, while Na_2O is not; but when SiO_2 and Na_2O are mixed, a glass with better properties results. In an excellent overview of the formation and properties of oxide glasses, Rawson (1991) points out that Zachariasen's first paper "is remarkable in that it is entirely speculative and qualitative – no new observations were reported". The experiments followed in the subsequent papers. The central hypothesis of Zachariasen's paper is that the vitreous (glassy) form of an oxide should not have a significantly higher internal energy than the crystalline form, and this required that both forms must contain the same kind of oxygen polyhedra that Lawrence Bragg had already determined in crystalline silicates. His central hypothesis led him to corollaries for a pure glass-former, such as that no oxygen must be linked to more than two A atoms, the oxygen polyhedra share corners with each other but not edges or faces, and each

(a)   **(b)**

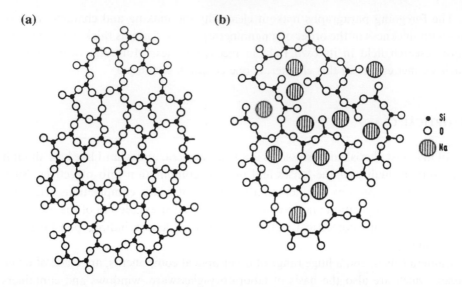

Figure 7.10. (a) Zachariasen's two-dimensional model of an A_2O_3 glass, after Zachariasen (1932). (b) Two-dimensional representation of a sodium silicate glass.

polyhedron must share at least three corners. In Figure 7.10(b), where a *glass-modifier* has been added to a *glass-former*, more oxygen become 'non-bridging', the viscosity goes down accordingly and other properties change too. Later experimental work by X-ray scattering, much of it involving Warren and his pupils at MIT, confirmed in detail the correctness of Zachariasen's model and generated statistical information about, for instance, the distribution of O–A–O angles. The gradual softening process of a glass when heated is characterised by a *glass transition temperature*, a temperature at which the glass viscosity falls to a critical level. The precise nature of the glass transition has led to much research and a huge journal literature; one of the key features of this transition is that a long anneal just below it slowly densifies the glass and also rids it of internal stresses. Annealing of optical glass is therefore a key part of the production of high-quality optical components.

Optical glasses, mostly oxide glasses, constitute a crucial class of materials. Their most important physical properties, apart from high transparency at all or most visible wavelengths, are the refractive index and the dispersion, the latter being the degree of variation of the refractive index with change of light wavelength. Another important property of all glasses, rarely spelled out explicitly, is their isotropy; the refractive index does not vary with direction of propagation, unlike many kinds of transparent crystals or polycrystals. These aspects are reviewed very clearly in an overview of optical glasses (Weber 1991). An important feature of glasses is that the refractive index and dispersion of glasses can be expressed by simple additive relationships involving just the percentages of different constituents (Huggins and Sun 1943); this makes glass design for optical purposes relatively straightforward. These properties are *not structure-sensitive*, although they do of course depend on density and are thus slightly affected by annealing.

At this point, a brief aside about the concept of 'structure-sensitivity' is in order. A German physicist, Adolf Smekal, in 1933, first divided properties into those that depend on the 'structure' – meaning crystal structure, grain structure and crystal defects of all kinds – and those that depend merely on composition. Plastic mechanical properties are prime instances of structure-sensitivity; physical properties such as thermal expansion, specific heat, elastic moduli, electron energy band gap, refractive index, dispersion are among those which are not structure-sensitive. Photoelastic properties (the induced birefringence of glass when stressed, used to measure stress distributions by optical means) are again not structure-sensitive. This concept of Smekal's has found resonance in the materials science community ever since; its thorough recognition has accompanied the conversion of physicists from their original aloofness towards all aspects of solids to their present intense involvement with solids.

Most glasses have dispersions which follow a standard relation with change of wavelength. Some glasses have anomalous dispersion, and these are useful for the design of highly colour-corrected lens sets, so-called apochromatic objectives.

There are also a range of other optical properties which cannot be discussed in detail here, for instance, non-linear optical properties... polarisation not proportional to applied electric field. These properties include 'photoinduced nonlinearities': if a standing wave is induced in an optical fibre by a laser beam, it can lead to a permanent modulation of the refractive index, so that a photo-induced diffraction grating is thereby created. (The first report of photorefractive behaviour was in a crystal, $LiNbO_3$, by Ashkin *et al.* 1966.) This kind of advanced optical engineering of materials has become a large field of expertise in recent decades; it involves both glasses and crystals, and is driven by the objective of manipulating optical signals in telecommunications without the need to keep changing from optical to electronic media.

A concise survey of where the broad domain of optical information processing had got to a few years ago is in a book issued by the European Commission (Kotte *et al.* 1989), while a good overview of non-linear optical materials is by Bloor (1994).

7.5.1 Optical fibres
One of the most spectacular modern developments in materials has been the design, improvement and exploitation of fine glass fibres for purposes of large-scale communication. Messages, above all telephone conversations, but also TV signals

and e-mail transmissions, are now converted into trains of optical pulses and sent along specially designed fibres, to be reconverted into electrical signals at the far end; hundreds of thousands of distinct messages can be sent along the same fibre, because the frequency of light is so high and the 'bandwidth' which decides the frequency range needed by one message becomes tiny compared to the available frequency when light is used, as distinct from microwaves which were standard vectors previously.

The fact that a glass fibre can 'capture' light and transport it over distances of a few meters – by repeated total internal reflection at the surface of the fibre – was known to French physicists in the 19th century, as is described in a fine popular book by Hecht (1999). In the 20th century, the various primary problems, such as leakage of light where adjacent fibres touch each other, had been solved, and bundles of fibres were applied to surgical instruments such as gastroscopes to examine defects in the stomach wall, and to face-plates to transport complex images over very short distances. The idea of using fibres for long-distance telecommunication was conceived and fostered in 1964–1966 by an imaginative and resolute Chinese immigrant to Britain, Charles K. Kao (b. 1933). He worked for Standard Telephone Laboratories in the east of England, a company which in the early 1960s concluded that none of the communications media used hitherto, up to and including millimetric microwaves, were capable of giving sufficient message capacity, and Kao's manager reluctantly concluded that "the only thing left is optical fibres", as reported in the careful historical account in Hecht's book. The reluctance stemmed from the daunting manufacturing tolerances required. When his manager was tempted away to a chair in Australia in 1964, pursuing this pipedream was left in the hands of Kao. There were two central problems: one was the need to arrange for a core and coating layer in the fibre, since uncoated fibres lost too much light at the imperfect surface; the other problem was absorption of light in the fibre, which in 1964 meant that in the best fibres, after 20 m, only 1% of the original light intensity remained.

Figure 7.11 shows the ways in which the first problem could be tackled – multimode and single mode transmission of light. In multimode transmission, the light follows many paths, in single mode transmission, there is only a single path. The main problem with multimode transmission is the gradual spreading-out of a sharp pulse; this can be ameliorated but not entirely cured by replacing the outer coating of low-refractive glass with a stepped layer of gradually changing refractive index. The difficulty with single-mode fibre is that the highly refractive core must have a diameter of the order of the light wavelength, around one micrometre, and this is technologically extremely difficult to achieve. Kao also began to study the theory of light absorption, and concluded that the only way to improve transparency in a dramatic way was to aim for purity. He and his assistant Hockham also looked

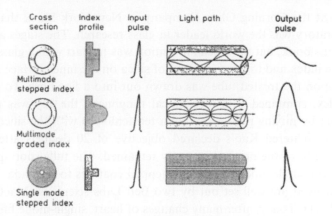

Cross section	Index profile	Input pulse	Light path	Output pulse

Multimode stepped index

Multimode graded index

Single mode stepped index

Figure 7.11. Optical fibres, showing light trajectories for different refractive index profiles (after MacChesney and DiGiovanni 1991).

at the communications theory in depth, and then Kao and Hockham (1966) published a very detailed study of a future light-optic communication channel, assuming that the practicalities of fibre design and manufacture could be solved. The paper attracted worldwide attention and brought others into the hunt, which had been Kao's objective. Many companies now became active, but many quickly became discouraged by the technical difficulties and dropped the research. Kao remained focused, and in the end his determination and collaboration with others brought success. A consequence was that the practical demonstration of optical communication was achieved by Kao's company and their close collaborators, the British Post Office. The chronological steps in this fascinating history are carefully set out by Hecht.

Kao can best be described as a 'product champion': various studies of the innovation process in industry have demonstrated the central importance of the product champion in driving a difficult innovation to the point of success. Many years ago (Cahn 1970) I published an analysis of the kinds of case history of innovations from which this conclusion eventually emerged in the 1970s. Kao would have been the perfect exemplar in this analysis, but he was still striving to convince the doubters when I wrote that paper.

Attempts by Kao and others to enhance transparency by chemically removing impurities from glass met with little success: the level of purity required was indeed comparable with that needed in silicon for integrated circuits. In the event, the required purification was achieved in the same way in which semiconductor-grade silicon is now manufactured, by going through the gas phase (silicon tetrachloride), which can be separated from the halides of impurity species because of differences in vapour pressures. This breakthrough was achieved by R.D. Maurer and his collaborators at the Corning Glass Company in New York State; that company's research laboratory was the world leader in glass research. The stages are set out in detail in Hecht's book, but the crucial invention was to start with a glass tube with a high refractive index and to deposit a soot of silica on the inner surface from the gas vector; thereupon the treated tube was drawn out into a fine fibre. To get the right refractive index, combined with mechanical toughness, the key was to codeposit some germania by mixing pure germanium tetrachloride with the silicon tetrachloride. This fibre bettered Kao's declared objective of 20 decibels attenuation per kilometre, i.e., 1% of the original intensity remained. The tube/soot approach also allowed excellent single-mode fibres with stepped coatings to be made. The detailed techniques involved are well set out by two Bell Labs scientists (MacChesney and DiGiovanni 1991). Today, after many changes of heart, single-mode fibres carry the bulk of optical messages, and transparency has been improved to better than 1 decibel per km at the preferred wavelength 'windows', 1.33 and 1.55 μm wavelength.

Efforts to improve the transparency of glass fibres even further are still continuing. In a very recent study (Thomas *et al.* 2000), the role of water – or, more specifically, the light-absorbent OH^- ion – in reducing the transparency of silica glass from the theoretical limit of about 0.2 decibel per km at 1.55 μm wavelength was examined. Water enters the glass through the use of a hydrogen/oxygen torch and this study measured the diffusion rate of water into the fibre; it turns out that the transparency is a function of radial position in the fibre, matching the decrease of OH^- ion concentration towards the centre line. This study indicates that further improvement in transparency should result from an abandonment of the hydrogen/oxygen torch.

Four other issues required to be addressed before optical communications could reach their destined perfection: a light source; a means of modulating light intensity in synchrony with an electrical signal, in the form of pulse modulation, was needed; a convenient form of amplification of the light intensity, repeaters in effect, was needed for long-distance signals through transatlantic cables and the like; and a reliable detector was needed to reconvert light into electronic signals. There is no space to go into details here: suffice it to say that heterostructured semiconductor lasers are the preferred light source, and after years of trouble with instabilities they are now reliable; there is a surfeit of possible modulators and detectors depending on magneto-optic, acousto-optic and other approaches (Nishizawa and Minikata 1996); light amplification has become possible by laser excitation of a length of fibre doped with erbium, as independently discovered in Southampton and Bell Labs in 1987. The engineering practicalities of the circuits have become extremely complex and expensive, because if millions of channels are to be passed along a single fibre with a micrometer-sized core, multiple wavelengths have to be used which requires rapidly tunable lasers and elaborate detectors. As Figure 7.12 shows, this level of 'multiplexing' channels had become possible by 1995.

Not only the number of messages that can pass along one fibre, but also the speed of transmission, has increased steadily over the past two decades; according to Sato (2000), in Japan this speed has increased by about an order of magnitude per decade, as a consequence of improved fibres and lasers and also improved networking hardware.

7.6. LIQUID CRYSTALS

One is inclined to think of "materials" as being solids; when editing an encyclopedia of materials some years ago, I found it required an effort of imagination to include articles on various aspects of water, and on inks. Yet one of the most important families of materials in the general area of consumer electronics are liquid crystals, used in inexpensive displays, for instance in digital watches and calculators. They have a fascinating history as well as deep physics.

Liquid crystals come in several varieties: for the sake of simplified illustration, one can describe them as collections of long molecules tending statistically to lie along a specific direction; there are three types, nematic, cholesteric and smectic, with an increasing measure of order in that sequence, and the variation of degree of alignment as the temperature changes is akin to the behaviour of spins in ferromagnets or of atomic order in certain alloys. The definite history of these curious materials goes back to 1888, when a botanist-cum-chemist, Friedrich Reinitzer, sent some cholesteric esters to a 'molecular physicist', Otto Lehmann.

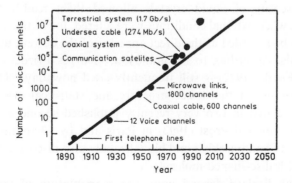

Figure 7.12. Chronology of message capacity showing exponential increase with time. The number of voice channels transmitted per fibre increases rapidly with frequency of the signalling medium. The three right-hand side points refer to optical-fibre transmission (after MacChesney and DiGiovanni 1991, with added point).

They can be considered the joint progenitors of liquid crystals. Reinitzer's compounds showed two distinct melting-temperatures, about 30° apart. Much puzzlement ensued at a time when the nature of crystalline structure was quite unknown, but Lehmann (who was a single-minded microscopist) and others examined the appearance of the curious phase between the two melting-points, in electric fields and in polarised light. Lehmann concluded that the phase was a form of very soft crystal, or 'flowing crystal'. He was the first to map the curious defect structures (features called 'disclination' today). Thereupon the famous solid-state chemist, Gustav Tammann, came on the scene. He was an old-style authoritarian and, once established in a prime chair in Göttingen, he refused absolutely to accept the identification of "flowing crystals" as a novel kind of phase, in spite of the publication by Lehmann in 1904 of a comprehensive book on what was known about them. Ferocious arguments continued for years, as recounted in two instructive historical articles by Kelker (1973, 1988). Lehmann, always eccentric and solitary, became more so and devoted his last 20 years to a series of papers on Liquid Crystals and the Theories of Life.

During the first half of this century, progress was mostly made by chemists, who discovered ever new types of liquid crystals. Then the physicists, and particularly theoreticians, became involved and understanding of the structure and properties of liquid crystals advanced rapidly. The principal early input from a physicist came from a French crystallographer, Georges Friedel, grandfather of the Jacques Friedel who is a current luminary of French solid-state physics. It was Georges Friedel who invented the nomenclature, nematic, cholesteric and smectic, mentioned above; as Jacques Friedel recounts in his autobiography (Friedel 1994), family tradition has it that this nomenclature "was concocted during an afternoon of relaxation with his daughters, especially Marie who was a fine Hellenist." Friedel grandpère recognised that the low viscosity of liquid crystals allowed them readily to change their equilibrium state when external conditions were altered, for instance an electric field, and he may thus be regarded as the direct ancestor of the current technological uses of these materials. According to his grandson, Georges Friedel's 1922 survey of liquid crystals (Friedel 1922) is still frequently cited nowadays. The very detailed present understanding of the defect structure and statistical mechanics of liquid crystals is encapsulated in two very recently published second editions of classic books, by de Gennes and Prost (1993) in Paris and by Chandrasekhar (1992) in Bangalore, India. (Chandrasekhar and his colleagues also discovered a new family of liquid crystals with disc-shaped molecules.)

Liquid crystal displays depend upon the reorientation of the 'director', the defining alignment vector of a population of liquid crystalline molecules, by a localised applied electric field between two glass plates, which changes the way in which incident light is reflected; directional rubbing of the glass surface imparts a directional memory to the glass and thence to the encapsulated liquid crystal. To apply the field, one uses transparent ceramic conductors, typically tin oxide, of the type mentioned above. Such applications, which are numerous and varied, have been treated in a book series (Bahadur 1991). The complex fundamentals of liquid crystals, including the different chemical types, are treated in the first volume of a handbook series (Demus *et al.* 1998). The linkage between the physics and the technology of liquid crystals is explained in very accessible way by Sluckin (2000). A particularly useful collection of articles covering both chemistry and physics of liquid crystals as well as their uses is to be found in the proceedings of a Royal Society Discussion (Hilsum and Raynes 1983). A more popular treatment of liquid crystals is by Collins (1990).

It is perhaps not too fanciful to compare the stormy history of liquid crystals to that of colour centres in ionic crystals: resolute empiricism followed by fierce strife between rival theoretical schools, until at last a systematic theoretical approach led to understanding and then to widespread practical application. In neither of these domains would it be true to say that the empirical approach sufficed to generate practical uses; such uses in fact had to await the advent of good theory.

7.7. XEROGRAPHY

In industrial terms, perhaps the most successful of the many innovations that belong in this Section is xerography or photocopying of documents, together with its offspring, laser-printing the output of computers. This has been reviewed in historical terms by Mort (1994). He explains that "in the early 1930s, image production using electrostatically charged insulators to attract frictionally charged powders had already been demonstrated." According to a book on physics in Budapest (Radnai and Kunfalvi 1988), this earliest precursor of modern xerography was in fact due to a Hungarian physicist named Pal Selenyi (1884–1954), who between the Wars was working in the Tungsram laboratories in Budapest, but apparently the same Zworykin who has already featured in Section 7.2.2, presumably during a visit to Budapest, dissuaded the management from pursuing this invention; apparently he also pooh-poohed a (subsequently successful) electron multiplier invented by another Hungarian physicist, Zoltan Bay (who died recently). If the book is to be believed, Zworykin must have been an early exponent of the "not invented here" syndrome of industrial scepticism.

Returning to Mort's survey, we learn that the first widely recognised version of xerography was demonstrated by an American physicist, Chester Carlson, in 1938; it was based on amorphous sulphur as the photosensitive receptor and lycopodium powder. It took Carlson 6 years to raise $3000 of industrial support, and at last, in 1948, a photocopier based on amorphous selenium was announced and took consumers by storm; the market proved to be enormously greater than predicted! Later, selenium was replaced by more reliable synthetic amorphous polymeric films; here we have another major industrial application of amorphous (glassy) materials. Mort recounts the substantial part played by John Bardeen, as consultant and as company director, in fostering the early development of practical xerography. A detailed account of the engineering practicalities underlying xerographic photo-copying is by Hays (1998). It seems that Carlson was severely arthritic and found manual copying of texts almost impossible; one is reminded of the fact that Alexander Graham Bell, the originator of the telephone, was professionally involved with hard-of-hearing people. Every successful innovator needs some personal driving force to keep his nose to the grindstone.

There was an even earlier prefiguration of xerography than Selenyi's. The man responsible was Georg Christoph Lichtenberg, a polymath (1742–1799), the first German professor of experimental physics (in Göttingen) and a name to conjure with in his native Germany. (Memoirs have been written by Bilaniuk 1970–1980 and by Brix 1985.) Among his many achievements, Lichtenberg studied electrostatic breakdown configurations, still today called 'Lichtenberg figures', and he showed in 1777 that an optically induced pattern of clinging dust particles on an insulator surface could be repeatedly reconfigured after wiping the dust off. Carlson is reported as asserting: "Georg Christoph Lichtenberg, professor of physics at Göttingen University and an avid electrical experimenter, discovered the first electrostatic recording process, by which he produced the so-called 'Lichtenberg figures' which still bear his name." Lichtenberg was also a renowned aphorist; one of his sayings was that anyone who understands nothing but chemistry cannot even understand chemistry properly (it is noteworthy that he chose not to use his own science as an example). His aphorism is reminiscent of a *New Yorker* cartoon of the 1970s in which a sad metallurgist tells his cocktail party partner: "I've learned a lot in my sixty years, but unfortunately almost all of it is about aluminum".

Just as the growth of xerographic copying and laser-printing, which derives from xerography, was a physicists' triumph, the development of fax machines was driven by chemistry, in the development of modern heat-sensitive papers most of which have been perfected in Japan.

7.8. ENVOI

The many and varied developments treated in this chapter, which themselves only scratch the surface of their theme, bear witness to the central role of functional materials in modern MSE. There are those who regard structural (load-bearing) materials as outdated and their scientific study as of little account. As they sit in their load-bearing seats on a lightweight load-bearing floor, in an aeroplane supported on load-bearing wings and propelled by load-bearing turbine blades, they can type their critiques on the mechanical keyboard of a functional computer. All-or-nothing perceptions do not help to gain a valid perspective on modern MSE. What is undoubtedly true, however, is that functional materials and their applications are a development of the postwar decades: most of the numerous references for this chapter date from the last 40 years. It is very probable that the balance of investment and attention in MSE will continue to shift progressively from structural to functional materials, but it is certain that this change will never become total.

REFERENCES

Adams, W.G. and Day, R.E. (1877) *Proc. Roy. Soc. Lond. A* **25**, 113.

Agullo-López, F. (1994) *MRS Bulletin* **19**(3), 29.

Amato, I. (1997) *Stuff: The Materials the World is Made of* (Basic Books, New York) p. 205.

Ames, I., d'Heurle, F.M. and Horstmann, R. (1970) *IBM J. Res. Develop.* **14**, 461.

Anon. (1998) Article on copper-based chip-making technology, *The Economist* (*London*) (June 6), 117.

Ashkin, A., Boyd, G.D., Dziedzic, J.M., Smith, R.G. and Ballman, A.A. (1966) *Appl. Phys. Lett.* **9**, 72.

Attardi, M.J. and Rosenberg, R. (1970) *J. Appl. Phys.* **41**, 2381.

Bachmann, K.J. (1995) *The Materials Science of Microelectronics* (VCH, Weinheim).

Bahadur, B. (ed.) (1991) *Liquid Crystals: Applications and Uses*, 3 volumes (World Scientific, Singapore).

Baibich, M.N. *et al.* (1988) *Phys. Rev. Lett.* **61**, 2472.

Baker, W.O. (1967) *J. Mater.* **2**, 915.

Bartha, L., Lassner, E., Schubert, W.-D. and Lux, B. (eds.) (1995) *The Chemistry of Non-Sag Tungsten* (Pergamon Press, Oxford).

Bednorz, J.G. and Müller, K.A. (1986) *Z. Phys. B* **64**, 189.

Bilaniuk, O.M. (1970–1980) in *Dictionary of Scientific Biography*, ed. Gillispie, C.C., vol. 7 (Charles Scribner's Sons, New York) p. 320.

Bloor, D. (1994) in *The Encyclopedia of Advanced Materials*, ed. Bloor, D. *et al.*, vol. 3 (Pergamon Press, Oxford) p. 1773.

Boyd, I.W. (1985) *Nature* **313**, 100.

Briant, C.L. (ed.) (1999) *Impurities in Engineering Materials: Impact, Reliability and Control* (Marcel Dekker, New York).

Brix, P. (1985) *Physikalische Blätter* **41**, 141.

Brown, R.G.W. and Pike, E.R. (1995) in *Twentieth Century Physics*, ed. Brown, L.M., Pais, A. and Pippard, B. vol. 3 (American Institute of Physics, Britol, Institute of Physics Publication and New York) p. 1385.

Burke, J.J. (1995) *Encyclopedia of Applied Physics*, vol. 12 (VCH Publishers, New York) p. 369.

Busch, G. (1991) *Condensed Matter News* **1**(2), 20.

Busch, G. (1993) *Condensed Matter News* **2**(1), 15.

Buschow, K.H.J. (1989) *J. Less-Common Metals* **155**, 307.

Cahn, R.W. (1970) *Nature* **225**, 693.

Cahn, R.W. (1983) *Nature* **302**, 294.

Casimir, H.B.G. (1983) *Haphazard Reality: Half a Century of Science*, Chapter 8 (Harper and Row, New York) p. 224.

Chandrasekhar, S. (1992) *Liquid Crystals*, 2nd edition (Cambridge University Press, Cambridge).

Chaplin, D.M., Fuller, C.S. and Pearson, G.L. (1954) *J. Appl. Phys.* **25**, 676.

Cho, A.Y. (1995) *MRS Bull.* **20**(4), 21.

Clogston, A.M., Hannay, N.B. and Patel, C.K.N. (1978) *J. Less-Common Metals* **62**, vii (also Raub, C., p. xi).

Colborn, R. *et al.* (eds.) (1966) Interview with Bernd Matthias, in *The Way of the Scientist* (Simon and Schuster, New York) p. 35.

Collins, P.J. (1990) *Liquid Crystals: Nature's Delicate Phase of Matter* (Princeton University Press, Princeton, NJ).

Crease, R.P. (1999) *Making Physics: A Biography of Brookhaven National Laboratory, 1946–1972* (University of Chicago Press, Chicago) p. 133.

Cross, L.E. and Newnham, R.E. (1986) History of Ferroelectrics, in *High-Technology Ceramics, Past, Present and Future*, ed. Kingery, W.D. (American Ceramic Society, Westerville, Ohio) p. 289.

Dash, W.C. (1958, 1959) *J. Appl. Phys.* **29**, 736; *ibid* **30**, 459.

De Gennes, P.G. and Prost, J. (1993) *The Physics of Liquid Crystals*, 2nd edition (Clarendon Press, Oxford).

Demus, D. *et al.* (eds.) (1998) *Handbook of Liquid Crystals, Vol. 1, Fundamentals* (Wiley-VCH, Weinheim).

De Rango, P. *et al.* (1991) *Nature* **349**, 770.

Devonshire, A.F. (1949) *Phil. Mag.* **40**, 1040.

Esaki, L. and Tsu, R. (1970) *IBM J. Res. Develop.* **4**, 61.

Fair, R.B. (editor) (1993) *Rapid Thermal Processing* (Academic Press, San Diego).

Faupel, F., Willecke, R. and Thran, A. (1998) *Mater. Sci. Eng.* **R22**, 1.

Fasor, G. (1997) *Science* **275**, 941.

Friedel, G. (1922) *Ann. Phys.* **18**, 273.

Friedel, J. (1994) *Graine de Mandarin* (Odile Jacob, Paris) p. 111.

Geballe, T.H. and Hulm, J.K. (1992) Superconducting Materials: An Overview, in *Concise Encyclopedia of Magnetic and Superconducting Materials*, ed. Evetts, J.E. (Pergamon Press, Oxford) p. 533.

Glanz, J. (1996) *Science* **271**, 1230.

Gleason, R.E. (2000) *How far will circuits shrink? Science Spectra*, issue 20, p. 32.

Goldsmid, H.J. (1964) *Thermoelectric Refrigeration* (Plenum Press, New York).

Goyal, A. (1995) *JOM* **47**(8), 55

Goyal, A. *et al.* (1999) *JOM* **51**(7), 19.

Grunberg, P. *et al.* (1986) *Phys. Rev. Lett.* **57**, 2442.

Gupta, L.C. (1999) *Proc. Indian Nat. Sci. Acad., Part A* **65A**, 767.

Haller, E.E. (1995) *J. Appl. Phys.* **77**, 2857.

Harvey, E.N. (1957) *History of Luminescence* (American Philosophical Society, Philadelphia).

Hays, D.A. (1998) in *Encyclopedia of Applied Physics*, vol. 23 (VCH Publishers, New York) p. 541.

Hecht, J. (1999) *City of Light: The Story of Fiber Optics* (Oxford University Press, Oxford).

Henisch, H.K. (1964) Electroluminescence, in *Reports on Progress in Physics*, vol. 27, p. 369.

Herman, F. (1984) *Physics Today*, June 1984, p. 56.

Herring, C. (1991) Solid State Physics, in *Encyclopedia of Physics*, ed. Lerner, R.G. and Trigg, R.L. (VCH Publishers, New York).

Hicks, L.D., Harman, T.C. and Dresselhaus, M.S. (1993) *Appl. Phys. Lett.* **63**, 3230.

Hilsum, C. and Raynes, E.P. (editors) (1983) *Liquid Crystals: Their Physics, Chemistry and Applications* (The Royal Society, London).

Hoddeson, L., Schubert, H., Heims, S.J. and Baym, G. (1992) in *Out of the Crystal Maze*, ed. Hoddeson, L. *et al.* (Oxford University Press, Oxford) p. 489.

Hodgson, S. and Wilkie, J. (2000) *Mater. World* **8**(8), 11.

Holton, G., Chang, H. and Jurkowitz, E. (1996) *Am. Sci.* **84**, 364.

Howson, M.A. (1994) *Contemp. Phys.* **35**, 347.

Huggins, M.L. and Sun, K.H. (1943) *J. Am. Ceram. Soc.* **26**, 4.

Ioffe, A.F. (1957) *Semiconductor Thermoelements and Thermoelectric Cooling* (English version) (Infosearch, London).

Jackson, K.A. (editor) (1996) in *Processing of Semiconductors, Materials Science and Technology: A Comprehensive Treatment*, vol. 16, ed. Cahn, R.W., Haasen, P. and Kramer, E.J. (VCH, Weinheim).

Joffe, A.F. and Stil'bans, L.S. (1959) *Rep. Progr. Phys.* **22**, 167.

Junod, P. (1959) *Helv. Phys. Acta* **32**, 567.

Känzig, W. (1991) *Condens. Mat. News* **1**(3), 21.

Kao, K.C. and Hockham, G.A. (1966) *Proc. IEE.* **113**, 1151.

Kasper, E., Herzog, H.J. and Kibbel, H. (1975) *Appl. Phys.* **8**, 199.

Kay, H.F. (1948) *Acta Crystallog.* **1**, 229.

Keith, S.T. and Quédec, P. (1992) Magnetism and magnetic materials, in *Out of the Crystal Maze*, Chapter 6, ed. Hoddeson, L. *et al.* (Oxford University Press, Oxford) p. 359.

Kelker, H. (1973) *Mol. Cryst. Liq. Cryst.* **21**, 1; (1988) *ibid* **165**, 1.

Keusin-Elbaum, L. *et al.* (1997) *Nature* **389**, 243.

Kotte, E.-U. *et al.* (1989) *Technologies of Light: Lasers, Fibres, Optical Information Processing, Early Monitoring of Technological Change* (Springer, Berlin).

Kraner, H.M. (1924) *J. Amer. Ceram. Soc.* **7**, 868.

Kraner, H.M. (1971) *Amer. Ceram. Soc. Bull.* **50**, 598.

Kulwicki, B.M. (1981) *PTC Materials Technology, 1955–1980*, in *Grain Boundary Phenomena in Electronic Ceramics – Advances in Ceramics*, vol. 1, ed. Levinson, L.M. (American Ceramic Society, Columbus, Ohio) p. 138.

Lenard, P., Schmidt, F. and Tomaschek, R. (1928) *Handbuch der Experimentalphysik*, vol. 23.

Levinson, L.M. (editor) (1981) *Grain Boundary Phenomena in Electronic Ceramics – Advances in Ceramics*, vol. 1 (American Ceramic Society, Columbus, Ohio).

Levinson, L.M. (1985), private communication.

Li, J., Duewer, F., Chang, H., Xiang, X.-D. and Lu, Y. (2000) *Appl. Phys. Lett.* (in press).

Lifshitz, E.M. and Kosevich, A.K. (1953) *Dokl. Akad. Nauk SSSR* **91**, 795.

Livingston, J.D. (1998) 100 Years of Magnetic Memories, *Sci. Amer.* (*November*) 80.

Loferski, J.J. (1995) Photovoltaic devices, in *Encyclopedia of Applied Physics*, vol. 13, ed. G.L. Trigg, p. 533.

MacChesney, J.H. and DiGiovanni, D.J. (1991) in *Glasses and Amorphous Materials*, ed. Zarzycki, J.; *Materials Science and Technology*, vol. 9, Cahn, R.W., Haasen, P. and Kramer, E.J. (VCH, Weinheim) p. 751.

Mahajan, S. and Sree Harsha, K.S. (1999) *Principles of Growth and Processing of Semiconductors* (McGraw-Hill, New York).

Megaw, H. (1945) *Nature* **155**, 484; **157**, 20.

Megaw, H. (1957) *Ferroelectricity in Crystals*, Methuen 1957 (London).

Miyayama, M. and Yanagida, H. (1988) Ceramic semiconductors: non-linear, in *Fine Ceramics*, ed. Saito, S. Elsevier (New York and Ohmsha, Tokyo) p. 275.

Moharil, S.V. (1994) *Bull. Mater. Sci. Bangalore* **17**, 25.

Mort, J. (1994) *Phys. Today* **47**(1), 32.

Mott, N.F. and Jones, H. (1936) *The Theory of the Properties of Metals and Alloys* (Clarendon Press, Oxford) p. 310.

Moulson, A.J. and Herbert, J.M. (1990) *Electroceramics: Materials, Properties, Applications* (Chapman and Hall, London).

Nagarajan, R. *et al.* (1994) *Phys. Rev. Lett.* **72**, 274.

Nakamura, S. (1996) *Japanese J. Appl. Phys.* **35**, L74–L76.

Néel, L. (1936) *Compt. Rend. Acad. Sci. Paris* **203**, 304.

Néel, L. (1948) *Annal. Phys.* **3**, 137.

Newnham, R.E. (1975) *Structure-Property Relations* (in a monograph series on *Crystal Chemistry of Non-Metallic Materials*) (Springer, Berlin).

Newnham, R.E. (1997) *MRS Bull.* **22**(5), 20.

Nishizawa, J. and Minakata, M. (1996) *Encyclopedia of Applied Physics*, vol. 15 (VCH Publishers, New York) p. 339.

Notis, M.R. (1986) in *High-Technology Ceramics, Past, Present and Future*, vol. 3, ed. Kingery, W.D. (American Ceramic Society, Westerville, Ohio) p. 231.

Paul, D. (2000) *Phys. World* **13**(2), 27.

Pfann, W.G. (1954) *Trans. AIME* **194**, 747.

Pfann, W.G. (1958, 1966) *Zone Melting*, 1st and 2nd editions (Wiley, New York).

Pippard, B. (1994) Obituary of John Bardeen, *Biograp. Mem. Fellows R. Soc.* **39**, 21.

Pippard, B. (1995) Electrons in solids, in *Twentieth Century Physics*, vol. 3, ed. Brown, L.M., Pais, A. and Pippard, B. (Institute of Physics Publications, Bristol and Amer. Inst. of Physics, New York) p. 1279.

Ponce, F.A. and Bour, D.P. (1997) *Nature* **386**, 351.

Radnai, R. and Kunfalvi, R. (1988) *Physics in Budapest* (North-Holland, Amsterdam) pp. 64, 74.

Rawson, H. (1991) in *Glasses and Amorphous Materials*, ed. Zarzycki, J.; *Materials Science and Technology*, vol. 9, ed. Cahn, R.W., Haasen, P. and Kramer, E.J. (VCH, Weinheim) p. 279.

Regain, B.O. and Grätzel (1991) *Nature* **353**, 637.

Reid, T.R. (1984) *The Chip* (Simon and Schuster, New York).

Roth, W.L. (1972) *J. Solid-State Chem.* **4**, 60.

Riordan, M. and Hoddeson, L. (1997) *Crystal Fire: The Birth of the Information Age* (W.W. Norton and Co., New York and London).

Sales, B.C. (1997) *Current Opinion in Solid State and Materials Science*, vol. 2, p. 284.

Sato, K.-I. (2000) *Phil. Trans. Roy. Soc. Lond. A* **358**, 2265.

Scaff, J.H. (1970) *Metall. Trans.* **1**, 561.

Schropp, R.E.I. and Zeeman, M. (1998) *Amorphous and Microcrystalline Silicon Solar Cells* (Kluwer Academic Publishers, Dordrecht).

Seitz, F. (1996) *Proc. Amer. Philo. Soc.* **140**, 289.

Seitz, F. and Einspruch, N.G. (1998) *Electronic Genie: The Tangled History of Silicon* (University of Illinois Press, Urbana and Chicago).

Shick, A.B., Ketterson, J.B., Novikov, D.L. and Freeman, A.J. (1999) *Phys. Rev. B* **60**, 15480.

Shull, C.G., Wollan, E.O. and Strauser, W.A. (1951) *Phys. Rev.* **81**, 483.

Simonds, J.L. (1995) *Phys. Today* (April), 26.

Slack, G.A. (1995) in *CRC Handbook of Thermoelectrics*, ed. Rowe, D.M. (Chemical Rubber Co. Boca Raton, FL) p. 470.

Sluckin, T.J. (2000) *Contemp. Phys.* **41**, 37.

Smekal, A. (1933) Aufbau der zusammenhängende Materie, in *Handbuch der Physik*, vol. 24 (part 2), p. 795.

Smit, J. and Wijn, H.P.J. (1959) *Ferrites* (Philips Technical Library, Eindhoven).

Snoek, J.L. (1936) *Physica* **3**, 463.

Sondheimer, E.H. (1999) Biographical memoir of Sir Alan Herries Wilson, *Biog. Mems. Fell. R. Soc. Lond.* **45**, 547.

Spear, W.E. (1974) in *Proc. Int. Conf. on Amorphous and Liquid Semiconductor*, ed. Stuke, J. and Brenig, W. (Taylor and Francis, London) p. 1.

Street, R.A. (1991) *Hydrogenated Amorphous Silicon* (Cambridge University Press, Cambridge).

Suits, C.G. and Bueche, A.M. (1967) Cases of research and development in a diversified company, in *Applied Science and Technological Progress* (no editor cited) (National Academy of Sciences, Washington, DC) p. 297.

Sze, S.M. (editor) (1991) *Semiconductor Devices: Pioneering Papers* (World Scientific, Singapore).

Thomas, G.A., Shraiman, B.I., Glodis, P.F. and Stephen, M.J. (2000) *Nature* **404**, 262.

Toigo, J.W. (2000) Avoiding a data crunch, *Sci. Amer.* **282**(5), 40.

TRACES (1968) *Technology in Retrospect and Critical Events in Science* (*TRACES*), Illinois Institute of Technology, Research Institute; published for the National Science Foundation (no editor or author named).

Valenzuela, R. (1994) *Magnetic Ceramics* (Cambridge University Press, Cambridge).

Verwey, E.J.W. and Heilmann, E.I. (1947) *J. Chem. Phys.* **15**, 174.

Weber, M.J. (1991) in *Glasses and Amorphous Materials*, ed. J. Zarzycki; *Materials Science and Technology*, vol. 9, ed. Cahn, R.W., Haasen, P. and Kramer, E.J. (VCH, Weinheim) p. 619.

Whall, T.E. and Parker, E.C.H. (1995) *J. Mater. Elect.* **6**, 249.

Wilson, A.H. (1931) *Proc. Roy. Soc. Lond. A* **133**, 458; **134**, 277.

Wilson, A.H. (1939) *Semi-conductors and Metals* (Cambridge University Press, Cambridge).

Wilson, A.H. (1980) *Proc. Roy. Soc. Lond. A* **371**, 39.

Yeack-Scranton, C.E. (1994) *Encyclopedia of Applied Physics*, vol. 10 (VCH Publishers, New York) p. 61.

Zachariasen, W.H. (1932) *J. Amer. Ceram. Soc.* **54**, 3841.

Ziman, J.M. (1960) *Electrons and Phonons* (Clarendon Press, Oxford) p. 396.

Zweibel, K. (1990) *Harnessing Solar Power: The Photovoltaics Challenge* (Plenum Press, New York).

Chapter 8
The Polymer Revolution

Chapter 8
The Polymer Revolution

8.1. BEGINNINGS

The early years, when the nature of polymers was in vigorous dispute and the reality of long-chain molecules finally came to be accepted, are treated in Chapter 2, Section 2.1.3. For the convenience of the reader I set out the sequence of early events here in summary form.

The understanding of the nature of polymeric molecules was linked from an early stage with the stereochemical insights due to van't Hoff, and the recognition of the existence of isomers. The main argument was between the followers of the notion that polymers are "colloidal aggregates" of small molecules of fixed molecular weight, and those, notably Staudinger, who insisted that polymers were long-chain molecules, covalently bound, of high but variable molecular weight. That argument was not finally settled until 1930. After that, numerous scientists became active in finding ever more ingenious ways of determining MWs and their distributions.

The discovery of stereoactive catalysts to foster the polymerisation of monomers transformed the study of polymers from an activity primarily to satisfy the curiosity of a few eccentric chemists into a large-scale industrial concern. These discoveries started in the 1930s with the finding, by ICI in England, that a combination of high pressure and oxygen served to create an improved form of polyethylene, and peaked in the early 1950s with the discoveries by Ziegler and Natta of low-pressure catalysts, initially applicable to polyethylene but soon to other products as well. In a separate series of events, Carothers in America set out to find novel synthetic fibres, and discovered nylon in the early 1930s. In the same period, chemists struggled with the difficult task of creating synthetic rubber.

After 1930, when the true nature of polymers was at last generally, recognised, the study of polymers expanded from being the province of organic specialists; physical chemists like Paul Flory and physicists like Charles Frank became involved. In this short chapter, I shall be especially concerned to map this broadening range of research on polymers.

A number of historically inclined books are recommended in Chapter 2. Here I will only repeat the titles of some of the most important of these. The best broad but concise overview is a book entitled *Polymers: The Origins and Growth of a Science* (Morawetz 1985); it covers events up to 1960. A very recent, outstanding book is *Inventing Polymer Science: Staudinger, Carothers and the Emergence of Macromolecular Chemistry* (Furukawa 1998). His last chapter is a profound consideration of

307

"the legacy of Staudinger and Carothers". These two books focus on the underlying science, though both also describe industrial developments. A British multiauthor book, *The Development of Plastics* (Mossman and Morris 1994), edited by specialists at the Science Museum in London, covers industrial developments, not least the Victorian introduction of parkesine, celluloid and bakelite. Published earlier is a big book classified by specific polymer families and types (e.g., polyesters. styrenes, polyphenylene sulfide, PTFE, epoxys, fibres and elastomers) and focusing on their synthesis and uses: *High Performance Polymers: Their Origin and Development* (Seymour and Kirshenbaum 1986). Still earlier was a fine book about the discovery of catalytic methods of making synthetic stereoregular polymers, which in a sense was the precipitating event of modern polymer technology (McMillan 1979).

8.2. POLYMER SYNTHESIS

For any of the many distinct categories of materials, extraction or synthesis is the necessary starting-point. For metals, the beginning is the ore, which has to be separated from the accompanying waste rock, then smelted to extract the metal which subsequently needs to be purified. Extractive metallurgy, in the 19th century, was the central discipline. It remains just as crucial as ever it was, especially since ever leaner ores have to be treated and that becomes ever more difficult; but by degrees extractive metallurgy has become a branch of chemical engineering, and university courses of materials science keep increasingly clear of the topic. There are differences: people who specialise in structural and decorative ceramics, or in glass, are more concerned with primary production methods... but here the starting-point is apt to be the refined oxide, as distinct from the raw material extracted from the earth.

The point of this digression is to place the large field of polymer chemistry, alternatively polymer synthesis, in some kind of perspective. The first polymers, in the 19th century, were made from natural precursors such as cotton and camphor, or were natural polymers in the first place (rubber). Also the objective in those early days was to find substitutes for materials such as ivory or tortoiseshell which were becoming scarce: 'artificial' was the common adjective, applied alike to polymers for billiard balls, combs, and stiff collars (e.g., celluloid), and to the earliest fibres ('artificial silk'). Bakelite was probably the first truly synthetic polymer, made from laboratory chemicals (phenol and formaldehyde), early in the twentieth century, invented independently by Leo Baekeland (1863–1944) and James Swinburne (1858–1958); bakelite was not artificial anything. Thereafter, and especially after ICI's perfection, in 1939, of the first catalyst for polymerising ethylene under high pressure, the classical methods of organic chemistry were used, and steadily improved. At first the task was simply to bring about polymerisation at all; soon, chemists began to focus on the equally important tasks of controlling the *extent* of polymerisation, and its stereochemical character. If one is to credit an introductory chapter (*Organic chemistry and the synthesis of well-defined polymers*) to a very recent text on polymer chemistry (Müllen 1999), even today "organic chemists tend to avoid polymers and are happy when 'polymers' remain at the top of their chromatography column. They consider polymers somewhat mysterious and the people who make them somewhat suspect. Polydisperse compounds (i.e., those with variable MWs) are not accepted as 'true' compounds and it is believed that a method of bond formation, once established for the synthesis of a small compound, can be extended without further complication toward polymer synthesis." Polymer specialists have become a chemical breed apart. As Müllen goes on to remark "While a synthesis must be 'practical' and provide sufficient quantities, the limitations of the synthetic method, with respect to the occurrence of side products and structural defects, must be carefully investigated, e.g., for establishing a reliable structure-property relationship". The situation was reminiscent of the difficulties encountered by the early semiconductor researchers who found their experimental materials too impure, too imperfect and too variable.

The 665 pages of the up-to-date text for which Müllen wrote cover an enormous range of chemical and catalytic techniques developed to optimise synthetic methods. One feature which sets polymer chemistry apart from traditional synthetic organic chemistry is the need to control mean MWs and the range of MWs in a polymeric product (the degree of 'polydispersity'). Such control is feasible by means of so-called 'living radical polymerisation' (Sawamoto and Kamigaito 1999); initiators are used to start the polymerisation reaction and 'capping reagents' to terminate it. The techniques of making polymers with almost uniform MWs are now so well developed that such materials have their own category name, 'model polymers', and they have extensive uses in developing novel materials, structures and properties and in testing concepts in polymer physics (Fettes and Thomas 1993). Quite generally, recent developments in polymerisation catalysis have made possible the precise control not only of molecular weight but also of co-monomer sequence and stereo-sequence (Kobayashi 1997).

A special form of polymerisation is in the solid state; in this way, single crystals of diacetylenes have been made, and this was the starting-point of the major developments now in progress with electrically conducting polymers. Yet another unexpected approach is the use of radiation to enhance polymerisation or cross-linking of polymers, for instance of rubbers during tire manufacture (Charlesby 1988).

Occasionally, a completely new family of polymers is discovered, and then the synthesizers have to start from scratch to find the right methods: an example is the family of dendrimers (Janssen and Meijer 1999), discovered in the 1980s, polymers which spread radially from a nucleus, with branching chains like the branches of a tree (hence the name, from the Greek word for a tree). Such polymers can be made with virtually uniform MWs, but at the cost of slow and extremely laborious synthetic methods.

The standard textbook of polymer science in the 1960s was that by Billmeyer (1962); of its 600 pages, 125 were devoted to polymerisation, i.e., to polymer chemistry. But this has changed: the important domain of polymer chemistry has become, by degrees, a branch of science almost wholly divorced from the rest of polymer science, with its own array of journals and conferences, and certainly not an integral part of materials science, and not treated in most general texts on polymer science. Accordingly, I will not treat it further in this chapter. The aspects of polymer science that form part of MSE nowadays are polymer processing and polymer physics.

8.3. CONCEPTS IN POLYMER SCIENCE

The whole of polymer science is constructed around a battery of concepts which are largely distinct from those familiar in other families of materials, metals in particular. This is the reason why I invited an eminent polymer scientist who was originally a physical metallurgist to write, for a textbook of physical metallurgy edited by me, a chapter under the title "A metallurgist's guide to polymers" (Windle 1996). The objective was to remove some of the mystery surrounding polymer science in the eyes of other kinds of materials scientists.

In outline form, here are some of the key concepts treated in that chapter. Polymers can be homopolymers (constituted of only one kind of monomer) or copolymers, constituted of (usually) two chemically different kinds of monomers. Copolymers, in turn, can be statistically mixed (random copolymers) or else made up of blocks of the two kinds of monomers... block copolymers or, if there are sidechains, graft copolymers; the lengths of the blocks can vary widely. Both kinds of polymer have variable MWs; the 'polydispersity' can be slight or substantial. The chains can be linear or branched, and linear chains can be stereotactic (with sidegroups arranged in a regular conformation), or disordered (atactic). According to the chemistry, a polymer can be resoftened by reheating (thermoplastic) or it can harden irreversibly when fully polymerised (thermoset).

Many polymers are amorphous, i.e., a kind of glass, complete with a glass transition temperature which is dependent on heating or cooling rate. Even crystalline polymers have a melting range depending on molecular weight. (It was these two features – variable MWs, and absence of a well-defined melting temperature – which stuck in the craw of early organic chemists when they contemplated polymers).

A polymer can consist of a three-dimensional, entangled array of chains of various lengths, which can be cross-linked to a greater or lesser degree. The chain lengths and cross-linking, together with the temperature, decide whether the material is rigid, fluid or – as an in-between condition – elastomeric, that is, rubber-like. Fluid polymers have a visco-elastic character that distinguishes their mechanical behaviour from fluids like water or molten metals. Elastomeric polymers are ultra-resilient and their elasticity is of almost wholly entropic origin; such materials become stiffer when heated, unlike non-polymeric materials.

Amorphous stereotactic polymers can crystallise, in which condition neighbour-ing chains are parallel. Because of the unavoidable chain entanglement in the amorphous state, only modest alignment of amorphous polymer chains is usually feasible, and moreover complete crystallisation is impossible under most circum-stances, and thus many polymers are semi-crystalline. It is this feature, semicrys-tallinity, which distinguished polymers most sharply from other kinds of materials. Crystallisation can be from solution or from the melt, to form spherulites, or alternatively (as in a rubber or in high-strength fibres) it can be induced by mechanical means. This last is another crucial difference between polymers and other materials. Unit cells in crystals are much smaller than polymer chain lengths, which leads to a unique structural feature which is further discussed below.

Most pairs of homopolymers are mutually immiscible, so that phase diagrams are little used in polymer science... another major difference between polymers on the one hand, and metals and ceramics on the other. Two-phase fields can be at lower or higher temperatures than single-phase fields... another unique feature.

Plastic deformation in polymers is not usually analysed in terms of dislocations, because crystallinity is not usually sufficiently perfect for this concept to make sense. Nevertheless, polymers do work-harden, like metals... indeed, strongly drawn fibres become immensely strong, because the intrinsic strength of the carbon–carbon backbone of a polymer chain then makes itself felt. Deformed polymers, especially amorphous ones, develop 'crazes', thin regions filled with nanosized voids; the fracture mechanics of polymers is intimately bound up with crazes, which are not known in other materials. Crazes propagate like cracks, but unlike cracks, they can support some load. As Windle puts it, "development of a craze is a potent, albeit localised, energy absorption mechanism which makes an effective contribution to resisting the propagation of a crack which follows it; a craze is thus both an incipient fracture *and* a toughening mechanism".

The methods used to characterise polymers are partly familiar ones like X-ray diffraction, Raman spectroscopy and electron microscopy, partly less familiar but widespread ones like neutron scattering and nuclear magnetic resonance, and partly unique to polymers, in particular, the many methods used to measure MWs and their distribution.

It is clear enough why polymers strike many materials scientists as very odd. However, since the 1930s, some physical chemists have made crucial contributions to the understanding of polymers; in more recent decades, many physicists have turned their attention wholly to polymer structures, and a number of metallurgists, such as the writer of the chapter referred to in this Section, have done likewise. As we will see in the next Section, some cross-fertilisation between polymer science and other branches of MSE has begun.

8.4. CRYSTALLINE AND SEMICRYSTALLINE POLYMERS

8.4.1 Spherulites

The most common form of crystallization in polymers is the *spherulite* (Figure 8.1(a) and (b)), which can grow from solution, melt or the solid amorphous form of a polymer. Spherulites do form in a number of inorganic systems, but only in polymers are they the favoured crystalline form. The first proper description of spherulites was by two British crystallographers, working in the chemical industry (Bunn and Alcock 1945); they used optical microscopy and X-ray diffraction to characterise the nature of the spherulites. In general, the individual polymer chains run tangentially (normal to the radius vector). The isothermal growth rate is found to be constant,

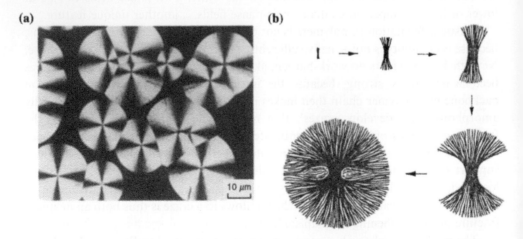

Figure 8.1. (a) Spherulites growing in a thin film of isotactic polystyrene, seen by optical microscopy with crossed polars (from Bassett 1981, after Keith 1963). (b) A common sequence of forms leading to spherulitic growth (after Bassett 1981). The fibres consist of zigzag polymer chains.

independently of the radius. The universality of this morphology has excited much theoretical analysis. A good treatment is that by Keith and Padden (1963), which draws inspiration from the then-new theory of freezing of alloys due to Chalmers and Rutter; the build-up of rejected impurities or solute leads to 'constitutional supercooling' (see ch. 9, sect. 9.1.1). Here, the 'impurities' are disordered (atactic) or branched chains. This leads to regular protuberances on growing metal crystal interfaces, while in polymers the consequence is the formation of fibrils, as seen schematically in Figure 8.1(b).

Spherulites are to be distinguished from dendrimers, which also have spherical form. A dendrimer is a single molecule of a special kind of polymer which spreads from a nucleus by repeated branching.

8.4.2 Lamellar polymer crystals

A very different morphology develops in a few polymers, grown from solution. Early experiments, in the 1930s and again the early 1950s, were with gutta-percha, a rather unstable natural polymer. The first report of such a crystal morphology from a well characterised, synthetic polymer was by Jaccodine (1955), who grew thin platelets from a solution of linear polyethylene, of molecular weight ≈10,000, in benzene or xylene. Figure 8.2 shows a population of such crystals. Jaccodine's report at once excited great interest among polymer specialists, and two years later, three scientists independently confirmed and characterised such polyethylene crystals (Till 1957, Keller 1957, Fischer 1957) and all showed by electron diffraction in an electron microscope that the polymer chains were oriented normal to the lamellar plane. They thereby started a stampede of research, accompanied by extremely vigorous disputes as to interpretation, which continues to this day. These monocrystal lamellae can

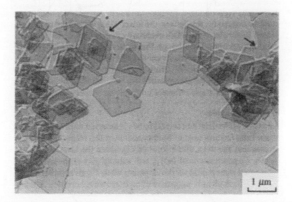

Figure 8.2. Lozenge-shaped monocrystals of polyethylene grown from solution by a technique which favors monolayer-type crystals. Electron micrograph (after Bassett 1981).

only be made with stereoregular polymers in which the successive monomers are arranged in an ordered pattern; Figure 8.3 shows the unit cell of a polyethylene crystal according to Keller (1968).

One of the active researchers on polymer crystals was P.H. Geil, who in 1960 reported nylon crystals grown from solution; in his very detailed early book on polymer single crystals (Geil 1963) he remarks that all such crystals grown from dilute solution consist of thin platelets, or lamellae, about 100 Å in thickness; today, a compilation of published data for polyethylene indicates that the thickness ranges between 250 and 500 Å (25–50 nm), increasing sharply with crystallization temperature. The exact thickness depends on the polymer, solvent, temperature, concentration and supersaturation. Such a crystal is much thinner than the length of a polymer chain of M.W. 10,000, which will be in excess of 1000 Å. The inescapable conclusion is that each chain must fold back on itself several times. As Keller put it some years later, "folding is a straightforward necessity as the chains have nowhere else to go". It has been known since 1933 that certain paraffins can crystallize with two long, straight segments and one fold, the latter occupying approximately five carbon atoms' worth of chain length. To make this surprising conclusion even harder

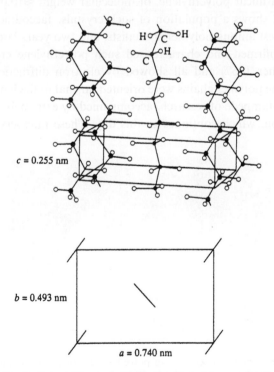

$c = 0.255$ nm

$b = 0.493$ nm

$a = 0.740$ nm

Figure 8.3. Unit cell of crystalline polyethylene, adapted from a figure by Keller 1968.

to accept than it intrinsically is, it soon became known that annealing of the thin crystals allowed them gradually to thicken; what this meant in terms of the comportment of the multiple folds was mysterious.

In the decade following the 1957 discovery, there was a plethora of theories that sought, first, to explain how a thin crystal with folds might have a lower free energy than a thick crystal without folds, and second, to determine whether an emerging chain folds over into an adjacent position or folds in a more disordered, random fashion... both difficult questions. Geil presents these issues very clearly in his book. For instance, one model (among several 'thermodynamic' models) was based on the consideration that the amplitude of thermal oscillation of a chain in a crystal becomes greater as the length of an unfolded segment increases and, when this as well as the energy of the chain ends is considered, thermodynamics predicts a crystal thickness for which the total free energy is a minimum, at the temperatures generally used for crystallization. The first theory along such lines was by Lauritzen and Hoffman (1960). Other models are called 'kinetic', because they focus on the kinetic restrictions on fold creation. The experimental input, microscopy apart, came from neutron scattering (from polymers with some of the hydrogen substituted by deuterium, which scatters neutrons more strongly), and other spectroscopies. Microscopy at that time was unable to resolve individual chains and folds, so arguments had to be indirect. The mysterious thickening of crystal lamellae during annealing is now generally attributed to partial melting followed by recrystallisation. The issue here is slightly reminiscent of the behaviour of precipitates during recrystallisation of a deformed alloy; one accepted process is that crystallites are dissolved when a grain boundary passes by and then re-precipitate.

The theoretical disputes gradually came to center on the question whether the folds are regular and 'adjacent' or alternatively are statistically distributed, as exemplified in Figure 8.4. The grand old man of polymer statistical mechanics, Paul Flory, entered the debate with rare ferocity, and the various opponents came together in a memorable Discussion of the Faraday Society (by then a division of the Royal Society of Chemistry in London). Keller (1979) attempted to set out the different points of view coolly (while his own preference was for the 'adjacent' model), but his attempted role as a peacemaker was slightly impeded by a forceful General Introduction in the same publication by his Bristol colleague Charles Frank, who by 1979 had converted his earlier concern with crystal growth of dislocated crystals into an intense concern with polymer crystals, and by even more extreme remarks by the aged Paul Flory, who was bitterly opposed to the 'adjacent' model. Frank included a "warning to show what bizarrely different models can be deemed consistent with the same diffraction evidence". He also delivered a timely reminder that applies equally to neutron scattering and X-ray diffraction: "All we can do is to make models and see whether they will fit the scattering data within experimental

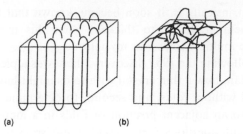

Figure 8.4. Schematic representation of chain folds in polymer single crystal. (a) regular adjacent reentry model; (b) random switchboard model.

error. If they don't, they are wrong. If they do, they are not necessarily right. You must call in all aids you can to limit the models to be tested." After the Discussion, Flory sent in the following concluding observations: "As will be apparent from perusal of the papers... denunciation of those who have the temerity to challenge the sacrosanct doctrine of regular chain folding in semicrystalline polymers is the overriding theme and motivation. This purpose is enunciated in the General Introduction, with a stridency that pales the shallow arguments mustered in support of chain folding with adjacent re-entry. The cant is echoed with monotonous iterations in ensuing papers and comments..." (Then, with regard to papers by some of the opponents of the supposed orthodoxy:) "The current trend encourages the hope that rationality may eventually prevail in this important area".

It is not often that discussion in such terms is heard or read at scientific meetings, and the 1979 Faraday Discussion reveals that disputatious passion is by no means the exclusive province of politicians, sociologists and littérateurs. Nevertheless, however painful such occasions may be to the participants, this is one way in which scientific progress is achieved.

The arguments continued in subsequent years, but it is beginning to look as though the enhanced resolution attainable with the scanning tunneling microscope may finally have settled matters. A recent paper by Boyd and Badyal (1997) about lamellar crystals of poly(dimethylsilane), examined by atomic force microscopy (Section 6.2.3) yielded the conclusion: "It can be concluded that the folding of polymer chains at the surface of polydimethylsilane single crystals can be seen at molecular scale resolution by atomic force microscopy. Comparison with previous electron and X-ray diffraction data indicates that polymer chain folding at the surface is consistent with the regular adjacent reentry model." The most up-to-date general overview of research on polymer single crystals is a book chapter by Lotz and Wittmann (1993).

Andrew Keller (1925–1999, Figure 8.5), who was a resolute student of polymer morphology, especially in crystalline forms, for many decades at Bristol University

Figure 8.5. Andrew Keller (1925–1999) (courtesy Dr. P. Keller).

in company with his mentor Charles Frank, was a chemist who worked in a physics department. In a Festschrift for Frank's 80th birthday (Keller 1991), Keller offered a circumstantial account of his key discovery of 1957 and how the special atmosphere of the Bristol University physics department, created by Frank, made his own researches and key discoveries possible. It is well worth reading this chapter as an antidote to the unpleasant atmosphere of the 1979 Faraday Discussion.

In concluding this discussion, it is important to point out that crystalline polymers can be polymorphic because of slight differences in the conformation of the helical disposition of stereoregular polymer chains; the polymorphism is attributable to differences in the weak intermolecular bonds. This abstruse phenomenon (which does not have the same centrality in polymer science as it does in inorganic materials science) is treated by Lotz and Wittmann (1993).

8.4.3 Semicrystallinity

The kind of single crystals discussed above are all made starting from solution. In industrial practice, bulk polymeric products are generally made from the melt, and such polymers (according to their chemistry) are either wholly amorphous or have 30–70% crystallinity. Indeed, even 'perfect' lamellar monocrystals made from solution have a little non-crystalline component, namely, the parts of each chain where they curl over for reentry at the lamellar surface. The difference is that in bulk polymers the space between adjacent lamellae gives more scope for random configuration of chains, and according to treatment, that space can be thicker or thinner (Figure 8.6). Attempts to distinguish clearly between the 'truly' crystalline regions and the disturbed space have been inconclusive; indeed, the terms under which a percentage of crystallinity is cited for a polymer are not clearly defined.

Perhaps the most remarkable polymeric configuration of all is the so-called shish-kebab structure (Figure 8.7). This has been familiar to polymer microscopists for decades. Pennings in the Netherlands (Pennings *et al.* 1970) first studied it systematically; he formed the structure by drawing the viscous polymer solution (a gel) from a rotating spindle immersed in the solution. Later, Mackley and Keller (1975) showed that the same structure could be induced in flowing solution with a longitudinal velocity gradient, and thereby initiated a sequence of research on controlled flow of solutions or melts as a means of achieving desired polymer morphologies. A shish-kebab structure consists of substantially aligned but non-crystalline chains, so arranged that at intervals along the fibre, a proportion of the chains splay outwards and generate crystalline lamellae attached to the fibre. Quite recently, Keller and Kolnaar (1997) discuss the formation of shish-kebab morphology in depth, but my impression is that even today no one really understands how and why this form of structure comes into existence, or what factors determine the periodicity of the kebabs along the shish.

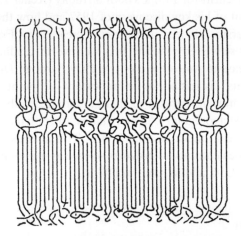

Figure 8.6. A diagrammatic view of a semicrystalline polymer showing both chain folding and interlamellar entanglements. The lamellae are 5–50 nm thick (after Windle 1996).

(a)   **(b)**

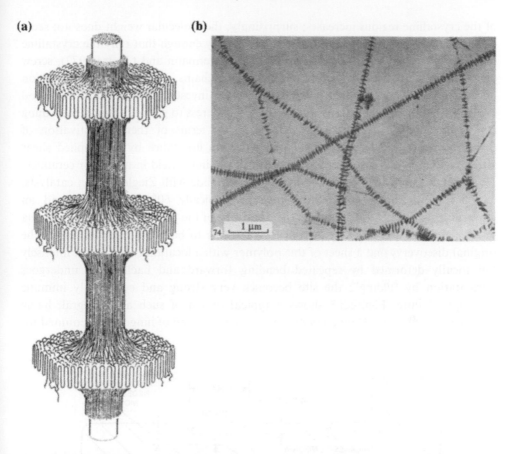

Figure 8.7. (a) Idealised view of a shish-kebab structure (after Pennings *et al.* 1970, Mackley and Keller 1975). (b) Shish kebabs generated in a flowing solution of polyethylene in xylene (after Mackley and Keller 1975).

8.4.4 Plastic deformation of semicrystalline polymers

Typically, a semicrystalline polymer has an amorphous component which is in the elastomeric (rubbery) temperature range – see Section 8.5.1 – and thus behaves elastically, and a crystalline component which deforms plastically when stressed. Typically, again, the crystalline component strain-hardens intensely; this is how some polymer fibres (Section 8.4.5) acquire their extreme strength on drawing.

The plastic deformation of such polymers is a major research area and has a triennial series of conferences entirely devoted to it. The process seems to be drastically different from that familiar from metals. A review some years ago (Young 1988) surveyed the available information about polyethylene: the yield stress is linearly related to the fraction of crystallinity, and it increases sharply as the thickness of the crystalline regions increases; surprisingly, the molecular weight does not seem to have any systematic effect. All this shows clearly enough that only the crystalline regions deform irreversibly. As early as 1972 (Petermann and Gleiter 1972), screw dislocations, with Burgers vectors parallel to the chains, were observed by electron microscopy in semicrystalline polyethylene; these investigators also obtained good evidence that these dislocations were activated by stress to generate slip steps. Young (1974) interpreted the measured yield stress in terms of thermal activation of dislocations at the edges of crystal platelets with assistance by the applied shear stress...an approach just like that current in examining yield in metals or ceramics.

Isotactic (sterically ordered) polypropylene, made with Ziegler–Natta catalysts, has become a major commodity polymer, typically 60% crystalline, and an important reason for this success is the discovery of the *polypropylene hinge* (Hanna 1990). It was found many years ago (there seems to be no documentation of the original discovery) that a sheet of this polymer with a local thin area, when intensely but locally deformed by repeated bending forward and backwards, undergoes "orientation by folding"; the site becomes very strong and completely immune to fatigue failure. Figure 8.8 shows a typical design of such an "integral, living polypropylene hinge". Hanna (1990) opines that this kind of hinge has accounted for

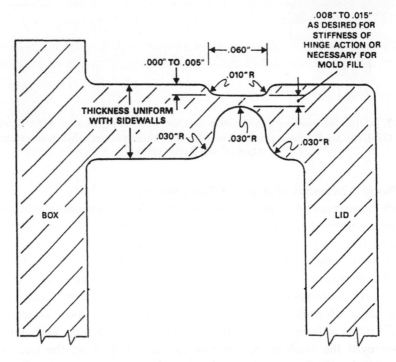

Figure 8.8. Design for a polypropylene hinge (modified from Hanna 1990).

much of the rapid growth of the industrial usage of polypropylene. It should be added that no interpretation has been offered for this unique immunity to fatigue failure.

The mechanical behavior of polymers, as well as many other topics in polymer engineering, are presented in an up-to-date way in a book by McCrum *et al.* (1998).

8.4.5 Polymer fibres

Leaving aside rayon and 'artificial silks' generally, the first really effective polymeric textile fibre was nylon, discovered by the chemist Wallace Hume Carothers (1896–1937) in the Du Pont research laboratories in America in 1935, and first put into production in 1940, just in time to make parachutes for the wartime forces. This was the first of several major commodity polymer fibres and, together with high-density polyethylene introduced about the same time and 'Terylene', polyethylene tereph-thalate, introduced in 1941 (the American version is Dacron), transformed the place of polymers in the materials pantheon.

The manufacture of nylon fibre involves a drawing step, rather like the drawing of an optical glass fibre (Section 7.5.1), which serves to align the chains. This form of drawing has been developed to the point, today, where immensely strong fibres with very intense chain alignment are routinely manufactured. It seems to have been Frank (1970) who originally analysed, from first principles, the strength and stiffness that might be expected of such products when strongly aligned. A schematic view of such a fibre is shown in Figure 8.9. The secret of obtaining a high elastic modulus is not only to achieve high alignment of the chains but also to minimise the volume of the intercrystalline tangles. Different treatments and different polymers generate different properties: thus nylon ropes, with large elastic extensibility, are used by mountaineers because they can absorb the high kinetic energy of a falling body without breaking, while terylene (dacron) cords with their very high modulus are used by archers for bowstrings.

The problems involved in orienting polymers for improved properties were first surveyed in a special issue of *Journal of Materials Science* (Ward 1971b). Another early survey of this important modern technology was a book edited by Ciferri and Ward (1979), while a recent authoritative account of the modern technology is by Bastiaansen (1997).

8.5. STATISTICAL MECHANICS OF POLYMERS

From about 1910 onwards, physical chemists began studying the characteristics of polymer solutions, measuring such properties as osmotic pressure, and found them

Figure 8.9. Diagram of the structure of a drawn polymer fibre. The Young's modulus of the crystallised portions is between 50 and 300 GPa, while that of the interspersed amorphous 'tangles' will be only 0.1–5 GPa. Since the strains are additive, the overall modulus is a weighted average of the two figures (after Windle 1996).

to be non-ideal; an outline of the stages is to be found in Chapter 16 of Morawetz (1985). The key event was the formulation, independently by the Americans Huggins (1942) and Flory (1942), of a statistical theory of the (Gibbs) free energy of mixed homopolymers in solution. (One of these papers was published in the *Journal of Physical Chemistry*, the other in the *Journal of Chemical Physics*). The theory was worked out on the understanding, which itself took a long time to gel, that polymer chains are highly flexible and can assume a great many alternative shapes in solution. This theory formed part of one of the most enduring of polymer texts, Flory's *Principles of Polymer Chemistry* (1953), which is still regularly cited today; it was followed by the same author's *Statistical Mechanics of Chain Molecules* (1969). Paul Flory (1910–1985) was stimulated to his crucial researches by William Carothers whom he joined at Du Pont in 1934 as a young physical chemist; he constituted part of that "restoration of the physicalist approach" to polymer science which is treated in the illuminating Chapter 5 of Furukawa's book on Staudinger and Carothers. Flory was awarded the Nobel Prize for Chemistry in 1974.

The Flory–Huggins equation has assumed a central place in the understanding of the mixing of different polymers, both in solution and in the melt. Any expression for a free energy must include enthalpy (internal energy) and entropy terms. The key conclusion is that the *configurational entropy* of mixing of polymer chains is very much smaller than that for individual atoms in a metallic solid solution. A crude way of explaining this is to point out that the constituent atoms in a polymer chain are linked inseparably together and thus have less freedom to rearrange themselves than the 'free' atoms in a metallic alloy; the difference is the greater, the higher the mean molecular weight of the polymer chains. The enthalpy term differs much less as between polymeric and metallic systems. The result is, in the words of Windle (1996), "For polymeric systems where the MWs of the chains are high, the enthalpic term (in the expression for free energy) will be very dominant. Given that, in bonding terms, like tends to prefer like, and thus the enthalpic term will usually be positive, solubility, or 'miscibility' as it is known in polymer parlance, will be unlikely. This is in accord with observation. *In general, dissimilar polymers are insoluble in each other.* There are, however, important and interesting exceptions." According as the constituent atoms of distinct chain types attract or repel each other, one can find polymer pairs in solution which mix at high temperatures but phase-separate below a critical temperature, or else be intersoluble at low temperatures and phase-separate as they are heated. It is fair to say, however, that solid-solution formation is rare enough that phase diagrams play only a modest role in polymer science, compared with their very central role in metallurgy and ceramics.

8.5.1 Rubberlike elasticity: elastomers

Rubber was a very major component of the polymer industry from its very beginning. From the beginning of the 20th century, attempts were made to make synthetic rubber, because the natural rubber industry was beset by severe economic fluctuations which made supplies unpredictable. A wide range of synthetic rubber-like materials were made from the late 1930s onwards, initially by the German chemical industry under ruthless pressure from Hitler. The German methods were known by some American companies and were taken over and quickly improved by those companies from 1942 onwards, once America had entered the War. The pressure for reliable rubber supplies in America can be attributed to the fact that in the late 1930s, the USA, with twice the population of Germany, manufactured 15 times as many automobiles. All these variegated rubbers – 'elastomers' in polymer language – were chemically distinct from natural rubber, polyisoprene; an elastomer chemically identical to natural rubber was successfully synthesised only in 1953, in the US; until then, heavy-duty truck tires, a particularly demanding product, could only be made from natural rubber, but thereafter all products could, if necessary, be manufactured from synthetics. The complicated story is told from a chemical viewpoint by Morawetz (1985) in his Chapter 8, and by Morris (1994) from a more political and economic viewpoint.

By the 1960s, a great range of synthetic rubbers were available to tire designers. David Tabor at the Cavendish Laboratory in Cambridge, whose research expertise was in friction between solids, formulated a hypothesis relating tire adhesion to the road surface to the resilience of the rubber (the degree to which it rebounds in shape after deformation); he took out a patent in 1960. This view soon became more elaborate, and adhesion was linked to hysteresis, the delay in resilience. Since highly hysteretic rubber generates much heat on cyclic deformation, it became necessary to use different elastomers for the tread and the tire sidewall where much of the heat is generated by flexure during each rotation of the wheel. For a time, this kind of tire construction became the orhodoxy. The subtle linkage between the viscoelastic properties of elastomers and tire properties is very clearly set out by Bond (1990), who put Tabor's ideas into effect.

Throughout the early stages of the synthetic rubber industry, there was essentially no understanding why rubbers have the extraordinary elastic extensibility which is the raison d'être of their many applications. The sequence of events which finally dispelled this ignorance is set out in Chapter 15 of Morawetz's admirable book. They began in Germany. The suggestion that the origin of rubberlike elasticity lay in configurational entropy, based on careful measurements of heat absorption and emission during stretching and retraction of rubber, was made in a key paper by Meyer *et al.* (1932). In 1934, W. Kuhn presented evidence that, contrary to Staudinger's conviction at that time, polymer chains in the rubbery or molten state are not rigid but are free to rotate at each bond, and in the same year, Guth and Mark (1934) put forward the essential feature of modern theory, relating rubberlike elasticity to the probability distribution of different degrees of curling of a long, flexible chain. (This is the same Herman Mark who featured in early research on metal single crystals, 12 years before, Section 4.2.1.) A completely straight chain has only one possible configuration, but the more curled up a chain is, i.e., the shorter the distance between its ends, the more distinct configurations are compatible with that distance. This means that a force will resist attempts to change a chain from a more probable to a less probable configuration, and that is the restoring force that causes a stretched rubber band to retract. Rubberlike elasticity is entropy made tangible.

For the behavior of the individual chain to be reflected in the behavior of the aggregate, neighbouring chains must be crosslinked at intervals, which is done by partial vulcanisation of rubber. It became clear that rubber progressively crystallises, reversibly, as it is stretched. Rubber can also be crystallised thermally, by cooling to the right temperature, and then the chains have no preferred orientation. If rubber is cooled below its glass transition, the chains cease to be flexible and rubberlike behavior ceases. An early exposition of the modern theory can be found in Chapter 3 of an influential little book by Treloar (1958). One of Treloar's figures (Figure 8.10), taken from a later book (Treloar 1970), refers to a rubber sample which has been thermally crystallized and then half of it has been heated enough to convert it back to the amorphous form; if the specimen is then kept at the right temperature, both parts stay metastably as they are, and on stretching only the amorphous part extends.

An idea of the present complexity of the statistical theory of rubberlike elasticity can be garnered from Chapter 7 of a recent book on *The Physics of Polymers*, by Strobl (1996).

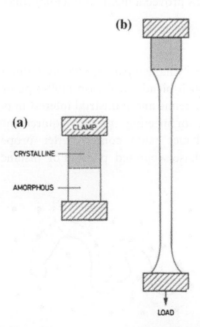

Figure 8.10. A sample of rubber treated to make it half crystalline, half amorphous. On stretching, measurable extension is restricted to the amorphous part (after Treloar 1970).

8.5.2 Diffusion and reptation in polymers

In Section 4.2.2 the central role of atomic diffusion in many aspects of materials science was underlined. This is equally true for polymers, but the nature of diffusion is quite different in these materials, because polymer chains get mutually entangled and one chain cannot cross another. An important aspect of viscoelastic behavior of polymer melts is 'memory': such a material can be deformed by hundreds of per cent and still recover its original shape almost completely if the stress is removed after a short time (Ferry 1980). This underlies the use of shrink-fit cling-film in supermarkets. On the other hand, because of diffusion, if the original stress is maintained for a long time, the memory of the original shape fades.

The principal way in which a polymer molecule can diffuse through a population of chains is by *reptation*, which can also be described as the Brownian diffusion of a polymer chain among fixed obstacles. This idea and its ramifications are due to de Gennes (1971) and Edwards (1976), and the process is schematically shown in Figure 8.11. The notion is that a chain is constrained by its neighbours, shown as dots (cross-sections of chains); the wriggling molecule is constrained to stay within a 'virtual tube' but it can move by a snake-like progression within that tube. The mobility, of course, diminishes as the chain becomes longer. The kinetics of the process and its relation to a traditionally defined diffusion constant are concisely set out by Léger and Viovi (1994). Reptation has proved a highly influential concept.

8.5.3 Polymer blends

Polymer 'alloys' are generally named *polymer blends* within the polymer community. In a recent overview of such blends, Robeson (1994) points out that "the primary reason for the surge of academic and industrial interest in polymer blends is directly related to their potential for meeting end-use requirements". He points out that, in general, miscible polymer pairs confer better properties, mechanical ones in particular, than do phase-separated pairs. For instance, the first commercial

Figure 8.11. Reptation of a polymer chain. The chain moves snake-like through its confining virtual tube.

miscible blend of synthetic polymers emerged in the early 1940s: poly(vinyl chloride) and butadiene-acrylonitrile copolymer (a form of rubber) were mixed in order to improve oxidative and ultraviolet stability of the rubber. Robeson cites an early survey of polymer blends in 1968 which listed only 12 miscible pairs, of which several were actually copolymers.

A copolymer, random or block, should not really be counted as an example of a miscible blend, because there is only a single population of polymer chains, albeit with variable composition along their lengths. Very important examples of such a block copolymer are the various forms of rubber-toughened polystyrene (PS). Polystyrene is in itself a cheap and strong mass polymer, but very brittle. It was found in the 1930s that the brittleness could be obviated by copolymerising PS with a synthetic elastomer (rubber) such as polybutadiene; the key product is ABS, acrylonitrile-butadiene-styrene copolymer, which was finally commercialised in 1953 after more than 10,000 laboratory experiments to get the chemistry right (Pavelich 1986). The interesting feature of such copolymers is that the rubber blocks on different chains dispose themselves adjacent to each other, so that chunks (often, microspheres) of rubber are dispersed regularly in the PS matrix. Unmodified PS fractures in tension at very small strains by crazing (see Section 8.3, above), while rubber-modified polystyrene can be elongated by $\approx 50\%$. Argon and Cohen (1989) showed that this large strain comes from a large number of minute crazes originating at interfaces between the glassy matrix and the more compliant inclusions; the crazing strain acts as a stress-relief mechanism, retarding fracture. The early development of rubber-toughened polymers was described in a book by Bucknall (1977).

The separation of the polybutadiene and polystyrene blocks into separate 'phases' poses an intriguing conceptual question. Can they really be considered as distinct phases in view of the fact that the blocks are linked together by covalent bonds in the same polymer chains? This poses a problem for established ideas, such as Findlay's phase rule that governs the form of phase diagrams. I do not know the answer. A very general treatment of the processing and properties of block copolymers with 'interphase' interfaces by Inoue and Maréchal (1997) includes a comparison of the structures of such products made from preexisting copolymeric chains with the same product made by dispersing homopolymers and then copolymerising them in situ in the solid state. This again underlines the fact that polymer science is replete with procedures and issues that have no parallel elsewhere in materials science.

Some very peculiar features have been discovered in the microstructures of copolymers. Thus, Hanna *et al.* (1993) showed that a *random* copolymer of two aromatic monomers has chains in which random but similar sequences of the two monomers on distinct chains 'find' each other and "come into register to form a layered structure with crystalline periodicity perpendicular to the chains but with no periodicity parallel to the chains". This is an early example of *self-assembly* in controlling polymer chain shape, a topic which has become very much to the fore in materials chemistry. Another recent paper (Percec *et al.* 1998) is entitled "Controlling polymer shape through the self-assembly of dendritic side-groups".

8.5.4 Phase transitions in polymers

In the preceding section, I asked how the phase rule should apply to the structure of block copolymers and confessed to puzzlement. Altogether, phase transitions in polymers are even more complex than in metals and ceramics, and a number of new principles are beginning to emerge. One thing is clear: in the polymer literature, one does not often see phase (equilibrium) diagrams, and I know of no collection of polymer phase diagrams (unlike the situation with metals and ceramics, where many thousands of diagrams have been collected and are in very frequent use by researchers).

One of the few investigators to have homed in on phase transitions in polymers, especially in two-component systems, and to introduce phase diagrams from time to time, is Hugo Berghmans in Belgium. An example of his work is in a paper by Aerts *et al.* (1993): here the polythene/diphenyl ether system is examined and the linkage between phase behavior and morphology is examined and a phase diagram established. One crucial point he emphasises is that the classical phase rule does not apply to such systems: a state with two liquid phases and one crystalline phase should be temperature-invariant according to the phase rule, but it is not so (the authors claim) because of the 'polydispersity' of the polymers, i.e., the fact that the molecular weight of each polymer shows a broad distribution. This is a variable which obviously has no analogue in metal alloys and ceramics.

The most striking treatment I know of phase transitions in polymers, and of metastability in particular, is by Keller and his coworkers. When Keller (1995) first addressed this issue, he pointed out that in polymers the state of ultimate equilibrium is hardly ever attained, and metastability is the rule. He even claimed the existence of stability inversion as crystallite size changes. His ideas were further developed in two papers written shortly before his death (Cheng and Keller 1998, Keller and Cheng 1998). One extraordinary observation presented and discussed here is that a single polymer crystal can have regions of different thicknesses and thus different degrees of metastability and also different melting temperatures. In another system, crystal thicknesses were shown to be 'quantised' as a function of changing crystallisation temperature. There is no space here to go further into these subtleties, but clearly there is enormous scope for research into the linked thermodynamic, kinetic and morphological aspects of phase transformations in polymers.

8.6. POLYMER PROCESSING

In no other branch of MSE, perhaps, is the role of processing in determining properties quite so intense as with polymers. Methods such as injection-molding, extrusion, drawing for 'ultimate properties', blow-molding, film-casting, each have to be controlled in fine detail to ensure the desired morphology and consequent properties, and the whole matter is further complicated by the fact that the viscoelastic properties of polymer melts depend not only on the chemical nature of the polymer in question, but also on the mean molecular weight and its distribution. Most (but by no means all) processing starts from the melt, but drawing of ultrastrong fibres takes place in the solid state. Casting, in the sense familiar from metals, plays little part, likewise, the sintering of powders. Computer modelling plays a particularly important part in improving processing technology; this is briefly discussed in Chapter 12.

Quite generally, the details of processing methods play an exceptionally central role in determining the resultant polymer properties; this is underlined by the title of the opening chapter in a major text on processing of polymers (Meijer 1997) – "Processing for properties". Properties are determined alike by the processing route and by the intrinsic chemical structure. This linkage is underlined by a famous polymer reference book, *Properties of Polymers* (van Krevelen 1990) which is devoted to the "correlation of properties with chemical structure, their numerical estimation and prediction from additive group contributions".

It is not feasible here to go in any detail into the history of processing methods; let it suffice to point out that that history goes back to the Victorian beginnings of polymer technology. Thus, as Mossman and Morris (1993) report, the introduction of camphor into the manufacture of parkesine in 1865 was asserted to make it possible to manufacture more uniform sheets than before. Processing has always been an intimate part of the gradual development of modern polymers.

Another important part of polymer science which I do not have space to consider in the detail it deserves is the theory of flow of viscoelastic polymeric melts – a topic closely linked to diffusion and, indeed, to processing. The science of fluid flow generally is the province of *rheology*. That discipline takes its name from the Greek… 'panta rhei', everything flows, a motto enunciated by the Greek philosopher Heraclitus. The term was introduced in 1929, when the first national society devoted to that field was founded in the USA. Since that time, much of the emphasis in rheology has been devoted to polymeric fluids and their peculiar behavior under stress (see, particularly, Ferry 1980). An outstanding treatment of the history of rheology, with vignettes of dozens of the founding fathers, and accounts of the schools of thoughts and disputes between them, has recently been published by Tanner and Walters (1998). These two books make excellent partners for the leading early treatment of the mechanical properties of solid polymers (Ward 1971a).

8.7. DETERMINING MOLECULAR WEIGHTS

At the end of the 1930s, the only generally available method for determining mean MWs of polymers was by chemical analysis of the concentration of chain end-groups; this was not very accurate and not applicable to all polymers. The difficulty of applying well tried physical chemical methods to this problem has been well put in a reminiscence of early days in polymer science by Stockmayer and Zimm (1984). The determination of MWs of a solute in dilute solution depends on the ideal, Raoult's Law term (which diminishes as the reciprocal of the MW), but to eliminate the non-ideal terms which can be substantial for polymers and which are independent of MW, one has to go to ever lower concentrations, and eventually one "runs out of measurement accuracy". The methods which were introduced in the 1940s and 1950s are analysed in Chapter 11 of Morawetz's book.

In the 1930s, one novel method was introduced by a Swedish chemist, The Svedberg, who invented the ultracentrifuge, an instrument in which a solution (of colloidal particles, proteins or synthetic polymers) is subjected to forces many times greater than gravity, and the equilibrium distribution of concentration (which may take weeks to attain) is estimated by measuring light absorption as a function of position along the length of the specimen chamber as the centrifuge spins. It took a long time for this approach to be widely used for polymers because of the great cost of the instrument; Du Pont acquired the first production instrument in 1937. Eventually it became a major technique and Svedberg (who himself was mainly concerned with proteins) earned a Nobel Prize. The theory that related equilibrium concentration gradients to molecular weight is the same as that put forward in Einstein's 1905 paper that was applied to Brownian motion and thus served to cement the atomic hypothesis (Section 3.1.1).

Two classical approaches for MWs of polymers, osmometry and viscometry, both go back to the early years of the 20th century: the former was plagued by technical difficulties with membranes, the latter, by long drawn-out arguments about the theory. Staudinger worked out his own theory of the relation between viscosity and MW, but on the assumption of rigid chains. Morawetz claims that "although the validity of Staudinger's 'law' proved later to have been an illusion, there can be little doubt that its acceptance at the time advanced the progress of polymer science". This is reminiscent of Rosenhain's erroneous views about amorphous layers at grain boundaries in metals, which nevertheless stimulated research on grain boundaries, mainly by those determined to prove him wrong. Motives in scientific research are not always impeccable. Viscometry has considerable drawbacks, including the fact that viscosities depend on chain shape, unbranched or branched.

An approach which began during the War was light scattering from polymer solutions. This again depended on an Einstein paper, this time dated 1910, in which he calculated scattering from density and compositional fluctuations. The technique was applied early to determine particle size in colloidal solutions, especially by Raman in India (e.g. Raman 1927), but its application to the more difficult problem of polymers awaited the input of the famous Dutch physical chemist Peter Debye (1884–1966), who in the 1940s had become a refugee in the USA. Stockmayer and Zimm describe in detail how Debye's theory (Debye 1944) opened the doors, by stages, to MW determination by light scattering.

The crowning development in MW determination was the invention of gel permeation chromatography, the antecedents of which began in 1952 and which was finally perfected by Moore (1964). A column is filled with pieces of cross-linked 'macroporous' resin and a polymer solution (gel) is made to flow through the column. The polymer solute permeates the column more slowly when the molecules are small, and the distribution of molecules after a time is linked not only to the average MW but also, for the first time with these techniques, to the vital parameter of MW distribution.

This brief outline of the gradual solution of a crucial characterisation dilemma in polymer science could be repeated for other aspects of characterisation; in polymer science, as in other parts of MSE, characterisation techniques and theories are crucial.

8.8. POLYMER SURFACES AND ADHESION

Most adhesives either are wholly polymeric or contain major polymeric constituents, and therefore the study of polymer surfaces is an important branch of polymer science, and it turns out that polymer diffusion is of the essence here. A great battery of characterisation techniques has been developed to study the structure of surfaces and near-surface regions in polymers, and the high activity in this field is attested by the fact that in 1995, a Faraday Discussion (volume 98) was held on *Polymers at Surfaces and Interfaces..* Not only adhesion depends on the nature of polymer surfaces. In Section 7.6 we saw that the functioning of liquid-crystal displays depends on glass plates coated with polyimide in contact with a liquid crystal layer, which induce alignment of the liquid-crystal 'director'. It has recently been proved that light brushing of the polyimide coating generates substantial chain alignment; such brushing had been found empirically to be necessary to prepare the glass plates for their function.

Adhesion generally requires the polymer(s) involved to be above their glass transition temperature, so that polymer diffusion (reptation) can proceed. Polymers can diffuse not only into other polymers but also, for instance, into slightly porous metal surfaces. The details have been effectively studied by Brown (1991, 1995): one approach is to use a diblock copolymer and deuterate one of the blocks, so that after interdiffusion the location of residual deuterium (heavy hydrogen) can be assessed. It turns out that according to the length of the chains, the adhesive layer fractures either by pullout or by 'scission' at the join between the blocks. Another aspect of the behaviour of adhesive layers depends on the energy required to develop and propagate crazes at the interface, which has been intensively studied by E.J. Kramer and others. When an adhesive has the right elastomeric character, it may be possible to generate very weak bonds by simple finger pressure, readily reversible without damage to the surface; this is the basis of the well-known Post-it^{TM} notes.

The broader issues of adhesion are beyond my scope here; a good source is a book by Kinloch (1987).

8.9. ELECTRICAL PROPERTIES OF POLYMERS

Until about twenty years ago, the concept of "electrical properties of polymers", or indeed of any organic chemicals, was equivalent to "dielectric properties"; organic conductors and semiconductors were unknown. Polymers were (and still are) used as dielectrics in condensers and to insulate cables, especially in demanding uses such as radar circuits, and latterly (in the form of polyimides) for dielectric layers in integrated circuits. The permittivity and loss factor (analogous to permeability and hysteresis in ferromagnets) are linked to structural relaxations in individual polymer molecules, and through this they are linked to mechanical hysteresis when a polymer is reversibly stressed. The variables need to be accurately measured at frequencies from main frequency (50 cycles/s) to microwave frequencies (up to 10^{11} cycles/s). The needed techniques were developed in America by Arthur von Hippel and in Britain by Willis Jackson, both of whom were early supporters of the concept of materials science. This early work, which included researches on polymers, was assembled in a renowned monograph (von Hippel 1954). This was supplemented by a different kind of book which has also achieved classic status, (McCrum *et al.* 1967), devoted to a discussion, side by side, of dielectric and mechanical forms of relaxation and hysteresis in polymers. The origins of the different kinds of relaxation were discussed in terms of the underlying molecular motional processes. An updated treatment of these matters is by Williams (1993).

In 1972, the first stable organic conductor was reported, one of the forms of TCNQ, TetraCyaNo-Quinodimethane. Its room-temperature conductivity was found to be close to that of metals like lead or aluminium; it is a one-dimensional property linked to the long shape of the molecules. Study of such organic conductors (dubbed 'synthetic metals') grew apace and the field soon had its own journal. Even before this, there was a short burst of research on organic superconductors (with very low critical temperatures), and the first (it was also the last) international conference on organic superconductors was held in 1969. The story of organic (non-polymeric) conductors and superconductors is outlined by Jérome (1986). A later concise view of this intriguing field, with a estimate of successes and failures, is by Campbell Scott (1997); he points out that around 1980, "the 'holy grail' became an air-stable polymer with the conductivity of copper. In retrospect, it is hard to believe that serious consideration was given to the use of plastics to replace wiring, circuit board connections, major windings, or solenoid coils." So it is probably fair to say that 'synthetic metals' have come and gone.

By the time the next overview of 'electrical properties of polymers' was published (Blythe 1979), besides a detailed treatment of dielectric properties it included a chapter on conduction, both ionic and electronic. To take ionic conduction first, ion-exchange membranes as separation tools for electrolytes go back a long way historically, to the beginning of the twentieth century: a polymeric membrane semipermeable to ions was first used in 1950 for the desalination of water (Jusa and McRae 1950). This kind of membrane is surveyed in detail by Strathmann (1994). Much more recently, highly developed polymeric membranes began to be used as electrolytes for experimental rechargeable batteries and, with particular success, for fuel cells. This important use is further discussed in Chapter 11.

About the time that 'synthetic metals' reached their apogee, twenty years ago, research began on semiconducting polymers. Today, at the turn of the century, such polymers have taken the center of the stage, and indeed promise some of the most important applications of polymers.

A completely separate family of conducting polymers is based on ionic conduction; polymers of this kind (Section 11.3.1.2) are used to make solid electrolyte membranes for advanced batteries and some kinds of fuel cell.

8.9.1 Semiconducting polymers and devices

The key concept in connection with semiconducting polymers is that of the *conjugated chain*. This is readily appreciated by examining a simplified diagram of the structure of poly(acetylene), C_nH_n (Figure 8.12), with the hydrogen atoms omitted. It can be seen that there is an alternation of single and double bonds. There are different ways of looking at the consequences of this conjugated configuration; one involves an examination of the electronic charge distribution in the bond orbitals (well explained, for instance, by Friend *et al.* 1999), but this falls outside my limits

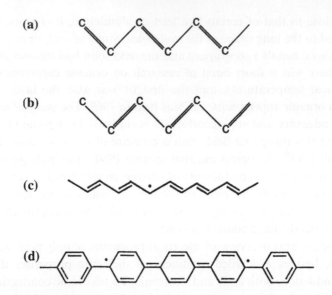

Figure 8.12. A conjugated chain in poly(acetylene). (a) changes to (b) when a charge passes along the backbone of the molecule. (c) and (d) show chains of poly(acetylene) and poly(para phenylene) respectively, each containing solitons (after Windle 1996).

here. Another way (after Windle 1996) is that one can visualise charge moving along the chain by the stepwise movement of double bonds from (say) right to left (going from (a) to (b) in the figure). The key factor, now, is that in equilibrium the double bond is shorter than the single one by about 0.003–0.004 nm (only 1–2%), but this is still very significant. The bond length cannot catch up with the movement of electrons, because the latter is much faster than the phonon-mediated process which allows the bond length to change. This mismatch between actuality and equilibrium in the bond lengths brings about strain and hence an energy band gap, allowing semiconducting behaviour. The band gap is modified if there are 'errors' along the chain, in the form of solitons (Figure 8.12(c) and (d)); such defects are brought about by doping; in polymers, dopants have to be used at per cent levels instead of parts per million, as in inorganic semiconductors. An electron or hole will bind itself to a soliton, forming a charged defect called a polaron. For such conjugated chains to operate well in semiconducting mode, the polymer needs to be, and remain, highly stereoregular.

One of the earliest observations of high conductivity in such a material was in a form of poly(acetylene) by a Japanese team (Shirakawa and Ikeda 1971). Perhaps one should date the pursuit of semiconducting polymer devices from that experiment. It soon became clear that conjugated polymers had a severe drawback; most of them are extremely stable against potential solvents; they cannot be forced into solution and furthermore are infusible (they decompose before they melt), hence the standard forms of polymer processing are unavailable. One way in which this was overcome was by starting with a single crystal of a monomer, diacetylene, and polymerising this in the solid state. However, cheapness is crucial to the success of polymer devices, in competition with other devices which have a headstart of decades, and further development awaited the invention of a synthetic trick (the 'Durham route', Edwards and Feast 1980), by which a precursor polymer which *is* soluble in common solvents was prepared cheaply and then heat-treated to produce poly(acetylene). More recently, the most useful semiconducting polymer, poly(phenylene vinylene), or PPV, has been made soluble by attaching appropriate sidechains to the phenylene rings. It can then be processed by spin-coating (in which a drop of solution is placed on a rapidly spinning substrate), which is a cheap way of preparing a thin uniform film. These processing tricks are surveyed by Friend (1994), who had set up two highly active research groups in Cambridge (one academic and one industrial), and also from a chemical perspective by Wilson (1998), who at that time was working with Friend.

By 1988, a number of devices such as a MOSFET transistor had been developed by the use of poly(acetylene) (Burroughes *et al.* 1988), but further advances in the following decade led to field-effect transistors and, most notably, to the exploitation of electroluminescence in polymer devices, mentioned in Friend's 1994 survey but much more fully described in a later, particularly clear paper (Friend *et al.* 1999). The polymeric light-emitting diodes (LEDs) described here consist in essence of a polymer film between two electrodes, one of them transparent, with careful control of the interfaces between polymer and electrodes (which are coated with appropriate films). PPV is the polymer of choice.

Friend *et al.* (1999) explain that polymeric LEDs have advanced so rapidly that they are now as efficient as the traditional tungsten-filament light bulb, and as efficient as the InGaN semiconductor lasers with their green light, announced at about the same time (Section 7.2.1.4). They also point out that, when a way is found to deposit polymeric LEDs on a polymer substrate instead of glass, they will become so cheap (especially if printing techniques can be used for deposition) that they will presumably make substantial inroads into the huge market for backlights in devices such as mobile telephones. If polymeric LEDs can be developed that will emit well-defined colours (at present they emit a broad wavelength range) then they will become candidates for full-color flat-screen displays, which is a market worth tens of billions of dollars a year.

The latest review of the status and prospects of 'polymer electronics' (Samuel 2000), by a young physicist working in Durham University, England, goes at length into the possibilities on the horizon, including the use of copolymer chains with a series of blocks with distinct functions, and the possible use of dendrimer molecules designed to "have the designed electronic properties at the core and linked by conjugated links to surface groups, which are selected to control the processing properties". Samuel also goes out of his way to underline the value of having "flexible electronics", based on flexible substrates which will not break.

Polymers have come a long way from parkesine, celluloid and bakelite: they have become functional as well as structural materials. Indeed, they have become both at the same time: one novel use for polymers depends upon precision micro-embossing of polymers, with precise pressure and temperature control, for replicating electronic chips containing microchannels for capillary electrophoresis and for microfluidics devices or micro-optical components.

REFERENCES

Aerts, L., Kunz, M., Berghmans, H. and Koningsveld, R. (1993) *Makromol. Chemie* **194**, 2697.

Argon, A. and Cohen, R.E. (1989) *Adv. Polymer Sci.* **90/91**, 301.

Bassett, D.C. (1981) *Principles of Polymer Morphology* (Cambridge University Press, Cambridge).

Bastiaansen, C.W.M. (1997) High-modulus and high-strength fibres based on flexible macromolecules, in *Processing of Polymers*, ed. Meijer, H.E.H.; *Materials Science and Technology, A Comprehensive Treatment*, vol. 18, eds. Cahn, R.W., Haasen, P. and Kramer, E.J. (VCH, Weinheim) p. 551.

Billmeyer, F.W. (1962) *Textbook of Polymer Science* (Wiley, Interscience, New York).

Blythe, A.R. (1979) *Electrical Properties of Polymers* (Cambridge University Press, Cambridge).

Bond, R. (1990) Tire adhesion: role of elastomer characteristics, in *Supplementary Volume 2 of the Encyclopedia of Materials Science and Engineering*, ed. Cahn, R.W. (Pergamon Press, Oxford) p. 1338.

Boyd, R.D. and Badyal, J.P.S. (1997) *Adv. Mater.* **9**, 895.

Brown, H.R. (1991) Adhesion between polymers, *Annu. Rev. Mater. Sci.* **21**, 463.

Brown, H.R. (1995) *Phys. World* (January) p. 38.

Bucknall, C.B. (1977) *Toughened Plastics* (Applied Science, London).

Bunn, C.W. and Alcock, T.C. (1945) *Trans. Faraday Soc.* **41**, 317.

Burroughes, J.H., Jones, C.A. and Friend, R.H. (1988) *Nature* **335**, 137.

Campbell Scott, J. (1997) *Sci.* **278**, 2071.

Charlesby, A. (1988) Radiation processing of polymers, in *Supplementary Volume 1 of Encyclopedia of Materials Science and Engineering*, ed. Cahn, R.W. (Pergamon Press, Oxford) p. 454.

Cheng, S.Z.D. and Keller, A. (1998) *Annu. Rev. Mater. Sci.* **28**, 533.

Ciferri, A. and Ward, I.M. (eds.) (1979) *Ultra-High Modulus Polymers* (Applied Science Publishers, London).

Debye, P.J.W. (1944) *J. Appl. Phys.* **25**, 338.

De Gennes, P.G. (1971) *J. Chem. Phys.* **55**, 572.

Edwards, J.H. and Feast, W.J. (1980) *Polymer Commun.* **21**, 595.

Edwards, S.F. (1976) The configuration and dynamics of polymer chains, in *Molecular Fluids*, eds. Balian, R. and Weill, G. (Gordon & Breach) London.

Ferry, J.D. (1980) *Viscoelasticity of Polymers*, 3rd edition (Wiley, New York).

Fettes, L.J. and Thomas, E.L. (1993) Model polymers for materials science, in *Structure and Properties of Polymers*, ed. Thomas, E.L. *Materials Science and Technology, A Comprehensive Treatment*, vol. 12, eds. Cahn, R.W., Haasen, P. and Kramer, E.J. (VCH, Weinheim) p. 1.

Fischer, E.W. (1957) *Z, Naturforsch.* **12a**, 753.

Flory, P.J. (1942) *J. Chem. Phys.* **10**, 51.

Frank, F.C. (1970) *Proc. Roy. Soc., (London)* **319A**, 127.

Friend, R.H. (1994) Conductive polymers, in *Encyclopedia of Advanced Materials*, vol. 1, eds. Bloor, D. *et al.* (Pergamon Press, Oxford) p. 467.

Friend, R., Burroughes, J. and Shimoda, T. (1999) *Phys. World* **12**(6), 35.

Furukawa, Yasu (1998) *Inventing Polymer Science: Staudinger, Carothers and the Emergence of Macromolecular Chemistry* (University of Pennsylvania Press, Philadelphia).

Geil, P.H.(1963) *Polymer Single Crystals* (Wiley, Interscience, New York).

Guth, E. and Mark, H. (1934) *Monatshefte Chem.* **65**, 93.

Hanna, R.D. (1990) Polypropylene, in *Handbook of Plastic Materials and Technology*, ed. Rubin, I.I. (Wiley, New York).

Hanna, S., Romo-Uribe, A. and Windle, A.H. (1993) *Nature* **366**, 546.

Huggins, M.L. (1942) *J. Phys. Chem.* **46**, 151.

Inoue, T. and Maréchal, P. (1997) Reactive processing of polymer blends: polymer–polymer interface aspects, in *Processing of Polymers*, ed. Meijer, H.E.H.; *Materials Science and Technology, A Comprehensive Treatment*, eds. Cahn, R.W., Haasen, P. and Kramer, E.J. (VCH, Weinheim) p. 429.

Jaccodine, R. (1955) *Nature* **176**, 305.

Janssen, H.M. and Meijer, E.W. (1999) Dendritic molecules, in *Synthesis of Polymers*, ed. Schlüter, A.-D. (Wiley-VCH, Weinheim) p. 403.

Jérome, D. (1986) *Phys. Bull. (London)* **37**, 171.

Jusa, W. and McRae, W.A. (1950) *J. Am. Chem. Soc.* **72**, 1044.

Keith, H.D. (1963) in *Physics and Chemistry of the Organic Solid State*, eds. Fox, D. *et al.* (Wiley, Interscience, Chichester).

Keith, H.D. and Padden, F.J. (1963) *J. Appl. Phys.* **34**, 2409.

Keller, A. (1957) *Phil. Mag.* **2**, 1171.

Keller, A. (1968) Polymer crystals. *Rep. Progr. Phys.* **31**, 623.

Keller, A. (1979) Organization of macromolecules in the condensed phase, *Faraday Discussions of the Chem. Soc. (London)* (68), 145.

Keller, A. (1991) in *Sir Charles Frank, OBE, FRS: An 80th Birthday Tribute*, eds. Chambers, R.G. *et al.* (Adam Hilger, Bristol) p. 265.

Keller, A. (1995) *Macromol. Symp.* **98**, 1.

Keller, A. and Cheng, S.Z.D. (1998) *Polymer* **39**, 4461.

Keller, A. and Kolnaar, H.W.H. (1997) Flow-induced orientation and structure formation, in *Processing of Polymers*, ed. Meijer, H.E.H.; *Materials Science and Technology, A Comprehensive Treatment*, eds. Cahn, R.W., Haasen, P. and Kramer, E.J. (VCH, Weinheim) p. 189.

Kinloch, A.J. (1987) *Adhesion and Adhesives* (Chapman & Hall, London).

Kobayashi, S. (ed.) (1997) *Catalysis in Precision Polymerisation* (Wiley, New York).

Lauritzen, J.I. and Hoffman, J.D. (1960) *J. Res. Natl. Bur. Standards* **A64**, 73.

Léger, L. and Viovy, J.L. (1994) Polymers: diffusion and reptation, in *Encyclopedia of Advanced Materials*, vol. 3, eds. Bloor, D. *et al.* (Pergamon, Oxford) p. 2063.

Lotz, B. and Wittmann, J.-C. (1993) Structure of Polymer Single Crystals, in *Structure and Properties of Polymers*, ed. Thomas, E.L.; *Materials Science and Technology, A Comprehensive Treatment*, eds. Cahn, R.W., Haasen, P. and Kramer, E.J. (VCH, Weinheim) p. 79.

Mackley, M.R. and Keller, A. (1975) *Phil. Trans. Roy. Soc., (London)* **278A**, 29.

McCrum, N.G., Read, B.E. and Williams, G. (1967) *Anelastic and Dielectric Effects in Polymeric Solids* (Wiley, London and New York) (Reprinted in 1991 by Dover).

McCrum, N.G., Buckley, C.P. and Bucknall, C.B. (1998) *Principles of Polymer Engineering*, 2nd edition, (Oxford University Press, Oxford).

McMillan, F.M. (1979) *The Chain Straighteners – Fruitful Innovation: The Discovery of Linear and Stereoregular Synthetic Polymers* (Macmillan, London).

Meijer, H.E.H. (1997) Processing for properties, in *Processing of Polymers*, ed. Meijer, H.E.H.; *Materials Science and Technology, A Comprehensive Treatment*, vol. 18, eds. Cahn, R.W., Haasen, P. and Kramer, E.J. (VCH, Weinheim) p. 3.

Meyer, K.H. Susich, G. von and Valkó, E. (1932) *Kolloid Z.* **59**, 208.

Moore, J.C. (1964) *J. Polymer Sci.* **2**, 835.

Morawetz, H. (1985) *Polymers: The Origins and Growth of a Science* (Wiley, New York, Constable, London) (Reprinted as a Dover edition in 1995).

Morris, P.J.T. (1994) Synthetic rubber: autarky and war, in *The Development of Plastics*, eds. Mossman, S.T.I. and Morris, P.J.T. (Royal Society of Chemistry, London).

Mossman, S.T.I. and Morris, P.J.T. (eds) (1994) *The Development of Plastics* (Royal Society of Chemistry, London).

Müllen, K. (1999). Organic chemistry and the synthesis of well-defined molecules, in *Synthesis of Polymers*, ed. Schlüter, A.-D. (Wiley-VCH, Weinheim) p. 1.

Pavelich, W.A. (1986) A path to ABS thermoplastics, in *High Performance Polymers: Their Origin and Development*, eds. Seymour, R.B. and Kirshenbaum, G.S. (Elsevier, New York) p. 125.

Pennings, A.J., van der Mark, J.M.A.A. and Kiel, A.M. (1970) *Kolloid Z. und Z. Polymere* **236**, 99.

Percec, V. *et al.* (1998) *Nature* **391**, 161.

Petermann, J. and Gleiter, H. (1972) *Phil. Mag.* **25**, 813; *J. Mater. Sci.* **8**, 673.

Raman, C.V. (1927) *Indian J. Phys.* **2**, 1.

Robeson, I.M. (1994) Polymer blends, in *Encyclopedia of Advanced Materials*, vol. 3, eds. Bloor, D. *et al.* (Pergamon, Oxford) p. 2043.

Samuel, I.D.W. (2000) *Phil. Trans. Roy. Soc., (London)* A **358**, 193.

Sawamoto, M. and Kamigaito, M. (1999) Living radical polymerisation, in *Synthesis of Polymers*, ed. Schlüter, A.-D. (Wiley-VCH, Weinheim) p. 163.

Seymour, R.B. and Kirshenbaum, G.S. (eds.) (1986) *High Performance Polymers: Their Origin and Development* (Elsevier, New York).

Shirakawa, H. and Ikeda, S. (1971) *Polymer J.* **2**, 231.

Stockmayer, W.H. and Zimm, B.H. (1984) When polymer science looked easy, *Annu. Rev. Phys. Chem.* **35**, 1.

Strathmann, H. (1994) Ion-exchange membranes, in *Encyclopedia of Advanced Materials*, vol. 2, eds. Bloor, D. *et al.* (Pergamon Press, Oxford) p. 1166.

Strobl, G. (1996) *The Physics of Polymers* (Springer, Berlin).

Tanner, R.I. and Walters, K. (1998) *Rheology: An Historical Perspective* (Elsevier, Amsterdam).

Till, P.H. (1957) *J. Polymer Sci.* **24**, 301.

Treloar, L.R.G. (1958) *The Physics of Rubberlike Elasticity* (Oxford University Press, Oxford).

Treloar, L.R.G. (1970) *Introduction to Polymer Science* (Wykeham Publications, London).

Van Krevelen, D.W. (1990) *Properties of Polymers*, 3rd edition (Elsevier, Amsterdam).

Von Hippel, A.R. (1954) *Dielectric Materials and Applications* (Wiley, New York).

Ward, I.M. (1971a) *Mechanical Properties of Solid Polymers* (Wiley, Interscience, New York).

Ward, I.M. (ed.) (1971b) Orientation phenomena in polymers, *J. Mat. Sci.* (special issue) **6**, 451.

Williams, G. (1993) Dielectric properties of polymers, in *Structure and Properties of Polymers*, ed. Thomas, E.L.; *Materials Science and Technology*, *A Comprehensive Treatment*, vol. 12, eds. Cahn, R.W., Haasen, P. and Kramer, E.J. (VCH, Weinheim) p. 471.

Wilson, L.M. (1998) Conducting polymers and applications, in *Processing of Polymers*, ed. Meijer, H.E.H.; *Materials Science and Technology*, *A Comprehensive Treatment*, vol. 18, eds. Cahn, R.W., Haasen, P. and Kramer, E.J. (VCH, Weinheim) p. 659.

Windle, A.H. (1996) A metallurgist's guide to polymers, in *Physical Metallurgy*, 4th edition, vol. 3, eds. Cahn, R.W. and Haasen, P. (North-Holland, Amsterdam) p. 2663.

Young, R.J. (1974) *Phil. Mag.* **30**, 85.

Young, R.J. (1988) *Materials Forum* (Australia) **11**, 210.

Chapter 9
Craft Turned into Science

Chapter 9
Craft Turned into Science

9.1. METALS AND ALLOYS FOR ENGINEERING, OLD AND NEW

In Section 3.2.1, something was said of the birthpangs of a new metallurgy early in the 20th century, and of the fierce resistance of the 'practical men' to the claims of 'metallography', which then meant 'science applied to metals'. In this chapter, I shall rehearse some examples, necessarily in a cursory fashion, of how the old metallurgy became new, and then go on to say something of the conversion of the old ceramic science into the new. The latest edition of my book on physical metallurgy (Cahn and Haasen 1996) has nearly 3000 pages and even here, some parepistemes receive only superficial treatment. It will be clear that this chapter cannot do more than scratch the surface if it is not to unbalance the book as a whole.

9.1.1 Solidification and casting

Metal objects can be shaped in one of three common ways: casting, plastic deformation, or the sintering of powder. For many centuries, shading back into prehistory, casting was a craft, with more than its due share of superstition. All kinds of magical additives, to the melt and to the mold, were sought to improve the soundness of cast objects; the memoirs of the great renaissance sculptor Benvenuto Cellini, for instance, are full of highly dramatic accounts of the problems in casting his statues and the magical tricks for overcoming them. Casting defects were a serious problem until well into this century. As recently as 1930, according to a memoir by Mullins (2000), the huge stern-post castings of heavy cruisers of the US Navy were apt to be full of defects and give poor service. Robert Mehl (see Section 3.2.1) then conceived the technique of gamma-ray radiography to detect defects in these large castings and, in the words of the memoir, "created a great sensation in engineering and practical metallurgical circles"; this was before the days of artificial radioisotopes.

Developments in casting since then fall into two categories, engineering innovations and scientific understanding of the freezing of alloys. It will come as no surprise to readers of this book that the two branches came to be linked. Among the engineering innovations I might mention are developments in molds – high-speed die-casting of low-melting alloys into metallic molds, casting into permanent ceramic molds – and then continuous casting of metallic sections, and 'thixocasting' (the use of a prolonged semi-solid stage to obviate casting defects). This is all set out in a classic text by Flemings (1974).

The understanding of the fundamentals of solidification is primarily the creation of Bruce Chalmers and his research school, first at Toronto University and from 1953 at Harvard. As it happens, I have an inside view of how this research came about. In 1947–1948, Chalmers (1907–1990; an English physicist turned metallurgist who had taken his doctorate with an eminent grower and exploiter of metal crystals, Neville Andrade in London) was head of metallurgy at the recently established Atomic Energy Research Establishment in Harwell, England, where I was a 'new boy'. In his tiny office he built a simple meccano contraption with which he studied the freezing of tin crystals, a conveniently low-melting metal, whenever he had a spare moment from his administrative duties. (I recall exploiting this obsession of his by getting him to sign, without even glancing at it, a purchase order for some hardware I needed.) He would suddenly decant the residual melt from a partly frozen crystal and examine what had been the solid/liquid interface. Its appearance was typically as shown in Figure 9.1 – a 'cellular' pattern – and when at his request I prepared an etched section from just behind the interface, its appearance was similar; this suggested that impurities might be concentrated at the cell boundaries. He was determined to get a proper understanding of what was going on, for which he needed more help, and so in 1948 he accepted an invitation to join the University of Toronto in Canada. Two famous papers in 1953 (Rutter and Chalmers 1953, Tiller *et al.* 1953) established what was happening. The second of these papers appeared in the first volume of *Acta Metallurgica*, a new journal of fundamental metallurgy which Chalmers himself had helped to create and was to edit for many years (see Section 14.3.2).

Figure 9.2 shows the essentials. The metal being solidified is assumed to contain a small amount of dissolved impurity. (a) shows a typical portion of a phase diagram,

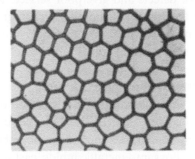

Figure 9.1. Decanted interface of cellularly solidified Pb–Sn alloy. Magnification ×150 (after Chadwick 1967).

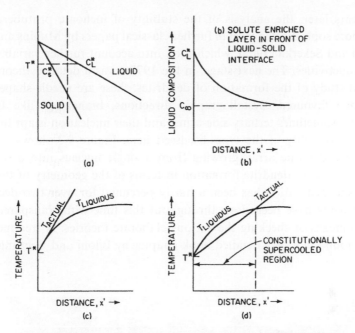

Figure 9.2. Constitutional supercooling in alloy solidification: (a) phase diagram; (b) solute-enriched layer ahead of the solid/liquid interface; (c) condition for a stable interface; (d) condition for an unstable interface.

while (b) shows a steady-state (but non-equilibrium) enhanced distribution of the corresponding solute, caused by the limited diffusion rate of the solute during continuous advance by the solid. (c) and (d) show the corresponding distribution of the *equilibrium* liquidus temperature ahead of the solid/liquid interface, related to the local solute content. What happens then depends on the imposed temperature gradient: when this is high, (c), solidification takes place by means of a stable plane front; if a protuberance transiently forms in the interface, it will advance into a superheated environment and will promptly melt back. If the temperature gradient is lower, (d), the situation represents what Chalmers called *constitutional supercooling*. Instabilities in the form of protuberances now develop because the impure metal in these 'bumps' is below its equilibrium freezing temperature; each protuberance rejects some solute to its periphery, leading to the configuration of Figure 9.1. It is straightforward to formulate a theoretical criterion for constitutional supercooling: the ratio of temperature gradient to growth rate has to exceed a critical value. Numerous studies in the years following all confirmed the correctness of this analysis, which constitutes one of the most notable postwar achievements of scientific metallurgy. An account in recollection of this research can be found in Chalmers's classic text (1974).

Some years later, the analysis of the stability of inchoate protuberances was taken to a more sophisticated level in further classical papers by Mullins and Sekerka (1963, 1964) and Sekerka (1965), which took into account further variables such as thermal conductivities. The next stage, in the 1970s, was a detailed theoretical and experimental study of the formation of dendrites; these are needle-shaped crystals growing along favoured crystallographic directions, branching (like trees) into secondary and sometimes tertiary side-arms, and their nucleation is apt to be linked to interfacial instability of the type discussed here. Figure 9.3 shows a computer simulation of a dendrite array growing from a single nucleus into a supercooled liquid. The analysis of dendrite formation in terms of the geometry of the rounded tips and of supersaturation has been a hardy perennial for over two decades, and many experiments have been done throughout this time with transparent organic chemicals as means of checking the various elaborate theories. A treatment of this field can be found in a very detailed book chapter by Biloni and Boettinger (1996).

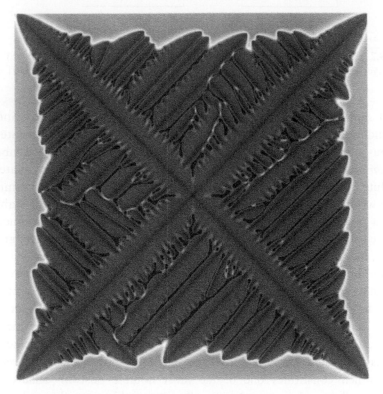

Figure 9.3. Computer simulation of dendrites growing into a Ni–Cu alloy with 41 at.% of Cu. The tints show local composition (courtesy W.J. Boettinger and J.A. Warren).

Earlier, a special issue of *Materials Science and Engineering* (Jones and Kurz 1984) to mark the 30th anniversary of the identification of constitutional supercooling includes 21 concise survey papers which constitute an excellent source for assessing the state of knowledge on solidification at that stage. Another source is a textbook (Kurz and Fisher 1984) published the same year.

The thixocasting mentioned above exploits dendritic solidification of alloys: a semi-solidified alloy is forged under pressure into a die; the dendrites are broken up into small fragments and a sound (pore-free) product is generated at a relatively low temperature, prolonging die-life. The array of related techniques of which this is one was introduced by Flemings and Mehrabian in 1971 and Flemings (1991) has recently reviewed them in depth.

Another major technical innovation in the casting field is the creation of non-brittle cast irons by doping with magnesium, causing the elemental graphite which is unavoidably present to convert from the embrittling flake form to harmless spherulites (rather like those described in Chapter 8 with respect to polymers). This work, perfected in the 1970s (Morrogh 1986), was an early example of nucleation control which has become very important in foundry work. A further example is the long-established 'modification' of Al–Si cast alloys by the addition of traces of sodium metal; the interpretation of this empirical method has given rise to decades of fundamental research. It is an example, not uncommon, of explanation after the event.

Such episodes of empirical discovery, followed only years later by explanation, were a major argument of the 'practical men' against the supposed uselessness of 'metallographists' (Section 3.2.1) but in fact the research leading to an explanation often smooths the way to subsequent, non-empirical improvements. A good recent instance of this was a study of the way in which grain-refining agents work in the casting of aluminium alloys. Fine particles of intermetallic compounds, TiB_2 and Al_3Ti, have long been used to promote heterogeneously catalysed nucleation from the melt of solid grains, on an empirical basis. Schumacher *et al.* (1998) have shown how a metallic glass based on aluminum can be used to permit analysis of the heterogeneous nucleation process: grain-refining particles are added to an Al–Y–Ni–Co composition which is cooled at about a million degrees per second to turn it into a metallic glass (in effect a congealed liquid). This is equivalent to stopping solidification of a melt at a very early stage, so that the interface between the nucleation catalyst and the crystalline Al-alloy nucleus, and the epitaxial fit between them, can be examined at leisure by electron microscopy: it was shown that nucleation is catalysed on particular crystal faces of an Al_3Ti crystallite which is itself attached to a TiB_2 particle. From this observation, certain methods of improving grain refinement were proposed. This is an impressive example of modern physical metallurgy applied to a practical task.

9.1.1.1 Fusion welding. One of the most important production processes in metallurgy is fusion welding, the joining of two metallic objects in mutual contact by melting the surface regions and letting the weld metal resolidify. Many different methods of creating the molten zone have been developed, but they all have in common a particular set of microstructural zones: primarily there is the fusion zone itself, then the heat-affected zone, a region which has not actually melted but has been unavoidably modified by the heat flowing from the fusion zone. In addition, internal stresses result from the thermal expansion and contraction acting on the rigidly held pieces that are being welded. The microstructure of the fusion zone in particular is sensitive to composition; in the case of steels, the carbon content has a particular influence.

Concise but very clear summaries of the microstructure of weld zones in steels are in book chapters by Honeycombe and Bhadeshia (1981, 1995), and by Porter and Easterling (1981), both of which also give references to more substantial treatments.

9.1.2 Steels

Steel, used for armour, swords and lesser civilian purposes, had been the aristocrat among alloys for the best part of a millennium. European, Indian and Japanese armorers vied with each other for the best product. The singular form of the word is appropriate, since for much of that time 'steel' meant a simple carbon–steel, admittedly with variable amounts of carbon remaining after crude pig iron has been refined to make steel. That refining process, steelmaking, has been slowly improved over the centuries, with major episodes in the nineteenth century, involving brilliant innovators like Bessemer, Siemens and Thomas in Britain, and leading to quite new processes in the 20th century, developed in many parts of the world (notably the USA, Austria and Japan). A good summary of the key technological events in the evolution of steel, together with a consideration of economic and social constraints, is a lecture by Tenenbaum (1976). A concise summary of the key events can also be found in a very recent book (West and Harris 1999); even a British prime minister, Stanley Baldwin, a member of an ironmaster's family, played a small part. By the end of the 19th century, 'steels' properly had to be discussed in the plural, because of the plethora of alloy steels which had begun to be introduced.

Lessons can be learned from the aristocrat among early steel products, the Japanese samurai sword, which reached its peak of perfection in the 13th century. This remarkable object consists of a tough, relatively soft blade joined by solid-state welding to a high-carbon, ultrahard edge, complete with a decorative pattern rather like the later Damascus steel. The most recent discussion of the samurai sword is to be found in an essay by Martin (2000), significantly titled *Stasis in complex artefacts.*

Martin points out the extremely complicated (and wholly empirical) steps which had evolved by long trial and error, involving multiple foldings and hammerings (which incidentally led to progressive carbon pickup from burning charcoal), followed by controlled water-quenching moderated by clay coatings of graded thickness. As Martin remarks: "The Japanese knew nothing of carbon. Neither did anyone else in the heyday of the sword: it was not identified as a separate material, an element, until the end of the 18th century. Nor did they know that they were adding this all-important material accidentally during the process of extraction of the iron from its ore, iron oxide (and more later, during hammering)." The clay-coating process had to be just right; the smallest error or peeling away of the coating would ruin the sword. So, as Martin emphasises, once everything at last worked perfectly, nothing must be changed in the process. "Having found a clay that works, in spite of (its) violent treatment, you treasure it. You lay hands on enough to last you through your career.... You will develop extreme caution in the surface finish of the steel to which you apply the slurry – not a hint of grease, not too smooth, a nice even oxide coating, but not a scale which could become detached.... The only way to achieve a success rate that can be lived with is to repeat each stage as exactly as possible." That represents craft at its highest level, but there is no science here. Once a craftsman has perfected a process, it must stay put. A scientific analysis, however, because it eventually allows an understanding of what goes on at each stage, allows individual features of a process to be progressively but rather rapidly improved. This change is essentially what began to happen in the late 19th century. It has to be admitted, though, that the classical Japanese sword, perfected empirically over centuries by superbly skilled and patient craftsmen, has never been bettered.

The scientific study of phase transformations in steel in the solid state during heat treatment, as a function of specimen dimensions and composition, then became a major branch of metallurgy; the way was shown by such classic studies as one by Davenport and Bain (1930) in America. This early study of the isothermal phase transformation of austenite (the face-centred cubic allotrope of iron), and the associated hardening of steel, was reprinted in 1970 by the American Society for Metals as one of a selection of metallurgical classics, together with a commentary placing this research in its historical context (Paxton 1970). This kind of research, including the study of the 'hardenability' of different steels in different sizes, is very well put in the perspective of the study of phase transformations generally in one of the best treatments published since the War (Porter and Easterling 1981).

After the Second World War, the technical innovations, both in steelmaking and in the physical metallurgy of steels, continued apace. A number of industrial research laboratories were set up around the world, of which perhaps the most influential was the laboratory of the US Steel Corporation in Pennsylvania, where some world-famous research was done, both technological and scientific. In the 1970s, a wave of optimism supported industrial metallurgy, especially in America, and university enrolments in metallurgy and MSE courses burgeoned (Figure 9.4). Then, by 1982, to quote a recent paper (Flemings and Cahn 2000), "newspapers, magazines and the television were full of stories about the non-competitiveness of the steel industry, the automotive industry, and a host of other related industries. Hiring of engineers by these industries came to a halt and a long period of 'downsizing' began. Students associated the materials departments with these distressed industries and enrollments dropped abruptly. By 1984, the reduced enrollments had worked their way through to the graduating class." This is very clear in Figure 9.4. Not only university courses felt the pinch; numerous industrial metallurgical laboratories, both ferrous and non-ferrous, were unceremoniously closed in America and in Europe, but not in Japan, where steelmaking and steel exploitation continued to make rapid progress. Since that time, steelmaking has acquired the unjust cachet of a 'smokestack' or 'rustbelt' industry.

Like all reactions, this one overshot badly. Steels are still by far the major class of structural metallic materials and the performance of steels, both high-grade alloy steels and routine carbon steels, has been steadily improved by the application of modern physical metallurgy and of modern process control. The most important development has been in microalloying – the evolution, via research, of steel types with small alloying additions, in fractions of 1%, and often also very low carbon contents. As a class, these are called high-strength low-alloy (HSLA) steels. One variant, used in large amounts for building work and bridges, is weathering steel, which is resistant to corrosion in the open, hence the name. A good account of this large and variegated new family of steels is by Gladman (1997). Other novel steel families, such as the dual-phase family (martensite in a matrix of ferrite), maraging steels (precipitation-hardened martensites, used where extreme strength is needed),

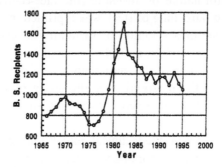

Figure 9.4. US bachelor's degrees in metallurgy and materials, numbers graduating 1966–1995 (after Flemings and Cahn 2000).

and a variety of tool steels for shaping and cutting tools, have been developed for special needs; much of this development has been done in the past two decades in the supposedly decaying smokestack plants, in spite of the gradual disappearance of research laboratories dedicated to steels.

Perhaps the most important innovation of all is in the thermomechanical control processes, involving closely controlled simultaneous application of heat and deformation, to improve the mechanical properties, especially of ultra-microalloyed compositions. Processes such as 'controlled rolling' are now standard procedures in steel mills.

The Nippon Steel Corporation in 1972 pioneered the use of 'continuous annealing lines', in which rolled steel sheet is heat-treated and quenched under close computerised control while moving. For this advanced process to give its best results, especially when the objective is to make readily shapable sheet for automobile bodies, steel compositions have to be tailored specifically for the process; composition and processing are seamlessly tied to each other. Today, dozens of these huge processing lines are in use worldwide (Ohashi 1988).

Part of the 'specific tailoring' of steel compositions to both the processing procedure and to the end-use is the steady move towards *clean steels*, alloys with, typically, less than 20 parts per million in all of undesired impurities, and especially of insoluble inclusions. Such steels are now standard for automobile bodies, drawn steel beverage cans, shadow masks for colour TV tubes, ball-bearings and gas piping. The elements that need specific control include P, C, S, N, H, Cu, Ni, Bi, Pb, Zn and Sn (many of these threaten to increase when scrap steel is used in steelmaking). It is noteworthy that carbon, once the defining constituent of steel, is now an element that needs to be kept down to a very low concentration for some applications. An account of 'high-purity, low-residual clean steels' and the methods of removing unwanted impurities is by Cramb (1999). Advanced modern methods of high-temperature chemistry, such as electroslag refining, are needed for such purification.

Two good general overviews of the design and processing of modern steels are by Pickering (1978, 1992).

To conclude this section, I want to return to the 'anti-smokestack' convulsion of the early 1980s. Figure 9.4 shows clearly that even after the shakeout in student numbers, numbers graduating remain above the levels of the 1960s and 1970s, which were a time of greater optimism. As the few comments here have shown, steel metallurgy, as a kind of indicator for metallurgy as a whole, is in rude good health; much has been achieved in recent decades, and there is more to do. I will conclude with a comment at the end of a recent survey article entitled *From the Schrödinger Equation to the Rolling Mill* (Jordan 1996): "The present time is one of unprecedented opportunities for alloy research, particularly for exciting basic science and its possible exploitation".

9.1.3 Superalloys

Superalloys as a class constitute the currently reigning aristocrats of the metallurgical world. They are the alloys which have made jet flight possible, and they show what can be achieved by drawing together and exploiting all the resources of modern physical and process metallurgy in the pursuit of a very challenging objective.

Steam turbines were patented by Charles Parsons in England in 1884 and in 1924, Ni–Cr–Mo steels were introduced to improve the performance of turbine rotors. These can be regarded as early precursors of superalloys. The modern gas turbine, a major enhancement of the steam turbine because combustion was no longer external to the turbine, was invented independently in Germany and Britain in 1939. The adjective 'modern' is needed here because simpler forms were developed much earlier. Old country houses open to visitors in Britain dating from the 17th century sometimes contain simple turbine wheels that turn in the warm updraft from a domestic fireplace and are linked to a rotating spit for roasting meat. In the early 1930s, turbochargers, essentially small gas turbines used to compress and heat incoming air, were developed to allow internal combustion (reciprocating) aero engines to work at high altitudes where the partial oxygen pressure is low, and they are used now to upgrade the acceleration of advanced automobile engines even at sea level. Propelling a plane entirely by means of a pure jet powered by a gas turbine was another challenge altogether, first met by Hans von Ohain in Germany and Frank Whittle in Britain about the time the Second World War began in 1939. Alloys had to be found to make the turbine blades, the disc on which they are mounted and the remaining hot constituents such as the combustion chamber, as well as the compressor blades at the front of the engine which do not become so hot. Since the first engines, the 'hot alloys' have been nickel-based and remain so today, 60 years later, though at intervals cobalt gets a look-in as a base metal when the African producers are not so embroiled in chaos that supplies are endangered. The operating temperature limit of superalloys increased from 700°C in 1950 to about 1050°C in 1996.

The evolution of superalloys has been splendidly mapped by an American metallurgist, Sims (1966, 1984), while the more restricted tale of the British side of this development has been told by Pfeil (1963). I have analysed (Cahn 1973) some of the lessons to be drawn from the early stages of this story in the context of the methods of alloy design; it really is an evolutionary tale... the survival of the fittest, over and over again. The present status of superalloy metallurgy is concisely presented by McLean (1996).

Around 1930, in America, presumably with the early superchargers in mind, several metallurgists sought to improve the venerable alloy used for electric heating elements, 80/20 nickel–chromium alloy (nichrome), by adding small amounts of titanium and aluminum, and found significant increase in creep resistance.

According to Pfeil's version of events, in Britain in the early 1940s, creep tests were at first made on ordinary commercial nichrome, but the results were not self-consistent; this was traced to differences in titanium and carbon content resulting from the use of titanium as a deoxidiser. A little later, a nickel–titanium additive with some aluminum was tried. The first superalloy, Nimonic 75, was made by 'doping' nichrome with controlled small amounts of carbon and titanium. From there, development continued on the hypothesis (which metallurgists had formulated in the 1930s but had been unable to prove) that creep resistance was conditional on precipitation-hardening. At this stage, in a British industrial laboratory in Birmingham, phase diagram work was thought essential, and the key to all superalloys was established by Taylor and Floyd (1951–1952) , at the time of what I have called the 'quantitative revolution': they found that age-hardening in the early superalloys was entirely due to the ordered intermetallic phases Ni_3Al and Ni_3Ti, or rather a mixed intermetallic, $Ni_3(Al, Ti)$, a phase they dubbed γ', gamma prime, as it is still called, dispersed in a more nickel-rich, disordered matrix, called gamma. A little later it became clear that the microstructure (Figure 9.5) was an epitaxial arrangement; both phases were of cubic crystallography and their cube axes were parallel (this was the epitaxial feature); also the structure was extremely fine in scale. The microstructure was reminiscent of the Widmanstätten structures studied by Barrett and Mehl in Pittsburgh in the 1930s (see Section 3.2.2 and Figure 3.16) but finer, and with one important difference: the lattice parameters (length of the sides of the cubic unit cells) of gamma and gamma prime were almost identical. This turned out to be the key to superalloy performance.

The gamma prime phase has the highly unusual characteristic, first discovered by Westbrook (1957), of becoming stronger with increasing temperature, up to

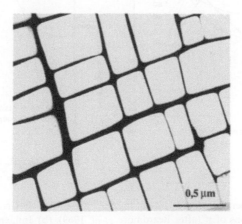

0,5 μm

Figure 9.5. Electron micrograph of a superalloy, showing ordered (gamma prime) cuboids dispersed epitaxially in a disordered (gamma) matrix (courtesy of Dr. T. Khan, Paris).

about 800°C. The reasons for this, closely linked to the geometry of dislocations in this ordered phase, have been argued over for decades and have at last been resolved at the end of the century – but the details do not matter here. As Figure 9.6(a) – taken from an important study, by Beardmore *et al.* (1969) – demonstrates, the γ/γ′ alloys, if they contain only about 50% of the disordered matrix, no longer show this anomaly, but they are as strong at room temperature as the ordered phase is at high temperature; this is the synergistic effect of the two phases together. Even more important is the quality of the fit between the two phases. Figure 9.6(b) shows that the creep-rupture life (the time to fracture under standardised creep conditions) rises to a very intense maximum when the lattice parameter mismatch is only a small fraction of 1%. In fact, it turned out that the creep resistance is best when (a) the parameter mismatch is minimal, and (b) the volume fraction of gamma prime is as high as feasible. (Decreasing the lattice mismatch from 0.2% to zero led to a 50-fold increase in the creep rupture life!) These insights come under the heading of 'phenomenological'. The conditions for optimum creep resistance are quite clear in terms of measurable variables, but *why* just this microstructure is so effective is still today the subject of vigorous discussion: the consensus seems to be that dislocations are constrained to stay in the narrow 'corridors' of the matrix and are prevented from crossing into the ordered cuboids, in part because the equilibrium dislocation configuration is quite different in the corridors and in the cuboids. We have here an example of a clear phenomenology and a disputed

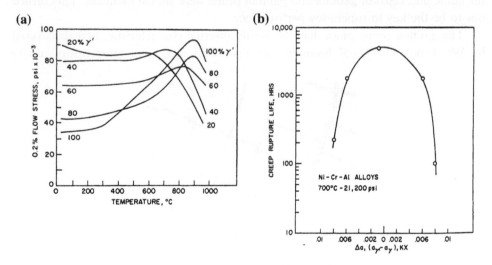

Figure 9.6. (a) The temperature dependence of the flow stress for a Ni–Cr–Al superalloy containing different volume fractions of γ′ (after Beardmore *et al.* 1969). (b) Influence of lattice parameter mismatch, in kX (effectively equivalent to Å) on creep rupture life (after Mirkin and Kancheev 1967).

aetiology to go with it (see footnote on page 206) – a common enough situation in materials science.

There is one other feature that distinguishes the microstructure of Figure 9.5, and that is its stability. Normally, a metallurgist would expect a population of tiny precipitates to coarsen progressively at high temperature. This crucial process, known as *Ostwald Ripening*, after the German physical chemist Wilhelm Ostwald whom we met in Chapter 2 and who first recognised it, arises because the solubility of a small sphere in the matrix is greater than that of a large sphere, so that the large precipitates will grow larger, the small will disappear. The kinetics of increase of average particle size, which turn out to be linear in time 1/3, depend on the interfacial energy, the diffusion rate of the solute in the matrix, and its solubility. The theory was developed more or less simultaneously by scientists in England, Germany and Russia, but the father of the theory is usually held to be Greenwood (1956) in England. The theory indicates that one way of reducing the rate of coarsening is to reduce the interfacial energy between the particles and the matrix, and in the case of superalloys, this energy is reduced to a negligible value by ensuring a very close match of lattice parameters. This helps to explain the form of the plot in Figure 9.6(b).

As we learn from Sims's reviews, many other improvements have been made to superalloys and to their exploitation in recent decades. Solid-solution strengthening, grain-boundary strengthening with carbides and other precipitates, and especially the institution, some twenty years ago, of clean processing which allows the many unwanted impurities to be avoided (Benz 1999) have all improved the alloys to the point where (McLean 1996) the best superalloys now operate successfully at a Kelvin temperature which is as much as 85% of the melting temperature; this shows that the prospect of significant further improvement is slight.

On top of this alloy development, turbine blades for the past two decades have been routinely made from single crystals of predetermined orientation; the absence of grain boundaries greatly enhances creep resistance. Metallic monocrystals have come a long way since the early research-centred uses described in Section 4.2.1.

All the different aspects of the processing and properties of superalloys, including monocrystals, are systematically set out in chapters of an impressive book (Tien and Caulfield 1989). The latest subtleties in the microstructural design of monocrystal superalloys are set out by Mughrabi and Tetzlaff (2000); among other new insights, it now appears that the optimum misfit between the two major phases is not exactly zero.

9.1.4 Intermetallic compounds

In Section 3.2.2, I briefly introduced the family of ordered intermetallic compounds, of which Cu_3Au was the first to be identified, early in the 20th century. We saw in the discussion of superalloys that such phases, Ni_3Al in particular, have a crucial role to play in modern metallurgy as constituents of multiphase heat-resistant alloys. Following the Second World War, moreover, resolute attempts have been pursued to develop single-phase intermetallics (as they are called for short) as engineering materials in their own right. A substantial fraction of published papers in physical metallurgy at present is devoted to intermetallics, in pursuit of what some regard as a hopeless dream and others perceive as a sober venture.

In the 1950s and 1960s, research focused on 'reversibly ordered' intermetallics, such as Cu_3Au, $CuAu$, $FeCo$, Fe_3Al, Ni_4Mo. The idea was to compare the properties, especially mechanical and electrical properties, of the same specimen in fully ordered, imperfectly ordered and disordered states, and these states could be produced by suitable heat-treatment and quenching (e.g., Stoloff and Davies 1966). Of those listed above, only Ni_4Mo has found appreciable use in high-temperature alloys. From the 1970s onwards, attention was drastically transferred to 'permanently ordered' alloys, alloys which are so strongly ordered that they remain so on heating until they melt, such as Ni_3Al, $NiAl$, $FeAl$, Ti_3Al, $TiAl$, Nb_3Al, and investigation focused on creep resistance (closely linked to the magnitude of the ordering energy) and also on the Achilles's heel of the entire family, brittleness at room temperature (Yamaguchi and Imakoshi 1990). The brittleness results partly from the difficulty of driving dislocations through the strongly bonded unlike atom pairs making up the crystal structure, and partly, as we now know, from 'environmental embrittlement' the passage of hydrogen, from water vapour, along grain boundaries. Once again, grain boundaries have proved to be a key concern in determining the behaviour of a new family of metallic materials. In these researches, all the sophisticated techniques of modern characterisation, processing and mechanical analysis are in constant use, and alloying has been systematically used both to reduce the brittleness and to enhance high-temperature strength. This field is unmistakably in the province of the 'new metallurgy'.

Nickel and iron aluminides have now been improved to the point where they are routinely used for a number of terrestrial applications, especially for components of furnaces (Deevi *et al.* 1997). These two families have also been critically evaluated in depth (Liu *et al.* 1997). The central hope of the large and international research community, however, is to improve lightweight intermetallics, especially TiAl, to the point where they can be used to make key components of jet engines, especially turbine discs and blades. Technologically, that stage seems to be within sight, in spite of the very limited ductility of TiAl, but in terms of expense, the very cost-conscious jet-engine industry is proving hard to convince. Another usage, TiAl blades for the rotors of automotive turbochargers (a kind of return to the first gas turbines of the 1930s but at a higher temperature) has required years of painstaking development, and is at last about to go into large-scale use, especially in Japan where those who finance such research have proved strikingly patient (see a group of 27 Japanese papers devoted to intermetallics, Yamaguchi 1996). A fine recent overview of the whole intermetallics field is a book by Sauthoff (1995). A cynical comment made by one industrial researcher some 30 years ago, that intermetallics are the materials of the future and always will be, is not being echoed so frequently now. The jury remains out.

9.1.5 High-purity metals

Repeatedly in this book, the important functions of 'dopants', intentional additives made in small amounts to materials, have been highlighted; the use of minor additives to the tungsten used to make lamp filaments is one major example. The role of impurities, both intentional and unintentional, in matters such as phase transformations, mechanical properties and diffusion, was critically reviewed in one of the early seminar volumes published by the American Society for Metals (Marzke 1955). But extreme purity was not considered; that came a little later.

In Chapter 7 the invention, by William Pfann at the Bell Telephone Laboratories, of zone-refining of silicon and germanium was outlined. This process, in which successive narrow molten zones are made to pass along a crystal so that dissolved impurities are swept along to one end where they can be cut off and discarded, made a huge impact at the time (1954) because it was rightly seen as one of the keys to the creation of the transistor. It was thus to be expected that metallurgists would wish to apply the technique to traditional metals with a view to improving their engineering properties, and this approach got under way in the late 1950s. By 1961, enough progress had been made, in North America and France, for a seminar to be organised in 1961 and its proceedings published the next year (Smith 1962). Pfann himself gave the inspirational opening talk, entitled "Why ultra-pure metals?" Both chemical and electrolytic methods of achieving extreme purity, and zone-refining methods, were treated, as well as the mechanical, electrical, thermoelectric properties of a range of metals (iron particularly) and their recovery and recrystallization after plastic deformation. It has to be admitted that nothing remotely comparable in importance with zone-refining of semiconductors was discovered.

Meanwhile, a group of researchers at the GE Corporate Research Laboratory, led by J.D. Cobine, had made a striking discovery. The company was interested in manufacturing an effective high-amperage sealed vacuum circuit breaker (power switch) for electrical utilities, to obviate fire hazard and to allow reduction of the gap between the electrodes and thus very rapid operation. Electrical engineers had been striving to perfect such a device ever since the 1920s, but it turned out that the operation of the switch released gases from the copper electrodes and this destroyed the vacuum in the sealed enclosure. In 1952, Cobine and his team zone-refined the copper from which the electrodes were to be made and found, to their astonishment, that the residual gas content in the resultant single crystals was less than one part in

10 million. A little later, GE's switchgear division used this copper for experimental sealed vacuum circuit breakers and the procedure was patented, from 1958 on, and led to a major industry. This was not sufficiently well known outside the world of electrical engineering to have found its way to the 1961 ASM Seminar. A detailed account of the sequence of events that led to this important breakthrough was published by two retired GE research directors in a little-known book which deserves to be widely read even today (Suits and Bueche 1967).

40 years later, ultra-pure copper is still being manufactured, in Japan, by a combination of electrolytic refining, vacuum-melting and floating-zone zone-refining (Kato 1995). The long-established 5N grade (i.e., 99.999% pure) is now replaced by 7N grade, that is, less than 0.1 part per million of (non-gaseous) impurities. Residual resistivity (at liquid helium temperature) is the best approximate way of estimating purity of such metals, since chemical analysis is approaching its limits. Industrially, this ultrapure copper is used in Japan for wires in hi-fi audio systems (it is actually claimed that its use improves the quality of sound reproduction!), and also as starting-material for lightly alloyed wires for various robotic and microcircuit uses. A more fundamental approach was taken by Abiko (1994) who continued the long-established tradition of purifying iron (by electrolytic refining) so as to establish a database of the properties and, again, to have a pure base for subsequent trace alloying. A few highly unconventional uses have been described, for instance, a German procedure for making highly reflective X-ray monochromator devices for synchrotron sources, using ultrapure beryllium monocrystals.

Independently of all this, for many years an isolated institute in East Germany (Dresden) carried out careful research on ultrapure refractory metals such as Mo, W, Nb (Köthe 1994); this was at a time when these heat-resistant metals were exciting more interest than they are now.

The upshot of all this research since 1954 is rather modest, with the exception of the GE research, which indicates that techniques and individual materials have to be married up; an approach which is crucial for one material may not be very productive for another. This is of course not to say that this 40-year programme of research was wasted. The initial presumption of the potential value of ultra-pure metals was reasonable; it is the obverse of the well-established principle that minor impurities and dopants can have major effects on the properties of metals.

9.2. PLASTIC FORMING AND FRACTURE OF METALS AND ALLOYS AND OF COMPOSITES

In this book, the process of plastic deformation and the related crystal defects have been discussed repeatedly. In Section 2.1.6, the distinction between continuum mechanics and atomic mechanics was set out; in Section 3.2.3.2, the early history of research on dislocations was outlined, Section 4.2.1 was devoted to the crucial role of metal crystals in studying plasticity, and in Section 5.1, the impact of quantitative approaches on the understanding of dislocations and their interactions was reported. If there were space, it would be desirable now to give a detailed account of one of the most active fields of research in the whole of MSE – the interpretation of yield stresses, strain-hardening, fatigue damage and creep resistance in terms of dislocation geometry and dynamics, and also of the related field of fracture mechanics. The study of plasticity is largely an exercise in what I have called atomic mechanics; the study of fracture, one in continuum mechanics. However, to avoid unbalancing the book, I can only find space for a bare outline of these fields, together with a brief discussion of the engineering use of plastic forming methods in what is sometimes unkindly called 'metal bashing'.

The resistance to plastic flow at ambient temperature is linked to the 'strength' of dislocation sources such as that illustrated in Figure 3.14, together with the operation of various obstacles to dislocation motion (dispersed particles, solutes and, indeed, other dislocations intersecting the moving ones). In some metals the Peierls force 'tying' a dislocation to the lattice is high enough to affect flow stresses as well. Other structural features, such as stacking-faults in close-packed metals and partial long-range order, also influence the motion of dislocations. All these interactions have been modelled and the 'constitutive equations' which emerged are used, inter alia, to draw deformation-mechanism maps (Section 5.1.2.2). The theme that has proved most obdurate to accurate modelling is strain-hardening, the gradual hardening of any metal as it is progressively deformed, because here dislocation dynamics have to be combined with a statistical approach. An outline history of some of these themes, especially the transition from monocrystal to polycrystal mechanics, has recently been published (Cahn 2000). Detailed facts and models are to be found in a comprehensive and authoritative volume (Mughrabi 1993), while the distinctive topic of fatigue damage after cycles of stressing in opposed directions, a most crucial theme in engineering practice, has been excellently treated by Suresh, in a book (1991) and also in the Mughrabi volume.

The termination of plastic deformation by fracture, or brittle fracture in the absence of plastic deformation, might be thought to be something that does not warrant much attention since fracture signals the end of usefulness. This would be a big mistake: the quantitative study of fracture, *and its avoidance*, has been one of the most fruitful fields since the Second World War. That field is nowadays called *fracture mechanics*, and it emerged from the ideas of A.A. Griffith (Section 5.1.2.1 and Figure 5.4), first applied in the 1920s to the statistically very variable fracture stress of glass fibers. Griffith, as we have seen, postulated a population of sharp surface cracks of varying depth, together with a simple but potent elastic analysis of how such a crack will magnify an applied tensile stress, to an extent depending simply on the crack depth. When the applied stress is large enough, but very much smaller than the theoretical intrinsic strength of the perfect crystal, Griffith's analysis shows that it is energetically favourable for the crack to advance explosively and lead to fracture, unless the material is capable of plastic deformation which will blunt the crack, impede its stress-magnifying function and thus arrest its spread. Although it took many years before the postulated 'Griffith cracks' were observed micrographically (Ernsberger 1963), his theory was widely accepted 'sight unseen', so to speak. Griffith's ideas, combined with statistical arguments, were particularly fruitful in interpreting the brittle fracture resistance of glass and ceramics. Because of the statistical distribution of crack sizes and the crucial role of the surface in initiating fracture, fracture resistance of strong, brittle fibres is a function of surface area and therefore of fibre diameter. The standard text is by Lawn and Wilshaw (1975, 1993).

There is in fact a continuous gradation from highly brittle materials to thoroughly ductile ones, those in which cracks cannot advance because they are at once blunted by local plastic deformation. In others, finely dispersed obstacles keep on diverting and finally arresting advancing cracks, all the time absorbing energy, that is, enhancing the work of fracture. The aim is always to secure as large a work of fracture as possible, together with a high yield stress. Section 9.5, below, cites examples of this strategy in action.

Fracture of plastically deformable metals and alloys, steels in particular, became a matter of great concern during the Second World War because large numbers of merchant ships were for the first time fabricated by welding (to save construction time) and some of them broke in two when buffeted by storms in cold winter waters. This was, to put the matter crudely, because a steel has a brittle/ductile transition in a critical temperature range, and welded ships (as distinct from rivetted ships) have no discontinuity to stop a running crack from propagating. The brittleness fostered by notches that locally enhance the applied stress was analysed by Orowan (1952), followed by a more sophisticated treatment of Griffith's model by Irwin (1957), called the stress-intensity approach, which then gave birth to the modern science of *fracture mechanics*. This allows the fracture risk of semi-brittle or plastic metals, complete with intrinsic defects, to be rationally assessed, and a stress limit for safe use to be calculated. The approach is crucial in engineering design today, for ships, bridges, pressure vessels in particular, and many other structures, and allows such structures to be designed with high confidence that fracture will be avoided under specified conditions of loading. The standard text is by Knott (1973).

Not all fracture is by crack propagation. Highly ductile materials stressed at high temperature will eventually break by the growth, through absorption of lattice vacancies, of plastic voids. This shades into the phenomenon of superplasticity, which was examined in Section 4.2.5.

There is a paradox at the heart of the design of very strong solids: in general, the higher the intrinsic strength (largely a matter of very strong covalent chemical bonds as found in ceramics), the more subject the solid is to premature fracture by crack propagation. The way round this paradox is to combine ultrastrong ceramic or graphite fibers with a relatively soft matrix, normally either of polymer or of metal, but latterly also of ceramic. (It is even possible to have a composite in which both phases are chemically identical, such as carbon/carbon composites, used for aircraft brakepads – graphite fibers dispersed in a vapour-generated amorphous carbon matrix.) The fibers both reinforce the matrix (like the steel rods in reinforced concrete) and act as efficient crack-arresters, so that the material can be fracture-resistant (tough) even if neither constituent is plastically deformable (as in rubber-toughened polymer blends, Section 8.5.3). This strategy has led to the large domain of *synthetic composite materials*, of which glass-reinforced epoxy resins were the first and best known, used from about 1950 on; wood, of course, is the archetype of a natural composite material. Again, there is no space here to do more than indicate the existence of this very important field, and to point to an early standard text (Kelly 1966, 1986), a detailed account of the microstructural design of fiber composites (Chou 1992), together with a recent text about the various kinds of reinforcing fibers (Chawla 1998) and another book focused on metal–matrix composites (Clyne and Withers 1993). The various types of fiber-reinforced composites have in common the theory that governs their mechanical behaviour: this quite elaborate body of micromechanics covers such features as pullout of fibers from the matrix, statistics of strength of fibers as a function of diameter, arresting and diversion of advancing cracks, anisotropy of properties in relation to orientation distribution of fibers. Kelly's book in its first (1966) edition was the first compilation of this body of general theory of composites. In a recent article (Kelly 2000) he illustrates modern applications of fibre-reinforced composites and also goes in some detail into the history of their development, with special emphasis on Britain.

The main reason why metals are the principal category of structural materials is that they are plastically deformable, which both prevents sudden, catastrophic fracture and allows the material to be plastically shaped to a desired form. Again, no room can be found here to discourse on the large field of metal-forming – rolling, forging, extrusion, wire-drawing, deep drawing of sheet. These techniques are mostly analysed by continuum mechanics, and in this way, for instance, the forces and power requirements in a planned rolling mill, say, or an extrusion press, can be estimated in advance of construction. I will point out only one aspect which cannot be encompassed by a continuum approach, and that is the generation of *deformation textures*. This is the statistical tendency of the millions of grains in a plastically deformed polycrystal to approximate to a single orientation, with statistical scatter. This is important for two reasons: the building up of texture can interfere with the further plastic deformation of the material, and if the material is elastically or in some other way anisotropic, then the properties of the resultant sheet, rod or wire, mechanical properties in particular, will be different in different directions. This two-way interaction between textures and plastic deformation has been very clearly explained in a recent standard text (Kocks *et al.* 1998). The deep drawing (shaping) of steel sheet to make automobile bodies depends on close control of texture if the sheet is not to crack at sites of locally intense deformation, and the drawing of aluminum alloy sheet into beverage cans, now a huge industry worldwide, is only feasible if the texture is so controlled (inter alia, by annealing, which changes the texture) that deformation is accurately isotropic; this has required many years of development work on the optimum composition and precise processing of aluminum alloy sheet (e.g., Hutchinson and Ekström 1990). Textures are determined by a variant of polycrystal x-ray diffraction, and in the 80 years since research on this began, a huge mass of information and interpretation has been accumulated. Two recent overviews are by Cahn (1991), and by Randle and Engler (2000) who write especially on microtextures, which are concerned with the statistics of misorientations across a population of grain boundaries. We saw in Section 7.2.2.4 how microtextures in superconducting ceramics determine the magnitude of the current that can be carried before superconduction is destroyed.

9.3. THE EVOLUTION OF ADVANCED CERAMICS

9.3.1 Porcelain

The production of ceramic containers, and of statuettes, is the oldest of man's major crafts, certainly older than metalworking. Containers were needed to store liquids and grain, i.e., for use; statuettes were made for religious ritual and also to please the eye of aristocrats. Thus from early in man's development as a technologist, utility and beauty were twin criteria. Often these criteria were combined, in the manufacture of decorated pots and statuettes, and the high point of this pursuit was undoubtedly Chinese porcelain, which was first made during the Tang dynasty (618–907 AD) and reached perfection during the Sung dynasty, in the mid-13th century (about the same time as the Japanese sword reached its apogee). Marco Polo in that century brought back a specimen of Chinese porcelain, still exhibited in Venice today. For a long time, Chinese porcelain (and to a much lesser extent, Japanese and Indian forms of porcelain) was assiduously exported to Europe; western demand was insatiable for this magical material, white, thin, strong, translucent and beautifully decorated. Europeans tried over and over to reproduce it, increasingly from the late 16th century onwards, when Florentine potters came close; French potters succeeded in making a somewhat inferior version of porcelain (the 'soft-paste' variety) at St. Cloud and Vincennes from about 1720 onwards. This material, however, tended to sag in the kiln and shape control was thus very difficult. True Chinese-style porcelain, known as 'hard-paste', was not made until 1709, by Johann Böttger (1682–1719), who was kept as a captive for many years by Augustus, the ruler of Saxony. Böttger was a self-proclaimed alchemist, and Augustus kept him in his castle for many years under repeated threat of death if he did not manufacture gold, the key 'arcanum' of his time. Böttger did not make gold directly, but he did make it indirectly for his demanding master: he found out, instead, how to make true porcelain, and a few years later Augustus set up in Meissen the first large-scale factory for manufacturing an alternative to the expensive Chinese imports. The money rolled in.

The difficulty in making porcelain was 2-fold: first, the needed ingredients were unknown (the Chinese were good at preserving their arcane secret) and second, the high temperatures needed (at least 1350°C) to 'fuse' the ingredients together could not be reached in the comparatively primitive European kilns. The captive Böttger was diverted from the quest for gold by a courtier, Count von Tschirnhaus (1651–1708), who was an early 'natural philosopher' with a special interest in using large lenses to concentrate sunlight. The high temperatures thus attainable led naturally to the idea of seeking to make porcelain, and the two men worked together on this, and eventually they were able to make kilns which could reach the necessary high temperatures without the use of lenses.

Tschirnhaus recognised that specially pure clay had to be used to prevent coloration of the product, and that something had to be added to make the clay fusible (i.e., turn it partly into a melt) to create the partly glassy, translucent body of porcelain. After years of experimentation, in January 1708 Böttger tried different proportions of white kaolin (nowadays called China clay) and alabaster, a calcium sulphate. With a low alabaster content of around 12%, beautiful porcelain resulted, and Böttger had saved his neck from the axe again. It proved, in the words of the time, to be 'white gold', especially after ways of decorating the surface in colours under a subsequently applied glaze had been developed. Kaolin has ever since been an essential constituent of porcelain, though the additions needed to make it fusible have varied somewhat.

The difficulties of the search and the appalling conditions under which it had to be conducted are memorably depicted in a recent book by Gleeson (1998). (Böttger died at the age of only 37 from the effects of the terrible conditions under which he had worked.) A more technical account, showing phase diagrams and placing the achievement in the context of attempts elsewhere in Europe, is by Kingery (1986), as part of a multi-volume study of the emergence of modern ceramic science. He sets out the consequences of the Tschirnhaus/Böttger triumph during the remainder of the 18th century. When that century began, chemistry played no part in ceramic craftsmanship. When Saxon porcelain was sent to France, the great chemist Réaumur (whom we have already met) analysed it and his analysis helped the French Sèvres pottery to make porcelain too, and somewhat later, Josiah Wedgwood (1730–1795) founded his famous pottery and introduced chemical methods to control raw materials better than before. In the 1780s he also introduced the first high-temperature pyrometer – actually pieces of ceramic of controlled composition which sagged at different temperatures (Dorn 1970–1980). Thus, a physical task (crude temperature measurement) was achieved by exploiting chemical expertise.

Kingery concludes: "By the end of the (18th) century chemical analysis and control of the constitution of bodies, glazes and raw materials was accepted. Ceramics had changed its role from that of an instigator of chemical studies to a net user of chemical studies." This view of things is consistent with the remark by Wachtman (1999) that "the long process of moving ceramics from a tradition-based craft to a science-based technology conducted under the direction of engineers was underway in the 1800s and has continued to the present day".

9.3.2 The birth of high-tech ceramics: lamps

Porcelain hardly comes under the rubric 'structural material', yet it is immensely strong. I recall an advertisement a few years ago showing a London doubledecker bus balanced on four inverted teacups under the tires. Once this feature of a high-grade ceramic came to be recognized, one path was open to more technological uses for this family of materials. But recognition of high compressive strength was certainly not the only factor in this development; the coming of the age of electricity at the end of the nineteenth century and the role of ceramics in helping that age along were even more important. This was much earlier than the developments in electronic and magnetic ceramics described in Chapter 7.

The title of this section is taken from the title of another essay by Kingery, one of the most eloquent and expert proponents of the central role of ceramics in MSE (for a sketch of his educational innovations, see Section 1.1.1). The essay is in a book series entitled *Ceramics and Civilization* (Kingery 1990).

The electrical age was built on the discovery in the early 1830s, independently by Joseph Henry (1797–1878) in America and Michael Faraday (1791–1867) in England, of electromagnetic induction, which led directly to the invention of the dynamo to generate electricity from steam-powered rotation. It came to fruition on New Year's Eve, 1879, when Thomas Edison (1847–1931) in rural New Jersey, after systematic and exhaustive experiments, made the first successful incandescent lamp, employing a carbonised filament made from some thread taken from Mrs. Edison's sewing cabinet. The lamp burned undimmed for 40 h, watched anxiously by Edison and some of his numerous collaborators. This lamp was ideal for domestic use, unlike the arc lamp perfected a few years previously which was only thought suitable for open-air use. Edison not only made the first successful filament lamp, he also organised the building of the first central electric power station, after a brief interval when dispute reigned over the relative merits of central and individual domestic generation of electricity. The Edison Electric Light Company, both to generate electricity and to sell the lamps to use it, was incorporated in 1878. Thereupon, a no-holds-barred race took place between robber barons of various types for power generation and lamp design and manufacture. By 1890, Edison had six major competitors. All this is recounted in splendid detail in a book by Cox (1979), published to celebrate the centenary of Edison's momentous success.

Edison's lamps were primitive, and their life was limited because of the fragility of the carbon filaments, the expense of hand manufacture and the inadequacy of contemporary vacuum pumps. The extraordinary lengths to which Edison went to find the best organic precursor for filaments, including the competitive trying-out of beard-hairs from two men, is retailed in a racy essay by Jehl (1995). Many alternatives, notably platinum and osmium, were tried, especially after Edison's patents ran out in the mid-1890s, until in 1911 General Electric put on sale lamps made with the 'non-sag' tungsten filaments developed by William Coolidge and they swept all before them. These filaments are still, today, made essentially by the same elaborate methods as used in 1911, using sintering of doped metal powder (see Section 9.4). An entire book was recently devoted to the different stages and aspects of manufacture of tungsten filaments (Bartha *et al.* 1995). Many manufacturers tried to break GE's patents and the lawyers and their advisers had a splendid time: my wife's father, a metallurgist, to whose memory this book is dedicated, sent his three children to boarding school on the proceeds of his work as expert witness in one such trial over lamp patents.

The complicated history of General Electric's progressive development of the modern incandescent lamp is clearly told in a book about the GE Research Laboratory (Birr 1957). In particular, this includes a summary of the crucial researches, experimental and (particularly) theoretical by a brilliant metallurgist turned physical chemist, Irving Langmuir (1881–1957). He examined in a fundamental way the kinetics of metal evaporation, the possible role of inert gas filling in counteracting this, and the optimum configurations of coiled (and coiled coil) filaments to reduce heat loss and thus electricity wastage from the filaments. Langmuir joined the Laboratory in 1909 and had essentially solved the design problems of incandescent lamps by 1913. We shall meet Langmuir again in Section 11.2.3, in his guise as physical chemist.

The 32-year interval between 1879 and 1911 saw a classic instance of challenge and response, in the battle between electric and gas lighting, and between two rival methods of electric lighting. Kingery, in his 1990 essay, describes the researches of Carl Auer, Baron von Welsbach, in Austria (1858–1929), who discovered how to improve 'limelight', produced when a flame plays on a block of lime, for domestic use. He discovered that certain rare-earth oxides generated a particularly bright incandescent light when heated with a Bunsen burner, and in 1866 he patented a mixture of yttria or lanthana with magnesia or zirconia, used to impregnate a loosely woven cotton fabric by means of a solution of salts of the elements concerned. He then spent years, Edison-fashion, in improving his ceramic mixture; in particular, he experimented with thoria, and found that the purer his sample was, the less efficiently did it illuminate. As so often in materials research, he tracked down these variations to contamination, in this instance with the oxide of cerium, and this oxide became the key to the commercial *Welsbach mantle*, marketed in 1890. Kingery remarks that "as far as I'm aware, the Auer incandescent gas mantle was the first sintered oxide alloy to be formed from chemically prepared raw materials". Its great incandescent capacity "put renewed life into gas light as a competitor with the newer electric lighting systems". Eventually, of course, electric lamps won the competition, but, as Kingery says, "for isolated and rural areas without electrification, the incandescent gas mantle remains the lighting system of choice" (using bottled gas).

In the 1890s, a third competitor arrived to challenge the electric filament lamp and the Welsbach gas mantle. This was the Nernst lamp. We have already briefly met the German chemist Walther Nernst (1864–1941) in Section 2.1.1. Nernst was acutely aware of the limitations of the filament lamp in its 1890 incarnation and especially of the poor vacuum pumps of the time, and decided to try to develop an electric lamp based, not on electronic conduction as in a metal, but on what we now know as ionic conduction. Of course at the time, so far as any chemist knew, ions were restricted to aqueous solutions of salts, so the mechanism of conduction must have been obscure. Nernst finally filed a patent in 1897 (just as Thomson announced the existence of the electron). His patent specified a conductor based on "such substances as lime, magnesia, zirconia, and other rare earths". (Recently, a small fragment of one of Nernst's surviving lamps was analysed for Kingery and found to be ≈88 wt% zirconia and 12 wt% yttria-group rare earths.) These ceramic 'glowers' did not conduct electricity sufficiently well at ambient temperature and had to be preheated by means of a platinum wire that encircled the glower; once the glower was operating, the preheater was automatically switched off and an overload surge protector was also built in. The need for preheating led to some delay in lighting up, and in later years Nernst, who had a mordant wit, remarked that the introduction of his lamp coincided with another major invention, the telephone, which "made it possible for the brokers at the Stock Exchange to ring up home when business was finished and ask their wives to switch on the light". Nernst's lamps were steadily improved

(Kingery 1990) and sold very widely, but they had to capitulate to the tungsten filament lamp after 1911. They had an effective commercial life of only 12 years.

The history of these three lamp types offers as good an example as I know of the mechanism of challenge and response in industrial design. Several more major electric lamp types have been introduced during the past century – one of them will be outlined in the next section – but competition did not eliminate any of them.

Kingery's 1990 essay also discusses another of Edison's inventions, the carbon granule microphone which he developed in 1877 for the new telephone, announced by Alexander Graham Bell the previous year (well before Nernst's lamp, in actual fact). Edison had in 1873 discovered the effect of pressure on electrical resistance in a carbon rheostat; building on that, he discovered that colloidal carbon particles made of 'lampblack' (soot from an oil lamp) had a similar characteristic and were ideal for operation behind an acoustic membrane. Telephones are still made today with carbon granules – a technology even longer-lived than tungsten filaments for lamps. This is one of many applications for different allotropic forms of carbon, which are often reckoned as ceramics (though carbon neither conducts electricity ionically nor is an insulator).

9.4. SINTERING AND POWDER COMPACTION

When prehistoric man made and fired clay pots, he relied (although he did not know it) upon the phenomenon of *sintering* to convert a loosely cohering array of clay powder particles steeped in water into a firmly cohering body. 'Sintering' is the term applied to the cohesion of powder particles in contact without the necessary intervention of melting. The spaces between the powder particles are gradually reduced and are eventually converted into open, interconnected pores which in due course become separate, 'closed' pores. The production of porcelain involves sintering too, but at a certain stage of the process, a liquid phase is formed and infiltrates the open pores – this is liquid-phase sintering. The efficacy of the sintering process is measured by the extent to which pores can be made to disappear and leave an almost fully dense ceramic.

Sintering is not restricted to clay and other ceramic materials, though for them it is crucial; it has also long been used to fabricate massive metal objects from powder, as an alternative to casting. For many years, furnaces could not quite reach the melting-point of iron, 1538°C, and the reduction of iron oxide produced iron powder which was then consolidated by heat and hammering. The great iron pillar of Delhi, weighing several tons, is believed to have been made by this approach. The same problem attended the early use of platinum, which melts at ≈1770°C. It was William Hyde Wollaston (1766–1828) in London who first proved that platinum was an element (generally accompanied by other elements of its group) and perfected a way of making 'malleable platinum' by precipitating the powder from solution and producing a cake, coherent enough to be heated and forged; this was reported just before Wollaston's death in 1828. The intriguing story of this metal and its 'colleagues' is concisely told in Chapter 8 of a recent book (West and Harris 1999). We have already seen that tungsten filaments for incandescent lamps were made from 1911 onwards by sintering of fine tungsten powder. Unlike the other historical processes mentioned here, these filaments were initially made by loose sintering, without the application of pressure, and it was this process which for many years posed a theoretical mystery. Sintered metal powders were not always made to be fully dense; between the Wars, sintered porous bronze, with communicating pores, was made in America to retain oil and thus create self-lubricating bearings. These early applications were reviewed by Jones (1937) and more recent uses and methods in accessible texts by German (1984) and by Arunachalam and Sundaresan (1991). These include discussions of sintering aided by pressure (pressure-sintering, especially the modern use of hot isostatic pressing (see Section 4.2.3)), methods which are much used in industrial practice.

Returning to history, a little later still, in 1925, the Krupp company in Germany introduced what was to become and remain a major product, a tough *cermet* (*ceramic–metal* composite) consisting of a mixture of sharp-edged, very hard tungsten carbide crystallites held together by a soft matrix of metallic cobalt. This material, known in Germany as 'Widia' (*Wie Dia*mant) was originally used to make wire-drawing dies to replace costly diamond, and later also for metal-cutting tools. Widia (also called cemented carbide) was the first of many different cermets with impressive mechanical properties.

According to an early historical overview (Jones 1960), the numerous attempts to understand the sintering process in both ceramics and metals fall into three periods: (1) speculative, before 1937; (2) simple, 1937–1948; (3) complex, 1948 onwards. The 'complex' experiments and theories began just at the time when metallurgy underwent its broad-based 'quantitative revolution' (see Chapter 5).

The elimination of surface energy provides the driving force for pressureless sintering. When a small group of powder particles is sintered (Figure 9.7), some of the metal/air surface is replaced by grain boundaries which have a lower specific energy; moreover, two surfaces are replaced by one grain boundary. The importance of the low grain-boundary energy in driving the sintering process is underlined by a beautiful experiment originally suggested by an American metallurgist, Paul Shewmon, in 1965 and put into effect by Herrmann *et al.* (1976). Shewmon was concerned to know whether the plot of grain-boundary energy vs angular misorientation, as shown in Figure 5.3 (dating from 1950), was accurate or whether there were in fact minor local minima in energy for specific misorientations, as later and more exact theories were predicting. He suggested that small metallic single-

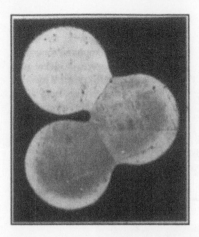

Figure 9.7. Metallographic cross-section through a group of 3 copper particles sintered at 1300 K for 8 h. The necks are occupied by grain boundaries (after Exner and Arzt 1996).

crystal spheres could be scattered on a single-crystal plate of the same metal and allowed to sinter to the plate; he predicted that each sphere would 'roll' into an orientation that would give a particularly low specific energy for the grain boundary generated by sintering. Herrmann and his coworkers made copper crystal spheres about 0.1 mm in diameter, simply by melting and resolidifying small particles. These spheres were then disposed on a copper monocrystal plate (with a surface parallel to a simple crystal plane) and heated to sinter them to the plate, as shown in Figure 9.8(a). (The same was done with silver also.) X-ray diffraction was then used to find the statistical orientation distribution of the sintered spheres, and it was found that after sufficiently long annealing (hundreds of hours at 1060°C) all the spheres, up to 8000 of them in one experiment, acquired accurately the same orientation, or one of two alternative orientations. The authors argued that if a 'cusp' of low energy exists at specific misorientations between a sphere and the plate, a randomly oriented sphere which has already begun to sinter, so that a grain boundary has been formed, will then reorient itself by means of atom flow as shown in Figure 9.8(b) until the misorientation has become such that the boundary energy reaches a local minimum. An actual sintered sphere is shown in Figure 9.8(c). Subsequent work has shown very clearly (Palumbo and Aust 1992), by a variety of experimental and simulation techniques, that indeed the energy of a grain boundary varies with misorientation not as shown in Figure 5.3, but as shown in the example of Figure 9.9. The energy 'cusps' arise for orientation relationships marked by the 'sigma numbers' indicated at the top of the graph, for which the atomic fit at the boundaries is particularly good.

This experiment is discussed here in some detail both because it casts light on the driving force for sintering and because it is a beautiful example of the ingenious

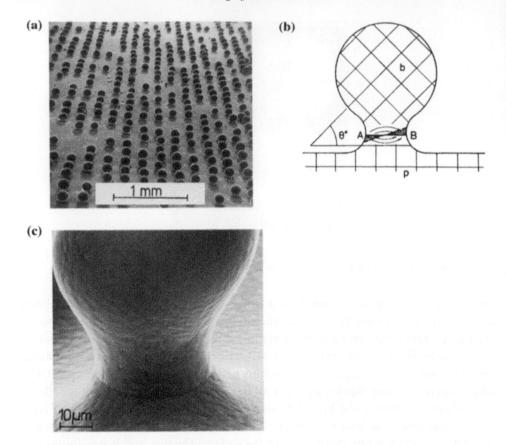

Figure 9.8. Sintering of single-crystal copper spheres to a single-crystal copper substrate. (a) experimental arrangement; (b) mechanism for rotation of an already-sintered sphere; (c) scanning electron micrograph of a sintered sphere (courtesy H. Gleiter).

approaches used by the 'new metallurgy' after the quantitative revolution of ≈1950, and further, because it serves to disprove David Kingery's assertion, quoted in Section 1.1.1, that "the properties and uses of metals are not very exciting". Finally, I urge the reader to note that the Herrmann experiment could equally well have been performed with a ceramic, and indeed a somewhat similar experiment was done a little later with polyethylene (Miles and Gleiter 1978), and the energy cusps which turned up were explained in terms of dislocation patterns. Attempts to reserve scientific fascination to a particular class of materials are doomed to disappointment. That is one reason why materials science flourishes.

Several of the early studies aimed at finding the governing mechanisms of sintering were done with metal powders. A famous study was by Kuczynski (1949) who also examined the sintering of copper or silver to single-crystal metal plates; but

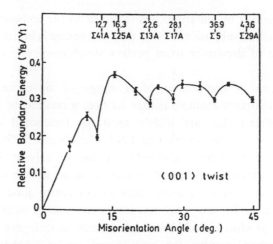

Figure 9.9. Relative boundary energy versus misorientation angle for boundaries in copper related by various twist angles about [1 0 0] (after Miura *et al.* 1990).

he was interested in sintering kinetics, not in orientations, and so he measured the time dependence of the radius of curvature, r, of the 'weld' interface between spheres and the plate. He then worked out the theoretical dependence of r on time, t, for a number of different rate-determining mechanisms, such as r^2 proportional to t for diffusional creep (see Section 4.2.5), r^5 proportional to t for volume diffusion of metal through the bulk, and r^7 proportional to t for metal diffusion along surfaces. Kuczynski claimed to have shown that volume diffusion was the preponderant mechanism. In the past half-century, Kuczynski's lead has been followed by numerous studies, of both metals and ceramics, (for instance an analysis by Herring (1950) of the effects of change of scale) and a number of research groups have been founded around the world to pursue both the theory and experimental testing of scaling and kinetic studies. Exner and Arzt (1996) survey these studies, which now suggest that surface diffusion and especially grain-boundary diffusion both play significant parts in the sintering process. This scaling approach to teasing out the truth is reminiscent of the use of the form of the observed grain-size dependence of creep rates to determine whether Nabarro–Herring (diffusional) creep is in operation.

   In the same year as Kuczynski's research was published, Shaler (1949), who had done excellent work on measuring surface energies and surface tensions on solid metals, argued that surface tension must play a major part in fostering shrinkage of powder compacts during sintering; his paper (Shaler 1949) led to a lively discussion, a feature of published papers in those more spacious days.

   The chemistry of ceramics plays a role in their behaviour during sintering. Non-stoichiometry of oxides has been found to play a major role in the extent to which a powder can be densified by sintering; this is linked to the emission of vacancies on the cationic and anionic sublattices from a pore. Sintering is better in anion-deficient ceramics. The role of departure from perfect stoichiometry is clearly set out by Reijnen (1970).

Sintering is now a component of a range of novel ceramic processing technologies: an important example is *tape casting*, a method of making very thin, smooth ceramic sheets that are widely used for functional applications. The technique was introduced in America in 1947: Hellebrand (1996) defines it as "a process in which a slurry of ceramic powder, binder and solvents is poured or 'cast' onto a flat substrate, then evenly spread, and the solvents subsequently evaporated". Sintering then follows. An enormous range of consumer goods, such as kitchen appliances, computers, TV sets, photocopiers, make use of such tapes. A variant, since 1952, is the production of laminated ceramic multilayers, used for various forms of miniaturised circuits: the multilayers act as 'skeletons' to hold the components and metallic interconnects.

9.4.1 Pore-free sintering

One aspect of sintering remains to be discussed, and that is the linkage between the efficiency of sintering and grain growth, that is, the migration of grain boundaries through a powder compact while sintering is in progress. The importance of this derives from the fact, first demonstrated at MIT by Alexander and Balluffi (1957) with respect to sintered copper, that pores lying on a grain boundary are eliminated while those situated in a grain interior remain. At about the same time, also at MIT, Kingery and Berg (1955), working with ceramics, pointed out that the ready diffusion of vacancies along grain boundaries, which according to Nabarro and Herring can be both sources and sinks for vacancies, provided a mechanism for shrinkage for powder compacts. These findings had a corollary: when grain boundaries sweep through a polycrystal, they can 'gather up' pores along their path provided they migrate slowly enough. This established the major link between grain growth and the late stage of sintering.

A brief word about grain growth, a major parepisteme in its own right, is in order here. This process is driven simply by the reduction of total grain-boundary energy (that is the ultimate driving force) and more immediately, by the usual unbalance of forces acting on three grain boundaries meeting along a line. Whether or not the microstructure responds to this ever-present pair of driving forces depends on the factors tending to hold the grain boundaries back; of these, the most important is the possible presence of an array of tiny dispersed particles which latch on to a moving boundary and slow it down or, if there are enough of them, stop it entirely. The reality of this effect has been plentifully demonstrated, and the modelling of grain growth, especially in the presence of such particles, is a 'growth industry' which I discuss further in Section 12.2.3.3. In the presence of a critical concentration of dispersed particles, most grain boundaries are arrested but a few still move, and this leads to abnormal or 'exaggerated' grain growth, and the creation of a few huge grains. In this connection, pores act like dispersed particles. The complicated circumstances of this process are surveyed by Humphreys and Hatherly (1995). When exaggerated grain growth takes place, any one location in a densifying powder compact is passed just once, rapidly, by a moving grain boundary, whereas normal grain growth ensures repeated slow passages of the myriad of grain boundaries in the compact, giving time for vacancies to 'evaporate' from pores and diffuse away along intersecting grain boundaries. To ensure adequate pore removal and hence densification it is necessary to ensure that normal, but not abnormal, grain growth operates, and that furthermore the migration of boundaries is slowed down as much as possible. The famous micrograph reproduced in Figure 9.10, from Burke (1996), of a densifying powder compact of alumina, demonstrates the sweeping up of pores by a moving grain boundary.

Burke, and also Suits and Bueche (1967), tell the history of the evolution of pore-free, and hence translucent, polycrystalline alumina, dating from the decision by Herbert Hollomon at GE (see Section 1.1.2) in 1954 to enlarge GE's research effort on ceramics. In 1955, R.L. Coble joined the GE Research Center from MIT and

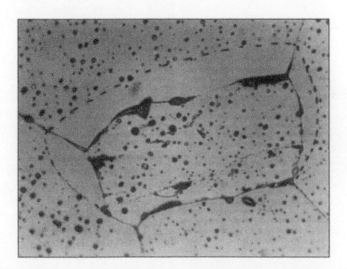

Figure 9.10. Optical micrograph of a powder compact of alumina at a late stage of sintering, showing pore removal along the path of a moving grain boundary. (The large irregular pores are an artefact of specimen preparation.) Grain boundaries revealed by etching. Micrograph prepared at GE in the late 1950s, and reproduced by Burke (1996) (reproduced by permission of GE).

began to study the mechanisms of the stages of sintering of alumina powder. The features outlined in the preceding paragraphs soon emerged and Coble then had the brilliant idea of braking migrating grain boundaries by 'alloying' the alumina with soluble impurities which might segregate to the boundaries and slow them down. Magnesia, at around 1% concentration, did the job beautifully. Figure 9.11 shows sintered alumina with and without magnesia doping. In 1956, a visiting member of GE's lamp manufacturing division chanced to see Coble's results with doped alumina and was struck by the near transparency of his sintered samples (there were no pores left to scatter light). From this chance meeting there followed the evolution of pore-free alumina, trademarked Lucalox, and its painstaking development as the envelope material for a new and very efficient type of high-pressure sodium-vapour discharge lamp. (Silica-containing envelopes were not chemically compatible with sodium vapour.) Burke, and Suits/Bueche, tell the tale in some detail and spell out the roles of the many GE scientists and engineers who took part. Nowadays, all sorts of other tricks can be used to speed up densification during sintering: for instance, the use of a population of rigorously equal-sized spherical powder particles ensures much better packing before sintering ever begins and thus there is less porosity to get rid of. But all this is gilt on the gingerbread; the crucial discovery was Coble's

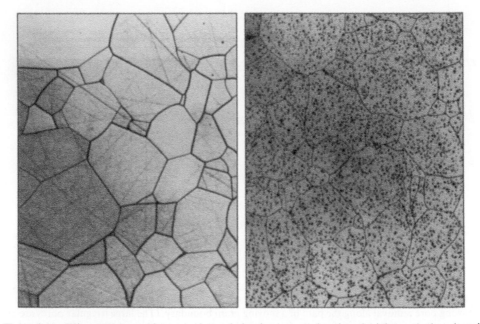

Figure 9.11. Microstructures of porous sintered alumina prepared undoped (right) and when doped with magnesia (left). Optical micrographs, originally 250× (after Burke 1996).

identification of how sintering actually worked, and that insight was then effectively exploited.

The Lucalox story is a prime specimen of a valuable practical application of a parepistemic study begun for curiosity's sake.

9.5. STRONG STRUCTURAL CERAMICS

Intrinsically, ceramics are immensely strong, because they are made up of mostly small atoms such as silicon, aluminum, magnesium, oxygen, carbon and nitrogen, held together by short, strong covalent bonds. So, individual bonds are strong and moreover there are many of them per unit volume. It is only the tiny Griffith cracks at free surfaces, and corresponding internal defects, which detract from this great potential strength of materials such as silicon nitride, silicon carbide, alumina, magnesia, graphite, etc. The surface and internal defects limit strength in tension and shear but have little effect on strength in compression, so many early uses of these materials have focused on loading in compression. Overcoming the defect-enhanced brittleness of ceramics has been a central concern of modern ceramists for much of the 20th century, and progress, though steady, has been very slow. This has allowed functional ("fine") ceramics, treated in Chapter 7, to overtake structural ceramics in recent decades, and the bulk of the international market at present is for functional ceramics. Japanese materials engineers made a good deal of the running on the functional side, and recently they have similarly taken a leading role in improving and exploiting load-bearing ceramics.

In the preceding section, we saw that removing internal defects, in the form of pores, made sintered alumina, normally opaque, highly translucent. Correspondingly, advanced ceramists in recent years have developed methods to remove internal defects, which often limit tensile strength more than do surface cracks. This program began 'with a bang' in the early 1980s, when Birchall *et al.* (1982) at ICI's New Science Group in England showed that "macro-defect-free" (MDF) cement can be used (for demonstration purposes) to make a beam elastically deformable to a much higher stress and strain than conventional cement (Figure 9.12). The cement was made by moulding in the presence of a substantial fraction of an 'organic rheological aid' that allowed the liquid cement mix to be rolled or extruded into a highly dense mass without pores or cracks. Next year, the same authors (Kendall *et al.* 1983, Birchall 1983) presented their findings in detail: the elastic stiffness was enhanced by removal of pores, and not only the strength but also the fracture toughness was greatly enhanced. Later, (Alford *et al.* 1987), they showed the same features with regard to alumina; in this latest publication, the authors also revealed some highly original indirect methods of estimating the sizes of the largest flaws present. At its

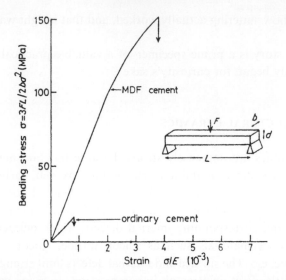

Figure 9.12. Bend strengths of ordinary and MDF cements (after Birchall *et al.* 1982).

high point, this approach to high-strength cements formed the subject-matter for an international conference (Young 1985).

The ICI group, with collaboration around the world, put a great deal of effort into developing this MDF approach to making ceramics strong in tension and bending, including the use of such materials to make bullet-resistant body armour. However, commercial success was not sufficiently rapid and, sadly, ICI closed down the New Science Group and the MDF effort. However, the recognition that the removal of internal defects is a key to better engineering ceramics had been well established. Thus, the experimental manufacture of silicon nitride for a new generation of valves for automotive engines deriving from research, led by G. Petzow, at the Powder Metallurgical Laboratory (which despite its name focuses on ceramics) of the Max-Planck-Institut für Metallforschung makes use of clean rooms, like those used in making microcircuits, to ensure the absence of dust inclusions which would act as stress-raising defects (Hintsches 1995). Petzow is quoted here as remarking that "old-fashioned ceramics using clay or porcelain have as much to do with the high-performance ceramics as counting on five fingers has to do with calculations on advanced computers".

The removal of pores and internal cracks is also of value where functional ceramics are concerned. Dielectrics such as are used in capacitors in enormous quantities, alumina in particular, have long been made with special attention to removing any pores because these considerably lower the breakdown field and therefore the potential difference that the capacitors can withstand.

Another mode of toughening – transformation-toughening – was invented a little earlier than MDF cement. The original idea was published, under the arresting title "Ceramic Steel?", by Garvie *et al.* (1975). These ceramists, working in Australia, focused on zirconia, ZrO_2, which can exist in three polymorphic forms, cubic, tetragonal or monoclinic in crystal structure, according to the temperature. Their idea exploits the fact that a martensitic (shear) phase transformation can be induced by an applied shear stress as well as by a change in temperature. Garvie and his colleagues proposed that by doping zirconia with a few percent of MgO, CaO, Y_2O_3 or CeO_2, the tetragonal or even the cubic form can be 'partially stabilised' so that the martensitic transformation to a thermodynamically more stable form cannot take place spontaneously but can do so if a crack advancing under stress unleashes an embryo of the stable structure and enables it to form a crystallite. This process absorbs energy from the advancing crack and thus functions as a crack arrester. The end result is that a crack is diverted along a tortuous path, or completely stopped, and this toughens the ceramic. The material is pre-aged to the point where partial transformation has taken place; if the treatment is just right, a peak level of toughness is attained. This brilliant idea led to a burst of research around the world, and transformation-toughened zirconia, or alumina provided with a dispersed toughened zirconia phase, became a favourite engineering material, especially for applications such as wire-drawing dies which have to be hard and tough. Figure 9.13 shows two micrographs of this kind of material. It is good to record that the Australians who invented the approach also retained the market in the early days and indeed much of it still today. The extensive literature on this kind of material is discussed in a chapter on toughening mechanisms in ceramic systems (Becher and Rose 1994) and in a recent review by Hannink *et al.* (2000), while the fracture mechanics of transformation-toughened zirconia is analysed by Lawn (1993, p. 225). A limitation is that toughening by this approach is not possible at high temperatures.

The principle behind transformation-toughened zirconia was originally developed, a few years earlier (Gerberich *et al.* 1971), for a steel, called TRIP – TRansformation-Induced Plasticity. (Hence the name proposed in 1975 for the novel form of zirconia... "ceramic steel".) The austenite phase is barely metastable and, where an advancing crack generates locally enhanced stress, martensite is formed locally and the fact that this requires energy causes the steel to be greatly toughened over a limited temperature range.

9.5.1 Silicon nitride

There is no space here to go into details of the many recent developments in ceramics developed to operate under high stresses at high temperatures; it is interesting that a

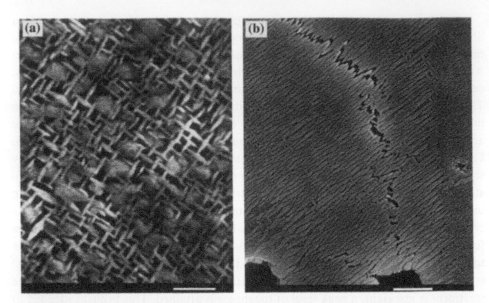

Figure 9.13. (a) Transmission electron micrograph of MgO-stabilised ZrO_2 aged to peak toughness. Tetragonal precipitates on cube planes are shown; the cubic matrix has been etched away with hydrofluoric acid. Bar = 0.5 μm. (b) Scanning electron micrograph of an overaged sample of MgO-stabilised ZrO_2 with coarsened precipitates, subjected to loading. Note the strong crack deflection and bridging. Bar = 2.5 μm (courtesy Dr. R.H.J. Hannink).

detailed memorandum on advanced structural ceramics and composites, issued by the US Office of Technology Assessment in 1986, remarks: "Ceramics encompass such a broad class of materials that they are more conveniently defined in terms of what they are not, rather than what they are. Accordingly, they may be defined as all solids which are neither metallic nor organic." I shall restrict myself to just one family of ceramics, the silicon nitrides (Hampshire 1994, Leatherman and Katz 1989); the material was first reported in 1857. Si_3N_4 has two polymorphs, of which one (β) is the stable form at high temperatures. The powder can be prefabricated and then hot-pressed (or hot isostatically pressed), or silicon powder can be sintered and then reacted with nitrogen, which has the advantage of preserving shape and dimensions and being a cheaper process. A range of additives is used to ensure good density and absence of porosity in the final product, and a huge body of research has been devoted to this ceramic since the War. In 1971/1972, two groups, one in Japan (Oyama, Kamigaito) and in England (Jack, Wilson) independently developed more complex variants of silicon nitrides, the 'sialons' (an acronym derived from Si–Al–O–N), complex materials some of which can be pressureless-sintered to full density. They are also fully presented in Hampshire's book chapter.

Silicon nitride has been used for some years to make automotive turbine rotors, because its low density, 3.2 g/cm^3, ensures low centrifugal stresses. As we saw in Section 9.1.4, now titanium aluminide, also very light, is beginning to be used instead. Since about 1995, silicon nitride inlet and exhaust valves have been used on an experimental basis in German cars, and have recorded very long lives. The low density means that higher oscillation frequencies are feasible, and there is no cooling problem because the material can stand temperatures as high as 1700°C without any problems. As is typical for structural ceramic components, this usage still seems to remain experimental, although a German car manufacturer has ceramic valves running effectively in some 2000 cars. Over recent years, there has again and again been hopeful discussion of the 'all-ceramic engine', either a Diesel version or, in the most hopeful form, a complete gas turbine; the only all-ceramic engine currently in production is a two-stroke version. The action on ceramic Diesel engines has now shifted to Japan (e.g., Kawamura 1999). Silicon nitride has the benefit not only of high temperature tolerance and low thermal conductivity but also of remarkably low friction for rotating or sliding components. The main problem is high fabricating cost (as mentioned above, clean-room methods are desirable), but present results indicate a significant reduction of fuel consumption with experimental engines and the benefits of the engine needing little or no cooling. Determined efforts seem to be under way to reduce production costs. (As with titanium aluminide, the cost per kilogram comes almost entirely from processing costs; the elements involved are all intrinsically cheap.) When, recently, silicon nitride production costs in Germany dropped to DM 10 per valve, the makers of steel valves reduced their price drastically (Petzow 2000). This is classic materials competition in action!

9.5.2 Other ceramic developments

I should add here a mention of a peculiar episode, still in progress, which is based on an attempt to extrapolate from the known properties of silicon nitride to those of a postulated carbon nitride, C_3N_4, which should theoretically (because of the properties a C—N bond should possess) be harder than diamond. This idea was first promulgated by Liu and Cohen (1989) and led to an extraordinary stampede of research. Within a few years, several hundred papers had been published, but no one has as yet shown unambiguously that the postulated compound exists; however, very high hardnesses have been measured in imperfect approximants to the compound. Two reviews of work to date are by Cahn (1996) (brief) and Wang (1997) (detailed). The theoretically driven search for superhard materials generally has been surveyed by Teter (1998) under the title 'Computational Alchemy'. This whole body of research, squarely nucleated by theoretical prediction, has bounced back and forth between experiment and theory; it may well be a prototype of ceramic research programmes of the future.

There is no room here to give an account of the many adventures in processing which are associated with modern 'high-tech' ceramics. The most interesting aspect, perhaps, is the use of polymeric precursors which are converted to ceramic fibres by pyrolysis (Section 11.2.5); another material made by this approach is glassy carbon, an inert material used for medical implants. The standard methods of making high-strength graphite fibres, from poly(acrylonitrile), and of silicon carbide from a poly(carbosilane) precursor, both developed more than 25 years ago, are examples of this approach. These important methods are treated in Chapters 6 and 8 of Chawla's (1998) book, and are discussed again here in Chapter 11.

Another striking innovation is the creation, in Japan, of ceramic composite materials made by unidirectional solidification in ultra-high-temperature furnaces (Waku *et al.* 1997). This builds on the metallurgical practice, developed in the 1960s, of freezing a microstructure of aligned tantalum carbide needles in a nickel–chromium matrix. An eutectic microstructure in $Al_2O_3/GdAlO_3$ mixtures involves two continuous, interpenetrating phases; this microstructure proves to be far tougher (more fracture-resistant) than the same mixture processed by sintering. The unidirectionally frozen structure is still strong at temperatures as high as 1600°C.

9.6. GLASS–CERAMICS

In Chapter 7, I gave a summary account of optical glasses in general and also of the specific kind that is used to make optical waveguides, or fibres, for long-distance communication. Oxide glasses, of course, are used for many other applications as well (Boyd and Thompson 1980), and the world glass industry has kept itself on its toes by many innovations, with respect to processing and to applications, such as coated glasses for keeping rooms cool by reflecting part of the solar spectrum. Another familiar example is Pilkington's float-glass process, a British method of making glass sheet for windows and mirrors without grinding and polishing: molten glass is floated on a still bed of molten tin, and slowly cooled – a process that sounds simple (it was in fact conceived by Alastair Pilkington while he was helping his wife with the washing-up) – but in fact required years of painstaking development to ensure high uniformity and smoothness of the sheet.

The key innovations in turning optical waveguides (fibres) into a successful commercial product were made by R.D. Maurer in the research laboratories of the Corning Glass Company in New York State. This company was also responsible for introducing another family of products, crystalline ceramics made from glass precursors – glass-ceramics. The story of this development carries many lessons for the student of MSE: It shows the importance of a resolute product champion who will spend years, not only in developing an innovation but also in forcing it through against inertia and scepticism. It also shows the vital necessity of painstaking perfecting of the process, as with float-glass. Finally, and perhaps most important, it shows the value of a carefully nurtured research community that fosters revealed talent and protects it against impatience and short-termism from other parts of the commercial enterprise. The laboratory of Corning Glass, like those of GE, Du Pont or Kodak, is an example of a long-established commercial research and development laboratory that has amply won its spurs and cannot thus be abruptly closed to improve the current year's profits.

The factors that favour successful industrial innovation have been memorably analysed by a team at the Science Policy Research Unit at Sussex University, in England (Rothwell *et al.* 1974). In this project (named SAPPHO) 43 pairs of attempted similar innovations – one successful in each pair, one a commercial failure – were critically compared, in order to derive valid generalisations. One conclusion was: "The responsible individuals (i.e., technical innovator, business innovator, chief executive, and – especially – product champion) in the successful attempts are usually more senior and have greater authority than their counterparts who fail".

The prime technical innovator and product champion for glass-ceramics was a physical chemist, S. Donald Stookey (b. 1915; Figure 9.14), who joined the Corning Laboratory in 1940 after a chemical doctorate at MIT. He has given an account of

Figure 9.14. S. Donald Stookey, holding a photosensitive gold-glass plate (after Stookey 1985, courtesy of the Corning Incorporated Department of Archives and Records Management, Corning, NY).

his scientific career in an autobiography (Stookey 1985). His first assigned task was to study photosensitive glasses of several kinds, including gold-bearing 'ruby glass', a material known since the early 17th century. Certain forms of this glass contain gold in solution, in a colourless ionised form, but can be made deeply colored by exposure to ultraviolet light. For this to be possible, it is necessary to include in the glass composition a 'sensitizer' that will absorb ultraviolet light efficiently and use the energy to reduce gold ions to neutral metal atoms. Stookey found cerium oxide to do that job, and created a photosensitive glass that could be colored blue, purple or ruby, according to the size of the colloidal gold crystals precipitated in the glass. Next, he had the idea of using the process he had discovered to create gold particles that would, in turn, act as heterogeneous nuclei to crystallise other species in a suitable glass composition, and found that either a lithium silicate glass or a sodium silicate glass would serve, subject to rather complex heat-treatment schedules (once to create nuclei, a second treatment to make them grow). In the second glass type, sodium fluoride crystallites were nucleated and the material became, what had long been sought at Corning, a light-nucleated opal glass, opaque where it had been illuminated, transparent elsewhere. This was trade-named FOTALITE and after a considerable period of internal debate in the company, in which Stookey took a full part, it began to be used for lighting fittings. (In the glass industry, scaling-up to make industrial products, even on an experimental basis, is extremely expensive, and much persuasion of decision-makers is needed to undertake this.) Patents began to flow in 1950.

A byproduct of these studies in heterogeneous nucleation was Stookey's discovery in 1959 of photochromic glass, material which will reversibly darken and lighten according as light is falling on it or not; the secret was a reversible formation of copper crystallites, the first reversible reaction known in a glass. This product is extensively used for sunglasses.

Stookey recounts how in 1948, the research director asked his staff to try and find a way of 'machining' immensely complex patterns of holes in thin glass sheets... a million holes in single plate were mentioned, with color television screens in mind. Stookey had an idea: he experimented with three different photosensitive glasses he had found, exposed plates to light through a patterned mask, crystallised them, and then exposed them to various familiar glass solvents. His lithium silicate glass came up trumps: all the crystallized regions dissolved completely, the unaltered glass was resistant. "Photochemically machinable" glass, trademarked FOTO-FORM, had been invented (Stookey 1953). Figure 9.15 shows examples of objects made with this material; no other way of shaping glass in this way exists. Stookey says of this product: "(It) has taken almost 30 years to become a big business in its own right; it is now used in complexly shaped structures for electronics, communications, and other industries (computers, electronic displays, electronic

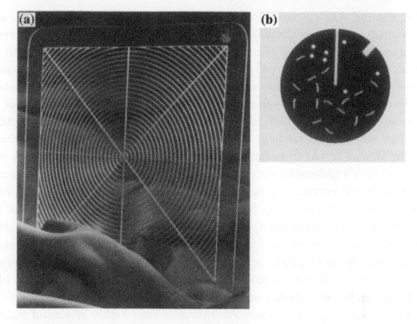

Figure 9.15. Photochemically machined objects made from FOTOFORM™ (after Stookey 1985, and a trade pamphlet, courtesy of the Corning Incorporated Department of Archives and Records Management, Corning, NY).

printers, even as decorative collectibles). Its invention also became a key event in the continuing discovery of new glass technology, proving that photochemical reactions, which precipitate mere traces (less than 100 parts per million) of gold or silver, can nucleate crystallization, which results in major changes in the chemical behavior of the glass."

In the late 1950s, a classic instance happened of accident favouring the prepared mind. Stookey was engaged in systematic etch rate studies and planned to heat-treat a specimen of FOTOFORM™ at 600°C. The temperature controller malfunctioned and when he returned to the furnace, he found it had reached 900°C. He knew the glass would melt below 700°C, but instead of finding a pool of liquid glass, he found an opaque, undeformed solid plate. He lifted it out, dropped it unintentionally on a tiled floor, and the piece bounced with a clang, unbroken. He realised that the chemically machined material could be given a further heat-treatment to turn it into a strong ceramic. This became FOTOCERAM™ (Stookey 1961). The sequence of treatments is as follows: heating to 600°C produces lithium metasilicate nucleated by silver particles, and this is differentially soluble in a liquid reagent; then, in a second treatment at 800–900°C, lithium disilicate and quartz are formed in the residual glass to produce a strong ceramic.

This was the starting-point for the creation of a great variety of bulk glass-ceramics, many of them by Corning, including materials for radomes (transparent to radio waves and resistant to rain erosion) and later, cookware that exploits the properties of certain crystal phases which have very small thermal expansion coefficients. Of course many other scientists, such as George Beall, were also involved in the development. Another variant is a surface coating for car windscreens that contains minute crystallites of such phases; it is applied above the softening temperature so that, on cooling, the surface is left under compression, thereby preventing Griffith cracks from initiating fracture; because the crystallites are much smaller than light wavelengths, the coating is highly transparent. As Stookey remarks in his book, glass-ceramics are made from perfectly homogeneous glass, yielding perfect reliability and uniformity of all properties after crystallisation; this is their advantage, photomachining apart, over any other ceramic or composite structure.

Stookey's reflection on a lifetime's industrial research is: "An industrial researcher must bring together the many strings of a complex problem to bring it to a conclusion, to my mind a more difficult and rewarding task than that of the academic researcher who studies one variable of an artificial system".

In today's ferocious competitive environment, even highly successful materials may have to give way to new, high-technology products. Recently the chief executive of Corning Glass, "which rivals Los Alamos for the most PhDs per head in the world" (Anon. 2000), found it necessary to sell the consumer goods division which includes some glass-ceramics in order to focus single-mindedly on the manufacture of the world's best glass fibres for optical communications. Corning's share price has not suffered.

From the 1960s onwards, many other researchers, academic as well as industrial, built on Corning's glass-ceramic innovations. The best overview of the whole topic of glass-ceramics is by a British academic, McMillan (1964, 1970). He points out that the great French chemist Réaumur discovered glass-ceramics in the middle of the 18th century: "He showed that, if glass bottles were packed into a mixture of sand and gypsum and subjected to red heat for several days, they were converted into opaque, porcelain-like objects". However, Réaumur could not achieve the close control needed to exploit his discovery, and there was then a gap of 200 years till Stookey and his collaborators took over. McMillan and his colleagues found that P_2O_5 serves as an excellent nucleating agent and patented this in 1963. Many other studies since then have cast light on heterogeneously catalysed high-temperature chemical reactions and research in this field continues actively. One interesting British attempt some 30 years ago was to turn waste slag from steel-making plant into building blocks ("Slagceram"), but it was not a commercial success. But at the high-value end of the market, glass-ceramics have been one of the most notable success stories of materials science and engineering.

REFERENCES

Abiko, K. (1994) in *Ultra High Purity Base Metals* (*UHPM-94*), ed. Abiko, K. *et al.* (Japan Institute of Metals, Sendai) p. 1.

Alexander, B.H. and Balluffi, R.W. (1957) *Acta Metall.* **5**, 666.

Alford, N.M., Birchall, J.D. and Kendall, K. (1987) *Nature* **330**, 51.

Anon. (2000) *The Economist*, 19 August, p. 65.

Arunachalam, V.S. and Sundaresan, R. (1991) Powder metallurgy, in *Processing of Metals and Alloys*, ed. Cahn, R.W.; *Materials Science and Technology*, vol 15, eds. Cahn, R.W., Haasen, P. and Kramer, E.J. (VCH, Weinheim) p. 137.

Bartha, L., Lassner, E., Schubert, W.D., and Lux, B. (1995) *The Chemistry of Non-Sag Tungsten* (Pergamon, Oxford).

Beardmore, P., Davies, R.G. and Johnston, T.L. (1969) *Trans. Met. Soc. AIME* **245**, 1537.

Becher, P.F. and Rose, L.R.F. (1994) Tougnening mechanisms in ceramic systems, in *Structure and Properties of Ceramics*, ed. Swain, M.V.; *Materials Science and Technology: A Comprehensive Treatment*, vol. 11, eds. Cahn, R.W., Haasen, P. and Kramer, E.J. (VCH, Weinheim) p. 409.

Benz, M.G. (1999) in *Impurities in Engineering Materials: Impact, Reliability and Control*, ed. Bryant, C.L. (Marcel Dekker, New York), p. 31.

Biloni, H. and Boettinger, W.J. (1996) in *Physical Metallurgy*, 4th edition, vol. 1, eds. Cahn, R.W. and Haasen, P. (North-Holland, Amsterdam) p. 669.

Birchall, J.D. (1983) *Phil. Trans. Roy. Soc. Lond. A* **310**, 31.

Birchall, J.D., Howard, A.J. and Kendall, K. (1982) *J. Mater. Sci. Lett.* **1**, 125.

Birr, K. (1957) *Pioneering in Industrial Research: The Story of the General Electric Research Laboratory* (Public Affairs Press, Washington, DC) pp. 33, 40.

Boyd, D.C. and Thompson, D.A. (1980) Glass, in *Kirk-Othmer Encyclopedia of Chemical Technology*, vol. 11, 3rd edition (Wiley, New York) p. 807.

Burke, J.E. (1996) *Lucalox*™ *Alumina: The Ceramic That Revolutionized Outdoor Lighting*, *MRS Bull.* **21/6**, 61.

Cahn, R.W. (1973) *J. Metals (AIME), February, p. 1.*

Cahn, R.W. (1991) Measurement and control of textures, in *Processing of Metals and Alloys*, ed. Cahn, R.W.; *Materials Science and Technology: A Comprehensive Treatment*, vol. 15, eds. Cahn, R.W., Haasen, P. and Kramer, E.J. (VCH, Weinheim) p. 429.

Cahn, R.W. (1996) *Nature* **380**, 104.

Cahn, R.W. (2000) Historical overview, in *Multiscale Phenomena in Plasticity* (NATO ASI) eds. Saada, *G. et al.* (Kluwer Academic Publishers, Dordrecht) p. 1.

Cahn, R.W. and Haasen, P. (eds.) (1996) *Physical Metallurgy*, 3 volumes, 4th edition (North-Holland, Amsterdam).

Chadwick, G.A. (1967) in *Fractional Solidification*, eds. Zief, M. and Wilcox, W.R. (Marcel Dekker, New York) p. 113.

Chalmers, B. (1974) *Principles of Solidification* (Wiley, New York).

Chawla, K.K. (1998) *Fibrous Materials* (Cambridge University Press, Cambridge).

Chou, T.W. (1992) *Microstructural Design of Fiber Composites* (Cambridge University Press, Cambridge).

Clyne, T.W. and Withers, P.J. (1993) *An Introduction to Metal Matrix Composites* (Cambridge University Press, Cambridge).

Cox, J.A. (1979) *A Century of Light* (Benjamin Company/Rutledge Books, New York).

Cramb, A.W. (1999) in *Impurities in Engineering Materials: Impact, Reliability and Control*, ed. Bryant, C.L. (Marcel Dekker, New York) p. 49.

Davenport, E.S. and Bain, E.C. (1930) *Trans. Amer. Inst. Min. Met. Engrs.* **90**, 117.

Deevi, S.C., Sikka, V.K. and Liu, C.T. (1997) *Prog. Mater. Sci.* **42**, 177.

Dorn, H. (1970–1980) Memoir of Josiah Wedgwood, in *Dictionary of Scientific Biography*, vol. 13, ed. Gillispie, C.C. (Scribner's Sons, New York) p. 213.

Ernsberger, F.M. (1963) Current status of the Griffith crack theory of glass strength, in *Progress in Ceramic Science*, vol. 3, ed. Burke, J.E. (Pergamon, Oxford) p. 58.

Exner, H.E. and Arzt, E. (1996) Sintering processes, in *Physical Metallurgy*, vol. 3, 4th edition, eds. Cahn, R.W. and Haasen, P. (North-Holland, Amsterdam) p. 2627.

Flemings, M.F. (1974) *Solidification Processing* (McGraw-Hill, New York).

Flemings, M.F. (1991) *Metall. Trans.* **22A**, 957.

Flemings, M.F. and Cahn, R.W. (2000) *Acta Mat.* **48**, 371.

Garvie, R.C., Hannink, R.H.J. and Pascoe, R.T. (1975) *Nature* **258**, 703.

Gerberich, W., Hemming, P. and Zackay, V.F. (1971) *Metall. Trans.* **2**, 2243.

German, R.M. (1984) *Powder Metallurgy Science* (Metal Powder Industries Federation, Princeton).

Gladman, T. (1997) *The Physical Metallurgy of Microalloyed Steels* (The Institute of Materials, London).

Gleeson, J. (1998) *The Arcanum: The Extraordinary True Story of the Invention of European Porcelain* (Bantam Press, London).

Greenwood, G.W. (1956) *Acta Met.* **4**, 243.

Hampshire, S. (1994) Nitride ceramics, in *Structure and Properties of Ceramics*, ed. Swain, M.V., *Materials Science and Technology: A Comprehensive Treatment*, vol. 11, eds. Cahn, R.W., Haasen, P. and Kramer, E.J. (VCH, Weinheim) p. 119.

Hannink, R.H.J., Kelly, P.M. and Muddle, B.C. (2000) *J. Amer. Ceram. Soc.* **83**, 461.

Hellebrand, H. (1996) Tape casting, in *Processing of Ceramics, Part I*, ed. Brook, R.J.; *Materials Science and Technology: A Comprehensive Treatment*, vol. 17A, eds. Cahn, R.W., Haasen, P. and Kramer, E.J. (VCH, Weinheim) p. 189.

Herring, C. (1950) *J. Appl. Phys.* **21**, 301.

Herrmann, G., Gleiter, H. and Bäro, G. (1976) *Acta Metall.* **24**, 343.

Hintsches, E. (1995) *MPG Spiegel* (No. 5, 20 November), p. 36.

Honeycombe, R.W.K. and Bhadeshia, H.K.D.H. (1981, 1995) *Steels: Microstructure and Properties* (Edward Arnold, London).

Humphreys, F.J. and Hatherly, M. (1995) *Recrystallization and Related Annealing Phenomena* (Pergamon, Oxford) p. 314.

Hutchinson, W.B. and Ekström, H.-E. (1990) *Mater. Sci. Tech.* **6**, 1103.

Irwin, G.R. (1957) *Trans. Amer. Soc. Mech. Eng., J. Appl. Mech.* **24**, 361.

Jehl, F. (1995) *Inventing electric light* (reprinted from a 1937 publication by the Edison Institute), in *The Faber Book of Science*, ed. Carey, J. (Faber and Faber, London) p. 169.

Jones, W.D. (1937) *Principles of Powder Metallurgy* (Edward Arnold, London).

Jones, W.D. (1960) *Fundamental Principles of Powder Metallurgy* (Edward Arnold, London) p. 442.

Jones, H. and Kurz, W. (eds.) (1984) *Solidification Microstructure: 30 Years after Constitutional Supercooling, Mat. Sci. Eng.* **65/1**.

Jordan, R.G. (1996) in *The Ray Smallman Symposium: Towards the Millenium, A Materials Perspective*, eds. Harris, R. and Ashbee, K. (The Institute of Materials, London) p. 229.

Kato, M. (1995) *JOM* **47/12**, 44.

Kawamura, H. (1999) *Key Eng. Mat.* **161–163**, 9.

Kelly, A. (1966) *Strong Solids* (Clarendon Press, Oxford). Third edition with N.H. Macmillan, 1986.

Kelly, A. (2000) Fibre composites: the weave of history, *Interdisciplinary Sci. Rev.* **25**, 34.

Kendall, K., Howard, A.J. and Birchall, J.D. (1983) *Phil. Trans. Roy. Soc. Lond. A* **310**, 139.

Kingery, W.D. (1986) The development of European porcelain, in *High-Technology Ceramics, Past, Present and Future*, vol. 3, ed. Kingery, W.D. (The American Ceramic Society, Westerville, Ohio) p. 153.

Kingery, W.D. (1990) An unseen revolution: the birth of high-tech ceramics, in *Ceramics and Civilization*, vol. 5 (The American Ceramic Society, Westerville, Ohio) p. 293.

Kingery, W.D. and Berg, M. (1955) *J. Appl. Phys.* **26**, 1205.

Knott, J.F. (1973) *Fundamentals of Fracture Mechanics* (Butterworths, London).

Kocks, U.F., Tomé, C.N. and Wenk, H.-R. (1998) *Texture and Anisotropy: Preferred Orientations in Polycrystals and their Effect on Materials Properties* (Cambridge University Press, Cambridge).

Köthe, A. (1994) in *Ultra High Purity Base Metals* (*UHPM-94*), eds. Abiko, K. *et al.* (Japan Institute of Metals, Sendai) p. 291.

Kuczynski, G.C. (1949) *Trans. Amer. Inst. Min. Metall. Engrs.* **185**, 169.

Kurz, W. and Fisher, D.J. (1984) *Fundamentals of Solidification* (Trans Tech, Aedermannsdorf).

Lawn, B. and Wilshaw, T.R. (1975) *Fracture of Brittle Solids*, 2nd edition, by Lawn alone, in 1993 (Cambridge University Press, Cambridge).

Leatherman, G.L. and Katz, R.N. (1989) Structural ceramics: processing and properties, in *Superalloys, Supercomposites and Superceramics*, eds. Tien, J.K. and Caulfield, T. (Academic Press, Boston) p. 671.

Liu, A.Y. and Cohen, M.L. (1989) *Phys. Rev. B* **41**, 10727.

Liu, C.T., Stringer, J., Mundy, J.N., Horton, L.L. and Angelini, P. (1997) *Intermetallics* **5**, 579.

Martin, G. (2000) Stasis in complex artefacts, in *Technological Innovation as an Evolutionary Process*, ed. Ziman, J. (Cambridge University Press, Cambridge) p. 90.

Marzke, O.T. (editor) (1955) *Impurities and Imperfections* (American Society for Metals, Cleveland, Ohio).

McLean, M. (1996) in *High-Temperature Structural Materials*, eds. Cahn, R.W., Evans, A.G. and McLean, M. (The Royal Society and Chapman & Hall, London).

McMillan, P.W. (1964, 1970) *Glass Ceramics*, 1st and 2nd editions (Academic Press, London).

Miles, M. and Gleiter, H. (1978) *J. Polymer Sci., Polymer Phys. Edition* **16**, 171.

Mirkin, I.L. and Kancheev, O.D. (1967) *Met. Sci. Heat Treat.* (1 & 2), 10 (in Russian).

Miura, H., Kato, M. and Mori, T. (1990) *Colloques de Phys. Cl* **51**, 263.

Morrogh, H. (1986) in *Encyclopedia of Materials Science and Engineering*, vol. 6, ed. Bever, M.B. (Pergamon Press, Oxford) p. 4539.

Mughrabi, H. (ed.) (1993) *Plastic Deformation and Fracture of Materials*, in *Materials Science and Technology: A Comprehensive Treatment*, vol. 6, eds. Cahn, R.W., Haasen, P. and Kramer, E.J. (VCH, Weinheim).

Mughrabi, H. and Tetzlaff, U. (2000) *Adv. Eng. Mater.* **2**, 319.

Mullins, W.W. and Sekerka, R.F. (1963) *J. Appl. Phys.* **34**, 323; (1964) *ibid* **35**, 444.

Mullins, W.W. (2000) Robert Franklin Mehl, in *Biographical Memoirs*, vol. 78 (National Academy Press, Washington, DC) in press.

Ohashi, N. (1988) in *Supplementary Volume 1 of the Encyclopedia of Materials Science and Engineering*, ed. Cahn, R.W. (Pergamon Press, Oxford) p. 85

Orowan, E. (1952) Fundamentals of brittle behavior in metals, in *Fatigue and Fracture of Metals, Symposium at MIT*, ed. Murray, W.M. (MIT and Wiley, New York).

Palumbo, G. and Aust, K.T. (1992) Special properties of Σ grain boundaries, in *Materials Interfaces: Atomic-level Structure and Properties*, eds. Wolf, D. and Yip, S. (Chapman & Hall, London) p. 190.

Paxton, H.W. (1970) *Met. Trans.* **1**, 3473.

Petzow, G. (2000) Private communication.

Pfeil, L.B. (1963) in *Advances in Materials Research in the NATO Countries*, eds. Brooks, H. *et al.* (AGARD by Pergamon Press, Oxford) p. 407.

Pickering, F.B. (1978) *Physical Metallurgy and the Design of Steels* (Applied Science Publishers, London).

Pickering F.B. (ed.) (1992) *Constitution and Properties of Steels*, in *Materials Science and Technology: A Comprehensive Treatment*, vol. 7, eds. Cahn, R.W., Haasen, P. and Kramer, E.J. (VCH, Weinheim).

Porter, D.A. and Easterling, K.E. (1981) *Phase Transformations in Metals and Alloys* (Van Nostrand Reinhold, New York).

Randle, V. and Engler, O. (2000) *Macrotexture, Microtexture and Orientation Mapping* (Gordon & Breach, New York).

Reijnen, P.J.L. (1970) Nonstoichiometry and sintering in ionic solids, in *Problems on Nonstoichiometry*, ed. Rabenau, A. (North-Holland, Amsterdam) p. 219.

Rothwell, R. *et al.* (1974) *Research Policy* **3**, 258.

Rutter, J.W. and Chalmers, B. (1953) *Can. J. Phys.* **31**, 15.

Sauthoff, G. (1995) *Intermetallics* (VCH, Weinheim).

Schumacher, P., Greer, A.L., Worth, J., Evans, P.V., Kearns, M.A., Fisher, P. and Green, A.H. (1998) *Mater. Sci. and Techn.* **14**, 394.

Sekerka, R.F. (1965) *J. Appl. Phys.* **36**, 264.

Shaler, A.J. (1949) *J. Metals (AIME)* **1/11**, 796.

Sims, C.T. (1966) *J. Metals (AIME)*, October, p. 1119.

Sims, C.T. (1984) A history of superalloy metallurgy for superalloy metallurgists, in *Superalloys 1984*, eds. Gell, M. *et al.* (Metallurgical Society of AIME, Warrendale) p. 399.

Smith, R.L. (ed.) (1962) *Ultra-high-purity Metals* (American Society for Metals, Metals Park, Ohio).

Stoloff, N.S. and Davies, R.G. (1966) The mechanical properties of ordered alloys, *Prog. Materi., Sci.* **13**, 1.

Stookey, S.D. (1953) Chemical machining of photosensitive glass, *Ind. Eng. Chem.* **45**, 115.

Stookey, S.D. (1961) Controlled nucleation and crystallization leads to versatile new glass-ceramics, *Chem. Eng. News* **39/25**, 116.

Stookey, S.D. (1985) *Journey to the Center of the Crystal Ball: An Autobiography* (The American Ceramic Society, Columbus, Ohio) 2nd edition in 2000.

Suits, C.G. and Bueche, A.M. (1967) Cases of research and development in a diversified company, in *Applied Science and Technological Progress* (a report by the National Academy of Sciences) (US Govt. Printing Office, Washington) p. 297.

Suresh, S. (1991) *Fatigue of Materials* (Cambridge University Press, Cambridge).

Taylor, A. and Floyd, R.W. (1951–52) *J. Inst. Metals* **80**, 577.

Tenenbaum, M. (1976) Iron and society: a case study in constraints and incentives, *Metal. Trans. A* **7A**, 339.

Teter, D.M. (1998) *MRS Bull.* **23/1**, 22.

Tien, J.K. and Caulfield, T. (eds.) (1989) *Superalloys, Supercomposites and Superceramics* (Academic Press, Boston).

Tiller, W.A., Jackson, K.A., Rutter, J.W. and Chalmers, B. (1953) *Acta Met.* **1**, 428.

Wachtman, J.B. (1999) The development of modern ceramic technology, in *Ceramic Innovations in the 20th Century*, ed. Wachtman, J.B. (American Ceramic Society, Westerville, Ohio) p. 3.

Waku, Y. *et al.* (1997) *Nature* **389**, 49.

Wang, E.G. (1997) *Progr. Mater. Sci.* **41**, 241.

West, D.R.F. and Harris, J.E. (1999) *Metals and the Royal Society*, Chapter 7 (IOM Communications Ltd, London).

Westbrook, J.H. (1957) *Trans. AIME* **209**, 898.

Yamaguchi, M. (eds.) (1996) Symposium on Intermetallics as new high-temperature structural materials, *Intermetallics* **4**, *S1*.

Yamaguchi, M. and Umakoshi, Y. (1990) The deformation behaviour of intermetallic superlattice compounds, *Prog. Mater. Sci.* **34**, 1.

Young, J.F. (1985) Very high strength cement-based materials, *Mater. Res. Soc. Symp. Proc.* **42**.

Chapter 10
Materials in Extreme States

Chapter 10
Materials in Extreme States

10.1. FORMS OF EXTREMITY

In this chapter I propose to exemplify the many categories of useful materials which depend on extreme forms of preparation and treatment, shape, microstructure or function. My subject-matter here should also include ultrahigh pressure, but this has already been discussed in Section 4.2.3. As techniques of preparation have steadily become more sophisticated over the last few decades of the twentieth century, materials in extreme states have become steadily more prevalent.

My chosen examples include rapid solidification, where the extremity is in cooling rate; nanostructured materials, where the extremity is in respect of extremely small grains; surface science, where the extremity needed for the field to develop was ultrahigh vacuum, and the development of vacuum quality is traced; thin films of various kinds, where the extremity is in one minute dimension; and quasicrystals, where the extremity is in the form of symmetry. Various further examples could readily have been chosen, but this chapter is to remain relatively short.

10.2. EXTREME TREATMENTS

10.2.1 Rapid solidification

The industrial technique now known as Rapid Solidification Processing (RSP) is unusual in that it owes its existence largely to a research programme executed in one laboratory for purely scientific reasons. The manifold industrial developments that followed were an unforeseen and welcome by-product.

The originator of RSP was Pol Duwez (1907–1984). This inspirational metallurgist was born and educated in Belgium, then found his way to Pasadena, California and spent the rest of his productive life there, at first at the Jet Propulsion Laboratory and then, 1952–1984, as a professor at the California Institute of Technology. Before he turned to the pursuits with which we are concerned here, he had a number of major discoveries to his credit, such as, in 1950, the identification and characterisation of the sigma phase, a deleterious, embrittling phase in a number of mostly ferrous alloys. This did much to kindle an enthusiasm for the study of intermetallic compounds. For what happened next, I propose to reproduce some sentences from a biographical memoir of Duwez (Johnson 1986a), from which the portrait (Figure 10.1) is also taken: "(From 1952) with several graduate students

Figure 10.1. Portrait of Pol Duwez in 1962 (after Johnson 1986a).

Duwez continued his systematic investigations of the occurrence of intermetallic phases. The work of Hume-Rothery, Mott and Jones, and others had begun to provide a fundamental basis for understanding the occurrence of extended (solid) solubility and intermetallic phases in binary alloys. These theoretical efforts were based on the electronic structure of metals. As these ideas developed, questions were raised regarding the apparent absence of complete solubility in the simple binary silver–copper system (though there was such complete solubility in the Cu–Au and Ag–Au systems). Duwez raised this issue in particular during discussions with students as early as 1955 and 1956. He suggested that perhaps the separation of silver–copper into two (solid) solutions could be avoided by sufficiently rapid cooling of a thin layer of melt. Two students, Ron Willems and William Klement, ultimately devised a method to perform the necessary experiments using a primitive apparatus consisting of a quartz tube containing a metal droplet and connected to a pressurised gas vessel. The droplet was melted using a flame, the pressure applied, and the liquid alloy propelled against a strip of copper. A homogeneous solid solution was obtained (Duwez *et al.* 1960a). The modern science of rapid quenching was born. What is most remarkable was Duwez's grasp of the significance of this event. Within a matter of weeks, a more sophisticated apparatus was built and a systematic study of noble metals was begun. Within months, the simple eutectic alloy system, silver-germanium, had been rapidly quenched to reveal a new metastable crystalline intermetallic phase (Duwez *et al.* 1960b). Duwez recognised this as a missing 'Hume-Rothery' phase. Shortly thereafter in an effort to look for other such phases, a gold–

silicon alloy was rapidly quenched from the melt to yield the first metallic glass (Klement *et al.* 1960)." Most metallic glasses since then follow that same formula... major constituent, a metal, minor constituent, a metalloid. A few years later, Duwez (1967) gave his own account of those first few productive years devoted to RSP; Johnson (1986a) lists 41 of Duwez's most important papers.

Many years later, an electronic-structure calculation for the three systems, Cu–Ag, Cu–Au and Ag–Au which had sparked Duwez's initial experiments, showed (Terakura *et al.* 1987) that the different behaviour in the three systems could be rigorously interpreted. It is a mark of the compartmentalisation of research nowadays that this paper makes no reference to Duwez, though two of the authors work in metallurgical laboratories. There was an ingenious attempt, even earlier, to interpret the anomaly of the Cu–Ag system: Gschneidner (1979) sought to associate a high Debye temperature with what he called "lattice rigidity"; silver has a higher Debye temperature than gold, and correspondingly Gschneidner found that a range of lanthanide (rare earth) metals dissolved more extensively in gold than in silver. These two papers are cited to show that the anomaly which prompted Duwez's initiative has indeed exercised the ingenuity of metallurgists and physicists.

It appears that there was an independent initiative in RSP by I.V. Salli in Russia in 1958 (Salli 1959), but it was not pursued.

Duwez and, in due course, a number of people working in industry (especially Allied-Signal in New Jersey) developed ever-improving devices for RSP; it is interesting that at first these became more complicated, and then again simpler, until the chill-block melt-spinner materialised in the late 1960s and was energetically exploited in the USA and Japan. In its final form, this is simply a jet of molten alloy impinging on a rapidly rotating, polished copper wheel, producing a thin ribbon typically 1–3 mm wide. Later, a variant was developed in which the bottom of the nozzle is held less than a millimetre from the wheel and the nozzle is in the form of a slit, so that a wide sheet (up to 20 cm wide) can be manufactured.

The evolution of RSP (melt-quenching) devices carries an intriguing lesson. As mapped out by Cahn (1993) in a historical overview, some devices were introduced well before Duwez started his researches, but purely as a cheap method of manufacturing shapes such as steel wires for tire cords; when the original technological objective was not achieved, interest in these devices soon waned. It was Duwez's team, sustained by its scientific curiosity, that carried this technical revolution through to completion. There have been two major consequences of his work on RSP: (1) The exploitation of metastable (supersaturated) metallic solid solutions, such as tool steels and light alloys which could be age-hardened particularly effectively because so much excess solute was available for pre-precipitation. (2) The study of metallic glasses in all their variety, which both created an extensive new field for experimental and theoretical research (Cahn 1980) and, in due course, offered major technological breakthroughs. On a larger view still, Duwez's work created the whole concept of non-equilibrium processing of materials (including techniques such as surface treatment by laser), which has just been surveyed (Suryanarayana 1999). There is also substantial coverage in Cahn's historical review and in the whole book in which it appeared in 1993. One of the topics covered there is the gradual development of techniques, both theoretical and experimental, for estimating the cooling rates in an RSP device. A rate of as much as a million degrees per second is feasible, compared with a very few thousand degrees per second in the best solid-state quench.

10.2.1.1 Metallic glasses. With regard to metallic glasses, which were so unexpected that for years Duwez was still sceptical of his own group's discovery, an explosion of research followed in the 1960s and 1970s, on such topics as the factors governing the ability to form such glasses (primarily, what compositions?), their plastic behavior, diffusion mechanisms, electrical conduction and, especially, ferromagnetic behavior of certain glasses (Spaepen and Turnbull 1984). Johnson, in his biographical memoir, says that in or about 1962, Duwez met the great Peter Debye at a conference and discussed the possibility of ferromagnetism in a metallic glass, in spite of the absence of a crystal lattice which would provide a vector for the spins to align themselves along. Debye must have been encouraging, for Duwez began to modify an early glass composition, $Pd_{80}Si_{20}$, by substituting Fe for some of the Pd, and in 1966 weak ferromagnetism was observed. Further substitutions eventually led to $Fe_{75}P_{15}C_{10}$ which was strongly ferromagnetic. A composition close to this is still used nowadays in transformer manufacture.

The use of ferrous glasses in making small 'distribution transformers' for stepping down voltages of several thousand volts to domestic voltages developed by degrees, and a technical history of this fascinating story has been published by DeCristofaro (1998). The point here is that core losses (magnetic and eddy-current losses) are much lower than in grain-oriented silicon-iron, which has held sway for a century; part of the reason is that the absence of magnetocrystalline anisotropy means that the coercive field for a magnetic glass can be particularly small. The upshot is that the power loss in a transformer is so much reduced that the slightly greater cost of the glass is acceptable.

The final development of metallic glasses is the discovery of 'bulk metallic glasses'. Since the 1960s, certain compositions, such as one in a Cu–Pd–Si ternary, had been found to require a cooling rate of only a few hundred degrees per second to bypass unwanted crystallisation during cooling. W.L. Johnson, one of Duwez's right-hand collaborators, and T. Masumoto and A. Inoue in Sendai, Japan, independently developed such compositions into complex mixtures, usually with four or five constituents, that had critical cooling rates of only 10° per second or sometimes even 1° per second... almost like siliceous oxide melts! Objects several centimetres thick could be made glassy. A Zr–Ti–Ni–Cu–Be mixture was the first, and Johnson has pursued this theme with pertinacity: one recent review is by him (Johnson 1996). These compositions are usually close to a deep eutectic, which is an established feature favouring glass formation. A so-called "confusion principle" also operates; not all the multiple diffusions needed for such a glass to crystallise can take place freely, and some sluggish diffusers will in effect stabilise the glass against crystallisation. Up to now, applications are fewer than might have been expected; the manufacture of golf clubs that are more forgiving of duff strokes than earlier clubs (because of the low damping in these glasses) is the most lucrative. A range of bulk glasses based on aluminium has been energetically developed by Masumoto and Inoue in Sendai, Japan, from 1990 onwards, and several of the early papers are listed in Johnson's 1996 overview. Inoue has also written a range of interesting reviews of the field. Inoue's team also pioneered the creation of ultrastrong aluminium-base metallic glasses reinforced by nanocrystalline crystallites through appropriate heat-treatment (e.g., Kim *et al.* 1991). Johnson and his many coworkers (e.g., Löffler *et al.* 2000) have shown by detailed physical analysis why bulk glasses inherently favour copious crystallization in the form of nanocrystalline grains. They are likely to have an important future as useful materials in the partly or wholly crystallised form.

10.2.1.2 Other routes to amorphisation. RSP is not the only way to make metallic glasses. One unexpected approach, discovered by Johnson and coworkers in 1983 and later reviewed by Johnson (1986b) is the *solid-state amorphisation reaction*. Here adjacent thin layers of crystalline elements are heated to interdiffuse them and the mixed zone then becomes amorphous, because crystallisation of a thermodynamically stabler intermetallic compound is kinetically inhibited. An alternative approach to amorphization exploits ball-milling, i.e., intense mechanical deformation of a (usually) metallic or intermetallic compound powder by impacting with tumbling steel or ceramic balls in a mill; this has lately become a major research field in its own right. Such amorphization was first observed by A.E. Yermakov in Russia in 1981. A good review is by Koch (1991). A theoretical study by Desré in 1994 has shown that when the mean grain size of a ball-milled powder has been reduced to a critical size, it will in effect 'melt' to form a thermodynamically stable glass. In fact, amorphization and true melting have been found to be intimately related (Cahn and Johnson 1986).

There is also a large body of research on crystal-to-glass transformation induced by nuclear irradiation, beginning with the observation by Bloch in 1962 that U_6Fe

was amorphised by fission fragments. The physics of this process is surveyed in great depth in relation to other modes of amorphization, and to theoretical criteria for melting, by Okamoto *et al.* (1999).

10.3. EXTREME MICROSTRUCTURES

10.3.1 Nanostructured materials

At a meeting of the American Physical Society in 1959, the Nobel prize-winning physicist, Richard Feynman, speculated in public about the likely effects of manipulating tiny pieces of condensed matter: "I can hardly doubt that when we have some control of the arrangements of things on a small scale, we will get an enormously greater range of possible properties that substances can have". A few years previously, in 1953, as we saw in Section 7.2.1.4, Lifshitz and Kosevich in Russia predicted quantum size effects in what have since come to be known as quantum wells and quantum dots, leading on to Esaki and Tsu's discovery of semiconducting 'superlattices' in 1970–1973. A little later, the pursuit of atomic clusters, predominantly of metals or semiconductors, took wing, because of an interest in the way properties, such as melting behavior, varies with cluster size for minute clusters. In 1988, a lengthy survey was published (Brus, Siegel et al. 1988) of both clusters and "cluster-assembled materials". The term in quotes was one of many synonyms in use at that time for polycrystalline solids made up of extremely small grains; recently, the international community interested in such materials has settled on "nanostructured materials" as the preferred term, with "nanophase materials" and "nanocrystalline materials" as backups. ("Nanostructures" is also sometimes used, but risks confusion with another burgeoning field, the production of minute mechanisms such as nano-electric motors, often from silicon monocrystals, which I do not discuss here; the term 'micromechanoelectrical' devices, or MEMs, is now often used for these. In 1959 Feynman offered a cash prize for the first electric motor less than 1/64 inch across, and it was not very long before he was called upon to make good his promise.)

Attention had been focused on nanostructured materials by a lecture delivered in Denmark by Herbert Gleiter (1981); in a recent outline survey of the field, Siegel (1996) describes this lecture as a 'watershed event'. A little later, Gleiter and Marquardt (1984) set forth some further ideas. Gleiter proposed that the kind of solid materials he envisaged could be made by evaporating substances into a space occupied by an inert gas at high pressure; nanoclusters would condense, be harvested without breaking the enclosure and be compressed by a piston to form a 'green' solid, which would then need further compaction by heat treatment. This for a while became the orthodox way of producing small samples for the study, primarily, of mechanical properties. Gleiter's view of the essential structure of these materials, when single phase, is shown in Figure 10.2: a substantial fraction of the atoms lies in the disordered grain boundaries. It was predicted that resistance to plastic deformation by dislocation motion would steadily increase as grain size is reduced, and this proved to be true, except that at the very smallest grain sizes there is often an inversion and strength again diminishes; this aspect is still a matter of frequent investigation. Such studies have also been made for 'nanocomposites': Figure 10.3 shows that nanostructured WC–Co 'cermet', now a commercial product used for cutting tools, is substantially harder than the same material with conventional grain size; the fine-grained cermet is also considerably tougher (more resistant to cracking).

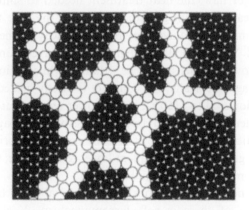

Figure 10.2. Schematic of the microstructure of a nanostructured single-phase material (after Gleiter 1996).

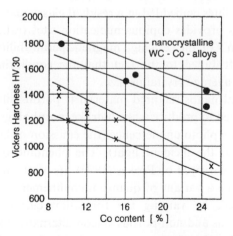

Figure 10.3. Hardness of WC–Co cermets with nanostructured and conventional grain sizes (after Gleiter 1996, reproduced from a report by Schlump and Willbrandt).

The most intriguing aspect of nanostructured metals and, especially, ceramics such as titania is that the very small grain size encourages Herring-Nabarro creep which, in turn is the precondition of superplastic forming under stress. The essential facts concerning this process are laid out in Section 4.2.5. Nanostructured ceramics can be plastically formed, in spite of extreme resistance to dislocation motion, and this has been plentifully documented in many studies. Examples are set out in Gleiter's own (1996) overview of nanostructured materials. The ability to form nano-ceramics to 'near net shapes' looks to have very promising industrial potential.

The exploitation of easy superplastic forming of nanostructured ceramics is hindered by one major flaw: the heat treatment needed to sinter a 'green' solid to 100% density also leads to grain growth, so that by the time the material is fully dense, it is no longer nanocrystalline. Very recently, a way has been found round this difficulty. Chen and Wang (2000), studying Y_2O_3, have found that a two-stage sintering process allows full density to be attained while grain growth is arrested during the second stage. Typically, the compact is briefly heated to 1310°C and the temperature is then lowered to 1150°C; if that lower temperature were applied from the start, complete densification would not be possible. The paper analyses various conceivable explanations, but it is not at present clear why a brief high-temperature anneal inhibits grain growth at a subsequent lower temperature; this valuable finding is likely to engender much consequential research.

A number of 'functional' properties can also be affected by nanocrystallinity. The most interesting of these is soft ferromagnetism. Yoshizawa *et al.* (1988) discovered that a bulk metallic glass (trade-named "Finemet") of composition $Fe_{73.5}Si_{13.5}B_9Cu_1Nb_3$, on partial crystallization, assumes a structure with nanometre-sized (5–20 nm) crystallites embedded in a residual glassy matrix. The small amount of copper in the glass provides copious nucleation sites (rather as copper does in glass-ceramics, Section 9.6); the very high magnetic permeability of such glass/crystal composites can be attributed to the fact that the equilibrium magnetic domain thickness exceeds the average crystallite size.

Another functional nanostructured material is porous silicon, monocrystalline silicon chemically etched to produce a fine hairlike morphology: this material, unlike unetched silicon, shows photoluminescence (the emission of light of a wavelength – variable – longer than the incident light). The phenomenon was discovered by Canham (1990) and is surveyed by Prokes (1996). Its mechanism is still under lively debate; it appears to be a variant of quantum confinement. Frohnhoff and Berger (1994) have succeeded, by varying the formation current density, in making superlattices with porous and non-porous silicon alternating; such superlattices can be tuned to reflect the photoluminescence and therefore enhance light emission. There is hope of exploiting porous silicon in light-emitting devices based on silicon chips, as part of 'optoelectronic' circuitry. The prospects of success in this have been discussed by Miller (1996).

The comparatively new field of nanostructured materials has its own journals (though the first one has now been merged with another, broader journal) and frequent conferences; it is a good example of a parepisteme which appears to be successful. The best single source of information about the many aspects of the field is a substantial multiauthor book edited by Edelstein and Cammarata (1996).

The original 'Gleiter method' of making nanostructured solids is fine for research but not a feasible commercial method of making substantial quantities, for instance of a nanostructured cermet such as Co–WC. A whole range of chemical methods has now been developed, as described in the Edelstein/Cammarata book. These methods are mostly dependent on colloidal precursors, often using the so-called sol–gel approach. A sol is a colloidal liquid solution, often in water; on evaporation or other treatment, a sol turns into a gelatinous 'gel' which in turn can be converted into a nanostructured solid. A range of organometallic colloidal precursors can be converted into oxide ceramics by such an approach. Spray pyrolysis or conversion via an 'aerosol' (a suspension of colloidal particles in air or other gas) offer other potentially large-scale routes to make nanostructured materials, and yet another route, chemically sophisticated, is by stabilising metal clusters with 'ligands', chemical radicals which bind to and coat the clusters to stabilise them against agglomeration. This approach allows a population of uniformly sized clusters to be made, but it is not appropriate for conversion into continuous solid materials.

Gleiter, who effectively created this field of research, has very recently surveyed its present condition in a magisterial overview (Gleiter 2000).

It must be added that in the opinion of some observers, the claims of what is coming to be called 'nanotechnology' are often exaggerated, and long-term hopes are sometimes presented as though they were present-day reality. A carefully nuanced critical view can be found, for example, in a review by an engineer, Dobson (2000), of a large book entitled *Nanotechnology*. To balance this, again, there are some sober overviews of what may be in prospect; an example is a survey of work currently in progress at Oak Ridge National Laboratory, in America (ORNL 2000). In this survey, an intriguing remark is attributed to Eugene Wong of the National Science Foundation in America: "The nanometre is truly a magical unit of length. It is the point where the smallest manmade thing meets nature".

10.3.2 Microsieves via particle tracks

Small holes are the negative correlative of small objects, and there is in fact an industrial product, considerably antedating Gleiter's initiative, which is based on such holes.

Two physicists, R.M. Walker and P.B. Price, working at the GE central laboratory in Schenectady, NY (see Section 1.1.2) discovered in 1961 that heavy fission fragments from uranium leave damage trails in insulators such as mica which, on subsequent chemical attack, act as preferential loci for rapid etching. A population of fission tracks in a thin cleaved sliver of mica can be converted into a population of holes of fairly uniform size; the mean size is determined by the duration of etching. Holes typically 3–4 μm across were formed. (This specific research was stimulated by a colleague at GE who needed a controllable, ultraslow vacuum leak.) Together with a third physicist, R.L. Fleischer, the discoverers developed this finding into a means of studying many features and processes, such as the age of geological specimens, the scale of radon seepage from radioactive rocks, and even features of petroleum deposits. The really unexpected development, however, came in 1962, when a cancer researcher in New York got wind of this research; he was just then needing an ultrafilter for blood which would hold back the larger, more rigid cancer cells while allowing other cells to pass through. GE's etched mica slivers proved to be ideal. This led to the setting-up of a dedicated small manufactory to make such filters; Fleischer found that sieves made with GE's own polycarbonate resin (used in automotive lighting) were stronger and more durable than those made with mica. A major medical product resulted which soon made GE a sales of some ten million dollars a year. When, 17 years later, the patents expired, other companies began to compete, and the total sales of microfilters, used to analyse aerosols, etc., as well as cancerous blood, now exceeds 50 million dollars per annum.

The antecedents and circumstances of this research program are spelled out in some detail by Suits and Bueche (1967), two former research directors of GE, and much more recently in a popular book by Fleischer (1998). Both publications analyse why a hard-headed industrial laboratory saw fit to finance such apparently 'blue-sky' research. Suits and Bueche say: "...the research did not arise from any direct or specific need of GE's businesses and was related to them only in a general way. Why, then, was the research condoned, supported and encouraged in an industrial laboratory? The answer is that a large company and a large laboratory can invest a small fraction of its funds in speculative ventures in research; these ventures promise, however tentatively, departures into entirely new businesses." This research met "no recognised pre-existent need"; indeed, to adopt my preferred word, it was a pure parepisteme. A recent historical study of a number of recent practical inventions, with a focus on high-temperature superconduction (Holton *et al.* 1996) concludes: "...above all, historical study of cases of successful modern research has repeatedly shown that the interplay between initially unrelated basic knowledge, technology and products is so intense that, far from being separate and distinct, they are all portions of a single, tightly woven fabric".

Fleischer, from his perspective 31 years later, points out that (as it turned out) track etching had been independently discovered in the late 1950s at Harwell Laboratory in England a little before GE did, but because the laboratory was then not commercially oriented, nothing was done to follow up the possibilities. In a hard-hitting analysis (pp. 171–176 of his book) Fleischer examines the gradual decay of this kind of industrial research in industry across the world ("even in Japan"), to be replaced by demands from American industrial executives that government should finance universities to undertake more of this kind of parepistemic research that had formerly been done in industrial laboratories, specifically in order to help industrial firms. Fleischer remarks that such pleadings are "if not actually hypocritical, at least futile. Is it reasonable to expect decision-makers in government to be eager to invest in science from which industry has withdrawn?" In my own country, Britain, in the face of the closure of ICI's New Materials Group and of the entire New Ventures laboratory of BP, one can only echo this bitter rhetorical question.

10.4. ULTRAHIGH VACUUM AND SURFACE SCIENCE

10.4.1 The origins of modern surface science
The earliest transistors (Section 7.2.1), starting at the end of the 1940s, were made of germanium; silicon only followed some years later. However, germanium transistors proved disconcertingly unreliable. The experience of manufacturers in those early days was forcefully put in a book by Hanson (1980): "It was wondrous that transistors worked at all, and quite often they did not. Those that did varied widely in performance, and it was sometimes easier to test them after production and, on that basis, find out what kind or electronic component they had turned out to be... It was as if the Ford Motor Company was running a production line so uncontrollable that it had to test the finished product to find out if it was a truck, a convertible or a sedan."

In an illuminating overview of the linkage between semiconductor problems and the genesis of surface science, Gatos (1994) describes the research on germanium surfaces performed at MIT and elsewhere in the early 1950s. The erratic performance of germanium transistors was gradually linked to the unstable properties of germanium surfaces, especially the solubility of germanium oxide in water; the electronic 'surface states' on Ge were thus unstable. In spite of prolonged studies of etching procedures intended to stabilise Ge surfaces, "their reliable and permanent stabilisation, indispensable in solid-state electronics, remained a moving target", to quote Gatos verbatim. "Naturally, the emphasis shifted from Ge to Si. The very thin surface oxide on Si was found to be chemically refractory and, thus, assured surface chemical stability". The manufacturer was now able to predetermine whether he was making a truck or a convertible!

According to Gatos, the needs of solid-state electronics, not least in connection with various compound semiconductors, were a prime catalyst for the evolution of the techniques needed for a detailed study of surface structure, an evolution which gathered pace in the late 1950s and early 1960s. This analysis is confirmed by the fact that Gatos, who had become a semiconductor specialist in the materials science and engineering department at M.I.T., was invited in 1962 to edit a new journal to be devoted specifically to semiconductor surfaces. As Gatos remarks in his historical overview, "it was clear to me that the experimental and theoretical developments achieved for the study of semiconductor surfaces were being rapidly transplanted to the study of the surfaces of other classes of materials". He thus insisted on a broader remit for the new journal, and *Surface Science*, under Gatos' editorship, first saw the light of day in 1964. Gatos' essay is the first in a long series of review articles on different aspects of surface science to mark the 30th anniversary of the journal, making up volumes 299/300 of *Surface Science*.

Other fields of surface study were of course developing: the study of catalysts for the chemical industry and the study of friction and lubrication of solid surfaces were two such fields. But in sheer terms of economic weight, solid-state electronics seems to have led the field.

Before 1950, it was impossible to examine the true structure of a solid surface, because, even if a surface is cleaned by flash-heating, the atmospheric molecules which constantly bombard a solid surface very quickly re-form an adsorbed monolayer, which is likely to alter the underlying structure. Assuming that all incident molecules of oxygen or nitrogen stick to the surface, a monolayer will be formed in 3×10^{-6} second at 1 Torr (=1 mm of mercury), that is, at 10^{-3} atmosphere; a monolayer forms in 3 s at 10^{-6} Torr, or 10^{-9} atmosphere; but a complete monolayer takes about an hour to form at 10^{-9} Torr. The problem was that in 1950, a vacuum of 10^{-9} Torr was not achievable; 10^{-8} Torr was the limit, and that only provided a few minutes' grace before an experimental surface became wholly contaminated.

The scientific study of surfaces, and the full recognition of how much a surface differs from a bulk structure, awaited a drastic improvement in vacuum technique. The next Section is devoted to a brief account of the history of vacuum.

10.4.2 The creation of ultrahigh vacuum

Early in the 17th century, there was still vigorous disagreement as to the feasibility of empty space; Descartes denied the possibility of a vacuum. The matter was put to the test for the first time by Otto von Guericke (1602–1686), a German politician who "devoted his brief leisure to scientific experimentation" (Krafft 1970–1980). He designed a crude suction pump using a cylinder and piston and two flap valves, and with this, after many false starts, he succeeded in his famous 1657 public experiment, in Magdeburg, of evacuating a pair of tightly fitting copper hemispheres to the point that two teams of horses could not drag them apart. The reality of vacuum had been publicly demonstrated.

In fact, though probably von Guericke did not know about it, the Florentine Evangelista Torricelli (1608–1647) had also established the pressure of the atmosphere by showing in 1643 that there was a limiting height of mercury that could be supported by that pressure in a closed tube; a working barometer followed the next year. This famous experiment indirectly demonstrated the existence of the "Torricellian vacuum" above the mercury in the closed tube, hence the use of Torricelli's name for the unit of gas pressure in a partial vacuum, the torr (equivalent to the pressure exerted by a mercury column of one millimetre height). In 1650, no less a scholar than Blaise Pascal showed that the height of the supported mercury column varied with altitude above sea-level.

In 1850, the Toepler pump was invented; this is a form of piston pump in which the reciprocating piston consists of mercury; it was followed in 1865 by the Sprengel pump, in which air is entrained away by small drops of mercury falling under gravity. In 1874, the first accurate vacuum gauge, the McLeod gauge, again centred around mercury columns, was devised. These and other dates are listed in a concise history of vacuum techniques (Roth 1976). The first rotary vacuum pump, the workhorse of rough vacuum, was not invented until 1905, by Wolfgang Gaede in Germany, and the first diffusion pump, invented by Irving Langmuir at GE, followed in 1916.

It is noteworthy that inventors well before Edison, notably the Englishman Joseph Swan who in some people's estimation was the true inventor of the incandescent lamp, found it impossible to make a stable lamp because the vacuum pumps at their disposal simply were not effective enough, and also took an inordinate time to produce even a modest vacuum. By the time Edison developed his carbon filament lamp in 1879, the Toepler and Sprengel pumps had been sufficiently developed to enable him to protect his filaments from oxidation, by vacua of around 0.1 torr or even better. In due course, 'getters' were invented; these were small pieces of highly reactive metal inside light bulbs, which were briefly flashed by an electric current to absorb residual oxygen and nitrogen. It was only from 1879 onwards that vacuum quality began to be taken seriously.

With the rotary and diffusion pumps in tandem, aided by a liquid-nitrogen trap, a vacuum of 10^{-6} Torr became readily attainable between the wars; by degrees, as oils and vacuum greases improved, this was inched up towards 10^{-8} Torr (a hundred-billionth of atmospheric pressure), but there it stuck. These low pressures were beyond the range of the McLeod gauge and even beyond the Pirani gauge based on heat conduction from a hot filament (limit $\approx 10^{-4}$ Torr), and it was necessary to use the hot-cathode ionisation gauge, invented in 1937. This depends on a hot-wire cathode surrounded by a positively charged grid, which in turn is enclosed in an ion-collecting 'shell'. Electrons travelling outwards from the cathode occasionally collide with a gas molecule, ionising them; the positive ions are picked up by the negatively charged collection shell, and their number measures the quality of the vacuum.

As we have seen, by 1950 it had become clear that no proper surface science could begin until a vacuum considerably 'harder' than 10^{-8} Torr could be attained. The 10^{-8} Torr limit was therefore a great frustration. Then, in 1947, Wayne Nottingham of MIT came up with the suggestion that the limit was illusory: he thought that the limit was not in pumping, but in measurement: Nottingham suggested that the electrons bombarding the positively charged grid would generate X-rays, which would release more photoelectrons from the collector. So the gauge would register a signal even if there were no gas molecules whatever in the gauge! Two years later, Robert Bayard and Daniel Alpert, at the Westinghouse Research Laboratory in Pittsburgh, invented a way of circumventing the problem, if it had been correctly diagnosed (Bayard and Alpert 1950). They switched the positions of the cathode and the collector. Now the collector was no longer a large cylinder but just a wire, offering a very slender target to the X-rays from the grid, so that the "null signal" would be negligible. The strategy worked, indeed it worked better than predicted, because the ion gauge could operate as a pump at very low pressures as well as being an indicator. The new Alpert gauge was isolated by means of a novel all-metal valve that did not require an organic sealing compound with its unavoidable characteristic vapour pressure, and the quality of the vacuum sailed to 5×10^{-10} Torr. This was now a new limit; Alpert, who is the recognised father of ultrahigh vacuum, constructed a mass spectrometer to analyse the residual atmosphere, and found that the new 5×10^{-10} Torr limit was due to atmospheric helium percolating through the pyrex glass enclosure. Thereafter, glass was avoided and the bulk of vacuum apparatus for ultrahigh vacuum (UHV) was henceforth made of welded metal, usually stainless steel, with soft metallic gaskets that require no lubricant, and fully metallic valves.

Such vessels can also be 'baked' at a temperature of several hundred degrees, to drive off any gas adsorbed on metal surfaces. The pumping function of an ion gauge was developed into efficient ionic pumps and 'turbomolecular pumps', supplemented by low-temperature traps and cryopumps. Finally, sputter-ion pumps, which rely on sorption processes initiated by ionised gas, were introduced. A vacuum of 10^{-11}–10^{-12} Torr, true UHV, became routinely accessible in the late 1950s, and surface science could be launched.

An early account of UHV and its requirements is by Redhead *et al.* (1962); an even earlier summary of progress in vacuum technology, with perhaps the first tentative account of UHV, was by Pollard (1959). A lively popular account is by

Steinherz and Redhead (1962), while advances in vacuum techniques from a specifically chemical viewpoint were discussed by Roberts (1960).

The various new vacuum pumps certainly made possible much faster and more efficient pumping, but the essential breakthrough came from two events: the recognition that the older ionic vacuum gauges were drastically inaccurate, and the further recognition that UHV systems needed to be made from metal, with little or no glass and no organic greases, and that the systems had to be bakeable.

The curious behavior of ion gauges acting also as pumps has had a recent counterpart. Cohron *et al.* (1996) studied the effect of low-pressure hydrogen on the mechanical behavior of the intermetallic compound Ni_3Al. They found, to their astonishment, that the ductility of the compound with their ion gauge turned off was 3–4 times higher than with the gauge functioning. They discovered that Langmuir and Mackey (1914) had first identified hydrogen dissociation on a hot tungsten surface, and proved that the embrittlement was due to atomic hydrogen 'manufactured' inside the gauge that then diffused along grain boundaries of the compound and embrittled them. So it seems that one must always be alert to the possibility of a measuring device that influences the very variable that it is meant to measure... a very apposite precaution in the days of quantum ambivalence.

10.4.3 An outline of surface science

My principal objective in Section 10.4 has been to underline the necessity for a drastic enhancement of a crucial experimental technology, the production of ultrahigh vacuum, as a precondition for the emergence of a new branch of science, and this enhancement was surveyed in the preceding Section. It would not be appropriate in this book to present a detailed account of surface science as it has developed, so I shall restrict myself to a few comments. The field has been neatly subdivided among chemists, physicists and materials scientists; it is an ideal specimen of the kind of study which has flourished under the conditions of the interdisciplinary materials laboratories described in Chapter 1.

UHV is necessary but not sufficient to ensure an uncontaminated surface. Certainly, the surface will not be contaminated by atoms arriving from the vacuum space, but such contamination as it had before the vacuum was formed has to be removed by bombardment with argon ions. This damages the surface structurally, and that has to be 'healed' by in situ heat treatment. That, however, allows dissolved impurities to diffuse to the surface and cause contamination from below. This problem has to be dealt with by many cycles of bombardment and annealing, until the internal contaminants are exhausted. This is a convincing example of Murphy's Law in action: one of the many corollaries of the Law is that "new systems generate new problems".

The first key technique (UHV apart) in surface science was low-energy electron diffraction (LEED). This was used for the first time by Davisson and Germer at Bell Labs in 1927; it did not then give much information about surfaces, but it did for the first time confirm the wave-particle duality in respect of electrons and thereby earned the investigators a Nobel Prize. The technique uses electrons typically at energies of 20–300 eV, which penetrate only one or two atom layers deep. The great difficulty is in interpreting the patterns obtained; the problems are well set out in a standard text by Woodruff and Delchar (1986); it is necessary to take account of multiple scattering. The early mystifications among LEED practitioners are explained in reminiscences by Marcus (1994). Not only the two-dimensional surface reconstruction as exemplified in Figure 6.9(b) in Chapter 6, but also the complications ensuing from domains, steps and defects at the surface need to be allowed for. One eminent practitioner, J.B. Pendry, in an opinion piece in *Nature* (Pendry 1984) under the title "Removing the black magic", claimed that proper surface crystallography had only existed since about 1974. Now, pictures obtained by scanning tunnelling microscopy offer a direct check on conclusions reached by LEED. The other key technique which is now used in conjunction with LEED is Auger electron spectrometry: here an ionising primary beam unleashes a cascade of electron energy transitions until an 'Auger electron' with an energy that constitutes a finger print of the element emitting it is released into the vacuum. The ranges of Auger electrons are so small that effectively the technique examines and identifies the surface monolayer of atoms. An early survey of this key technique is by Rivière (1973).

One other technique has become central in surface research: this is X-ray photoelectron spectrometry, earlier known as ESCA, 'electron spectroscopy for chemical analysis'. Photoelectrons are emitted from a surface irradiated by X-rays. The precautions which have to be taken to ensure accurate quantitative analysis by this much-used technique are set out by Seah (1980).

It is now clear that surface defects, steps in particular, and two-dimensional crystallographic restructuring of surfaces, are linked: there is a phenomenon of reconstruction-linked faceting. Surface steps, particularly on vicinal crystal faces (faces close to but not coinciding with low-index planes) are important for various electronic devices; in particular, the migration of steps and thus the instability of surface morphology needs to be understood. The elaborate complexity of current understanding of surface steps has just been surveyed by Jeong and Williams (1999).

As remarked above, surface science has come to be partitioned between chemists, physicists and materials scientists. Physicists have played a substantial role, and an excellent early overview of surface science from a physicist's perspective is by Tabor (1981). An example of a surface parepisteme that has been entirely driven by physicists is the study of the *roughening transition*. Above a critical temperature but still well below the melting temperature, many smooth surfaces begin to become rough. This was first theoretically predicted in the famous 1951 paper by Burton, Cabrera and Frank on the theory of crystal growth (see Section 3.2.3.3): roughening is in essence due to the prevalence of vacancies at surfaces and the consequential enhanced probability of creating additional defects near an existing defect; diffusing vacancies and adatoms will begin to cluster above the roughening temperature, forming growing mounds. In the mid-1970s, the roughening transition was shown to be also linked, improbable though it may seem, to a two-dimensional metal-insulator transition. The story of theory and experiment relating to this curious phenomenon can be found in a review article by Pontikis (1993).

Nevertheless, chemists have played the biggest role by far. A particular reason for this is that chemists need catalysts to accelerate many reactions used in chemical manufacturing, in particular the cracking of petroleum into fractions; this has been a major field of research, focused on surface behavior, ever since Johann Döbereiner (1780–1849) in 1823 discovered that platinum sponge (very fine particles) catalysed the combination of hydrogen and oxygen. Some of these catalysts are colloidal (nanostructured) particles, in some cases even metallic glass particles, but the most important catalysts nowadays are zeolites. These are typically crystalline alumino-silicates with the formal composition $M_xO_y \cdot Al_2O_3 \cdot pSiO_2 \cdot qH_2O$. They have structural tunnels – internal surfaces – as shown in Figure 10.4; these admit some reactants but not others and can thus function as highly selective catalysts.

Crucial though they are industrially, I do not propose to discuss catalysts further here. My reason is that I do not regard them as materials. Up to this point, I have not sought to define what I mean by a 'material', but this is a convenient point to attempt such a definition. In my conception, a material is a substance which is then further processed, shaped and combined with others to make a useful object. Something like a lubricant, fertiliser, food, drug, ink or catalyst by that definition is not a material, because it is used 'as is'. Like all definitions, this is untidy at the edges: thus a drug may be combined with another substance to ensure slow release to the bodily tissues, and that auxiliary substance is then a material, and the status of cooked foods by my definition gives plentiful scope for casuistry.

(a)   **(b)**   **(c)**

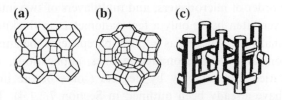

Figure 10.4. Outline structures of (a) zeolite A, (b) its homologue faujasite, (c) the channel network of the 'tubular' zeolite ZSM-5.

An excellent, accessible overview of what surface scientists do, the problems they address and how they link to technological needs is in a published lecture by a chemist, Somorjai (1998). He concisely sets out the function of numerous advanced instruments and techniques used by the surface scientist, all combined with UHV (LEED was merely the first), and exemplifies the kinds of physical chemical issues addressed – to pick just one example, the interactions of co-adsorbed species on a surface. He also introduces the concept of 'surface materials', ones in which the external or internal surfaces are the key to function. In this sense, a surface material is rather like a nanostructured material; in the one case the material consists predominantly of surfaces, in the other case, of interfaces.

A further field of research is linked to the influence of the surface state on a range of bulk properties: a recent example is the demonstration of enhancement of ductility of relatively brittle materials such as pure chromium and the intermetallic NiAl by careful removal of mechanical damage from their surfaces. A further large field of research is the design and properties of surface coatings, with objectives such as oxidation resistance (notably for superalloys in jet engines), ultrahardness and reduction of friction. This is a domain cultivated by materials engineers, as is the study of *tribology*, which comes from the Greek word for 'rubbing' and includes the study of friction as well as the rate and mechanisms of wear when one surface rubs against another under load. Tribology, an increasingly elaborate and important field, links closely with the study of lubrication. Tribology has become a beautiful exemplar of the marriage of engineering and science. The notable classic of this field is a text by Bowden and Tabor (1954), while a more recent concise overview is by Furey (1986). The history of tribology is surveyed by Dowson (1979).

10.5. EXTREME THINNESS

10.5.1 Thin films

Thin metallic or semiconducting films, almost invariably deposited on a substrate, come essentially in three forms: monolayers or ultrathin films; continuous films with thicknesses of the order of micrometers; and multilayers of two interleaving species, each successive layer often being only a few nanometers in thickness. This form of material was originally investigated as a 'pure' parepisteme, beginning with metallic films and going on later to semiconducting ones; applications, which are nowadays extremely varied, arrived only by degrees (some of the important ones in microelectronics have already been outlined in Section 7.2.1.4). Today, thin films have their own major journals and conferences. The subject is clearly linked to surface science, particularly so the study of the initial, monolayer films.

Much interest attaches to the mechanisms of thin film deposition, and these in turn are linked to the mechanisms of epitaxial growth (see below). The very early stages, up to and including monolayer growth, used to be investigated largely by Auger electron spectrometry: the completion of the first layer is revealed by a bend in the plot of signal intensity versus time of deposition, and LEED helped to identify the nature of the initial deposit; progressively, electron microscopy, both by transmission and by scanning microscopy, has gradually taken over. This kind of research has been closely linked with the investigation of chemisorption. The early work on monolayers is very competently surveyed by Rhead (1983).

The workhorse methods used for depositing thin films are thermal evaporation and sputtering. The second (evocatively named) method allows much more exact control than does evaporation: it involves bombarding a target consisting of the material(s) to be deposited with high-energy noble-gas ions, causing atoms of the target to spring out and hit the substrate. One starts with UHV and then bleeds in small pressures of the bombarding gas, which does not contaminate the substrate surface. For the most complete control, especially when semiconductor films are in question, molecular-beam methods and atomic layer epitaxy, as outlined in Section 7.2.1.4, are now used. The subtleties of sputtering are surveyed by Kinbara (1997).

In recent years, it has been established that bombarding the substrate (as distinct from a target) directly with noble-gas ions while a film is being deposited can greatly enhance the quality of adhesion between substrate and deposit, and controlling the direction of the bombarding ions can influence the crystallographic orientation of the deposit as well as its microstructure. This whole family of effects, now widely exploited, is surveyed by Rossnagel and Cuomo (1988).

Of the many properties of films in their successive stages, those most commonly studied nowadays are the magnetic, electrical and mechanical ones. The magnetic properties and uses of thin films, especially multilayers, have been outlined in Section 7.4 and need not be repeated here; however, it is worth pointing out an excellent survey of magnetic multilayers (Grünberg 2000). Electrical properties have been covered by Coutts (1974).

The mechanical properties, especially the internal stresses set up by interaction of substrate and deposit, have a close bearing on the behavior of metallic interconnects (electrical conductors) in integrated circuits. Such interconnects suffer from more diseases than does a drink-sodden and tobacco-crazed invalid, and stress-states play roughly the role of nicotine poisoning. A very good review specifically of stresses in films is by Nix (1989).

On the broad subject of thin films generally, a well-regarded early text is by an Indian physicist, Chopra (1969), while a very broad, didactic treatment of thin films in all their aspects is by Ohring (1992). A recent survey of the effect of structure on properties of thin films relevant to microelectronics is by Machlin (1998).

10.5.1.1 Epitaxy. There is often a sharp orientation relationship between a single-crystal substrate and a thin-film deposit, depending on the crystal structures and lattice parameters of the two substances. When such a relationship exists, the deposit is said to be *in epitaxy* with the substrate. The simplest relationship is parallel orientation, and this is common in semiconductor heterostructures, but more complex relationships are often encountered.

The word 'epitaxy' was introduced by a French mineralogist, L. Royer, who discovered the phenomenon (Royer 1928); the term, based on Greek, literally means 'arrangement on'. In the early years, the phenomenon was most commonly studied by evaporating metal films on to cleaved alkali halide monocrystals; before UHV was introduced, epitaxial studies were of course restricted to contaminated substrate surfaces. From the beginning, the crucial role of lattice misfit (the mismatch of lattice parameters of the two substances, whether or not they had the same crystal structure) in governing the appearance of epitaxy was fully recognised. A limiting misfit not more than 15% is often quoted as the empirical rule; this is reminiscent of Hume-Rothery's 15% rule governing extensive solid solubility between two isostructural metals (Section 3.3.1.1). A famous stage in the prolonged study of the factors governing the appearance of epitaxy was the publication of a group of papers by F.C. Frank (of crystal-growth fame) and his South African collaborator, J.H. van der Merwe (1949, 1950). They worked out the implications of the hypothesis that growth of an epitaxial deposit depends on the initial growth of a monolayer strained elastically to fit the substrate.

Figure 10.5 shows the three recognised forms of thin-film growth; epitaxy seems to depend on the initial operation of monolayer growth, as shown in Figure 10.5(a). Frank and van der Merwe analysed this in terms of the various surface and interfacial energies involved, including a term attributable to the elastically strained monolayer. These forms of initial growth, and coalescence of growth islands at a later stage, are crucial components of epitaxial growth, as are the defects (such as dislocation arrays) which are formed if the strain becomes too large. There is a detailed discussion of these stages and the factors governing them, and the many crystallographic forms of epitaxy, for metallic thin films, in a fine review by Pashley (1991), who played a major part in the early electron microscopic study of the phenomenon. The conditions governing epitaxy of semiconductors, with special reference to molecular-beam epitaxy, is treated for example by Bachmann (1995).

Oriented ultrathin overgrowth of a polymer on a non-polymeric substrate is the latest combination of materials to show epitaxy. The most recent, remarkable form of this phenomenon is the formation of an array of parallel polymer chains on a substrate by depositing monomers and then polymerising them in situ. The Japanese discoverer of this phenomenon (Sano 1996) has called it 'polymerisation-induced

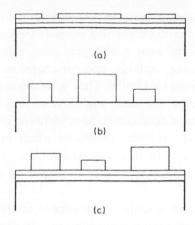

(a)

(b)

(c)

Figure 10.5. The three modes of growth of films: (a) Frank and van der Merwe's monolayer (two-dimensional) mode; (b) the Volmer-Weber three-dimensional mode; (c) the Stranski-Krastanov mode involving two-dimensional growth followed by three-dimensional growth.

epitaxy'. This only works with properly crystallizable polymers; atactic polymer chains cannot be aligned in this way. Sano points out that this is a way of aligning (effectively, crystallising in a two-dimensional manner) polymers whose monomers are soluble in appropriate solvents, even though the polymer itself is not.

Another recently discovered form of epitaxy is 'graphoepitaxy' (Geis *et al.* 1979). Here a non-crystalline substrate (often the heat-resistant polymer polyimide, with or without a very thin metallic coating) is scored with grooves or pyramidal depressions; the crystalline film deposited on such a substrate can have a sharp texture induced by the geometrical patterns. More recently, this has been tried out as an inexpensive way (because there is no need for a monocrystalline substrate) of preparing oriented ZnS films for electroluminescent devices (Kanata *et al.* 1988).

10.5.1.2 Metallic multilayers. In Section 7.4, we have met the recent discovery of multilayers of two kinds of metal, or of a metal and a non-metal, that exhibit the phenomenon of giant magnetoresistance. This discovery is one reason why the preparation and exploitation of such multilayers have recently grown into a major research field.

The original motivation for the preparation of regular metallic multilayers of carefully controlled periodicity was the need for X-ray reflectors, both to calibrate unknown X-ray wavelengths and to function as large and efficient monochromators, especially for 'soft' X-rays of wavelengths of several Å. This was first done by

Deubner (1930) and analysed in detail in a famous paper by DuMond and Youtz (1940). A typical modern multilayer for this purpose would be of W/Si.

The methods of growing such multilayers with rigorously regular spacing, involving especially sputtering methods, and for characterising them, are critically discussed by Greer and Somekh (1991). They also discuss some unexpected uses which have been discovered for such multilayers, in particular, their use for measuring very small diffusion coefficients: here, diffusion of a component from one layer to its neighbour leads to fuzzy interfaces which in turn leads to reduced intensities of reflected X-rays. In this way, diffusivities (for example, in metallic glasses) have been measured much smaller than can be examined by any other technique.

Strength as well as elastic modulus anomalies in multilayers for critical repeat distances caused great excitement a few years ago; it now seems that the elastic anomalies were the result of faulty experimental methods, but the strength enhancement, as well as enhancements of fracture toughness, for very small periodicities seem to be genuine and are beginning to find applications. The motion of dislocations is progressively inhibited as the thickness of individual layers is reduced. Two plots in Figure 10.6 illustrate these trends. In Figure 10.6(a), it can be seen that an Ag/Cr multilayer of wavelength 20 nm is much harder than would be predicted from the rule of mixtures applied to the measured hardnesses of individual layers. Figure 10.6(b) shows a measure of the temperature dependence of fracture toughness (resistance to the spread of cracks) of mild steel, ultrahigh-carbon steel and a laminated (multilayered) composite of the two kinds of steel. Each plot shows a transition temperature from ductile to brittle behavior; this transition is at a very low temperature for the tough composite. The maximum toughness is also much the largest for the multilayered material.

An intriguing recent review of "size effects in materials due to microstructural and dimensional constraints" with a focus on mechanical properties, including those of multilayers, is by Arzt (1998).

10.6. EXTREME SYMMETRY

10.6.1 Quasicrystals

In 1982, at the National Bureau of Standards near Washington, DC, an Israeli crystallographer, Daniel Shechtman, walked in a state of high excitement into the office of his colleague, my namesake John Cahn, to show him a photograph just like the one shown here as Figure 10.7. The pattern was made from an alloy foil (Al–Mn) rapidly quenched from the melt (and in a metastable condition), and shows fivefold symmetry. Every first-year undergraduate of materials science knows that no crystal

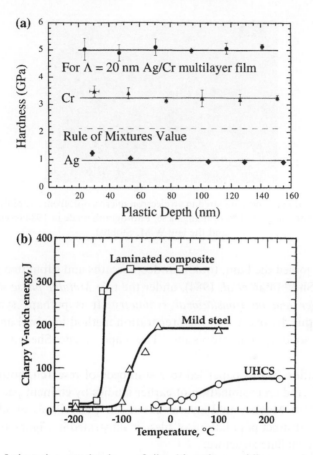

Figure 10.6. (a) Indentation nanohardness of silver/chromium multilayers and single films of the constituent metals, as a function of depth affected by plastic deformation. (b) Charpy impact energies, a measure of fracture toughness, of three materials, as a function of test temperature: they are mild steel, ultrahigh-carbon steel and a composite of the two kinds of steel (courtesy Dr. J. Wadsworth) (Fig. 10.6(b) is from Kum *et al.* (1983)).

can have fivefold symmetry, because this is incompatible with periodic stacking of atoms. Shechtman claimed this was a new kind of *quasiperiodic* material; the term *quasicrystal* came soon after (Levine and Steinhardt 1984, in a paper entitled *Quasicrystals: a new class of ordered structures*).

John Cahn was irate; what he had been shown was manifest nonsense, and he was sure that a publication making such a claim would relegate both of them, Shechtman and Cahn, to the nether regions of demonstrated crankiness. It took two more years of experimental work, and a good deal of reading of earlier theoretical speculation, before Shechtman and Cahn, together with two French crystallogra-

Figure 10.7. Diffraction pattern, prepared in an electron microscope, from a rapidly solidified foil of an Al–Mn alloy containing 14 at.% of manganese. Photograph made in 1984 (courtesy A.L. Greer and the late W.M. Stobbs).

phers who had joined the hunt, took four deep breaths and submitted a paper about their findings (Shechtman *et al.* 1984), under the title *Metallic phase with long-range orientation order and no translational symmetry*. It is perhaps symbolic of the strangeness of this discovery that the preparation method involved another extreme feature, rapid solidification processing. The paper made Shechtman an instant celebrity.

The publication of this paper led to a stampede of research, both experimental and theoretical, and an examination of earlier studies by eminent people like Roger Penrose and Alan Mackay in England about the possibilities of filling space by 'tiling' with two distinct populations of tiles, as illustrated in Figure 10.8. This is the basis of quasicrystalline structure.

It took a long time before everyone accepted the reality of quasicrystallinity. No less a celebrity than Linus Pauling took a hard line, and published a paper in *Nature* (Pauling 1985) insisting, erroneously as was finally proved some time later, that the pattern was caused by an array of minute crystals in twinned arrangement.

A great deal of theory was introduced in contemplation of these remarkable materials; the ancient Greek golden section, mathematical Fibonacci series, six-dimensional crystallography... these were three concepts which proved to be relevant to quasicrystals. An early study by Frank and Kasper (1958) – this is the second time that Charles Frank has appeared in this chapter – following the time-hallowed analysis of crystal chemistry in terms of atomic sizes, proved to be important in predicting which alloy systems would generate quasicrystals, and many of the alloys which proved to be convertible to quasicrystals had related Frank–Kasper true crystal structures.

The fivefold symmetry discovered by Shechtman is modelled in terms of the stacking of icosahedra and the term 'icosahedral symmetry' is sometimes used.

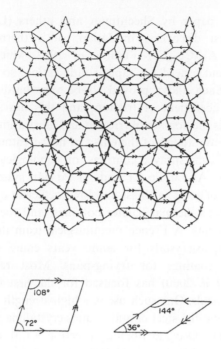

Figure 10.8. Two kinds of rhombi, acute and obtuse, arranged by 'matching rules' to generate a two-dimensional quasiperiodic tiling (courtesy S. Ranganathan).

A little later (Bendersky 1985, Chattopadhyay *et al.* 1985) decagonal (tenfold) symmetry was discovered in other Al-transition metal compounds; quasiperiodic layers are stacked periodically in the third dimension. Since then, one or other of these forms of quasicrystal have been identified in many different compositions. A detailed review of the decagonal type is by Ranganathan *et al.* (1997).

A good, accessible overview of quasicrystals, written only a few years after their discovery, is by Ranganathan (1990); Indian metallurgists played a major part in the early research. Many other published reviews require considerable mathematical sophistication before they can be understood by the reader.

Interest in physical properties of quasicrystals is growing. Thus, a recent comment (Thiel and Dubois 2000) analyses the implications of the fact that decagonal quasicrystals have very much higher electrical resistivity, by orders of magnitude, than do their constituent metals, and moreover that resistivity decreases with rising temperature. For one thing, it seems that the concentration of highly mobile 'free' electrons is much lower in such quasicrystals than in normal metals.

For the first 15 years after the discovery, quasicrystals were studied purely as a compelling scientific issue. Just recently, applications have begun to appear.

According to a recent paper by Shechtman and others (Lang *et al.* 1999) this followed from the first discovery, in Japan, of a thermodynamically stable quasicrystalline phase, $Al_{65}Cu_{20}Fe_{15}$ (Tsai *et al.* 1987) which could therefore be prepared in bulk, without a need for rapid solidification, so that properties could readily be measured. Such alloys proved to be very hard (and brittle) as well as having a very low coefficient of friction, and this has suggested tribological applications (tribology is the domain of friction and lubrication). The paper by Lang *et al.* reports on systematic studies of the effect of heat-treatment of quasicrystalline alloys, in the form of plasma-sprayed coatings, on tribological properties. It turned out that coatings in the Al–Cu–Fe–Cr and Al–Pd–Mn systems showed the most promising combination of good hardness and low coefficient of friction. Others have proposed such coatings as thermal barrier coatings, and they are being tested on aircraft turbine components. A French metallurgical team that had been studying the fundamentals of quasicrystals for some years came up with an apposite application, non-stick coatings for frying-pans. Most recently, a conference proceedings (Dubois *et al.* 2000) has focused for the first time on technological applications of quasicrystals. One such use is a high-strength steel used for surgical tools and electric shavers; the steel contains quasicrystalline precipitates which are particularly stable against Ostwald ripening (Section 9.1.3), apparently because of the low interfacial energy between precipitates and matrix, and thus resistant to overaging. A range of aluminum alloy strengthened by quasicrystalline precipitates has also been developed.

The many papers in this proceedings are partitioned into very abstruse theoretical analyses of structure and stability of quasicrystals on the one hand, and practical studies of surface structures, mechanical properties and potential applications. The subject shows signs of becoming as deeply divided between theorists and practical investigators, out of touch with each other, as magnetism became in the preceding century.

10.7. EXTREME STATES COMPARED

Virtually every reference at the end of this chapter is to post-war publications, and the majority are to papers published during the past 15 years. This shows, clearly enough, that 'extreme materials' are recent features of materials science and engineering (MSE), and there is every indication that the focus on materials of the kind discussed in this chapter will continue to develop. Individual approaches come and go – thus, rapid solidification processing, the oldest of the approaches discussed here, seems to have passed its apogee – while others go from strength to strength:

thus, a 'nanotechnology initiative' has recently been instituted at the highest level of US government. It is not perhaps too much of an exaggeration to claim that in this chapter, we can see something of the future of MSE.

REFERENCES

Arzt, E. (1998) *Acta Mater.* **46**, 5611.

Bachmann, K.J. (1995) *The Materials Science of Microelectronics*, Chapter 6 (VCH, New York).

Bayard, R.T. and Alpert, D. (1950) *Rev. Sci. Instr.* **21**, 571.

Bendersky, L. (1985) *Phys. Rev. Lett.* **54**, 2422.

Bowden, F.P. and Tabor, D. (1954) *The Friction and Lubrication of Solids*, revised reprint (Oxford University Press, Oxford).

Brus, L.E., Siegel, R.W. *et al.* (1988) *J. Mater. Res.* **4**, 704.

Cahn, R.W. (1980) *Contemp. Phys.* **21**, 43.

Cahn, R.W. (1993) in *Rapidly Solidified Alloys*, ed. Liebermann, H.H. (Marcel Dekker, New York) p. 1.

Cahn, R.W. and Johnson, W.L. (1986) *J. Mater. Res.* **1**, 724.

Canham, L.T. (1990) *Appl. Phys. Lett.* **57**, 1056.

Chattopadhyay, K. *et al.* (1985) *Scripta Metall.* **19**, 767.

Chen, I.-Wei and Wang, X.-H. (2000) *Nature* **404**, 168.

Chopra, K.L. (1969) *Thin Film Phenomena* (McGraw-Hill, New York).

Cohron, J.W., George, E.P., Heatherly, L., Liu, C.T. and Zee, R.H. (1996) *Intermetallics* **4**, 497.

Coutts, T.J. (1974) *Electrical Conduction in Thin Metal Films* (Elsevier, Amsterdam).

DeCristofaro, N. (1998) Amorphous metals in electric-power distribution applications, *MRS Bull.* **23**(5), 50.

Deubner, W. (1930) *Ann. Phys.* **5**, 261.

Dobson, P.J. (2000) Book review: Nanotechnology: opportunities missed, *Contemp. Phys.* **41**, 159.

Dowson, D. (1979) *Tribology* (Longman, London).

Dubois, J.M., Thiel, P.A., Tsai, A.-P. and Urban, K. (2000) *Quasicrystals*, proc. vol. 553 (Materials Research Society, Warrendale, PA).

DuMond, J. and Youtz, J.P. (1940) *J. Appl. Phys.* **11**, 357.

Duwez, P. (1967) *Trans. Amer. Soc. Metals* **60**, 607.

Duwez, P., Willens, R.H. and Klement Jr., W. (1960a) *J. Appl. Phys.* **31**, 1136.

Duwez, P., Willens, R.H. and Klement Jr., W. (1960b) *J. Appl. Phys.* **31**, 1137.

Edelstein, A.S. and Cammarata, R.C. (eds.) (1996) *Nanomaterials: Synthesis, Properties and Applications* (Institute of Physics Publishing, Bristol and Philadelphia).

Fleischer, R.L. (1998) *Tracks to Innovation* (Springer, New York, Berlin).

Frank, F.C. and van der Merwe, J.H. (1949) *Proc. Roy. Soc. Lond. A* **198**, 205; **200**, 125; (1950), *idem, ibid,* **201**, 261.

Frank, F.C. and Kasper, J.S. (1958) *Acta Cryst.* **17**, 184.

Frohnhoff, S. and Berger, M.G. (1994) *Adv. Mater.* **6**, 963.

Furey, M.J. (1986) *Tribology*, in *Encyclopedia of Materials Science and Engineering*, vol. 7, ed. Bever, M.B. (Pergamon, Oxford) p. 5145.

Gatos, H.C. (1994) *Surface Sci.* **299/300**, 1.

Geis, M.W., Flanders, D.C. and Smith, H.I. (1979) *Appl. Phys. Lett.* **35**, 71.

Gleiter, H. (1981) in *Deformation of Polycrystals: Mechanisms and Microstructures*, ed. Hansen, N. *et al.* (Risø National Lab., Roskilde, Denmark) p. 15.

Gleiter, H. and Marquardt, P. (1984) *Z. Metallkde.* **75**, 263.

Gleiter, H. (1996) in *Physical Metallurgy*, vol. 1, Chapter 9, 4th edition, eds. Cahn, R.W. and Haasen, P. (North-Holland, Amsterdam) p. 908.

Gleiter, H. (2000) Nanostructured materials: basic concepts and microstructure, *Acta Mater.* **48**, 1.

Greer, A.L. and Somekh, R.E. (1991) Metallic multilayers, in *Processing of Metals and Alloys*, ed. Cahn, R.W. *Materials Science and Technology: A Comprehensive Treatment*, vol. 15 (VCH, Weinheim) p. 329.

Grünberg, P. (2000) Layered magnetic structures in research and application, *Acta Mater.* **48**, 239.

Gschneidner, Jr., K.A. (1979) in *Theory of Alloy Phase Formation*, ed. Bennett, L.H. (The Metallurgical Society of AIME, New York) p. 1.

Hanson, D. (1980) *The New Alchemists* (Little Brown and Co, New York).

Holton, G., Chang, H. and Jurkowitz, E. (1996) *Amer. Scientist* **84**, 364.

Jeong, H.-C. and Williams, E.D. (1999) *Surface Sci. Rep.* **34**, 171.

Johnson, W.L. (1986a) *Int. J. Rapid Solidification* **1**, 331.

Johnson, W.L. (1986b) *Progr. Mat. Sci.* **30**, 81.

Johnson, W.L. (1996) *Current Opinion in Solid State and Materials Science*, vol. 1, p. 383.

Kanata, T. *et al.* (1988) *J. Appl. Phys.* **64**, 3492.

Kim, Y.H., Inoue, A. and Masumoto, T. (1991) *Mater. Trans. Japan Inst. Metals* **32**, 331.

Kinbara, A. (1997) Sputtering, in *Encyclopedia of Applied Physics*, vol. 19 (VCH Publishers, New York) p. 437.

Klement Jr., W., Willens, R.H. and Duwez, P. (1960) *Nature* **187**, 869.

Koch, C.C. (1981) Mechanical milling and alloying, in *Processing of Metals and Alloys*, eds. Cahn, R.W., Haasen, P. and Kramer, E.J.; *Materials Science and Technology: A Comprehensive Treatment*, vol. 15 (VCH, Weinheim) p. 193.

Krafft, F. (1970–1980) Article on Otto von Guericke, in *Dictionary of Scientific Biography*, vol. 5, ed. Gillispie, C.C. (Charles Scribner's Sons, New York) p. 574.

Kum, D.W., Oyama, T., Wadsworth, J. and Sherby, O.D. (1983) *J. Mod. Phys.* **31**, 173.

Lang, C.I., Shechtman, D. and González, E. (1999) *Bull. Mater. Sci. (India)* **22**, 189.

Langmuir, I. and Mackey, G.M.J. (1914) *J. Amer. Chem. Soc.* **36**, 417, 1708.

Levine, D. and Steinhardt, P.J. (1984) *Phys. Rev. Lett.* **53**, 2477.

Löffler, J.F., Bossuyt, S., Glade, S.C., Johnson, W.L., Wagner, W. and Thiyagarajan, P. (2000) *Appl. Phys. Lett.* **77**, 525.

Machlin, E.S. (1998) *The Effects of Structure on Properties of Thin Films* (GiRo Press, Croton-on-Hudson).

Marcus, P.M. (1994) *Surface Sci.* **299/300**, 447.

Miller, D.A.B. (1996) Silicon integrated circuits shine, *Nature* **384**, 307.

Nix, W.D. (1989) *Metall. Trans.* **20A**, 2217.

Ohring, M. (1992) *The Materials Science of Thin Films* (Academic, San Diego).

Okamoto, R., Lam, N.Q. and Rehn, L.E. (1999) Physics of crystal-to-glass transformations, *Solid State Phys.* **52**, 1.

ORNL (2000) *Report on current nanotechnology programs at Oak Ridge National Laboratory*, on the Worldwide Web at www.ornl.gov/ORNLReview/rev32_3/brave.htm.

Pashley, D.W. (1991) The epitaxy of metals, in *Processing of Metals and Alloys*, ed. Cahn, R.W.; *Materials Science and Technology: A Comprehensive Treatment*, vol. 15 (VCH, Weinheim) p. 290.

Pauling, L. (1985) *Nature* **15**, 317, 512.

Pendry, J.B. (1984) *Nature* **302**, 504.

Pollard, J. (1959) Progress in vacuum technology, *Rep. Prog. Phys.* **22**, 33.

Pontikis, V. (1993) Crystal surfaces: melting and roughening, in *Supplementary Volume 3 of the Encyclopedia of Materials Science and Engineering*, ed. Cahn, R.W. (Pergamon press, Oxford), p. 1587.

Prokes, S.M. (1996) Porous silicon nanostructures, in *Nanomaterials: Synthesis, Properties and Applications*, eds. Edelstein, A.S. and Cammarata, R.C. (Institute of Physics Publishing, Bristol and Philadelphia) p. 439.

Ranganathan, S. (1990) Quasicrystals, in *Supplementary Volume 2 of the Encyclopedia of Materials Science and Engineering* ed. Cahn, R.W. (Pergamon press, Oxford) p. 1205.

Ranganathan, S. *et al.* (1997) *Progr. Mater. Sci.* **41**, 195.

Redhead, P.A., Hobson, J.P. and Kornelsen, E.V. (1962) Ultrahigh vacuum, *Adv. Electronics and Electron Phys.* **17**, 323.

Rhead, G.E. (1983) *Contemp. Phys.* **24**, 535.

Rivière, J.C. (1973) *Contemp. Phys.* **14**, 513.

Roberts, M.W. (1960, August) High-vacuum techniques, *J. Roy. Inst. Chem.* 275.

Rossnagel, S.M. and Cuomo, J.J. (1988) *Vacuum* **38**, 73.

Roth, A. (1976) *Vacuum Technology* (North-Holland, Amsterdam) p. 10.

Royer, L. (1928) *Bull. Soc. Franç. Mineralogie* **51**, 7.

Salli, I.V. (1959) see *Chem. Abstr.* **53**, 1053; *Met. Abstr.* **26**, 492.

Sano, M. (1996) *Adv. Mater.* **8**, 521.

Seah, M.P. (1980) *Surface and Interface Anal.* **2**, 222.

Shechtman, D., Blech, I., Gratias, D. and Cahn, J.W. (1984) *Phys. Rev. Lett.* **53**, 1951.

Siegel, R.W. (1996, December) Creating nanophase materials, *Scientific American* **275**, 42.

Somorjai, G.A. (1998) From surface science to surface technologies, *MRS Bull.* **23/5**, 11.

Spaepen, F. and Turnbull, D. (1984) Metallic glasses, *Annu. Rev. Phys. Chem.* **35**, 241.

Steinherz, H.A. and Redhead, P.A. (1962) Ultrahigh vacuum, *Scientific American* **206/3**, 78.

Suits, C.G. and Bueche, A.M. (1967) Cases of research and development in a diversified company, in *Applied Science and Technological Progress* (a report by the National Academy of Sciences), US Government Printing Office, Washington, DC, p. 297

Suryanarayana, C. (ed.) (1999) *Non-equilibrium Processing of Materials* (Pergamon, Oxford).

Tabor, D. (1981) *Contemp. Phys.* **22**, 215.
Terakura, K., Oguchi, T., Mohri, T. and Watanabe, K. (1987) *Phys. Rev. B* **35**, 2169.
Thiel, P.A. and Dubois, J.M. (2000) *Nature* **406**, 570.
Tsai, A.P., Inoue, A. and Masumoto, T. (1987) *Jpn. J. Appl. Phys.* **29**, L1505.
Woodruff, D.P. and Delchar, T.A. (1986) *Modern Techniques of Surface Science* (Cambridge University Press, Cambridge).
Yoshizawa, Y., Oguma, S. and Yamauchi, K. (1988) *J. Appl. Phys.* **64**, 6044.

Chapter 11
Materials Chemistry and Biomimetics

Chapter 11
Materials Chemistry and Biomimetics

11.1. THE EMERGENCE OF MATERIALS CHEMISTRY

Chemistry has featured repeatedly in the earlier parts of this book. In Section 2.1.1, the emergence of physical chemistry is mapped, followed by a short summary of the status of solid-state chemistry in Section 2.1.5. The key ideas of phase equilibria and metastability are set forth in Section 3.1.2, with special emphasis on Willard Gibbs. The linkage between crystal structure, defects in crystals and equilibria in chemical reactions is outlined in Section 3.2.3.5, while crystal chemistry is treated at some length in Section 3.2.4. Chemical analysis features in Sections 6.2.2.3 and 6.3. The chemistry of magnetic ceramics is outlined in Section 7.3, while liquid crystals are presented in Section 7.6. The huge subject of polymer chemistry is briefly introduced in Section 8.2, and the field of glass-ceramics is explained in Section 9.6; this last can be regarded as an expression of high-temperature chemistry. The outline of surface science in Sections 10.4.1 and 10.4.3 includes some remarks about its chemical aspects.

Clearly, chemistry plays as large a part in the evolving science of materials as do physics and metallurgy. Nevertheless, when materials science arrived as a concept in the late 1950s, no chemist would have dreamed of describing himself as a materials chemist, though the term 'solid-state chemist' was just making its appearance at that time. Since then, in the 1980s, materials chemistry has arrived as a recognised category, and the term appears in the titles of several major journals.

We can get an idea of the gradual development of *solid-state chemistry* from a fine autobiographical essay by one of the greatest modern exponents of that science, the Indian Rao (1993). He remarks: "When I first got seriously interested in the subject in the early 1950s it was still in its infancy". He traces it through its stages, including a period of very intense emphasis on the chemical consequences of crystal defects as studied by electron microscopy; he refers to a book (Rao and Rao 1978) he co-authored on phase transitions, a topic which he claims had been neglected by solid-state chemists until then and had perhaps been too much the exclusive domain of metallurgists. He also remarks: "Around 1980, it occurred to me that there was need for greater effort in the synthesis of solid materials, not only to find novel ways of making known solids, but also to prepare new, novel metastable solids by unusual chemical routes". He goes on to point out that "the tendency nowadays is to avoid brute-force methods and instead employ methods involving mild reaction conditions. Soft chemistry routes are indeed becoming popular...". This interest led to yet another book (Rao 1994). His notable book on solid-state chemistry as a whole (Rao and Gopalakrishnan 1986, 1997) has already been discussed in Chapter 2.

So, by the 1990s, Professor Rao had been active in several of the major aspects which, *together*, were beginning to define materials chemistry: crystal defects, phase transitions, novel methods of synthesis. Yet, although he has been president of the Materials Research Society of India, he does not call himself a materials chemist but remains a famous solid-state chemist. As with many new conceptual categories, use of the new terminology has developed sluggishly.

As materials chemistry has developed, it has come to pay more and more attention to that archetypal concern of materials scientists, microstructure. That concern came in early when the defects inherent in non-stoichiometric oxides were studied by the Australian J.S. Anderson and others (an early treatment was in a book edited by Rabenau 1970), but has become more pronounced recently in the rapidly growing emphasis on self-assembly of molecules or colloidal particles. This has not yet featured much in books on materials chemistry, but an excellent recent popular account of the broad field has a great deal to say on self-assembly (Ball 1997). The phenomenon of graphoepitaxy outlined in Section 10.5.1.1 is a minor example of what is meant by self-assembly.

A notable chemist, Peter Day, has recently published an essay under the challenging title *What is a material*? (Day 1997). He makes much of the point that the properties of, say, a molecular material are not determined purely by the characteristics of the molecules but also by their interaction in a continuous solid, and that chemists have to come to terms with this if they wish to be materials chemists. If they do, they can hope to synthesise materials with very novel properties. He also puts emphasis, as did Rao, on the benefits of 'chimie douce', soft chemistry, in which very high temperatures are avoided. For instance, he points out, "to deposit thin films...., selectively decomposing carefully designed organometallic molecules has proved a notable advance over the 'engineering' approach of flinging atoms at a cold surface in ultrahigh vacuum". There is scope for a great deal of discussion in the wording of that sentence.

In the words of a recent paper on MSE education (Flemings and Cahn 2000), "chemistry departments have historically been interested in individual atoms and molecules, but increasingly they are turning to condensed phases". A report by the National Research Council (of the USA) in 1985 highlighted the opportunities for chemists in the materials field, and this was complemented by the NRC's later analysis (MSE 1989) which, inter alia, called for much increased emphasis on materials synthesis and processing. As a direct consequence of this recommendation, the National Science Foundation (of the USA) soon afterwards issued a formal call for research proposals in materials synthesis and processing (Lapporte 1995), and by that time it can be said that materials chemistry had well and truly arrived, in the

United States at least. The huge field of inorganic materials synthesis is not further discussed in this chapter, but the interested reader will benefit from reading a survey entitled "Inorganic materials synthesis: learning from case studies" (Roy 1996).

11.1.1 Biomimetics

The emphasis on microstructure as a major variable in materials chemistry has been strengthened by the emergence of yet another subdiscipline, that of *biomimetics*. This is simply, in the words of one practitioner, J.F.V. Vincent, retailed by another (Jeronimidis 2000), "the abstraction of good design from nature". (Vincent himself (1997) gave a lecture on "stealing ideas from nature"). Biomimetics seems to have begun as a study of strong and tough materials (skeletons, defensive starfish spines, mollusc shells) in order to mimic their microstructure in man-made materials. Such mimicry necessarily involves chemical methods, to the extent that a recent major text is entitled *Biomimetic Materials Chemistry* (Mann 1996). (In fact, the term 'biomimetic chemistry' was used as early as 1979 as the title of a symposium organised by the American Chemical Society, Dolphin *et al.* 1980.) An exceptionally illuminating presentation of a range of strong and tough biological materials, incorporating both those found in a range of quite distinct creatures and those specific to one taxum, is by Weiner *et al.* (2000). Two examples of the striking features discussed in this paper: echinoderm spines are essentially single crystals of calcite (or dolomite), but their readiness to cleave under stress is obviated by the division of the single crystal into mosaic domains that are very slightly mutually misoriented; this is a highly specific feature. On the other hand, the formation of a very tough structure via a sequence of multilayers in mutually crossed orientations is widespread in zoology: whether the material is based on aragonite in abalone shells, or on chitin in beetle wingcases, the basic principle is the same, and such structures always contain thin layers of biopolymers. As Calvert and Mann (1988) early recognised, "biological mineralisation demonstrates the possibility of growing inorganic minerals locally on or in polymer substrates". A very recent, detailed examination of an ultratough marine shell, that of the conch *Strombus gigas*, which has three hierarchical levels of aragonite lamellae separated by ultrathin organic layers, is by Kamat *et al.* (2000).

Quoting just a few of the chapter headings in Mann's book (a), and also in Elices' (2000) even more recent book (b), conveys the flavour of the subdiscipline: (a) Biomineralisation and biomimetic materials chemistry; biomimetic strategies and materials processing; template-directed nucleation and growth of inorganic materials; biomimetic inorganic–organic composites; organoceramic nanocomposites. (b) Structure and mechanical properties of bone; biological fibrous materials; silk fibres – origins, nature and consequences of structure. These headings indicate several things: a strong focus on synthesis and preparative methods; self-arrangement; and the thorough mixing of normally quite distinct categories of materials. Biomimetics is succeeding in breaking down almost every historical barrier between fields of MSE. Since 1993, there has been a journal entitled *Biomimetics*.

The book by Ball introduced above includes chapters both on "Only natural: biomaterials" and on "Spare parts: biomedical materials". The first of these is really about biomimetics (terminology is still somewhat in flux), the second is about the even larger field of artificial materials for use in the human body. This category includes such items as artificial heart-valves (polymeric or carbon-based), synthetic blood-vessels, artificial hips (metallic or ceramic), medical adhesives, collagen, dental composites, polymers for controlled slow drug delivery. There is plainly a link between biomimetics and biomedical materials, but whereas a biomimetic engineer seeks to make materials for non-biological uses under inspiration from the natural world, the biomedical engineer has to work hand-in-glove with surgeons and physicians, and must never forget such crucial considerations as the compatibility of synthetic surfaces with blood or the wear resistance of artificial hip joints. I have no room here for further details, and the interested reader is referred to Williams (1990).

11.1.2 Self-assembly, alias supramolecular chemistry

To get a feel for the kind of new issues that weigh on materials chemists nowadays, a brief account of the topic of self-assembly will serve well.

Chemists deal primarily with molecules but, as they concern themselves increasingly with condensed matter, they are brought face-to-face with the means of tying 'saturated' molecules, or other small particles, together by weaker bonds, such as hydrogen bonds or van der Waals bonds. This craft was originally dubbed *supramolecular chemistry* by pioneers such as the Nobel-prize-winning French chemist Lehn (1995). But that term seems to be playing hide-and-seek with *self-assembly*. A very recent paper (Nangia and Desiraju 1998) lays it down that "supramolecular chemistry is the chemistry of the *inter*molecular bond and is based on the theme of mutual recognition; such recognition is characterised by chemical and geometrical complementarity between interacting molecules". A very recent overview of the field from a materials science viewpoint (Moore 2000) emphasises 'design from the bottom-up' as the essence of the skills involved here. Whereas 'supramolecular chemistry' properly only applies to the assembly of molecules, 'self-assembly' can also include the assembly of larger units... so I prefer the latter term.

From the way the field has developed during the last few years, two quite distinct kinds of self-assembly are emerging. One kind focuses on the 'self' part of the nomenclature and relies entirely on the inherent forces acting between particles. A good example is the formation of colloidal pseudocrystals from small polymeric spheres, as outlined in Section 2.1.4; a recent set of reviews of this rather mysterious process is by Grier (1998). A subvariant of this is to coat the spheres with nickel and encourage them to align themselves in various configurations by applying a field. Another is the self-organised growth of nanosized arrays of iron crystallites on a copper bilayer deposited on a (1 1 1) face of platinum (Figure 11.1); here the source of organisation is the spontaneously regular array of dislocations resulting from strain-relief between the copper and the platinum which have different lattice constants, defects which in turn act as heterogeneous nucleation sites for iron crystallites when iron is evaporated onto the film (Brune *et al.* 1998). Yet another example of this approach is self-assembly of polymers by relying on interaction between dendritic side-branches (Percec *et al.* 1998).

Special attention has been paid recently to methods of creating 'photonic crystals', microstructured materials in which the dielectric constant is periodically modulated in three dimensions on a length scale comparable to the wavelength of the electromagnetic radiation to be used, whether that is visible light or a UHF radio wave; obviously the periodicity is much greater than that in natural ('real') crystals. One of the many techniques tried out is the use of interfering laser beams sent in four precisely chosen different directions into a layer of photoresist polymer (as used in microcircuit technology); highly exposed photoresist is rendered insoluble, other regions can be etched away, generating a regular array of holes (Campbell *et al.* 2000). An even more intriguing approach is that by Blanco *et al.* (2000) in which an

200 Å

Figure 11.1. Scanning tunnelling microscope image of a periodic array of Fe islands nucleated on the regular dislocation network of a Cu bilayer deposited on a platinum (1 1 1) face (after Brune *et al.* 1998).

opalescent structure of lightly sintered small silica spheres, in regular array, is infiltrated with silicon, followed by removal of the silica template. A very recent survey of the many ways in which photonic crystals can be made, together with an outline of their use in optical communications systems (for instance, in enabling light beams effectively to bend sharply rather than gradually) is by Parker and Charlton (2000).

This takes us to the second type of self-assembly which relies on some form of *template*, a pattern imposed on a surface that will act as a guide to further molecules or particles that are deposited subsequently – so, on a pedantic view, this is aided assembly rather than self-assembly. The current guru of this approach is the chemist George Whitesides of Harvard University: two papers of his illustrate his preferred approaches (Kim *et al.* 1995, Aizenberg *et al.* 1999). In the first paper, moulding in capillaries is described: a pattern, typically of grooves one or two nanometers wide and a fraction of a nanometer in depth, is made by photolithography (as practised in microcircuit fabrication) and then reproduced in negative by casting with an elastomeric polymer. The channel pattern is then filled with a 'prepolymer', e.g., some form of monomer solution, relying on capillarity to fill the grooves accurately. The polymer is cured and the elastomeric mould then peeled off. The second paper describes an even more elaborate process: self-assembled monolayers (SAMs) are patterned on a metallic substrate by microcontact printing with an elastomeric 'stamp'. A suitable chemical is used as "ink". The unexposed areas are then passivated with an appropriate wash, and the whole immersed in calcium chloride solution. Only the unpassivated SAM regions react to deposit calcite crystals, which thus form an array in regular positions and of regular sizes. In developing this technique, the investigators relied on information from an earlier study of biomineralisation. Neither of these papers proposes a specific use for these patterns; what is done here (as often in self-assembly research up to now) is 'technology push', the identification of a sophisticated technique; the market pull of a particular need is confidently expected to arrive later.

Colloidal crystals can be grown by a templated approach too. Thus van Blaaderen and Wiltzius (1997) have shown that allowing colloidal spheres to deposit under gravity on to an array of suitably spaced artificial holes in a plate quickly generates a single 'crystalline' layer of colloidal spheres, and a thick crystal will then grow on this basis.

Addadi and Weiner (1999) have concisely and critically reviewed these various strategies and have added their own variant – the use of biological templates, for instance bacterium surfaces to assist self-assembly. Here, self-assembly and biomimetics join forces productively.

One intriguing technique of manufacturing a regular array of sharp electrodes sitting in an insulating matrix, useful for flat-screen displays, relies on a mix between spontaneous and templated self-assembly. Hill *et al.* (1996) used a regular eutectic array, formed *spontaneously* by directional freezing, of single-crystal tungsten fibres (300–1000 nm in diameter, about 10^7 fibres/cm^2) in an oxide matrix such as UO_2, the assembly etched so that the fibres stand proud of the surface. Silica evaporated on to the array forms cones that act as shadow-masks for the subsequent deposition of a metallic film on the surface; the silica is then removed and the end-result is an array of free-standing vertical metallic needles in an insulator surrounded by a non-contacting 'gridded' superficial ring of metal film. If these conducting rings are made anodic, then 100 volts suffices to induce field-emission of electrons from the nearby electrodes. This is a beautiful example of a combined physical/chemical processing strategy, reminiscent of techniques used in microcircuitry, designed by a group of materials scientists.

Yet another variant of self-assembly relies on the repulsion between blocks of suitably constituted block copolymers, leading to fine-scale patterns of organisation. One very recent description of this approach is by de Rosa *et al.* (2000). Details of this kind of approach as cultivated at Oak Ridge National Laboratory can also be found on the internet (ORNL 2000).

11.2. SELECTED TOPICS IN MATERIALS CHEMISTRY

In this Section, I shall briefly exemplify some topics that illustrate how the needs of materials science and engineering have shaped chemists' approaches to synthesis and processing.

11.2.1 Self-propagating high-temperature reactions

In the 19th century, the steel rails of streetcars (trams) were welded in situ by packing a mixture of ferric oxide and aluminium powder between the rails to be joined and initiating a strongly exothermic reaction between the two powders by local heating; the reaction produces molten iron which achieves the weld. This is (gasless) *combustion synthesis*. This approach was generalised by a Russian chemist, A.G. Merzhanov, who began publishing accounts of the synthesis of compounds in this way inside a sealed 'bomb'; his first account of the synthesis of high-melting carbides, nitrides and borides was published in 1972 (Merzhanov and Borovinskaya 1972). The technique spread rapidly through Soviet industry.

The technique, now named *self-sustaining high-temperature synthesis* (SHS) – on the grounds that long names drive out short ones – was later taken up in the West, and has gradually become more sophisticated. The synthesis of TiC$_x$ by Holt and Munir (1986) marks the beginning of detailed analysis of heat generation and disposal, and brought in the practice of the use of inert diluents to limit temperature excursions. Figure 11.2 shows schematically how temperature varies with time in such a process. Much research has been done on the elimination of porosity in the product, often by the application of high pressure for short periods after the reaction is over. The technique, now named XD to mystify the reader, was applied by the Martin-Marietta Corporation in America to create alloys dispersion-hardened by fine intermetallic or ceramic particles; the constituent elements of the ceramic are mixed with a metal or alloy powder (Brubacher *et al.* 1987). The field received a major review by Munir and Anselmi-Tamburini (1989).

A particularly striking recent application was by Deevi and Sikka (1997): they developed an industrial process for casting intermetallics, especially nickel alumi-nides, so designed (by modifying the furnace-loading sequence) that the runaway temperature rise which had made normal casting particularly dangerous was avoided.

11.2.2 Supercritical solvents

In 1873, Johannes van der Waals (1837–1923) presented his celebrated doctoral thesis to the University of Leiden in the Netherlands, under the title "On the continuity of the liquid and gaseous states": here he established a simple molecular interpretation of the observed fact that a *critical temperature* exists for a particular gas below which a gas can be condensed to a two-phase system of vapour and liquid, whereas above it there can only be a homogeneous fluid phase ('fluid' strictly being neither vapour nor liquid). His equation of state for a gas, gradually improved, played a major part in the early understanding of gases and, for instance, helped his countryman Heike Kamerlingh Onnes to work out his method of liquefying helium.

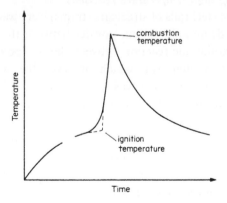

Figure 11.2. Schematic representation of the temperature profile associated with the passage of a gasless combustion front.

Soon after van der Waals' thesis was published, Hannay and Hogarth (1879) discovered that a *supercritical fluid*, SCF (i.e., a fluid above the critical temperature) can readily dissolve non-volatile solids. Nothing followed from this for a century; it was taken up again only in the 1970s. Now chemical engineers, in particular, are very actively examining the scope for the use of supercritical solvents in dissolving reactants and controlling their reactions in solution. A thorough overview has recently been published (Eckert *et al.* 1996). Supercritical carbon dioxide (critical temperature, 31°C), in particular, is finding growing use as a solvent; the solvent is easily removed, without causing environmental hazards in the way that organic solvents may do. (It is a delightful irony that CO_2, so often decried as an environmental hazard in its own right, is perceived as benign in its context as a solvent.)

A SCF is highly compressible compared with a normal liquid and accordingly, solubilities (and reaction rates in solution) can change rapidly for small changes in temperature and pressure. In this way, fine control can be exercised in synthesis of products and their physical form. One technique involves rapid depressurisation of a SCF containing a solute of interest; small particles are then precipitated because of the large supersaturation associated with the rapid loss of density in the highly compressible fluid phase. Methods are rapidly being developed to enhance further the solubility of a range of solids in SCF CO_2, in particular, by addition of co-solvents, surfactants especially.

11.2.3 Langmuir–Blodgett films

Benjamin Franklin's observations on the calming effect of oil films on turbulent water (Franklin 1774) has been described as the first recorded experiment in surface chemistry. Franklin noticed that a teaspoonful of oil covered about half an acre of water, which suggests how very thin the surface layer of oil must have been. A century later, Franklin was followed by a remarkable, self-taught German girl, Agnes Pockels, who from the age of 18, in her home, began a series of surface-chemical investigations which finally impressed the great Lord Rayleigh when she drew her work to his attention (Pockels 1891). She introduced, among other techniques, the use of a liquid trough for measuring the properties of thin surface films on liquids. Rayleigh (1899) finally proposed that the films she had studied were monomolecular in thickness.

Enter, now, Irving Langmuir (Figure 11.3), the remarkable American metallurgist/physical chemist whom we have met before in connection with incandescent lamps. During the First World War, he turned some of his attention from metallic surfaces to liquid surfaces, and by 1919, he was ready to read a paper to the Faraday Society in London, describing how he set about making films of fatty acids of

Irving Langmuir (1881–1957)

Katharine Burr Blodgett
(1898–1979)

Figure 11.3. Portraits of Irving Langmuir (1881–1957) and Katharine Blodgett (1898–1979) (after Gaines 1983).

varying molecular weights on water and gave evidence that Rayleigh was indeed correct, and furthermore that the molecules in the surface films were oriented with their chains normal to the surface. (These are 'amphiphilic' molecules, hydrophilic at one end and hydrophobic at the other.) In 1917 (Langmuir 1917), he had invented the film balance which allowed a known stress to be applied to a surface film until it was close-packed and could not be compressed further; in this way, he determined the true diameter of his chain molecules, and incidentally one of his measurements more or less tallied with Agnes Pockels' estimate. Later, in 1933, he published a paper, the very first to be printed in the then new *Journal of Chemical Physics* (see Section 2.1.1) which covered, inter alia, the behaviour of thin films adsorbed on a liquid surface. In the years between 1917 and 1933, Langmuir had been largely taken up with surface studies relevant to radio valves (tubes).

His assistant from 1920 on was a young chemist, Katharine Blodgett (Figure 11.3). In 1934, she published a classic paper on monomolecular fatty-acid films which she was able to transfer sequentially from water to a glass slide, so that multilayer films were thereby created (Blodgett 1934). In a concise historical note on these "Langmuir–Blodgett films", (which served as introduction to a major conference on these films, published in the same issue of *Thin Solid Films*), Gaines (1983) advances evidence that this research probably issued from an interest at GE in lubricating the bearings of electricity meters. The superb fundamental work of this pair was always, it seems, nourished (perhaps one should say, lubricated) by severely practical industrial concerns.

During the remainder of the 1930s, Langmuir and Blodgett carried out a brilliant series of studies on multilayer films of a variety of chemicals, supplemented by studies in Britain, especially at the ill-fated Department of Colloid Science in Cambridge (Section 2.1.4). Then the War came, and momentum was lost for a couple of decades. After that, L–B films came back as a major topic of research and have been so ever since (Mort 1980). It is current practice to refer to *molecular films*, made by various techniques (Swalen 1991), but the L–B approach remains central.

Molecular films are of intense current concern in electronics. For instance, diacetylenes and other polymerisable monomer molecules have been incorporated into L–B films and then illuminated through a mask in such a way that the illuminated areas become polymerised, while the rest of the molecules can be dissolved away. This is one way of making a resistance for microcircuitry. L–B films have also found a major role in the making of gas-sensors (Section 11.3.3).

A review of what has come to be called *molecular electronics* (Mirkin and Ratner 1992) includes many striking discoveries, such as a device based on azobenzene (Liu *et al.* 1990) that undergoes a stereochemical transition, trans-to-cis, when irradiated with ultraviolet light, but reverts to trans when irradiated with visible light. The investigators in Japan found that L–B films of their molecules can be used for a short-term memory system, but a chemical conversion to a related compound generates a film which can serve as a longterm memory. Electrochemical oxidation of the L–B film can erase memory completely, so this kind of film has all the key features of a memory system.

It will be clear that L–B films are intrinsically linked to self-assembly of molecules, and this has been recognised in the title of a recent overview book (Ulman 1991), *An Introduction to Ultrathin Organic Films from Langmuir–Blodgett to Self-Assembly: An Overview*.

11.2.4 Colossal magnetoresistance: the manganites

In 1993/1994, several papers from diverse laboratories appeared, all reporting a remarkable form of magnetoresistance, that is, a large change of electrical resistivity resulting from the application of a magnetic field, quite distinct from the so-called 'giant magnetoresistance' found in multilayers of metallic and insulating films (Sections 3.3.3, 7.4, 10.5.1.2). Two of the first papers were by Jin *et al.* (1993), reporting from Bell Laboratories, and from von Helmholt *et al.* (1994), reporting from Siemens Research Laboratory and the University of Augsburg, in Germany. The phenomenon (Figure 11.4) required low temperatures and a very high field. The first paper reported on $La_{0.67}Ca_{0.33}MnO_x$, the second on $La_{0.67}Ba_{0.33}MnO_x$.

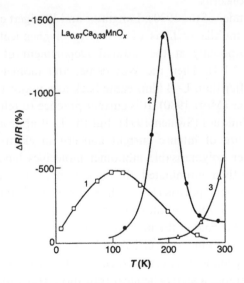

Figure 11.4. Three plots of $\Delta R/R$ curves for a La–Ca–Mn–O film: (1) as deposited; (2) heated to 700°C for 30 min in an oxygen atmosphere; (3) heated to 900°C for 3 h in oxygen (after Jin *et al.* 1993, courtesy of *Science*).

Such compounds have the cubic perovskite crystal structure, or a close approximation to that structure. Perovskites, much studied both by solid-state chemists and by earth scientists, have an extraordinary range of properties. Thus $BaTiO_3$ is ferroelectric, $SrRuO_3$ is ferromagnetic, $BaPb_{1-x}Bi_xO_3$ is superconducting. Several perovskitic oxides, e.g. ReO_3, show metallic conductivity. Goodenough and Longo (1970) long ago assembled the properties of perovskites known at that time in a wellknown database, but the new phenomenon, which soon came to be called *colossal magnetoresistance* (CMR) to distinguish it from giant magnetoresistance (GMR) of multilayers, came as a complete surprise.

The 1993/1994 papers unleashed a flood of papers during the next few years, both reporting on new perovskite compositions (mostly manganates) showing CMR, and also trying to make sense of the phenomenon. A good overview of the first 4 years' research, already citing 64 papers, is by Rao and Cheetham (1997). The ideas that have been put forward are very varied; suffice it to say that CMR seems to be characteristic of compounds in a heterogeneous condition, split into domains with different degrees of magnetisation, of electrical conductivity, with regions differently charge-ordered. So, though these perovskites are not made as multilayers, they behave rather as though they had been. A relatively accessible discussion of some of the current theoretical ideas is by Littlewood (1999).

The goldrush of research on perovskites showing CMR is reminiscent of similar goldrushes when the rare-earth ultrastrong permanent magnets were discovered, when the oxide ('high-temperature') superconductors were first reported and when the scanning tunnelling microscope was announced – all these within the last 30 years. For instance, the $Fe_{14}Nd_2B$ permanent-magnet compound discovered in the mid-1980s led to four independent determinations of its crystal structure within a few months. It remains to be seen whether the manganite revolution will lead to an outcome as useful as the other three cited here.

Another feature of this goldrush is instructive. The usefulness of CMR is much reduced by the requirement for a very high field and low temperature (though the first requirement can be bypassed, it seems, with CMR-materials of different crystal structure, such as pyrochlore type (Hwang and Cheong 1997). The original discovery in perovskite, in 1993/1994, was made by physicists, much of the research immediately afterwards was conducted by solid-state chemists; people in materials science departments were rather crowded out. An exception is found in a paper from the Cambridge materials science department (Mathur *et al.* 1997), in which a bicrystal of $La_{0.67}Ca_{0.33}MnO_3$, made by growing the compound epitaxially on a bicrystal substrate, and so patterned that the current repeatedly crosses the single grain boundary, is examined. Such a device displays large magnetoresistance in fields very much smaller than an ordinary polycrystal or monocrystal show, though the peak temperature is still well below room temperature. The investigators express the view that a similar device using a superconducting perovskite with a high critical temperature may permit room-temperature exploitation of CMR. This is very much a materials scientist's approach to the problem, centred on microstructure.

11.2.5 Novel methods for making carbon and ceramic materials and artefacts

At the start of this Chapter, an essay by Peter Day was quoted in which he lauds the use of 'soft chemistry', exemplifying this by citing the use of organometallic precursors for making thin films of various materials used in microelectronics. The same approach, but without the softness, is increasingly used to make ceramic fibres: here, 'ceramic' includes carbon (sometimes regarded as almost an independent state of matter because it is found in so many forms).

This approach was first industrialised around 1970, for the manufacture on a large scale of strong and stiff carbon fibres. The first technique, pioneered at the Royal Aircraft Establishment in Britain, starts with a polymer, polyacrylonitrile, containing carbon, hydrogen and nitrogen (Watt 1970). This is heated under tension and pyrolysed (i.e., transformed by heat) to turn it into essentially pure carbon; one of the variables is the amount of oxygen in the atmosphere in which the fibre is processed. During pyrolysis, sixfold carbon rings are formed and eventually turn into graphitic fragments which are aligned in different ways with respect to the fibre axis, according to the final temperature. Carbonisation in the range 1300–1700°C produces the highest fracture strength, while further heat-treatment above 2000°C maximises the elastic stiffness at some cost to strength. Figure 11.5 shows the structure of PAN-based fibres schematically, with thin graphite-like layers. An alternative source of commercial carbon fibres, used especially in Japan, is pitch made from petroleum, coal tar or polyvinyl chloride; the pitch is spun into fibre, stabilised by a low-temperature anneal, and then pyrolysed to produce a graphitic structure.

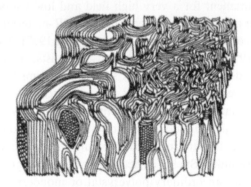

Figure 11.5. Model of structure of polyacrylonitrile-based carbon fibre (after Johnson 1994).

Similar techniques are used to make massive graphitic material, called *pyrolytic graphite*; here, gaseous hydrocarbons are decomposed on a heated substrate. Further heating under compression sharpens the graphite orientation so that a near-perfect graphite monocrystal can be generated ('highly oriented pyrolytic graphite', HOPG). HOPG is used, inter alia, for highly efficient monochromators for X-rays or thermal neutrons. An early account of this technique is by Moore (1973). A different variant of the process generates *amorphous* or *glassy carbon*, in which graphitic structure has vanished completely. This has proved ideal for one kind of artificial heart valve. Yet another product made by pyrolysis of a gaseous precursor is a carbon/carbon composite: bundles of carbon fibre are impregnated by pyrolytic graphite or amorphous carbon to produce a tough material with excellent heat conduction. These have proved ideal for brake-pads on high-performance aeroplanes, fighters in particular. When one takes these various forms of carbon together with the fullerenes to be described in the next Section and the diamonds discussed elsewhere in this book, one can see that carbon has an array of structures which justify its description as an independent state of matter!

Turning now to other types of ceramic fibre, the most important material made by pyrolysis of organic polymer precursors is silicon carbide fibre. This is commonly made from a poly(diorgano)silane precursor, as described in detail by Riedel (1996) and more concisely by Chawla (1998). Silicon nitride fibres are also made by this sort of approach. Much of this work originates in Japan, where Yajima (1976) was a notable pioneer.

Another approach for making ceramic artefacts which is rapidly gaining in adherents is more of a physical than a chemical character. It is coming to be called *solid freeform fabrication*. The central idea is to deposit an object of complex shape by projecting tiny particles under computer control on to a substrate. In one of several versions of this procedure (Calvert *et al.* 1994), a ceramic slurry (in an immiscible liquid) is ejected by small bursts of gas pressure from a microsyringe attached on a slide which is fixed to a table with x-y drive. The assembly is computer-driven by a stepper motor. The technique has also been used for nylon objects (ejecting a nylon precursor) and for filled polymeric resins. Such a technique, however, only makes economic sense for objects of high intrinsic value. A fairly detailed account of this approach as applied to metal powders has been published by Keicher and Smugersky (1997).

11.2.6 Fullerenes and carbon nanotubes

"Carbon is really peculiar" is one of the milder remarks by Harold Kroto (1997) in his splendid Nobel lecture. The 1996 Nobel Prize for chemistry was shared by Kroto in Brighton with Richard Smalley and Robert Curl in Texas, for the discovery of (buckminster)-fullerene, C_{60} and C_{70}, in 1985. These three protagonists all delivered Nobel lectures which were printed in the same journal issue. Kroto's lecture, which goes most fully into the complicated antecedents and history of the discovery, is entitled "Symmetry, space, stars and C_{60}". Stars come into the story because Kroto and astronomer colleagues had for years before 1985 made spectroscopic studies of interstellar dark clouds, had identified some rather unusual carbon-chain molecules with 5–9 carbon atoms, and had then joined forces with the Americans (using advanced techniques involving lasers contributed by the latter) in seeking to use streams of laser-induced tiny carbon clusters to recreate the novel interstellar molecules. They succeeded... but the mass spectra of the molecules also included a mysterious strong peak corresponding to a much larger molecule with 60 carbon atoms, and another weaker peak for 70 atoms. These proved to be the spherical molecules of pure carbon which won the Nobel Prize, called 'fullerenes' for short after Buckminster-Fuller, an architect who was famed for his part-spherical 'geodesic domes'. The discovery was first reported by Kroto *et al.* (1985).

The spherical fullerenes, of which C_{60} and C_{70} are just the two most common versions (they go down to 20 carbon atoms and up to 600 carbon atoms or perhaps even further, and some are even spheres within spheres, like Russian dolls), are a new collective allotrope of carbon, in addition to graphite and diamond. The 'magic-number' fullerenes, C_{60} and C_{70}, turn out to form strain-free spheres consisting of mixed hexagons (as in graphite sheets) and pentagons, Figure 11.6. Later, Krätschmer *et al.* (1990) established that substantial percentages of the fullerenes were formed in a simple carbon arc operating in argon, and a copious source of the molecules was then available from the soot formed in the arc, leading at once to a deluge of research. Krätschmer succeeded soon after in crystallising C_{60} from solution in benzene. The crystals are a classic example of a 'rotator phase', so called because molecules (or radicals) in the crystal are very weakly bonded, here by van der Waals forces, and thus rotate freely without moving away from their lattice sites. On severe cooling, the rotation stops. Rotator phases are also known as 'plastic

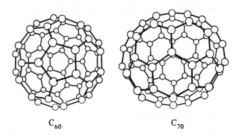

C_{60} C_{70}

Figure 11.6. Two fullerene molecules, C_{60} and C_{70}.

crystals' because they will flow under remarkably small stresses, on account of very high self-diffusivity; the study of this kind of crystal has become a well-established parepisteme of solid-state chemistry (Parsonage and Staveley 1978).

After 1990, the chemistry of fullerenes was studied intensively by teams all over the world; a summary account of what was initially found can be found in a survey by Kroto and Prassides (1994). The internal diameter of a C_{60} sphere is about 0.4 nm, large enough to accommodate any atom in the periodic table, and a number of atoms have in fact been accommodated there to form proper compounds. Kroto and Prassides describe these 'endohedral complexes' as "superatoms with highly modified electronic properties, opening up the way to novel materials with unique chemical and physical properties". Turning from chemistry to fundamental physics, another striking paper was published recently in *Nature*: Arndt *et al.* (1999) were able to show that a molecular beam of C_{60} undergoes optical diffraction in a way that clearly demonstrates that these heavy moving 'particles' evince wavelike properties, as originally proposed by de Broglie for subatomic particles. They are the heaviest 'particles' to have demonstrated wave characteristics.

The hoped-for applications of fullerenes have not materialised as yet. A cartoon published in America soon after the discovery shows a hapless hero sinking into a vat full of buckyballs (another name for fullerenes) with their very low friction. It is not known how the hero managed to escape...

Applications can be more realistically hoped for from a variant of fullerenes, namely, *carbon nanotubes*. These were discovered, in two distinct variants, on the surface of the cathode of a carbon arc, by a Japanese carbon specialist, Iijima (1991), and Iijima and Ichihashi (1993). These tubes consist of rolled-up graphene sheets (the name for a single layer of the normal graphite structure) with endcaps. Iijima's first report was of multiwalled tubes (Russian dolls again), but his second paper reported the discovery of single-walled tubes, about 1 nm in diameter, capped by well-formed hemispheres with C_{60} structure. (The multiwalled tubes are capped by far more complex multiwall caps). Printed alongside Iijima's second paper in *Nature* was a similar report by an American team (Bethune *et al.* 1993). It seems that *Nature* has established a speciality in printing adjacent pairs of papers independently reporting the same novelty: this also happened in 1951 with growth spirals on polytypic silicon carbide (Verma and Amelinckx) and earlier, in 1938, with pre-precipitation zones in aged Al-Cu alloys (Guinier, Preston) – see Chapter 3 for details of both these episodes.

Interest has rapidly focused on the single-walled, capped tubes, as shown in Figure 11.7. They can currently be grown up to ≈100 μm in length, i.e., about 100,000 times their diameter. As the figure shows, there are two ways of folding a graphene sheet in such a way that the resultant tube can be seamlessly closed with a C_{60} hemisphere... one way uses a cylinder axis parallel to some of the C—C bonds in

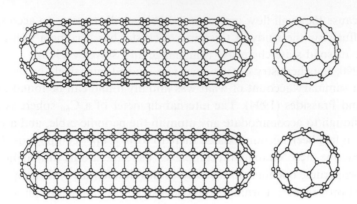

Figure 11.7. Two types of single-walled carbon nanotubes.

the sheet, the other, an axis normal to the first. The distinction is important, because the two types turn out to have radically different electrical properties.

Research on nanotubes has been so intensive that the first single-author textbook has already been published (Harris 1999), following an earlier multiauthor overview (Dresselhaus *et al.* 1996). In addition to discussing the mechanism of growth of the different kinds of nanotubes, he also discusses the many precursor studies which almost – but not quite – amounted to discovery of nanotubes. He also has a chapter on 'carbon onions', multiwalled carbon spheres first observed in 1992 (and again reported in *Nature*); these seem to be multiwalled versions of fullerenes and the reader is referred to Harris's book for further details. Just one feature about the onions that merits special attention is that the onions are under extreme internal pressure, as shown by the sharp diminution of lattice spacings in the inner regions of the onion. When such an onion is irradiated at high temperature with electrons, the core turns into diamond (Banhart 1997). For good measure, Harris also provides a historical overview of the spherulitic form of graphite in modified cast irons (see Section 9.1.1). His book also contains a fascinating chapter on chemistry inside nanotubes, achieved by uncapping a tube and sucking in reactants. One promising approach is to use a single-walled nanotube as a template for making ultrafine metallic nanowires.

Harris has this to say on the breadth of appeal of nanotubes: "Carbon nanotubes have captured the imagination of physicists, chemists and materials scientists alike. Physicists have been attracted to their extraordinary electronic properties, chemists to their potential as 'nanotest-tubes' and materials scientists to their amazing stiffness, strength and resilience".

An even more up-to-date account of the current state of nanotube research from physicists' perspective is in an excellent group of articles published in June 2000

(McEwen *et al.* 2000). One feature which is explained here is the fact that one of the structures in Figure 11.7 has metallic conductivity, the other is a semiconductor, because of the curious energy band structure of nanotubes. The metallic version is beginning to be applied for two purposes: (a) as flexible tips for scanning tunnelling microscopes (Section 6.2.3) (Dai *et al.* 1996), (b) as highly efficient field-emitting electrodes. In this second capacity, arrays of tubes have been used for lamps... electrons are emitted, accelerated and impinge on a phosphor screen. Now the extremely challenging task of using such nanotube arrays for display screens has been initiated, and one such display has been shown in Korea; one of the papers in the recent publication says: "In the extremely competitive display market there will be only a few winners and undoubtedly many losers".

Carbon nanotubes mixed with ruthenium oxide powder, and immersed in a liquid electrolyte, have been shown by a Chinese research group to function as 'supercapacitors' with much larger capacitance per unit volume than is normally accessible (Ma *et al.* 2000).

Nanotubes have also been found to be promising as gas sensors, for instance for N_2O, and in particular – this could prove to be of major importance – as storage devices for hydrogen. The capacity of both kinds of nanotubes to absorb various gases at high pressure was first found in 1997, and very recently, a Chinese team has established that one hydrogen atom can be stored for every two carbon atoms, using a 'chemically treated' population of nanotubes, a high capacity. Moreover, most of this absorbed gas can be released at room temperature by reducing the pressure; this seems to be the most valuable feature of all. The current position is reviewed by Dresselhaus *et al.* (1999).

The other striking feature of nanotubes is their extreme stiffness and mechanical strength. Such tubes can be bent to small radii and eventually buckled into extreme shapes which in any other material would be irreversible, but here are still in the elastic domain. This phenomenon has been both imaged by electron microscopy and simulated by molecular dynamics by Iijima *et al.* (1996). Brittle and ductile behaviour of nanotubes in tension is examined by simulation (because of the impossibility of testing directly) by Nardelli *et al.* (1998). Hopes of exploiting the remarkable strength of nanotubes may be defeated by the difficulty of joining them to each other and to any other material.

A distinct series of studies is focused on improved methods of growing nanotubes; Hongjie Dai in the 2000 group of papers focuses on this. In a recent research paper (Kong *et al.* 1998) he reports on the synthesis of individual single-walled nanotubes from minute catalyst islands patterned on silicon wafers – a form of templated self-assembly. The latest approach returns towards the 1985 technique: an anonymous report (ORNL 2000) describes an apparatus in which a pulsed laser locally vaporises ('ablates') a graphite target containing metal catalyst. A 'bubble' of

10^{16} carbon and metal atoms streams away through hot argon gas and they then combine to form single-wall nanotubes with high efficiency.

The foregoing is merely a very partial summary of a major field of materials science, into which chemistry and physics are indissolubly blended.

11.2.7 *Combinatorial materials synthesis and screening*

In the early 1990s, a new technique of investigation was introduced in the research laboratories of pharmaceutical companies – combinatorial chemistry. The idea was to generate, by automated techniques, a collection of hundreds or even thousands of compounds, in tiny samples, of graded compositions or chemical structure, and to bioassay them, again by automated techniques, to separate out promising samples. The choice of chemicals was determined by experience, crystallographic information on bond configuration, and inspired guesswork. A little later, this approach was copied by chemists to seek out effective homogeneous and heterogeneous catalysts for specific gas-phase reactions (Weinberg *et al.* 1998); this account cites some of the earlier pharmaceutical papers. Weinberg is technical director of a start-up company called Symys Technologies in Silicon Valley, founded with the objective of applying the above-mentioned approach to solid-state materials. After initial hesitation, the approach is also beginning to be tried by a number of major materials laboratories such as Bell Labs, and by an active group at the Lawrence Berkeley National Laboratory led by Xiao-Dong Xiang.

The main approach of materials scientists who wished to exploit this approach has been to deposit an array of tiny squares of material of systematically varying compositions, on an inert substrate, originally by sequential sputtering from multiple targets through specially prepared masks which are used repeatedly after 90° rotations. The array is then *screened* by some technique, as automated as possible to speed things up, to separate the sheep from the goats. Perhaps the first report of such a search was by Xiang *et al.* (1995), devoted to a search for new superconducting ceramics, with a sample density of as much as 10,000 per square inch. A four-point probe was used to screen the samples. New compositions were found, albeit not with any particularly exciting performance.

A slightly later example of this approach was a search for an efficient new luminescent material (Danielson *et al.* 1997a, b, Wang *et al.* 1998), using about 10 target materials mixed in greatly varying proportions. Screening in this instance was simple, since the entire array could be exposed to light and the 'winners' directly identified; in fact an automated light-measuring device was used to record the performance of each sample automatically. In this way, Sr_2CeO_4 was identified out of a combinatorial 'library' of more than 25 000 members; it gives a powerful blue–white emission and responds well to X-ray stimulation. In the *Science* paper, the authors show how a consequential test with Ba and Ca oxides was done to see whether a mixed oxide with Sr might perform even better. The array of samples was arranged in an equilateral triangle looking just like a ternary diagram; the pure Sr compound was unambiguously the best. This luminescence search was used as the text of an early survey of the combinatorial approach, under the slightly optimistic title "High-speed materials design" (Service 1997).

Xiang and his many collaborators went on to develop the initial approach in a major way. The stationary masks were abandoned for a technique using precision shutters which could be moved continuously under computer control during deposition; sputtering was replaced by pulsed laser excitation from targets. Figure 11.8 schematically shows the mode of operation. The result is a continuously graded thin film instead of separate samples each of uniform composition; Xiang calls the end-result a *continuous phase diagram* (CPD). Composition and structure at any point can be checked by Rutherford back-scattering of ions, and by an x-ray microbeam technique using synchrotron radiation, respectively, after annealing at a modest temperatures to interdiffuse the distinct, sequentially deposited layers. This approach to making a continuously variable thin film was originally tried by Kennedy *et al.* (1965), curiously enough in the same laboratory as Xiang's present research. At that time, deposition techniques were too primitive for the approach to be successful. Xiang's group (unpublished research) has tried out the technique by making a CPD of binary Ni–Fe alloys and testing magnetic characteristics for comparison with published data. More recently (Yoo *et al.* 2000), CPDs were used to locate unusual phase transitions in an extensive series of alloyed perovskite manganites of the kind that show colossal magnetoresistance (Section 11.2.4); this

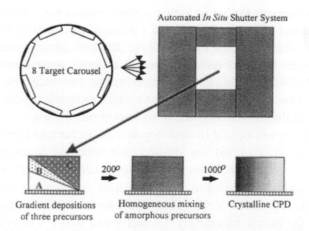

Figure 11.8. Schematic layout of procedure for creating a continuous phase diagram (courtesy X.-D. Xiang, after Yoo *et al.* 2000).

seems to be the first published account of the use of CPDs to examine hitherto unknown phenomena. Moreover, this important study revealed the compositions at which phase changes took place; this implies that 'continuous phase diagrams' can be used to locate the loci of phase transitions in, say, ternary systems at some specified temperature, and thus help to determine isothermal phase equilibria. This would be a considerable technical advance in materials science.

Xiang (1999) has recently published a critical account of the whole field of what he calls *combinatorial materials synthesis and screening*, a phrase which I have chosen to provide the title of this section.

The recent burst of research on the combinatorial approach is not, however, the first. Thirty years ago, a scientist at the laboratories of RCA (the Radio Corporation of America), Joseph Hanak, wrote a precocious paper on what he called the "multiple sample concept" in materials research (Hanak 1970), essentially the same notion. Some 25 papers by Hanak followed during the 1970s, reporting on the application of his concept to a variety of problems, for instance electroluminescence (Hanak 1977) and solar cells. Subsequently, attention lapsed, though a Japanese group in 1988 pursued combinatorial study of oxides. The leader of that group, H. Koinuma, has just published an account of recent Japanese work on the combinatorial approach (Koinuma *et al.* 2000); it includes details of a systematic survey of ZnO doped with variable amounts of transition metals to determine solubility limits and optical properties.

11.3. ELECTROCHEMISTRY

Electricity and chemistry are linked in two complementary ways: the use of chemical reactions to produce electricity is one, and the use of electricity to induce chemical reactions is the other. The first of these large divisions encompasses primary and secondary batteries and fuel cells; the second includes some forms of extractive metallurgy and of large-scale chemical manufacture and such processes as water purification. In between, there are phenomena which include local electric currents as an incidental; metallic corrosion is the most important of these.

Electrochemistry can be said to have begun with the famous experiments in 1791 by Luigi Galvani (1737–1798): he showed that touching a dissected frog's leg with metal under certain conditions caused the muscle to undergo spasm. Galvani thought his observations pointed to a 'nervous fluid', perhaps a form of 'life force'. His countryman, Alessandro Volta (1745–1827) reexamined the matter and finally concluded that the muscle was merely a detector and that the stimulus could come from two dissimilar metals separated by a poor conductor (Volta 1800). He capitalised on his insight by creating the world's first primary battery, a 'pile' (in

French, a battery is still called 'une pile') of metals and paper disks moistened with brine, in the sequence silver–paper–zinc–silver–paper–zinc etc. Volta's pile only worked for a day or two before the paper dried out, but it marked the beginning of electrochemistry. The next year, William Cruikshank in England designed the first of many variants of a 'trough battery', in which metal plates were dipped into a suitable aqueous solution (ammonium chloride initially). In 1807, Sir Humphry Davy at the Royal Institution in London used three large trough batteries in his famous experiments to separate sodium and potassium from their salts, in the forms of slightly damp, fused soda and potash (Davy 1808). Previously, in 1800, Nicholson and coworkers had been the first to demonstrate chemical reactions resulting from the passage of an electric current when they found that gas bubbles were formed when a drop of water shorted the top of a voltaic pile; they identified the bubbles as hydrogen and oxygen, on the purported basis of smell!

After Cruikshank, there was a steady succession of gradually improving primary batteries (by 'primary', I mean batteries which are not treated as rechargeable); by stages, the power and endurance of such batteries was enhanced, and in 1836, Frederic Daniell designed a battery with two vessels separated by a semipermeable biological membrane, to prevent polarization by gas bubbles. This was the first of a succession of constant-voltage standards. All these are explained and illustrated in a fine historical overview by King (1962). The first dry battery was the 1868 Leclanché cell, using a carbon electrode in a pasty mixture of MnO_2 and other constituents, with a zinc electrode separated from the rest by a semipermeable ceramic cylinder. In chemical terms, the modern primary dry battery relies on much the same process. The first secondary (or storage) battery was announced in 1859 (Planté 1860): by electrolysing sulphuric acid with lead electrodes, he generated a layer of lead oxide on lead; then the charging primary battery was removed and the lead-acid battery was able to return its charge. 140 years later, after endless improvements to the composition and microstructure of the lead grid (even preferred crystallographic orientation of the lead has recently been found to be vital in improving the longevity of such grids), Planté's approach is still used in every automobile. In 1860, dynamo-generated mains electricity, as primary source of charge for lead-acid batteries, was still two decades away.

Electrochemistry in the modern sense really began with Michael Faraday's experiments in the 1830s, using a giant primary battery made specifically for Faraday's laboratory in London. Williams (1970–1980), in a major essay on Faraday, interprets Faraday's motivation for these experiments as being his desire to prove that electricity from different sources, electrostatic generators, voltaic cells, thermocouples, dynamos and electric fishes was the same entity; Williams estimates that Faraday was successful in this quest. In the process, by establishing quantitative measures for 'quantity of electricity' indifferently from diverse sources, Faraday established his two laws of electrochemistry: (1) the chemical effect is proportional to the quantity of electricity which has passes into solution, and (2) the amounts of different substances deposited or dissolved by a fixed quantity of electricity are proportional to their equivalent weights. The way Williams puts it, Faraday had proved that "(electricity) was the force of chemical affinity"; Much later, von Helmholtz argued that these experiments of Faraday's had shown that "electricity must be particulate". This research, which put electrochemistry firmly on the map, shows Faraday at his most inspired.

In addition to the various early European electrochemists, there was one important American participant, Robert Hare Jr. (1781–1858), whose life is treated by Westbrook (1978). As Westbrook explains, when Hare (who, though largely selftaught, eventually became a professor at the new University of Pennsylvania) began research, science in America was still "in an emergent state", and the first scientific journal "with national pretensions" had only come into being in 1797. In 1818, he designed his own efficient version of a voltaic trough, which he called the *calorimotor* (not calorimeter), because he was still a believer in the caloric theory of heat and thought of a voltaic trough as accumulating heat as well as electricity, both to be regarded in particulate terms. So, his apparatus was to be seen as a 'heat mover'. A later, further improved version of his pile was now called a 'deflagrator' (he was addicted to curious names) because by striking an arc, he could cause burning, or 'deflagration'. In 1822, Hare with a friend, made what seem to have been the first demonstrations of electric light from a deflagrator. He also showed clearly, with use of a mercury cathode, the separation of metallic calcium from an aqueous $CaCl_2$ solution (Ca was obtained from its amalgam), putting to rest uncertainties remaining from Davy's earlier attempt (Hare 1841). He went on to design an electric arc furnace with which he achieved a number of 'firsts', including CaC_2 synthesis and metal spot-welding.

11.3.1 Modern storage batteries

Batteries, both primary and secondary, have become very big business indeed, which moreover is growing rapidly. Salkind (1998) in a concise overview of the entire domain of battery types and technologies, estimates that in 1996, the world market in the two types of battery combined totalled ≈ 33 billion dollars, and that the ratio of secondary to primary battery sales is steadily edging upwards. In spite of its poor charge density per unit mass, the lead-acid battery still accounts for more than a quarter of the total, because it costs so much less than its rivals and lasts well.

Newer batteries can be divided into small rechargeable batteries for consumer electronics, cell-phones and laptop computers primarily, and larger advanced storage systems. The field of research on battery concepts and materials has recently expanded dramatically. A very detailed overview of battery materials has been published very recently (Besenhard 1999).

Increasing numbers of advanced batteries for all purposes depend on ionically conducting solid electrolytes, so it will be helpful to discuss these before continuing. It should be remembered that any battery can be described as an 'electron pump', and the role of the electrolyte is to block the passage of electrons, letting ions through instead.

11.3.1.1 Crystalline ionic conductors. 'Superionic' conductors have already been briefly introduced in Section 7.2.2.2. They have been known for quite a long time, and a major NATO Advanced Study Institute on such conductors was held as early as 1972 (van Gool 1973). Of course, all ionic crystals are to a greater or lesser extent ionically conducting – usually they are cationic conductors, because cations are smaller than anions. Superionic conductors typically have ionic conductivities 10^{11} times higher than do 'ordinary' ionic crystals such as KCl or AgCl.

Certain ionically well-conducting crystals, ZrO_2 for instance, have long been exploited for such applications as sensors (see below) and, long ago, for early electric lamps (Section 9.3.2); nowadays, the compound is stabilised against allotropic transformations by adding yttria, Y_2O_3. Every mole of the dopant, moreover, brings with it an extra vacancy, which enhances ionic conductivity. This brings zirconia into the domain of ionic *super*conductors which have exceptionally large ionic mobilities, generally because of very high equilibrium vacancy concentrations which permit the ions bordering those vacancies to diffuse very fast, with or without applied electric fields. The materials chemistry of stabilised zirconia, used in the form of thin films less than 100 μm in thickness, has become very sophisticated. The interface between the zirconia and the complex electrodes now used affects the ionic conductivity, so that the microstructure of the interface has become a vital variable (Drennan 1998).

Beta-alumina, mentioned in Section 7.2.2.2, is just the best known and most exploited of this family. They have been developed by intensive research over more than three decades since Yao and Kummer (1967) first reported the remarkably high ionic conductivity of sodium beta-alumina. Many other elements have been used in place of sodium, as well as different crystallographic variants, and various processing procedures developed, until this material is now poised at last to enter battery service in earnest (Sudworth *et al.* 2000).

11.3.1.2 Polymeric ionic conductors. One of the most unexpected developments in recent decades in the whole domain of electrochemistry has been the invention of and gradual improvements in ionically conducting polymeric membranes, to the point where they have become *the* key components of advanced batteries and fuel cells. A comparison between the conductivity of an advanced member of this category and of two ionic superconductors is shown in Figure 11.9.

The original motive for developing such polymers was for the chemical function of ion-exchange membranes, for such purposes as water desalination or softening. This kind of usage was already well established at the beginning of the 1960s. At about that time, the GE Laboratory in Schenectady began research on ionically conducting polymers for use in the fuel cells that were to be used as power sources in the American 'moon shots'; the 'product champion' was a chemist, W. Thomas Grubb who, in the words of Koppel (1999) "got an inspiration from an unlikely source, the common water softener". The story is spelled out in much greater detail in an essay by Suits and Bueche (1967); in 1955, Grubb took out a patent on his sulfonated polystyrene resin and a version of this polymeric electrolyte, in conjunction with an improved way of attaching platinum electrocatalyst developed by Leonard Niedrach, also of GE, eventually was used in the fuel cells for the American Gemini moon shots in the early 1960s. This kind of membrane is now commonly called a PEM, a *proton exchange membrane*, because the ions of interest in this connection are hydrogen ions. Industrially important polymers are cation conductors.

Later, Du Pont in America developed its own ionically conducting membrane, mainly for large-scale electrolysis of sodium chloride to manufacture chlorine, Nafion®, (the US Navy also used it on board submarines to generate oxygen by electrolysis of water), while Dow Chemical, also in America, developed its own even more efficient version in the 1980s, while another version will be described below in connection with fuel cells. Meanwhile, Fenton *et al.* (1973) discovered the first of a

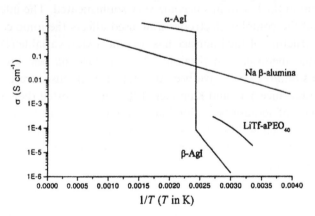

Figure 11.9. Conductivity vs temperature plot for two ionically conducting crystals and for a polymer electrolyte, LiTf-aPEO$_{40}$, which is based on amorphous poly(ethylene) oxide (after Ratner 2000).

series of polymers suitable specifically for batteries, based on dissolution of a salt in amorphous poly(ethylene) oxide, used in sheets of the order of 100 μm thick. Further development of membranes for battery use is concisely described by Scrosati and Vincent (2000); a number of quite different polymers and polymer composites have been developed; it has become a major branch of materials chemistry.

11.3.1.3 Modern storage batteries (resumed). The most advanced batteries to exploit superionic conductors have used beta-alumina. For some years, the sodium-sulphur battery held sway; here the electrodes are of molten sulphur and of molten sodium (the battery only functions at high temperature) and the electrolyte is of beta-alumina with sodium; that is, the electrodes are liquid and the electrolyte solid, standing tradition on its head. For a while, Ford Motor Company hoped to use this approach as a power source for automobiles; in the 1970s and 1980s much research was done on this system, but eventually it was abandoned for what Sudworth *et al.* (2000) call "a variety of technical and economic reasons". It seems that it has been replaced very recently by a sodium/nickel chloride battery, called ZEBRA, again using beta-alumina electrolyte; this well developed concept is peculiar in that the nickel chloride electrode has a liquid electrolyte incorporated, in contact with the solid electrolyte; this seems to be the first system of this type. Sudworth *et al.* indicate that vehicles have covered over 2 million kilometers with this kind of storage battery.

However, the battery system that has caused most excitement in recent years, and an enormous amount of associated research (see, e.g., dozens of papers in a recent MRS symposium, Ginley *et al.* 1998) is the Sony lithium ion battery for consumer electronics, introduced commercially in 1995 after many years of research and development. Without going into extensive details, this consists of a $LiCoO_2$ cathode and a Li anode, both intercalated in a specially developed carbon form (the anode consists of 'lithiated graphite', LiC_6; there is no free metallic lithium present). The electrolyte in the latest form of the battery is a newly developed, Li^+-conducting polymer, consisting of an amorphous matrix and salt-enriched crystalline regions; the conduction mechanism is still not properly understood. The Li^+ ions shuttle between two energy states in the two electrodes, and the battery gives a cell voltage of 3.8 V. The electrode chemistry is extremely complex, and alternative electrode strategies are being energetically researched; even computer simulation of electrochemical systems is being extensively applied in the search for improvements (e.g., Ceder *et al.* 1998).

The Sony cell is rapidly outstripping all other batteries for such uses as laptop computers, especially since the electrode design has overcome danger of fire which held back earlier versions of the battery. It has an energy density of >200 watt-hours/kg, compared with 35 for a modern lead-acid battery (and compared with 12,000 watt-hours/kg for gasoline!) Nevertheless, the Li battery in its latest form is the only one to date which exceeds the minimum battery characteristics officially set for automobile use. For the ZEBRA battery mentioned above, an energy density of 90 watt-hours/kg has been quoted.

It is interesting that one researcher on the lithium batteries, Manthiram (1999) of the University of Texas at Austin, found that to make progress in his group's researches, it was necessary to train students from various relevant disciplines, especially chemistry and physics, in an interdisciplinary materials science course before they acquired the right attitudes to make progress. The way he put it was: "It is difficult to achieve the research goals with graduate students having prior degrees in any of the traditional disciplines".

The great disadvantage of any battery, however advanced, for automobile power trains, is the long time required to charge a battery, and in my view this will be decisive. Here, fuel cells have an enormous advantage over batteries, and so I turn to fuel cells next.

11.3.2 Fuel cells

A fuel cell is simply a device with two electrodes and an electrolyte for extracting power from the oxidation of a fuel without combustion, converting the power released directly into electricity. The fuel is usually hydrogen. The principle of a fuel cell was first demonstrated by Sir William Grove in London in 1839 with sulphuric acid and platinum gauze as an electrocatalyst, and thereafter there were very occasional attempts to develop the principle, "not all of which were based on sound scientific principles", as one commentator put it.

The father of the modern fuel cell is Francis Thomas Bacon (known as Tom Bacon, 1904–1992), a descendant of Sir Nicholas Bacon, Elizabeth the First's Lord Keeper of the Great Seal and father of the 'original' Francis Bacon. From 1937 onwards, Tom Bacon became fascinated by the potential of fuel cells, and applied his considerable engineering skills to successive designs. He used nickel electrodes, highly pressurised hydrogen and a concentrated potassium hydroxide electrolyte and a temperature typically around 100°C, and the conditions he favoured gradually became more severe. He was faced with endless obstacles in the form of hostile research directors and unreliable financial backers. Fortunately he had a modest private income which throughout his life freed him from the tyranny of the money-men.

After the War, Tom Bacon worked for a while in the ill-fated Department of Colloid Science which we met in Chapter 2. His laboratory space there was taken away from him and he moved to the adjacent metallurgy laboratory and then again to the nearby chemical engineering department. In his own person, Tom Bacon worked in all the relevant departments in Cambridge University. All these stages are described in a biographical memoir by Williams (1994). Finally, Bacon obtained reasonably steadfast government support and by 1959 he was able to demonstrate a properly engineered 6 kW 40-cell device; the hydrogen electrode was of porous nickel, the oxygen electrode, eventually, of preoxidised nickel. At this stage, British Government support was withdrawn, but Pratt and Whitney in America became very interested, put some 1000 engineers on the project and by the mid-1960s an American fuel cell based on Bacon's design powered the Apollo moonshots, producing copious by-product water as a bonus. President Lyndon Johnson put his arm round Bacon's shoulders and said "Without you, Tom, we wouldn't have gotten to the moon".

The other main approach at the time was a fuel cell based on GE's ionically conducting polymer (Section 11.3.1.2), and this was used in the Gemini moonshots which preceded the Apollo programme. There were many teething troubles but fuel cells proved their worth in the space programme. The stages of this programme are described in Koppel's (1999) book.

Apart from Bacon's 'alkaline' fuel cell and the polymeric membrane cell, other variants are phosphoric acid cell, a molten carbonate cell and (greatly favored by many investigators) the solid oxide fuel cell, using stabilised zirconia as electrolyte and complex compound electrodes. These are all outlined in an encyclopedia article by Steele (1994), and the current design of the oxide fuel cell is described by Singhal (2000). There has been an enormous amount of gradual optimisation and Steele claims that the latest version has operated at $\approx 900°C$ with little degradation for more than 32,000 h. Both hydrogen and natural-gas fuels have been used, with very high generation efficiencies. Numerous cells are connected to form an industrial unit.

I suspect that the final competition for large-scale application will be between solid-oxide and polymeric-membrane versions, and that the former may well win out for stationary power sources, while the latter will be the victor for automotive uses, particularly since the operating temperature with polymeric electrolyte is so much lower and very little start-up time is needed. A detailed discussion of the design and merits of the different designs is in a book by Kordesch and Simader (1996), which pays special attention to the phosphoric acid cell. Another detailed review of the alternatives for the "electric option" for powering automobiles is by Shukla *et al.* (1999); they conclude, intriguingly, that a 50 kW polymer electrolyte fuel cell stack, together with a "supercapacitor" or a battery bank for short bursts of extra power, would be a viable arrangement. This takes us naturally to the experience of the most successful company currently active in this field.

The achievements of a small Canadian startup company, Ballard Power Systems, in Vancouver, are the main reason for my view that polymeric-membrane cells have the automotive market at their feet. The stages of the company's achievements, founded by Geoffrey Ballard, are fascinatingly described in Koppel's book, which also goes in considerable detail into the industrial battles between the rival configurations. The Ballard company by degrees improved the polymeric membrane; since the Du Pont and Dow membranes were too expensive and the prices would not come down, the company developed and then began to manufacture its own improved membrane, and also – in collaboration with Johnson Matthey, the precious-metal firm – found ways of using platinum electrocatalyst in ever more efficient physical forms, reducing the amount needed by a factor of ten. Ballard cells, using compressed hydrogen, powered a fleet of municipal buses in Vancouver as early as 1993. Finally, the company made common cause with a major automobile manufacturer and it looks as though a thoroughly practical automobile fuel cell is very close. A recent critical overview strikes an upbeat note (Appleby 1999).

As of 2000, it also looks as though more and more electric utilities are becoming interested in fuel cell stacks as local 'microgenerators' to top up power from large power stations, without the need for long-distance transmission of electricity and its attendant expense and power losses.

Storage of the fuel is the Achilles' heel of all fuel cells. Hydrogen is still the preferred fuel, methanol is another though even here the preference is for an on-board apparatus for 'reforming' the chemical to create hydrogen. Hydrogen can be effectively stored as compressed gas, liquid (here the difficulties are the low density and thus large volume of a supply of LH, and also the large amount of energy irreversibly used in liquefaction) or in the form of a reversibly formed hydride; hydrogen can be released by slight heating. Research on metal hydrides is now a major field of materials chemistry, but as yet the attainable ratio of hydrogen to metal is not quite sufficient and this form of hydrogen storage has to contend with excessive weight. However, magnesium hydride looks distinctly promising (Schwarz 1999), as does the reversible storage of hydrogen in carbon nanotubes (Dresselhaus *et al.* 1999). As with batteries, the speed, simplicity and cost of 'refuelling' will probably be the limiting factor in the development of automobiles driven by fuel cells, but this may not be a major consideration where microgenerators are concerned.

11.3.3 Chemical sensors

Electrochemistry plays an important role in the large domain of sensors, especially for gas analysis, that turn the chemical concentration of a gas component into an electrical signal. The longest-established sensors of this kind depend on superionic conductors, notably stabilised zirconia. The most important is probably the oxygen sensor used for analysing automobile exhaust gases (Figure 11.10). The space on one side of a solid-oxide electrolyte is filled with the gas to be analysed, the other side

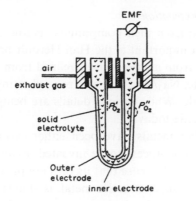

Figure 11.10. Gas sensor to monitor oxygen content of exhaust gases from automobile engines (after Fray 1990).

with a gas of standard composition, and the cell potential is measured. This kind of cell is much more sensitive at low concentrations than at high (Fray 1990). Similar cells can be designed to measure other gases such as CO_2 and SO_2 (Yamazoe and Miura 1999). Hydrogen can be analysed, for instance, by exposing SnO_2, a conductor, to oxygen, thereby creating a chemisorbed layer of high resistivity; then reducing this by hydrogen: the resistivity is related to the hydrogen concentration. To distinguish between different reducing gases, dopants such as La_2O_3 can be added to the SnO_2. To show the amount of materials chemistry that has gone into this kind of instrumentation, reference can be made to an overview of the dozens of devices developed to measure just one impurity gas, sulphur dioxide, many using molten salt electrolytes (Singh and Bhoga 1999). Other sensors are based on changes in resistivity or on MOSFET-type transistors, many are used for analysing solutions rather than gases; here the drain current depends on ion concentration. The subject is too vast to attempt any further classification here.

A subset of sensors is designed to function as *smart materials*; these are devices that function both as sensors and as actuators (Newnham 1998). An example is a smart shock absorber for automobiles, designed in Japan; this is a multilayer ferroelectric system in which sensed vibrations lead to a correcting signal acting on another part of the multilayer stack. The ferroelectric mount for the tip of a scanning tunneling microscope also functions as a smart material, in keeping the tip at a predetermined distance from the sample being examined. Magnetostriction and electrostriction are other responses used in certain smart materials. The foregoing are based on sensors for physical rather than chemical properties, but there is no reason why chemical sensors should not come to be incorporated in control systems, for instance to keep constant the concentration of an aqueous solution or of a gas in a gas mixture.

11.3.4 *Electrolytic metal extraction*

Many metals are extracted from their compounds, as found in ores, by electrolytic processes. By far the most important is the Hall-Héroult process, invented in 1886, for producing aluminium from alumina, itself refined from bauxite ore. Alumina is dissolved in molten cryolite, Na_3AlF_6, and electrolysed, using carbon anodes and the aluminium itself as cathode. While various details are being steadily improved, the basic process is still the same today.

Since 1886, many other metals have been either extracted or else refined by electrolytic means. The latest process to be invented involves titanium metal. This metal is intrinsically cheap in the sense that its ores are plentiful in the earth's crust; the high cost of titanium, a highly reactive metal, is almost entirely due to the very elaborate pyrometallurgical production process used; this is the Kroll process, introduced in 1940. An effective electrolytic process has been sought for decades. Now, it appears, an effective method has been developed (Chen *et al.* 2000): TiO_2 powder is made the cathode of a bath of molten $CaCl_2$ whose cation can form a more stable oxide, CaO. The oxygen in the TiO_2 is ionised and dissolves in the salt, leaving titanium metal behind. The approach is simple, has worked well on a kilogram scale, and may well prove to be cheap. If it is fully proved, it is likely to have a revolutionary effect on the scope of titanium in practical metallurgy.

11.3.5 *Metallic corrosion*

In economic terms, the study and prevention of metallic corrosion is one of the most important fields of materials science and engineering. Methods of study have been developed throughout the twentieth century. Perhaps the first major text to assemble the many insights gained was that by the Cambridge metallurgist Ulick Evans (1889–1980) (1937, 1945). Evans made it very clear that the operation of localised electrolytic microcells play a dominant role in corrosion. One form of such localised electrolysis was what Evans called "differential aeration": different rates of supply of oxygen to the centre and periphery of a water drop on metal suffice to set up a potential difference and thus a corrosive current. This particular concept was much discussed and disputed in the 1930s, and a recent overview of corrosion (Schütze 2000) makes no mention of it. This is typical of this disputatious field. However, the centrality of electrochemistry in corrosion is not in doubt, and the first chapter in Schütze's book is devoted to a description of the macroscopic experimental methods used to mimic the localised electrolytic processes in rusting steel and other corroding metals.

Corrosion is fought partly by developing alloys with a built-in proclivity to form protective oxide layers, such as 'stainless steels', and partly by designing protective coatings. A form of protection particularly closely linked to electrochemistry is cathodic or anodic protection. In one form of this strategy, a coating is designed to dissolve preferentially ('sacrificially') instead of the underlying metal: the use of zinc coatings on steel is the most familiar and long-established form of this approach. Another way is to pass an externally sourced current between the item to be protected, whether a ship or a buried pipeline, and an adjacent sacrificial piece of another metal. This form of protection has become a widespread technology; it is fully described by Juchniewicz *et al.* (2000).

Ultramodern techniques are being applied to the study of corrosion: thus a very recent initiative at Sandia Laboratories in America studied the corrosion of copper in air 'spiked' with hydrogen sulphide by a form of combinatorial test, in which a protective coat of copper oxide was varied in thickness, and in parallel, the density of defects in the copper provoked by irradiation was also varied. Defects proved to be more influential than the thickness of the protective layer. This conclusion is valuable in preventing corrosion of copper conductors in advanced microcircuits. This set of experiments is typical of modern materials science, in that quite diverse themes... combinatorial methods, corrosion kinetics and irradiation damage...are simultaneously exploited.

To keep this book in some kind of balance, no further treatment of corrosion and its prevention – or of high-temperature dry corrosion – is feasible here, important though these themes are.

REFERENCES

Addadi, L. and Weiner, S. (1999) *Nature* **398**, 461.
Aizenberg, J., Black, A.J. and Whitesides, G.M. (1999) *Nature* **398**, 495.
Appleby, A.J. (1999) The electrochemical engine for vehicles, *Sci. Amer.* **281**(1) (July) 58.
Arndt, M. *et al.* (1999) *Nature* **401**, 680.
Ball, P. (1997) *Made to Measure: New Materials for the 21st Century* (Princeton University Press, Princeton, NJ).
Banhart, F. (1997) *Physikalische Blätter* **53**, 33.
Besenhard, J.G. (editor) (1999) *Handbook of Battery Materials* (Wiley-VCH, Weinheim).
Bethune, D.S. *et al.* (1993) *Nature* **363**, 605.
Blanco, A. *et al.* (2000) *Nature* **405**, 437.
Blodgett, K. (1934) *J. Amer. Chem. Soc.* **56**, 495.
Brubacher, J.M., Christodoulou, L. and Nagle, D.C. (1987) *US Patent* 4 710 348.
Brune, H., Giovannini, M., Bromann, K. and Kern, K. (1998) *Nature* **394**, 451.
Calvert, P.D. and Mann, S. (1988) *J. Mater. Sci.* **23**, 3801.
Calvert, P.D. *et al.* (1994) in *Proc. Solid Freeform Fabrication Symposium*, ed. Marcus, H.L. *et al.* (University of Texas Press, Austin) p. 50.
Campbell, M., Sharp, D.N., Harrison, M.T., Denning, R.G. and Turberfield, A.J. (2000) *Nature* **404**, 53.

Ceder, G. *et al.* (1998) *Nature* **392**, 694.

Chawla, K.K. (1998) *Fibrous Materials* (Cambridge University Press, Cambridge) pp. 132, 211.

Chen, G.Z., Fray, D.J. and Farthing, T.W. (2000) *Nature* **407**, 361.

Dai, H., Hafner, J.H., Rinzler, A.G., Colbert, D.T. and Smalley, R.E. (1996) *Nature* **384**, 147.

Danielson, E. *et al.* (1997a) *Science* **279**, 837.

Danielson, E. *et al.* (1997b) *Nature* **389**, 944.

Davy, Humphry (1808) *Phil. Trans. Roy. Soc. Lond.* **98**, 1.

Day, P. (1997) What is a material? in *New Trends in Materials Chemistry*, ed. Catlow, C.R.A. (Kluwer Academic Publishers, Dordrecht) p. 1.

Deevi, S.C. and Sikka, V.K. (1997) *Intermetallics* **5**, 17.

De Rosa, C., Park, C., Thomas, E.L. and Lotz, B. (2000) *Nature* **405**, 433.

Dolphin, D., McKenna, C., Murakami, Y. and Tabushi, I. (1980) *Biomimetic Chemistry, Advances in Chemistry Series* #191 (American Chemical Society, Washington, DC).

Drennan, J. (1998) *J. Mater. Synthesis and Processing* **6**, 181.

Dresselhaus, M.S., Dresselhaus, G. and Eklund, P.C. (1996) *Science of Fullerenes and Carbon Nanotubes* (Academic Press, San Diego).

Dresselhaus, M.S., Williams, K.A. and Eklund, P.C. (1999) *MRS Bulletin* **24**(11), 45.

Eckert, C.A., Knutson, B.L. and Debenedetti, P.G. (1996) *Nature* **383**, 313.

Elices, M. (editor) (2000) *Structural Biological Materials* (Pergamon Press, Oxford).

Evans, U.R. (1937) *Metallic Corrosion, Passivity and Protection* (Edward Arnold, London), 2nd edition, 1945.

Fenton, D.E., Parker, J.M. and Wright, P.V. (1973) *Polymer* **14**, 589.

Flemings, M.C. and Cahn, R.W. (2000) *Acta Mater.* **48**, 371.

Franklin, B. (1774) *Phil. Trans. Roy. Soc. Lond.* **64**, 445.

Fray, D.J. (1990) Gas sensors, in *Suppl. Vol. 2 to Encyclopedia of Materials Science and Engineering*, ed. Cahn, R.W. (Pergamon Press, Oxford) p. 927.

Gaines, Jr., G.L. (1983) *Thin Solid Films* **99**, ix.

Ginley, D.S. *et al.* (eds.) (1998) in *Materials for Electrochemical Storage and Energy Conversion II – Batteries, Capacitors and Fuel Cells*, MRS Symp. Proc., vol. 496, (Warrendale, PA).

Goodenough, J.B. and Longo, J.M. (1970) in *Landolt–Börnstein Tables*, New Series III/4a (Springer, Berlin).

Grier, D.G. (ed.) (1998, October) *Directed self-assembly of colloidal materials*, *MRS Bull.* **23**(10), 21.

Hanak, J.J. (1970) *J. Mater. Sci.* **5**, 964.

Hanak, J.J. (1977) *J. Luminescence* **15**, 349.

Hannay, J.B. and Hogarth, J. (1879) *Proc. Roy. Soc. Lond. A* **29**, 324

Hare, R. (1841) *Amer. Phil. Soc. Trans.* **7**.

Harris, P.J.F. (1999) *Carbon Nanotubes and Related Structures: New Materials for the Twenty-first Century* (Cambridge University Press, Cambridge).

Hill, D.N., Lee, J.D., Cochran, J.K. and Chapman, A.T. (1996) *J. Mater. Sci.* **31**, 1789.

Holt, J.B. and Munir, Z.A. (1986) *J. Mater. Sci.* **21**, 251.

Hwang, H.Y. and Cheong, S.-W. (1997) *Nature* **389**, 942.

Iijima, S. (1991) *Nature* **354**, 56.

Iijima, S. and Ichihashi, T. (1993) *Nature* **363**, 603.

Iijima, S., Brabec, C., Maiti, A. and Bernholc, J. (1996) *J. Chem. Phys.* **104**, 2089.

Jeronimidis, G. (2000) *Structure-property relationships in biological materials*, in *Structural Biological Materials*, ed. Elices, M. (Pergamon Press, Oxford) p. 3.

Jin, S. *et al.* (1993) *Science* **264**, 413.

Johnson, D.J. (1994) Carbon fibres, in *Encyclopedia of Advanced Materials*, vol. 1, ed. Bloor, D. (Pergamon Press, Oxford) p. 342.

Juchniewicz, R., Jankowski, J. and Darowicki, K. (2000) in *Corrosion and Environmental Degradation*, vol. 1. ed. M. Schütze (Wiley-VCH, Weinheim) p. 383.

Kamat, S., Su, X., Bellarini, R. and Heuer, A.H. (2000) *Nature* **405**, 1036.

Keicher, D.M. and Smugeresky, J.E. (1997) *JOM*, May, p. 51.

Kennedy, K, Stefansky, T., Davy, G., Zackay, C.F. and Parker, E.R. (1965) *J. Appl. Phys.* **36**, 3808.

Kim, E., Xin, Y. and Whitesides, G.M. (1995) *Nature* **376**, 581.

King, W.J. (1962) *US National Museum Bull.* (*Smithsonian*) (228), 223.

Koinuma, H., Ayer, H.N. and Matsumoto, Y. (2000) *Sci. Technol. Adv Mater.* (Japan) **1**, 1.

Kong, J., Soh, H.T., Cassell, A.M., Quayte, C.F. and Dai, H. (1998) *Nature* **395**, 878.

Koppel, T. (1999) *Powering the Future: The Ballard Fuel Cell and the Race to Change the World* (Wiley, Toronto).

Kordesch, K. and Simader, G. (1996) *Fuel Cells and their Applications* (VCH, Weinheim).

Krätschmer, W., Lamb, L.D., Fostiropoulos, K. and Hiffman, D.R. (1990) *Nature* **347**, 354.

Kroto, H. (1997) *Rev. Mod. Phys.* **69**, 703.

Kroto, H.W., Heath, J.R., O'Brien, S.C., Curl, R.F. and Smalley, R.E. (1985) *Nature* **318**, 162.

Kroto, H.W. and Prassides, K. (1994) *Fullerenes*, in *Encyclopedia of Advanced Materials*, vol. 2, ed. D. Bloor *et al.* (Pergamon Press, Oxford) p. 891.

Langmuir, I. (1917) *J. Amer. Chem. Soc.* **39**, 1848.

Langmuir, I. (1933) *J. Chem. Phys.* **1**, 3.

Lapporte, S.J. (1995) *Adv. Mater.* **7**, 687.

Lehn, J.-M. (1995) *Supramolecular Chemistry: Concepts and Perspectives* (VCH, Weinheim).

Littlewood, P. (1999) *Nature* **399**, 529.

Liu, Z.F., Hashimoto, K. and Fujishima, A. (1990) *Nature* **347**, 658.

Ma, Renzhi *et al.* (2000) *Science in China* (*Series E, Technological Sciences*) **43**, 178.

McEwen, P.L., Schönenberger, C., Forró, L., Dai, H., de Heer, W.A. and Martel, R. (2000) *Physics World* (*London*) **13**(6), 37.

Mann, S. (1996) *Biomimetic Materials Chemistry* (VCH, New York).

Manthiram, A. (1999) Unpublished lecture at Pennsylvania State University, August.

Mathur, N.D. *et al.* (1997) *Nature* **387**, 266.

Merzhanov, A.G. and Borovinskaya, I.P. (1972) *Doklady Akad. Nauk SSSR* **204**, 366.

Mirkin, C.A. and Ratner, M.A. (1992) *Annu. Rev. Phys. Chem.* **43**, 719.

Moore, A.W. (1973) in *Chemistry and Physics of Carbon*, ed. Walker, P.L. Jr. and Thrower, P.A. (Marcel Dekker, New York) p. 69.

Moore, J.S. (ed.) (2000) Supramolecular materials, a group of papers, *MRS Bull.* **25**(4), 26.

Mort, J. (1980) *Science* **208**, 819.

MSE (1989) *Materials Science and Engineering for the 1990s*, Report of the Committee on Materials Science and Engineering from the National Research Council (National Academy Press, Washington, DC).

Munir, Z.A. and Anselmi-Tamburini, U. (1989) *Mater. Sci. Rep.* **3**, 277.

Nangia, A. and Desiraju, G.R. (1998) *Acta Cryst. A* **54**, 934.

Nardelli, M.B., Yakobson, B.I. and Bernholc, J. (1998) *Phys. Rev. Lett.* **81**, 4656.

Newnham, R.E. (1998) *Acta Cryst. A* **54**, 729.

ORNL (2000) *Report on current nanotechnology programs at Oak Ridge National Laboratory*, on the Worldwide Web at www.ornl.gov/ORNLReview/rev32_3/brave.htm.

Parker, G. and Charlton, M. (2000) *Phys. World* **13**(8), 29.

Parsonage, L.G. and Staveley, N.A.K. (1978) *Disorder in Crystals* (Oxford University Press, Oxford) p. 512.

Percec, V. *et al.* (1998) *Nature* **391**, 161.

Planté, G. (1860) *Compt. Rend. Acad. Sci. (Paris)*, **50**, 640.

Pockels, A. (1891) *Nature* **43**, 437.

Rabenau, A. (ed.) (1970) *Problems of Nonstoichiometry* (North-Holland, Amsterdam).

Rao, C.N.R. (1993) *Bull. Mater. Sci. (India)*, **16**, 405.

Rao, C.N.R. (1994) *Chemical Approaches to the Synthesis of Inorganic Materials* (Wiley, New York).

Rao, C.N.R. and Rao, K.J. (1978) *Phase Transitions in Solids* (McGraw-Hill, New York).

Rao, C.N.R. and Gopalakrishnan, J. (1986, 1997) *New Directions in Solid State Chemistry* (Cambridge University Press, Cambridge).

Rao, C.N.R. and Cheetham, A.K. (1997) *Adv. Mater.* **9**, 1009.

Ratner, M.A. (2000), Polymer Electrolytes: Ionic Transport Mechanisms and Relaxation Coupling, *MRS Bull.* **25**(3), 31.

Rayleigh, Lord (1899) *Phil. Mag.* **48**, 321.

Riedel, R. (1996) in Advanced ceramics from inorganic polymers, Processing of Ceramics, ed. Brook, R.J.; *Materials Science and Technology* vol. 17B, ed. Cahn, R.W. *et al.* (VCH, Weinheim), p. 2.

Roy, R. (1996) *J. Mater. Education* **18**, 267.

Salkind, A.J. (1998) in *Materials for Electrochemical Storage and Energy Conversion II – Batteries, Capacitors and Fuel Cells*, ed. D.S. Ginley *et al.*, MRS Symp. Proc. vol. 496, (Warrendale, PA) p. 3.

Schütze, M. (ed.) (2000) *Corrosion and Environmental Degradation* (Wiley-VCH, Weinheim) 2 volumes.

Schwarz, R.B. (1999) *MRS Bull.* **24**(11), 40.

Scrosati, B. and Vincent, C.A. (2000) Polymer electrolytes: the key to lithium polymer batteries, *MRS Bull.* **25**(3), 28.

Service, R.F. (1997) *Science* **277**, 474.

Shukla, A.K., Avery, N.R. and Muddle, B.C. (1999) *Current Sci. (India)* **77**, 1141.

Singh, K. and Bhoga, S.S. (1999) *Bull. Mater. Sci. (India)*, **22**, 71.

Singhal, S.C. (2000) Science and technology of solid-oxide fuel cells, *MRS Bull.* **25**(3), 16.

Steele, B.C.H. (1994) Fuel cells, in *The Encyclopedia of Advanced Materials*, vol. 2, ed. Bloor, D. *et al.* (Pergamon Press, Oxford) p. 885.

Sudworth, J.L. *et al.* (2000) Toward commercialization of the beta-alumina family of ionic conductors, *MRS Bull.* **25**(3), 22.

Suits. C.G. and Bueche, A.M. (1967) Cases of research and development in a diversified company, in *Applied Science and Technological Progress* (US Govt. Printing Office, Washington, DC) p. 297.

Swalen, J.D. (1991) *Annu. Rev. Mater. Sci.* **21**, 373.

Ulman, A. (1991) *An Introduction to Ultrathin Organic Films from Langmuir-Blodgett to Self-Assembly: An Overview* (Academic Press, Boston).

Van Blaaderen, A. and Wiltzius, P. (1997) *Adv. Mater.* **9**, 833.

Van Gool, W. (ed.) (1973) *Fast Ion Transport in Solids: Solid-State Batteries and Devices* (North-Holland, Amsterdam).

Vincent, J. (1997) *RSA Journal* (Royal Society of Arts, London) September, p. 16.

Volta, A. (1800) *Phil. Mag.* **7**, 289.

Von Helmholt, R. *et al.* (1994) *Phys. Rev. Lett.* **71**, 2331.

Wang, J. *et al.* (1998) *Science* **279**, 1712.

Watt, W. (1970) *Proc. Roy. Soc. Lond. A* **319**, 5.

Weinberg, W.H., Jandeleit, B., Self, K. and Turner, H. (1998) *Curr. Opinion Solid State Mater. Sci.* **3**, 104.

Weiner, S., Addadi, L. and Wagner, H.D. (2000) *Mater. Sci. Eng. C* **11**, 1.

Westbrook, J.H. (1978) in *Selected Topics in the History of Electrochemistry*, eds. Dubpernell, G.P. and Westbrook, J.H.; *Proc. Electrochem. Soc.* **78**(6), 100.

Williams, D. (1990) *Concise Encyclopedia of Medical and Dental Materials* (Pergamon Press, Oxford).

Williams, K.R. (1994) Memoir of F.T. Bacon, in *Biographical Memoirs of Fellows of the Royal Society* (The Royal Society, London) p. 3.

Williams, L. Pearce (1970–1980) Michael Faraday, in *Dictionary of Scientific Biography*, vol. 3, ed. Gillispie, C.C. (Charles Scribner's Sons, New York) p. 527.

Xiang, X.-D. *et al.* (1995) *Science* **268**, 1738.

Xiang, X.-D. (1999) *Annu. Rev. Mater. Sci.* **29**, 149.

Yajima, S. (1976) *Phil. Trans. Roy. Soc. Lond. A* **294**, 419.

Yamazoe, N. and Miura, N. (1999) Gas sensors using solid electrolytes, *MRS Bull.* **24**(6), 37

Yao, Y.F.Y. and Kummer, J.T. (1967) *J. Inorg. Nucl. Chem.* **29**, 2453.

Yoo, Y.-K., Duewer, F., Yang, H., Yi, D., Li, J.-W. and Xiang, X.-D. (2000) *Nature* **406**, 704.

Chapter 12
Computer Simulation

Chapter 12
Computer Simulation

12.1. BEGINNINGS

In late 1945, a prototype digital electronic computer, the Electronic Numerical Integrator and Calculator, (ENIAC) designed to compute artillery firing tables, began operation in America. There were many 'analogue computers' before ENIAC, there were primitive digital computers that were not programmable, and of course 19th-century computers, calculating engines, were purely mechanical. It is sometimes claimed that 'the world's first fully operational computer' was EDSAC, in Cambridge, England, in 1949 (because the original ENIAC was programmed by pushing plugs into sockets and throwing switches, while EDSAC had a stored electronic program). However that may be, computer simulation in fact began on ENIAC, and one of the first problems treated on this machine was the projected thermonuclear bomb; the method used was the Monte Carlo (MC) approach.

The story of this beginning of computer simulation is told in considerable detail by Galison (1997) in an extraordinary book which is about the evolution of particle physics and also about the evolving nature of 'experimentation'. The key figure at the beginning was John von Neumann, the Hungarian immigrant physicist whom we have already met in Chapter 1. In 1944, when the Manhattan Project at Los Alamos was still in full swing, he recognised that the hydrodynamical issues linked to the behaviour of colliding shock-waves were too complex to be treated analytically, and he worked out (in Galison's words) "an understanding of how to transform coupled differential equations into difference equations which, in turn, could be translated into language the computer could understand". The computer he used in 1944 seems to have been a punched-card device of the kind then used for business transactions. Galison spells out an example of this computational prehistory. Afterwards, within the classified domain, von Neumann had to defend his methods against scepticism that was to continue for a long time. Galison characterises what von Neumann did at this time as "carving out... a zone of what one might call *mesoscopic physics* perched precariously between the macroscopic and the microscopic".

In view of the success of von Neumann's machine-based hydrodynamics in 1944, and at about the time when the fission bomb was ready, some scientists at Los Alamos were already thinking hard about the possible design of a fusion bomb. Von Neumann invited two of them, Nicholas Metropolis and Stanley Frankel, to try to model the immensely complicated issue of how jets from a fission device might initiate thermonuclear reactions in an adjacent body of deuterium. Metropolis linked up with the renowned Los Alamos mathematician, Stanislaw Ulam, and they began to sketch what became the Monte Carlo method, in which random numbers were used to decide for any particle what its next move should be, and then to examine what proportion of those moves constitute a "success" in terms of an imposed criterion. In their first public account of their new approach, Metropolis and Ulam (1949) pointed out that they were occupying the uncharted region of mechanics between the classical mechanician (who can only handle a very few bodies together) and the statistical mechanician, for whom Avogadro's huge Number is routine. The "Monte Carlo" name came from Ulam; it is sometimes claimed that he was inspired to this by a favourite uncle who was devoted to gambling. (A passage in a recent book (Hoffmann 1998) claims that in 1946, Ulam was recovering from a serious illness and played many games of solitaire. He told his friend Vázsonyi: "After spending a lot of time trying to estimate the odds of particular card combinations by pure combinatorial calculations, I wondered whether a more practical method than abstract thinking might not be to lay the cards out say one hundred times and simply observe and count the number of successful plays... I immediately thought of problems of neutron diffusion and other questions of mathematical physics...").

What von Neumann and Metropolis first did with the new technique, as a try-out, with the help of others such as Richard Feynman, was to work out the neutron economy in a fission weapon, taking into account all the different things – absorption, scattering, fission initiation, each a function of kinetic energy and the object being collided with – that can happen to an individual neutron. Galison goes on to spell out the nature of this proto-simulation. Metropolis's innovations, in particular, were so basic that even today, people still write about using the "Metropolis algorithm".

A simple, time-honoured illustration of the operation of the Monte Carlo approach is one curious way of estimating the constant π. Imagine a circle inscribed inside a square of side a, and use a table of random numbers to determine the cartesian coordinates of many points constrained to lie anywhere at random within the square. The ratio of the number of points that lies inside the circle to the total number of points within the square $\approx \pi a^2 / 4 a^2 = \pi / 4$. The more random points have been put in place, the more accurate will be the value thus obtained. Of course, such a procedure would make no sense, since π can be obtained to any desired accuracy by the summation of a mathematical series... i.e., analytically. But once the simulator is faced with a complex series of particle movements, analytical methods quickly become impracticable and simulation, with time steps included, is literally the only possible approach. That is how computer simulation began.

Among the brilliant mathematicians who developed the minutiae of the MC method, major disputes broke out concerning basic issues, particularly the question whether any (determinate) computer-based method is in principle capable of generating an array of truly random numbers. The conclusion was that it is not, but that one can get close enough to randomness for practical purposes. This was one of the considerations which led to great hostility from some mathematicians to the whole project of computer simulation: for a classically trained pure mathematician, an approximate table of pseudo-random numbers must have seemed an abomination! The majority of theoretical physicists reacted similarly at first, and it took years for the basic idea to become acceptable to a majority of physicists. There was also a long dispute, outlined by Galison: "What *was* this Monte Carlo? How did it fit into the universally recognised division between experiment and theory – a taxonomic separation as obvious to the product designer at Dow Chemical as it was to the mathematician at Cornell?" The arguments went on for a long time, and gradually computer simulation came to be perceived as a form of experiment: thus, one of the early materials science practitioners, Beeler (1970), wrote uncompromisingly: "A computer experiment is a computational method in which physical processes are simulated according to a given set of physical mechanisms". Galison himself thinks of computer simulation as a hybrid "between the traditional epistemic poles of bench and blackboard". He goes in some detail into the search for "computational errors" introduced by finite object size, finite time steps, erroneous weighting, etc., and accordingly treats a large-scale simulation as a "numerical experiment". These arguments were about more than just semantics. Galison asserts baldly that "without computer-based simulation, the material culture of late-20th century microphysics (the subject of his book) is not simply inconvenienced – it does not exist".

Where computer simulation, and the numerical 'calculations' which flow from it, fits into the world of physics – and, by extension, of materials science – has been anxiously discussed by a number of physicists. One comment was by Herman (1984), an early contributor to the physics of semiconductors. In his memoir of early days in the field, he asserts that "during the 1950s and into the 1960s there was a sharp dichotomy between those doing formal solid-state research and those doing computational work in the field. Many physicists were strongly prejudiced against numerical studies. Considerable prestige was attached to formal theory." He goes on to point out that little progress was in fact made in understanding the band theory of solids (essential for progress in semiconductor technology) until "band theorists rolled up their sleeves and began doing realistic calculations on actual materials (by computer), and checking their results against experiment".

Recently, Langer (1999) has joined the debate. He at first sounds a distinct note of scepticism: "...the term 'numerical simulation' makes many of us uncomfortable. It is easy to build models on computers and watch what they do, but it is often unjustified to claim that we learn anything from such exercises." He continues by examining a number of actual simulations and points out, first, the value of obtaining *multiscale* information "of a kind that is not available by using ordinary experimental or theoretical techniques". Again, "we are not limited to simulating 'real' phenomena. *We can test theories by simulating idealised systems for which we know that every element has exactly the properties we think are relevant* (my emphasis)". In other words, in classical experimental fashion we can change one feature at a time, spreadsheet-fashion.

These two points made by Langer are certainly crucial. He goes on to point out that for many years, physicists looked down on instrumentation as a mere service function, but now have come to realise that the people who brought in tools such as the scanning tunnelling microscope (and won the Nobel Prize for doing so) "are playing essential roles at the core of modern physics. I hope" (he concludes) "that we'll be quicker to recognise that computational physics is emerging as an equally central part of our field". Exactly the same thing can be said about materials science and computer simulation.

Finally, in this Introduction, it is worthwhile to reproduce one of the several current definitions, in the Oxford English Dictionary, of the word 'simulate': "To imitate the conditions or behaviour of (a situation or process) by means of a model, especially for the purpose of study or training; specifically, to produce a computer model of (a process)". The Dictionary quotes this early (1958) passage from a text on high-speed data processing: "A computer can simulate a warehouse, a factory, an oil refinery, or a river system, and if due regard is paid to detail the imitation can be very exact". Clearly, in 1958 the scientific uses of computer simulation were not yet thought worthy of mention, or perhaps the authors did not know about them.

12.2. COMPUTER SIMULATION IN MATERIALS SCIENCE

In his early survey of 'computer experiments in materials science', Beeler (1970), in the book chapter already cited, divides such experiments into four categories. One is the *Monte Carlo approach*. The second is the dynamic approach (today usually named *molecular dynamics*), in which a finite system of N particles (usually atoms) is treated by setting up $3N$ equations of motion which are coupled through an assumed two-body potential, and the set of $3N$ differential equations is then solved numerically on a computer to give the space trajectories and velocities of all particles as function of successive time steps. The third is what Beeler called the *variational approach*, used to establish equilibrium configurations of atoms in (for instance) a crystal dislocation and also to establish what happens to the atoms when the defect moves; each atom is moved in turn, one at a time, in a self-consistent iterative process, until the total energy of the system is minimised. The fourth category of 'computer experiment' is what Beeler called a *pattern development calculation*, used to simulate, say, a field-ion microscope or electron microscope image of a crystal defect (on certain alternative assumptions concerning the true three-dimensional configuration) so that the simulated images can be compared with the experimental one in order to establish which is in fact the true configuration. This has by now become a widespread, routine usage. Another common use of such calculations is to generate predicted X-ray diffraction patterns or nuclear magnetic resonance plots of specific substances, for comparison with observed patterns.

Beeler defined the broad scope of computer experiments as follows: "Any conceptual model whose definition can be represented as a unique branching sequence of arithmetical and logical decision steps can be analysed in a computer experiment... The utility of the computer... springs mainly from its computational speed." But that utility goes further; as Beeler says, conventional analytical treatments of many-body aspects of materials problems run into awkward mathematical problems; computer experiments bypass these problems.

One type of computer simulation which Beeler did not include (it was only just beginning when he wrote in 1970) was finite-element simulation of fabrication and other production processes, such as for instance rolling of metals. This involves exclusively continuum aspects; 'particles', or atoms, do not play a part.

In what follows, some of these approaches will be further discussed. A very detailed and exhaustive survey of the various basic techniques and the problems that have been treated with them will be found in the first comprehensive text on "computational materials science", by Raabe (1998). Another book which covers the principal techniques in great mathematical detail and is effectively focused on materials, especially polymers, is by Frenkel and Smit (1996).

One further distinction needs to be made, that between 'modelling' and 'simulation'. Different texts favour different usages, but a fairly common practice is to use the term 'modelling' in the way offered in Raabe's book: "It describes the classical scientific method of formulating a simplified imitation of a real situation with preservation of its essential features. In other words, a model describes a part of a real system by using a *similar* but *simpler* structure." Simulation is essentially the putting of numbers into the model and deriving the numerical end-results of letting the model run on a computer. A simulation can never be better than the model on which it relies.

12.2.1 Molecular dynamics (MD) simulations

The simulation of molecular (or atomic) dynamics on a computer was invented by the physicist George Vineyard, working at Brookhaven National Laboratory in New York State. This laboratory, whose 'biography' has recently been published (Crease 1999), was set up soon after World War II by a group of American universities, initially to foster research in nuclear physics; because radiation damage (see Section 5.1.3) was an unavoidable accompaniment to the accelerator experiments carried out at Brookhaven, a solid-state group was soon established and grew rapidly. Vineyard was one of its luminaries. In 1957, Vineyard, with George Dienes, wrote an influential early book, *Radiation Damage in Solids*. (Crease comments that this book "helped to bolster the image of solid-state physics as a basic branch of physics".) In 1973, Vineyard became laboratory director.

In 1972, some autobiographical remarks by Vineyard were published at the front of the proceedings of a conference on simulation of lattice defects (Vineyard 1972). Vineyard recalls that in 1957, at a conference on chemistry and physics of metals, he explained the then current analytical theory of the damage cascade (a collision sequence originating from one very high-energy particle). During discussion, "the idea came up that a computer might be applied to follow in more detail what actually goes on in radiation damage cascades". Some insisted that this could not be done on a computer, others (such as a well-known, argumentative GE scientist, John Fisher) that it was not necessary. Fisher "insisted that the job could be done well enough by hand, and was then goaded into promising to demonstrate. He went off to his room to work; next morning he asked for a little more time, promising to send me the results soon after he got home. After two weeks... he admitted that he had given up." Vineyard then drew up a scheme with an atomic model for copper and a procedure for solving the classical equations of state. However, since he knew nothing about computers he sought help from the chief applied mathematician at Brookhaven, Milton Rose, and was delighted when Rose encouragingly replied that 'it's a great problem; this is just what computers were designed for'. One of Rose's mathematicians showed Vineyard how to program one of the early IBM computers at New York University. Other physicists joined the hunt, and it soon became clear that by keeping track of an individual atom and taking into account only near neighbours (rather than all the N atoms of the simulation), the computing load was roughly proportional to N rather than to N^2. (The initial simulation looked at 500 atoms.) The first paper appeared in the *Physical Review* in 1960. Soon after, Vineyard's team conceived the idea of making moving pictures of the results, "for a more dramatic display of what was happening". There was overwhelming demand for copies of the first film, and ever since then, the task of making huge arrays of data visualisable has been an integral part of computer simulation. Immediately following his mini-autobiography, Vineyard outlines the results of the early computer experiments: Figure 12.1 is an early set of computed trajectories in a radiation damage cascade.

One other remark of Vineyard's in 1972, made with evident feeling, is worth repeating here: "Worthwhile computer experiments require time and care. The easy understandability of the results tends to conceal the painstaking hours that went into conceiving and formulating the problem, selecting the parameters of a model,

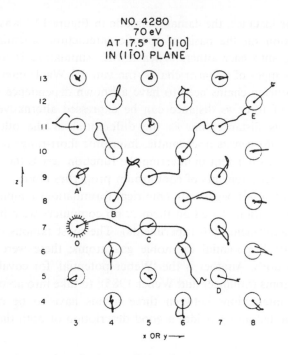

Figure 12.1. Computer trajectories in a radiation damage cascade in iron, reproduced from
Erginsoy *et al.* (1964).

programming for computation, sifting and analysing the flood of output from the
computer, rechecking the approximations and stratagems for accuracy, and out of it
all synthesising physical information". None of this has changed in the last 30 years!

Two features of such dynamic simulations need to be emphasised. One is the
limitation, set simply by the finite capacity of even the fastest and largest present-day
computers, on the number of atoms (or molecules) and the number of time-steps
which can be treated. According to Raabe (1998), the time steps used are 10^{-14}–
10^{-15} s, less than a typical atomic oscillation period, and the sample incorporates
10^3–10^9 atoms, depending on the complexity of the interactions between atoms. So,
at best, the size of the region simulated is of the order of 1 nm^3 and the time below
one nanosecond. This limitation is one reason why computer simulators are forever
striving to get access to larger and faster computers.

The other feature, which warrants its own section, is the issue of interatomic
potentials.

12.2.1.1 Interatomic potentials. All molecular dynamics simulations and some MC
simulations depend on the form of the interaction between pairs of particles (atoms

or molecules). For instance, the damage cascade in Figure 12.1 was computed by a dynamics simulation on the basis of specific interaction potentials between the atoms that bump into each other. When a MC simulation is used to map the configurational changes of polymer chains, the van der Waals interactions between atoms on neighbouring chains need to have a known dependence of attraction on distance. A plot of force vs distance can be expressed alternatively as a plot of potential energy vs distance; one is the differential of the other. Figure 12.2 (Stoneham *et al.* 1996) depicts a schematic, interionic short-range potential function showing the problems inherent in inferring the function across the significant range of distances from measurements of equilibrium properties alone.

Interatomic potentials began with empirical formulations (empirical in the sense that analytical calculations based on them... no computers were being used yet... gave reasonable agreement with experiments). The most famous of these was the Lennard-Jones (1924) potential for noble gas atoms; these were essentially van der Waals interactions. Another is the 'Weber potential' for covalent interactions between silicon atoms (Stillinger and Weber 1985); to take into account the directed covalent bonds, interactions between three atoms have to be considered. This potential is well-tested and provides a good description of both the crystalline and

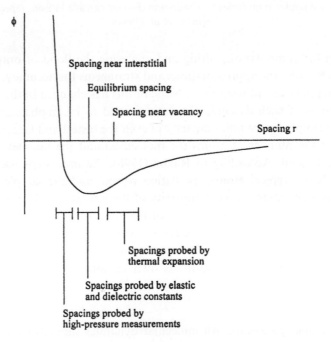

Figure 12.2. A schematic interionic short-range potential function, after Stoneham *et al.* (1996).

the amorphous forms of silicon (which have quite different properties) and of the crystalline melting temperature, as well as predicting the six-coordinated structure of liquid silicon. This kind of test is essential before a particular interatomic potential can be accepted for continued use.

In due course, attempts began to *calculate* from first principles the form of interatomic potentials for different kinds of atoms, beginning with metals. This would quickly get us into very deep quantum-mechanical waters and I cannot go into any details here, except to point out that the essence of the different approaches is to identify different simplifications, since Schrödinger's equation cannot be solved accurately for atoms of any complexity. The many different potentials in use are summarised in Raabe's book (p. 88), and also in a fine overview entitled "the virtual matter laboratory" (Gillan 1997) and in a group of specialised reviews in the *MRS Bulletin* (Voter 1996) that cover specialised methods such as the Hartree–Fock approach and the embedded-atom method. A special mention must be made of density functional theory (Hohenberg and Kohn 1964), an elegant form of simplified estimation of the electron–electron repulsions in a many-electron atom that won its senior originator, Walter Kohn, a Nobel Prize for Chemistry. The idea here is that all that an atom embedded in its surroundings 'knows' about its host is the local electron density provided by its host, and the atom is then assumed to interact with its host exactly as it would if embedded in a homogeneous electron gas which is everywhere of uniform density equal to the local value around the atom considered.

Most treatments, even when intended for materials scientists, of these competing forms of quantum-mechanical simplification are written in terms accessible only to mathematical physicists. Fortunately, a few 'translators', following in the tradition of William Hume-Rothery, have explained the essentials of the various approaches in simple terms, notably David Pettifor and Alan Cottrell (e.g., Cottrell 1998), from whom the formulation at the end of the preceding paragraph has been borrowed.

It may be that in years to come, interatomic potentials can be estimated experimentally by the use of the atomic force microscope (Section 6.2.3). A first step in this direction has been taken by Jarvis *et al.* (1996), who used a force feedback loop in an AFM to prevent sudden springback when the probing silicon tip approaches the silicon specimen. The authors claim that their method means that "force-distance spectroscopy of specific sites is possible – mechanical characterisation of the potentials of specific chemical bonds".

12.2.2 *Finite-element simulation*
In this approach, continuously varying quantities are computed, generally as a function of time as some process, such as casting or mechanical working, proceeds, by 'discretising' them in small regions, the finite elements of the title. The more

complex the mathematics of the model, the smaller the finite elements have to be. A good understanding of how this approach works can be garnered from a very thorough treatment of a single process; a recent book (Lenard *et al.* 1999) of 364 pages is devoted entirely to hot-rolling of metal sheet. The issue here is to simulate the distribution of pressure across the arc in which the sheet is in contact with the rolls, the friction between sheet and rolls, the torque needed to keep the process going, and even microstructural features such as texture (preferred orientation). The modelling begins with a famous analytical formulation of the problem by Orowan (1943), numerous refinements of this model and the canny selection of acceptable levels of simplification. The end-result allows the mechanical engineering features of the rolling-mill needed to perform a specific task to be estimated.

Finite-element simulations of a wide range of manufacturing processes for metals and polymers in particular are regularly performed. A good feeling for what this kind of simulation can do for engineering design and analysis generally, can be obtained from a popular book on supercomputing (Kaufmann and Smarr 1993).

Finite-element approaches can be supplemented by the other main methods to get comprehensive models of different aspects of a complex engineering domain. A good example of this approach is the recently established Rolls-Royce University Technology Centre at Cambridge. Here, the major manufacturing processes involved in superalloy engineering are modelled: these include welding, forging, heat-treatment, thermal spraying, machining and casting. All these processes need to be optimised for best results and to reduce material wastage. As the Centre's then director, Roger Reed, has expressed it, "if the behaviour of materials can be quantified and understood, then processes can be optimised using computer models". The Centre is to all intents and purposes a virtual factory. A recent example of the approach is a paper by Matan *et al.* (1998), in which the rates of diffusional processes in a superalloy are estimated by simulation, in order to be able to predict what heat-treatment conditions would be needed to achieve an acceptable approach to phase equilibrium at various temperatures. This kind of simulation adds to the databank of such properties as heat-transfer coefficients, friction coefficients, thermal diffusivity, etc., which are assembled by such depositories as the National Physical Laboratory in England.

12.2.3 Examples of simulations of a material
12.2.3.1 Grain boundaries in silicon.
The prolonged efforts to gain an accurate understanding of the fine structure of interfaces – surfaces, grain boundaries, interphase boundaries – have featured repeatedly in this book. Computer simulations are playing a growing part in this process of exploration. One small corner of this process is the study of the role of grain boundaries and free surfaces in the

process of melting, and this is examined in a chapter of a book (Phillpot *et al.* 1992). Computer simulation is essential in an investigation of how much a crystalline solid can be overheated without melting *in the absence of surfaces and grain boundaries* which act as catalysts for the process; such simulation can explain the asymmetry between melting (where superheating is not normally found at all) and freezing, where extensive supercooling is common. The same authors (Phillpot *et al.* 1989) began by examining the melting of imaginary crystals of silicon with or without grain boundaries and surfaces (there is no room here to examine the tricks which computer simulators use to make a model pretend that the small group of atoms being examined has no boundaries). The investigators finish up by distinguishing between mechanical melting (triggered by a phonon instability), which is homogeneous, and thermodynamic melting, which is nucleated at extended defects such as grain boundaries. The process of melting starting from such defects can be neatly simulated by molecular dynamics.

The same group (Keblinski *et al.* 1996), continuing their researches on grain boundaries, found (purely by computer simulation) a highly unexpected phenomenon. They simulated twist grain boundaries in silicon (boundaries where the neighbouring orientations differ by rotation about an axis normal to the boundary plane) and found that if they introduced an amorphous (non-crystalline) layer 0.25 nm thick into a large-angle crystalline boundary, the computed potential energy is lowered. This means that an amorphous boundary is thermodynamically stable, which takes us back to an idea tenaciously defended by Walter Rosenhain a century ago!

12.2.3.2 *Colloidal 'crystals'*.

At the end of Section 2.1.4, there is a brief account of regular, crystal-like structures formed spontaneously by two differently sized populations of hard (polymeric) spheres, typically near 0.5 nm in diameter, depositing out of a colloidal solution. Binary 'superlattices' of composition AB_2 and AB_{13} are found. Experiment has allowed 'phase diagrams' to be constructed, showing the 'crystal' structures formed for a fixed radius ratio of the two populations but for variable volume fractions in solution of the two populations, and a computer simulation (Eldridge *et al.* 1995) has been used to examine how nearly theory and experiment match up. The agreement is not bad, but there are some unexpected differences from which lessons were learned.

The importance of these pseudo-crystals is that their periodicities are similar to those of visible light and they can thus be used like semiconductors in acting on light beams in optoelectronic devices.

12.2.3.3 *Grain growth and other microstructural changes*.

When a deformed metal is heated, it will *recrystallise*, that is to say, a new population of crystal grains will replace the deformed population, driven by the drop in free energy occasioned by the removal of dislocations and vacancies. When that process is complete but heating is continued then, as we have seen in Section 9.4.1, the mean size of the new grains gradually grows, by the progressive removal of some of them. This process, *grain growth*, is driven by the disappearance of the energy of those grain boundaries that vanish when some grains are absorbed by their neighbours. In industrial terms, grain growth is much less important than recrystallisation, but it has attracted a huge amount of attention by computer modellers during the past few decades, reported in literally hundreds of papers. *This is because the phenomenon offers an admirable testbed for the relative merits of different computational approaches.*

There are a number of variables: the specific grain-boundary energy varies with misorientation if that is fairly small; if the grain-size *distribution* is broad, and if a subpopulation of grains has a pronounced preferred orientation, a few grains grow very much larger than others. (We have seen, Section 9.4.1, that this phenomenon interferes drastically with sintering of ceramics to 100% density.) The metal may contain a population of tiny particles which seize hold of a passing grain boundary and inhibit its migration; the macroscopic effect depends upon both the mean size of the particles and their volume fraction. All this was quantitatively discussed properly for the first time in a classic paper by Smith (1948). On top of these variables, there is also the different grain growth behaviour of thin metallic films, where the surface energy of the metal plays a key part; this process is important in connection with failure of conducting interconnects in microcircuits.

There is no space here to go into the great variety of computer models, both two-dimensional and three-dimensional, that have been promulgated. Many of them are statistical 'mean-field' models in which an average grain is considered, others are 'deterministic' models in which the growth or shrinkage of every grain is taken into account in sequence. Many models depend on the Monte Carlo approach. One issue which has been raised is whether the simulation of grain size *distributions* and their comparison with experiment (using stereology, see Section 5.1.2.3) can be properly used to prove or disprove a particular modelling approach. One of the most disputed aspects is the modelling of the limiting grain size which results from the pinning of grain boundaries by small particles.

The merits and demerits of the many computer-simulation approaches to grain growth are critically analysed in a book chapter by Humphreys and Hatherly (1995), and the reader is referred to this to gain an appreciation of how alternative modelling strategies can be compared and evaluated. A still more recent and very clear critical comparison of the various modelling approaches is by Miodownik (2001).

Grain growth involves no phase transformation, but a number of such transformations have been modelled and simulated in recent years. A recently published overview volume relates some experimental observations of phase transformations to simulation (Turchi and Gonis 2000). Among the papers here is one describing some very pretty electron microscopy of an order–disorder transformation by a French group, linked to simulation done in cooperation with an eminent Russian-emigré expert on such transformations, Armen Khachaturyan (Le Bouar *et al.* 2000). Figure 12.3 shows a series of micrographs of progressive transformation, in a Co–Pt alloy which have long been studied by the French group, together with corresponding simulated patterns. The transformation pattern here, called a 'chessboard pattern', is brought about by internal stresses: a cubic crystal structure (disordered) becomes tetragonal on ordering, and in different domains the unique fourfold axis of the tetragonal form is constrained to lie in orthogonal directions, to accommodate the stresses. The close agreement indicates that the model is close to physical reality… which is always the objective of such modelling and simulation.

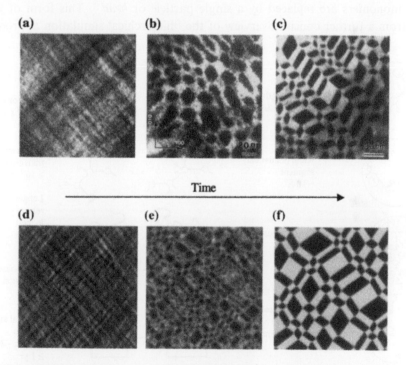

Figure 12.3. Comparison between experimental observations (a–c) and simulation predictions (d–f) of the microstructural development of a 'chessboard' pattern forming in a $Co_{39.5}Pt_{60.5}$ alloy slowly cooled from 1023 K to (a) 963 K, (b) 923 K and (c) 873 K. The last of these was maintained at 873 K to allow the chessboard pattern time to perfect itself (Le Bouar *et al.* 2000) (courtesy Y. Le Bouar).

12.2.3.4 Computer-modelling of polymers.

The properties of polymers are determined by a large range of variables – chemical constitution, mean molecular weight and molecular weight distribution, fractional crystallinity, preferred orientation of amorphous regions, cross-linking, chain entanglement. It is thus no wonder that computer simulation, which can examine all these features to a greater or lesser extent, has found a special welcome among polymer scientists.

The length and time scales that are relevant to polymer structure and properties are shown schematically in Figure 12.4. Bearing in mind the spatial and temporal limitations of MD methods, it is clear that a range of approaches is needed, including quantum-mechanical 'high-resolution' methods. In particular, configurations of long-chain molecules and consequences such as rubberlike elasticity depend heavily on MC methods, which can be invoked with "algorithms designed to allow a correspondence between number of moves and elapsed time" (from a review by Theodorou 1994). A further simplification that allows space and time limitations to weigh less heavily is the use of *coarse-graining*, in which "explicit atoms in one or several monomers are replaced by a single particle or *bead*". This form of words comes from a further concise overview of the "hierarchical simulation approach to

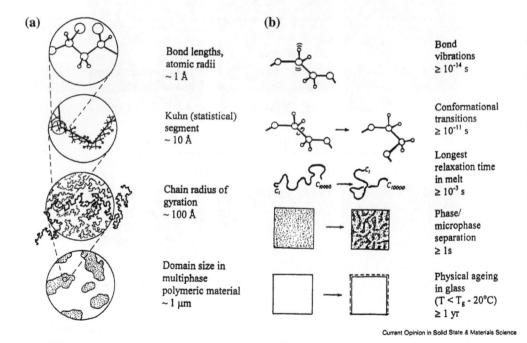

Current Opinion in Solid State & Materials Science

Figure 12.4. Hierarchy of length scales of structure and time scales of motion in polymers. T_g denotes the glass transition temperature. After Uhlherr and Theodorou (1998) (courtesy Elsevier Science).

structure and dynamics of polymers" by Uhlherr and Theodorou (1998); Figure 12.4 also comes from this overview. Not only structure and properties (including time-dependent ones such as viscosity) of polymers, but also configurations and phase separations of block copolymers, and the kinetics of polymerisation reactions, can be modelled by MC approaches. One issue which has recently received a good deal of attention is the configuration of block copolymers with hydrophobic and hydrophilic ends, where one constituent is a therapeutic drug which needs to be delivered progressively; the hydrophobically ended drug moiety finishes up inside a spherical micelle, protected by the hydrophilically ended outer moiety. Simulation allows the tendency to form micelles and the rate at which the drug is released within the body to be estimated.

The voluminous experimental information about the linkage between structural variables and properties of polymers is assembled in books, notably that by van Krevelen (1990). In effect, such books "encapsulate much empirical knowledge on how to formulate polymers for specific applications" (Uhlherr and Theodorou 1998). What polymer modellers and simulators strive to achieve is to establish more rigorous links between structural variables and properties, to foster more rational design of polymers in future.

A number of computer modelling codes, including an important one named 'Cerius 2', have by degrees become commercialised, and are used in a wide range of industrial simulation tasks. This particular code, originally developed in the Materials Science Department in Cambridge in the early 1980s, has formed the basis of a software company and has survived (with changes of name) successive takeovers. The current company name is Molecular Simulations Inc. and it provides codes for many chemical applications, polymeric ones in particular; its latest offering has the ambitious name "Materials Studio". It can be argued that the ability to survive a series of takeovers and mergers provides an excellent filter to test the utility of a published computer code.

Some special software has been created for particular needs, for instance, lattice models in which, in effect, polymer chains are constrained to lie within particular cells of an imaginary three-dimensional lattice. Such models have been applied to model the spatial distribution of preferred vectors ('directors') in liquid-crystalline polymers (e.g., Hobdell *et al.* 1996) and also to study the process of solid-state welding between polymers. In this last simulation, a 'bead' on a polymer chain can move by occupying an adjacent vacancy and in this way diffusion, in polymers usually referred to as 'reptation', can be modelled; energies associated with different angles between adjacent bonds must be estimated. When two polymer surfaces inter-reptate, a stage is reached when chains wriggle out from one surface and into the contacting surface until the chain midpoints, on average, are at the interface (Figure 12.5). At that stage, adhesion has reached a maximum. Simulation has shown that

(a)

(b)

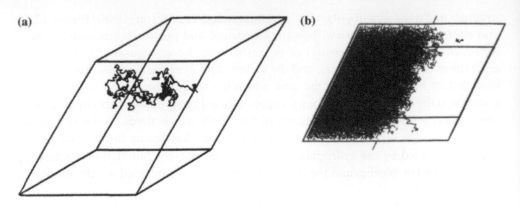

Figure 12.5. (a) Lattice model showing a polymer chain of 200 'beads', originally in a random configuration, after 10,000 Monte Carlo steps. The full model has 90% of lattice sites occupied by chains and 10% vacant. (b) Half of a lattice model containing two similar chain populations placed in contact. The left-hand side population is shown after 50,0000 Monte Carlo steps; the short lines show the location of the original polymer interface (courtesy K. Anderson).

the number of effective 'crossovers' achieved in a given time varies as the square root of the mean molecular weight, and has also allowed strength to be predicted as a function of molecular weight, temperature and time (K. Anderson and A.H. Windle, private communication). The procedure for the MC model involved in this simulation is described by Haire and Windle (2001).

Another family of simulations concerns polymeric fibres, which constitute a major industrial sector. Y. Termonia has spent many years at Du Pont modelling the extrusion and tensile drawing of fibres from liquid solution. A recent publication (Termonia 2000) uses an extreme form of coarse-graining to simulate the process; the model starts with an idealised oblique set of mutually orthogonal straight chains with regular entanglement points. This system is deformed at constant temperature and strain rate, by a succession of minute length increments, moving the top row of entanglement points in the draw direction while leaving the bottom row unmoved. The positions of the remaining entanglement points are then readjusted by an iterative relaxation procedure which minimises the net residual force acting on each site. After each strain increment and complete relaxation, each site is 'visited' by an MC lottery (in the author's words), and four processes are allowed to occur, breakage of interchain bonds, slippage of chains through entanglements, chain breakage and final network relaxation. Then the cycle restarts. All this is done for chains of different lengths, i.e., different molecular weights. The model shows how low molecular weights lead to poor drawability and premature fracture, in accord with experiment. Termonia's extrapolation of this simulation, in the same book, to ultrastrong spider web fibres, where molecular weight *distribution* plays a major part in determining mechanical properties, shows the power of the method.

12.2.3.5 Simulation of plastic deformation. The modelling of plastic deformation in metals presents in stark form the problems of modelling on different length scales. Until recently, dislocations have either been treated as flexible line defects without consideration of their atomic structure, or else the Peierls–Nabarro force tying a dislocation to its lattice has been modelled in terms of a relatively small number of atoms surrounding the core of a dislocation cross-section. There is also a further range of issues which has exercised a distinct subculture of modellers, attempting to predict the behaviour of a polycrystal from an empirical knowledge of the behaviour of single crystals of the same substance. This last is a huge subject (see Section 2.1.6), and I can do no more here than to cite a concise coverage of the issues in Chapter 17 of Raabe's (1998) book and also to refer to *the* definitive treatment in detail, representing many years of work (Kocks *et al.* 1998).

Very recently, the modelling of plastic deformation in terms of the motion, interaction and mutual blocking of dislocations moving under applied stress has entered the mesoscale. Two papers have applied MD methods to this task; instead of treating dislocations as semi-macroscopic objects, the motion of up to 100 million individual atoms around the interacting dislocation lines has been modelled; this is feasible since the timescale for such a simulation can be quite brief. One paper is by Bulatov *et al.* (1998), the other by Zhou *et al.* (1998); each has received an illuminating discussion in the same issues of their respective journals. These two studies follow a first attempt by L.P. Kubin and others in 1992. As P. Gumbsch points out in his discussion of the Zhou paper, these atomistic computations generate such a huge amount of information (some 10^4 configurations of 10^6 atoms each) that "one of the most important steps is to discard most of it, namely, all the atomistic information not directly connected to the *cores* of the dislocations. What is left is a physical picture of the atomic configurations in such a dislocation intersection and even some quantitative information about the stresses required to break the junction". This kind of information overload is a growing problem in modern super simulation, and knowing what to discard, as well as turning pages of numbers into readily assimilable visual displays, are central parts of the simulator's skill.

Baskes (1999) has discussed the "status role" of this kind of modelling and simulation, citing many very recent studies. He concludes that "modelling and simulation of materials at the atomistic, microstructural and continuum levels continue to show progress, but prediction of mechanical properties of engineering materials is still a vision of the future". Simulation cannot (yet) do everything, in spite of the optimistic claims of some of its proponents.

This kind of simulation requires massive computer power, and much of it is done on so-called 'supercomputers'. This is a reason why much recent research of this kind has been done at Los Alamos. In a survey of research in the American national laboratories, the then director of the Los Alamos laboratory, Siegfried Hecker (1990) explains that the laboratory "has worked closely with all supercomputer vendors over the years, typically receiving the serial No. 1 machine for each successive model". He goes on to exemplify the kinds of problems in materials science that these extremely powerful machines can handle.

12.3. SIMULATIONS BASED ON CHEMICAL THERMODYNAMICS

As we have repeatedly seen in this chapter, proponents of computer simulation in materials science had a good deal of scepticism to overcome, from physicists in particular, in the early days. A striking example of sustained scepticism overcome, at length, by a resolute champion is to be found in the history of CALPHAD, an acronym denoting CALculation of PHAse Diagrams. The decisive champion was an American metallurgist, Larry Kaufman.

The early story of experimentally determined phase diagrams and of their understanding in terms of Gibbs free energies and of the Phase Rule was set out in Chapter 3, Section 3.1.2. In that same chapter, Hume-Rothery's rationalisation of certain features of phase diagrams in terms of atomic size ratios and electron/atom ratios was outlined (Section 3.3.1.1). Hume-Rothery did use his theories to predict limited features of phase diagrams, solubility limits in particular, but it was a long time before experimentally derived values of free energies of phases began to be used for the prediction of phase diagrams. Some tentative early efforts in that direction were published by a Dutchman, van Laar (1908), but thereafter there was a void for half a century. The challenge was taken up by another Dutchman, Meijering (1957) who seems to have been the first to attempt the calculation of a complete ternary phase diagram (Ni–Cr–Cu) from measured thermochemical quantities from which, in turn, free energies could be estimated. Meijering's work was particularly important in that he recognised that for his calculation to have any claims to accuracy he needed to estimate a value for the free energy (as a function of temperature) of face-centred cubic chromium, a notional crystal structure which was not directly accessible to experiment. This was probably the first calculation of the lattice stability of a potential (as distinct from an actual) phase.

At the time Meijering published his research, Larry Kaufman was working for his doctorate at MIT with a charismatic steel metallurgist, Professor Morris Cohen, and they undertook some simple equilibrium calculations directed at practical problems of the steel industry. From the end of the 1950s, Kaufman directed his efforts at developing these methods, sustained by two other groups, one in Stockholm, Sweden, run from 1961 by Mats Hillert and another in Sendai, Japan, led by T. Nishizawa. Hillert and Kaufman had studied together at MIT and their interaction over the years was to be crucial to the development of CALPHAD.

At a meeting at Battelle Memorial Institute in Ohio in 1967, Kaufman demonstrated some approximate calculations of binary phase diagrams using an ideal solution model, but met opposition particularly from solid-state physicists who preferred to use first-principles calculations of electronic band structures instead of thermodynamic inputs. For some years, this became the key battle between competing approaches. At about this time, Kaufman began exchanging letters with William Hume-Rothery concerning the best way to represent thermodynamic equilibria. Thereupon, Hume-Rothery, in his capacity as editor of *Progress in Materials Science*, invited Kaufman to write a review about lattice stabilities of phases, which appeared (Kaufman 1969) shortly after Hume-Rothery's death in 1968. Shortly before his death, Hume-Rothery had written to Kaufman to say that he was "not unsympathetic to any theory which promises reasonably accurate calculations of phase boundaries, and saves the immense amount of work which their experimental determination involves", but that he was still sceptical about Kaufman's approach. This extract comes from a full account of the history of CALPHAD in Chapter 2 of a recent book (Saunders and Miodownik 1998). In his short overview, Kaufman took great trouble to counter Hume-Rothery's reservations, and he also gave a fair account of the competing band-theoretical approaches. The imperative need to account for the competition in stability between alternative phases, actual and potential, was central to Kaufman's case.

For the many ensuing stages in CALPHAD's history, including the incorporation of CALPHAD Inc. in 1973, and the practice of organising meetings at which different approaches to formulating Gibbs energies could be reconciled, Saunders and Miodownik's history chapter must be consulted. Effective international cooperation got under way in 1973, with involvement of a number of national laboratories, and numerous published computer codes from around the world such as Thermocalc and Chemsage are now in regular use. From 1976 on, physicists were encouraged to attend CALPHAD meetings in order to assess the feasibility of merging data obtained by thermochemistry with those calculated by first-principles electron-theory methods (Pettifor 1977). CALPHAD's own journal began publication in 1977. Kaufman is still, today, a key participant at the regular CALPHAD meetings.

Kaufman and Bernstein (1970) brought out the first book on phase diagram calculations; there was not another comprehensive treatment till Saunders and Miodownik's book came out in 1998. This last covers the ways of obtaining the thermodynamic input data, ways of dealing with complications such as atomic

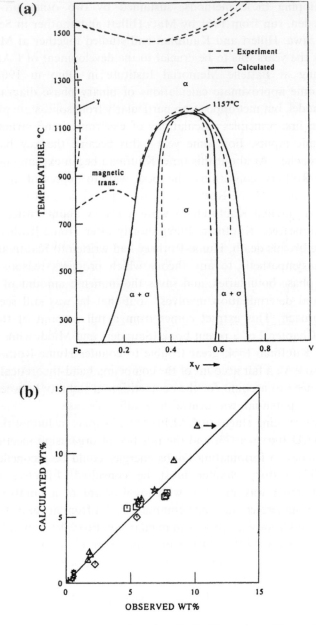

Figure 12.6. (a) Calculated $(\alpha + \sigma)$ phase boundary for Fe–V together with experimental boundaries (After Spencer and Putland 1973). (b) Comparison between calculated and experimental values of the concentration of Al, V and Fe in the two phases in Ti–6Al–4V alloy (after Saunders and Miodownik 1998).

ordering and ferromagnetism, and (in particular) many forms of application of CALPHAD. It also describes the crucial methods of critically *optimising* thermodynamic data for incorporation in internationally agreed databases. Figure 12.6(a) shows an early example of a simple binary phase diagram obtained by experiment together with calculated phase boundaries. Figure 12.6(b) shows a plot of observed versus calculated solute concentrations in a standard titanium alloy. It has been an essential part of the gradual acceptance of CALPHAD methods that theory and experiment have been shown to agree well, and progressively better so as methods improved. Indeed, *in all computer modelling and simulation, confidence can only come gradually from a steady series of such comparisons between simulation and experiment*.

Even when the CALPHAD approach had been widely accepted as valid, there was still the problem of the reluctance of newcomers to start using it. Hillert (1980) proposed looking at thermodynamics as a game and to consider how one can learn to play that game well. He went on: "For inspiration, one may first look at another game, the game of chess. The rules are very simple to learn, but it was always very difficult to become a good player. The situation has now changed due to the application of computers. Today there are programmes for playing chess which can beat almost any expert. It seems reasonable to expect that it should also be possible to write programmes for 'playing thermodynamics', programmes which should be almost as good as the very best thermodynamic expert." In other words... for novices, tried and tested commercial software, whether for thermodynamics or for MC or MD, should take the sting out of taking the plunge into computer simulation, and perhaps in the fullness of time such simulations will move out of specialised journals and coteries of specialists and find their way increasingly into mainline journals, and students with first degrees in materials science will be entirely at ease with this kind of activity. Perhaps we are already there.

Today, thermodynamic simulation has broadened out far beyond the calculation of binary and ternary (and even quaternary) phase diagrams. For instance, as explained in the last chapter of Saunders and Miodownik's book, methods have recently been developed to combine diffusional simulations with phase stability simulations in order to obtain estimates of the kinetics of phase transformation. A recent text issued by SGTE, the Scientific Group Thermodata Europe (Hack 1996) includes 24 short chapters instancing applications, in particular, to processing issues. Two examples from Sweden, relating to solidification, include the calculation of solidification paths for a multicomponent system and (severely practical) calculations directed towards the prevention of clogging by premature freezing in a continuous casting process. Another chapter discusses the formulation of a Co–Fe–Ni binder phase for use with a dispersion of tungsten carbide 'hard metal'. A chapter by Per Gustafson in Sweden hinges on computer development of a high-speed (cutting) steel. This last application is reminiscent of a protracted programme of research at Northwestern University (where materials science started in 1958) by Gregory Olson (e.g. Kuehmann and Olson 1998) to design steels for specially demanding purposes by a sophisticated computer-optimisation program, including extensive use of CALPHAD; some further remarks about this program can be found in Section 4.3.

Very recently, a very detailed report from two groups attending a 1997 meeting on 'Applications of Computational Thermodynamics' has been published (Kattner and Spencer 2000) with presentations of many applications to practical problems, with emphasis on processing methods, including processing of semiconductors and microcircuits. One process modelled here is the deposition of a compound semiconductor from an organometallic precursor, a 'soft chemistry' approach discussed in the preceding chapter.

The CALPHAD approach has been treated here at some length because its history illustrates the strengths and limitations of computer modelling and simulation. The strengths clearly outweigh the limitations, and this is becoming increasingly true throughout the broad spectrum of applications of computers in materials science and engineering.

REFERENCES

Baskes, M.I. (1999) *Current Opinion in Solid State and Mater Sci.* **4** 273.
Beeler, Jr., J.R. (1970) *The role of computer experiments in materials research*, in *Advances in Materials Research, Vol. 4*, ed. Herman, H. (Interscience, New York) p. 295.
Bulatov, V. *et al.* (1998) *Nature* **391**, 669 (see also p. 637).
Cottrell, A (1998) *Concepts in the Electron Theory of Alloys* (IOM Communications, London).
Crease, R.P. (1999) *Making Physics: A Biography of Brookhaven National Laboratory* (University of Chicago Press, Chicago).
Eldridge, M.D., Madden, P.A., Pusey, P.N. and Bartlett, P. (1995) *Molecular Physics* **84**, 395.
Erginsoy, C., Vineyard, G.H. and Engler, A. (1964) *Phys. Rev. A* **133**, 595.
Frenkel, D. and Smit, B. (1996) *Understanding Molecular Simulations: From Algorithms to Applications* (Academic Press, San Diego).
Galison, P. (1997) *Image and Logic: A Material Culture of Microphysics*, Chapter 8 University of Chicago Press, Chicago.
Gillan, M.J. (1997) *Contemp. Phys.* **38**, 115.
Hack, K. (editor) (1996) *The SGTE Casebook: Thermodynamics at Work* (The Institute of Materials, London).
Haire, K.R. and Windle, A.H. (2001) *Computational and Theoretical Polymer Science* **11**, 227.
Hecker, S.S. (1990) *Metall. Trans. A* **21A** 2617.

Herman, F. (1984, June) *Phys. Today* 56.

Hillert, M. (1980) in *Conference on the Industrial Use of Thermochemical Data*, ed. Barry, T. (Chemical Society, London) p. 1.

Hobdell, J.R., Lavine, M.S. and Windle, A.H. (1996) *J. Computer-Aided Mater. Design* **3**, 368.

Hoffman, P. (1998) *The Man Who Loved Only Numbers* (Fourth Estate, London) p. 238.

Hohenberg, P. and Kohn, W. (1964) *Phys. Rev. B* **136**, 864.

Humphreys, F.J. and Hatherly, M. (1995) *Recrystallization and Related Annealing Phenomena*, Chapter 9 (Pergamon, Oxford).

Jarvis, S.P., Yamada, H., Yamamoto, S.-I., Tokumoto, H. and Pethica, J.B. (1996) *Nature* **384**, 247.

Kattner, U.R. and Spencer, P.J. (eds.) (2000) *Calphad* **24**, 55.

Kaufman, L. (1969) *Progr. Mat. Sci.* **14**, 55.

Kaufman, L. and Bernstein, H. (1970) *Computer Calculations of Phase Diagrams* (Academic Press, New York).

Kaufmann III, W.J. and Smarr, L.L. (1993) *Supercomputing and the Transformation of Science*, Chapter 6 (Scientific American Library, New York).

Keblinski, P., Wolf, D., Phillpot, S.R. and Gleiter, H. (1996) *Phys. Rev. Lett.* **77**, 2965.

Kocks, U.F., Tomé, C.N. and Wenk, H.-R. (eds.) (1998) *Texture and Anisotropy: Preferred Orientations in Polycrystals and their Effect on Materials Properties* (Cambridge University Press, Cambridge).

Kuehmann, C.J. and Olson, G.B. (1998/5) Gear steels designed by computer, in *Advanced Materials and Processes*, p. 40.

Langer, J. (1999) *Phys. Today* (July), 11.

Le Bouar, Y., Loiseau, A. and Khachaturyan, A. (2000) in (2000) *Phase Transformations and Evolution in Materials*, eds. Turchi, P.E.A. and Gonis, A. (The Minerals, Metals and Materials Society, Warrendale, PA) p. 55.

Lenard, J.G., Pietrzyk, M. and Cser, L. (1999) *Mathematical and Physical Simulation of the Properties of Hot-Rolled Products* (Elsevier, Amsterdam).

Lennard-Jones, J.E. (1924) *Proc. Roy. Soc. (Lond.) A* **106**, 463.

Matan, N. *et al.* (1998) *Acta Mater.* **46**, 4587.

Meijering, J.L. (1957) *Acta Met.* **5**, 257.

Metropolis, N. and Ulam, S. (1949) Monte Carlo method, *Amer. Statist. Assoc.* **44**, 335.

Miodownik, M.A. (2001) Article on *Normal Grain Growth*, in *Encyclopedia of Materials* (Pergamon, Oxford).

Orowan, E. (1943) *Proc. Inst. Mech. Eng.* **150**, 140.

Pettifor, D.G. (1977) *Calphad* **1**, 305.

Phillpot, S.R., Yip, S. and Wolf, D. (1989) *Comput. Phys.* **3**, 20.

Phillpot, S.R., Yip, S., Okamoto, P.R. and Wolf, D. (1992) Role of interfaces in melting and solid-state amorphization, in *Materials Interfaces: Atomic-Level Structure and Properties*, ed. Wolf, D. and Yip, S. (Chapman and Hall, London) p. 228.

Raabe, D. (1998) *Computational Materials Science* (Wiley-VCH, Weinheim).

Saunders, N. and Miodownik, A.P. (1998) *CALPHAD: Calculation of Phase Diagrams – A Comprehensive Guide* (Pergamon, Oxford).

Smith, C.S. (1948) *Trans. Metall. Soc. AIME* **175**, 15.

Spencer, P.J. and Putland, F.H. (1973) *J. Iron and Steel Inst.* **211**, 293.

Stillinger, F.H. and Weber, T.A. (1985) *Phys. Rev. B* **31**, 5262.

Stoneham, M., Harding, J. and Harker, T. (1996) Interatomic potentials for atomistic simulations, *MRS Bull.* **21**(2), 29.

Termonia, Y. (2000) Computer model for the mechanical properties of synthetic and biological polymer fibres, in *Structural Biological Materials*, ed. Elices, M. (Pergamon, Oxford) p. 269.

Theodorou, D.N. (1994) Polymer structure and properties: modelling, in *Encyclopedia of Advanced Materials*, vol. 3, ed. Bloor, D. *et al.* (Pergamon Press, Oxford) p. 2052.

Turchi, P.E.A. and Gonis, A. (editors) (2000) *Phase Transformations and Evolution in Materials* (The Minerals Metals and Materials Society, Warrendale, PA).

Uhlherr, A. and Theodorou, D.N. (1998) *Current Opinion in Solid State and Mater Sci.* **3**, 544.

Van Krevelen, D.W. (1990) *Properties of Polymers*, 3rd edition (Elsevier, Amsterdam).

Van Laar, J.J. (1908) *Z. Phys. Chem.* **63**, 216; **64**, 257.

Vineyard, G.H. (1972), in *Interatomic Potentials and Simulation of Lattice Defects*, ed. Gehlen, P.C., Beeler, J.R. Jr. and Jaffee, R.I. (Plenum Press, New York) pp. xiii, 3.

Voter, A.F. (editor) (1996) *Interatomic potentials for atomistic simulations*, *MRS Bull.* **21**(2), (February) 17.

Zhou, S.J., Preston, D.L., Londahl, P.S. and Beazley, D.M. (1998) *Science* **279**, 1525. (see also p. 1489).

Chapter 13
The Management of Data

Chapter 13
The Management of Data

13.1. THE NATURE OF THE PROBLEM

As I write this, one of my grandchildren has just asked me which is the 'heaviest' metal. I did not have any listing available at home, and I guessed uranium or tungsten. My grandson Daniel preferred to believe that gold was the heaviest, i.e., densest. Now, sitting in my office, I cast about for a convenient listing of densities, and chose the *Metals Reference Book*, fourth edition, 1967, volume 3. It turns out that neither my grandson nor I were quite right: gold and tungsten have virtually the same density, uranium is marginally less dense, but several noble metals (platinum, osmium, and others) are even denser. Of course, I had to skim the whole column listing densities of pure metals and look for the highest value; with a suitable computerised listing, I could have asked the computer "Which of these elements is densest?" – if a computerised way of putting such a question is available – and (perhaps) got an answer in the blink of an eye. All this is a very minor example of the problems of data retrieval.

Of course, densities of pure metals do not change over time, though the available precision may improve a little over the decades. However, many more complex data do change significantly as new experiments are done, and new materials or material systems come along constantly and so entirely new data flood the literature. Materials scientists, like chemists, physicists and engineers, need means of finding these. Those means are called *databases*. This brief chapter surveys how databases are assembled and used, with special attention to materials.

13.2. CATEGORIES OF DATABASE

13.2.1 *Landolt–Börnstein, the* International Critical Tables *and their successors*
Initially, natural philosophers communicated by letter, and in this way measurements of physical and chemical quantities were slowly spread. Then scientific journals began to develop, slowly at first; 200 years ago, there were some 50 of these in the world; data were then spread through journals, for instance, the *Philosophical Transactions of the Royal Society of London*. Attempts to gather scattered data in lists began in earnest in 1883, when Hans Heinrich Landolt and Richard Börnstein in Germany published the first volume of their *Physikalische-Chemische Tabellen*, running to 261 pages. This was well received, and up to 1950, 25 further volumes

491

appeared with broad titles such as *Crystals, Fusion Equilibria and Interfacial Phenomena, Optical Constants*. The 26 volumes came out in six successive editions until 1969, whereupon the publisher, Springer, decided to change to a more flexible form and started the *New Series*, devoted to a great variety of specialised themes. 129 volumes and subvolumes are listed in the comprehensive 1996 catalogue, with several more planned, and in 2000 all of these were available on the internet, and access was offered free of charge until the entire *New Series*, some 140,000 pages, is on line. Many of the 129 volumes were a long way from relevance to materials – for instance, some are devoted to astronomy – but quite a number are directly related to MSE. This series is unique in its longevity and consistency.

By way of example, Volume 26 in Group III (Crystal and Solid State Physics) is devoted to *Diffusion in Solid Metals and Alloys*; this volume has an editor and 14 contributors. Their task was not only to gather numerical data on such matters as self- and chemical diffusivities, pressure dependence of diffusivities, diffusion along dislocations, surface diffusion, but also to exercise their professional judgment as to the reliability of the various numerical values available. The whole volume of about 750 pages is introduced by a chapter describing diffusion mechanisms and methods of measuring diffusivities; this kind of introduction is a special feature of "Landolt–Börnstein". Subsequent developments in diffusion data can then be found in a specialised journal, *Defect and Diffusion Forum*, which is not connected with Landolt–Börnstein.

Other early tabulations of numerical data were the French *Tables Annuelles de Constantes et Données Numériques* which appeared for some decades after 1920, and the British *Tables of Physical and Chemical Constants*, masterminded by the National Physical Laboratory and known affectionately as "Kaye and Laby" after the editors, which appeared annually in single volume form from 1911 to 1966. These last two, like Landolt–Börnstein, appeared regularly, in successive editions.

Something rather different was the set of 7 volumes of the *International Critical Tables* masterminded by the International Union of Pure and Applied Physics, edited by Edward Washburn, and given the blessing of the International Research Council (the predecessor of the International Council of Scientific Unions, ICSU). This appeared in stages, 1926–1933, once only; when Washburn died in 1934, "the work died with him". This last quotation comes from a lively survey of the history of ICSU (Greenaway 1996); this book has an entire chapter devoted to "Data, and Scientific Information".

It was not until the mid-1960s that Harrison Brown (later ICSU President) called attention to the absence of any successor to the *International Critical Tables*, and was asked by ICSU to make recommendations. This led to ICSU's creation of *CODATA*, following on from ICSU's earlier World Data Centers, devoted to specific sciences such as metereology. This body is more of a gadfly and organiser than publisher of databases, and for example by 1969 it had published an "International Compendium of Numerical Data Projects", and set up a task group on computer usage. *CODATA* became closely involved with the National Academy of Sciences in Washington, DC. In 1984, ICSU created another body, the International Council for Scientific and Technical Information, ICSTI, to be devoted to getting databases to the right user. This is, and has long been, a central problem, because there are now so many databases scattered around the world and most materials scientists know only a very few of them.

The *Metals Reference Book*, mentioned at the beginning of this chapter, published in Britain in successive editions from 1949, and edited by a metallurgist, C.J. Smithells, is a good example of a specialised database kept up to date by periodic new editions. It eventually extended to well over 1000 pages. A somewhat similar compilation focused on polymers is D.W. van Krevelen's book, *Properties of Polymers*, already mentioned in Chapter 8. This more discursive book is now in its third edition, 1990. An example of an even more specialised database is a book by R. Hultgren and 4 others, *Selected Values of the Thermodynamic Properties of the Elements* (American Society for Metals, 1973), which was of great importance, inter alia, for the early efforts in calculation of phase diagrams (see Section 12.3). A more recent compilation of measurements of high-temperature thermodynamic quantities for alloys, measured by high-temperature calorimetry and also valuable for the calculation of phase diagrams, is by Kleppa (1994).

For many materials scientists *the* database for which they automatically reach when a problem arises like the one with which I opened this chapter is the *Handbook of Chemistry and Physics*, now in its 81st edition, with over 2500 pages of densely packed information. This Handbook was first published in 1914 (a few years were missed because of wars), at the instigation of Arthur Friedman, a mechanical engineer and entrepreneur; one of his companies was the Chemical Rubber Company, CRC, in Cleveland, Ohio, which supplied laboratory items in rubber. The CRC published the Handbook from the start, and still does... hence the Handbook's nickname, *The Rubber Bible*. In the early years, Friedman used the *Handbook* as a promotional device for the sale of such items as rubber stoppers.

Information about this splendid compilation came to me from a chemist, Robert Weast (1985), who was editor from 1952 until 1988... 37 years! He also informed me that the creation (jointly by the American Chemical Society and the American Institute of Physics) of the *Journal of Physical and Chemical Reference Data*, which began publication in 1972, was encouraged by the results of a survey which indicated how widely the 'Rubber Bible' was used. Weast describes this journal as "a truly outstanding source of critically evaluated data". In saying this, he underlined the crucial role of editors' and contributors' critical judgment in selecting data for such compilations. David Lide, the editor of the journal, in 1989 succeeded Robert

Weast as editor of the *Rubber Bible*. Although the *Rubber Bible* is not primarily addressed to materials scientists, yet it has proved of great utility for them.

Database construction has now become sufficiently widespread that the ASTM (the American Society for Testing and Materials... a standards organisation) has issued a manual on the building of databases (ASTM 1993); it incorporates advice on computer practice.

An interesting question is what motivates researchers to choose a particular substance for precise measurement of some physical, chemical or mechanical characteristic. I consulted a well-known physicist, Guy White, who works for the Division of Applied Physics of CSIRO in Australia (White 1991). He is concerned with thermophysical measurements, thermal conductivity and thermal expansion in particular. He told me that his choice of materials for thermophysical measurements "were probably dictated by a combination of curiosity, availability and 'simplicity' plus, when opportunity offered, the benefit of a chat with an interested theorist". For instance, the availability of large crystals of certain substances from a British firm prompted their use for thermal expansion measurements. It went further than that, indeed. As White pointed out, "many dilatometers in common use have significant systematic errors in determining linear thermal expansion as evidenced by round-robin measurements on reproducible materials". So high-quality reference materials are needed to check and calibrate precision dilatometers; substances like oxygen-free copper and semiconductor-grade silicon were used for that purpose. Similar round-robin measurements of the lattice parameter of semiconductor-grade silicon powder were used many years ago to test the reliability of different X-ray diffraction instruments and to compare the accuracy of photographic and direct measurement methods. This kind of procedure also picks out the most conscientious operators.

13.2.2 Crystal structures

Crystal structure determination began, as we saw in Chapter 3, in 1912, and was initially rather slow to get under way. By 1929, however, enough crystal structures had been determined to stimulate the creation of a specialist journal, *Strukturbericht*, which continued after the War and until the mid-1980s as *Structure Reports*, published by the International Union of Crystallography. There were also compendia of crystal structures in book form, the best known being a series of books by R. Wyckoff, *The Structure of Crystals*, which began to appear in 1931. Many metallic crystal structures were included in the *Metals Reference Book*. Other specialised books appeared, for instance *Crystal Data*, intended primarily for mineralogists, and the *Powder Diffraction File* in its many successive formats, which listed the lattice spacings and intensities of the lines in powder diffraction patterns from many different substances.

Structure Reports were eventually replaced by structural databases "as curators of the world's primary crystal structure data". This description comes from a splendid overview of "The Development, Status and Scientific Impact of Crystallographic Databases" by Allen (1998). According to Allen, in 1998 the five principal crystallographic databases included about 288,000 entries. Of these, more than two thirds were organic crystal structures, proteins and nucleic acids, some exceedingly complex, the majority included in the Cambridge Crystallographic Database (which got under way in the late 1960s). Two other databases (Inorganic Crystal Structure Database, and CRYSTMET) deal with inorganic and specifically metallic crystal structures (including intermetallic compounds). Allen discusses in some detail the difficult but indispensable task of "validating" crystal structures and bond lengths; inter alia, this involves seeking out simple typesetting errors in the accounts of primary data. The Cambridge database is used extensively by biochemists and, particularly, pharmaceutical firms engaged in drug development: secondary data on bond lengths and interbond angles can prove very useful. These "research applications" are discussed in detail by Allen.

Another recent database, still in evolution, is the Linus Pauling File (covering both metals and other inorganics) and, like the Cambridge Crystallographic Database, it has a "smart software part" which allows derivative information, such as the statistical distribution of structures between symmetry types, to be obtained. Such uses are described in an article about the file (Villars *et al.* 1998). The Linus Pauling File incorporates other data besides crystal structures, such as melting temperature, and this feature allows numerous correlations to be displayed.

13.2.3 Max Hansen and his successors: phase diagram databases

In 1936, Springer in Berlin published a book by a German metallurgist, Max Hansen, which was devoted to a critical assembly of known binary metallic phase diagrams (Hansen 1936). We saw in Chapter 3 that the accurate determination of phase diagrams, such as Fe–C and Cu–Sn, began at the end of the 19th century and so Hansen's pathbreaking book brought together some 40 years' research. The book covered 828 systems and contained 456 phase diagrams, with about 5500 references to the literature. This at once shows that there were several references for each system about which enough was known to publish a diagram. It was Hansen's innovation to exercise his critical faculty on the many instances where different investigators differed as to a liquidus, solidus, eutectic, peritectic, eutectoid or peritectoid temperature, or compositions of solid solutions or eutectic mixtures. His diagrams were 'optimised' diagrams. After the War, Hansen's book was revised (with the help of a coauthor, K. Anderko), translated into English and published in America (Hansen and Anderko 1958). Now there were 1382 systems and 750 diagrams. Some diagrams were buttressed by more than 100 references, with many explicit critical judgments as to the most reliable values of disputed quantities. My personal copy of this book, which I was given as a gift in 1958, has almost fallen apart from frequent use. Subsequently, there were two addendum volumes, in 1965 and 1969, including many revised and improved diagrams.

The crucial role that phase diagrams play in materials science, and the consequential need for compilations like Hansen's, has been memorably portrayed in a lecture by Massalski (1989).

From 1977, the GE Corporate Research Center in New York State began its own private collection of binary metallic phase diagrams, and from 1990 Massalski in Pittsburgh began editing a series of such assessments for the American Society of Metals (now ASM International). Later, he extended his remit to ternary systems. Earlier, in England, Hume-Rothery and his pupil Raynor had brought out, through the Institute of Metals, a number of individual critically assessed phase diagrams for the most important binary systems. There has also been a variety of Russian compilations.

From 1988, Effenberg in Stuttgart, Germany, began the enormous task of masterminding the publication of optimised *ternary* metallic phase diagrams. At the time of writing, some 3500 systems have been scrupulously covered in 18 volumes, and there is still far to go. In 1995, Villars and colleagues brought out a rival ternary compilation which included even more systems.

Further, since 1993, Effenberg has edited the "Red Book", annual summaries of developments in the world literature of phase diagrams.

In 1986, APDIC, the Alloy Phase Diagram International Commission, with the participation of 18 national bodies, was set up "to safeguard the quality of phase diagram evaluations and to provide globally the best possible coordination of the major phase diagram projects". This phrasing is taken from a short article by Effenberg (2001) which provides a complete listing, with bibliographic details, of all the various compilations, and also goes out of its way to emphasise how small a proportion of all possible ternary systems (not to mention quaternaries) have been determined (even partially).

This 66-year programme, to date, of phase diagram evaluation and publication is probably the most judgment-intensive operation ever undertaken in the history of materials science. Reliability, reproducibility and accuracy must all be assessed, and moreover some detailed features of phase diagrams as published are often inconsistent with elementary thermodynamic principles: Okamoto and Massalski (1993) have provided guidelines for avoiding this kind of error. Usually, the many stages of optimising a phase diagram and resolving disparities between different experimenters are passed over in silence. To get an idea of the complexity of the process, a paper by Murray (1985) about the steps in the optimisation of the Cu–Ti phase diagram can be recommended; here, both direct experiment and the result of CALPHAD theory are brought together and assessed.

In addition to all the metallic phase diagrams, a series of volumes devoted to ceramic systems have been published since 1964 by the American Ceramic Society and is still continuing. The original title was *Phase Diagrams for Ceramists*; now it is named *Phase Equilibria Diagrams*. Some 25,000 diagrams, binary and ternary mostly, have been published to date. There is no compilation for polymeric systems, since little attention has been devoted to phase diagrams in this field up to now.

In addition to printed compilations, more and more of the information is available on CD-ROM and latterly also on-line on the internet. This last is a feature of the service provided by MSI, Materials Science International Services in Stuttgart. This organisation, under the working name of MSIT® Workplace (http://www.msiwp.com), provides information on the entire corpus of phase diagram compilations.

It is a reflection on present-day priorities in industry that the research laboratory of a great company, Metallgesellschaft in Frankfurt-am-Main, Germany, where Hansen began work on his epoch-making book, was closed down a few years ago to save money. This laboratory was initially directed, from 1918 onwards, by Jan Czochralski, the Pole whom we met in Section 4.2.1 and who gave his name to the present-day process for growing silicon crystals, and subsequently by Georg Sachs and, after he had been driven from Germany by the Nazis in 1935, by Erich Schmid, all highly distinguished figures. The manifold achievements of the laboratory are described in a book issued on the occasion of the company's centenary, when the laboratory was still going strong (Wassermann and Wincierz 1981).

A long history of distinguished contributions does not suffice, nowadays, to save a research institution from casual destruction.

13.2.4 Other specialised databases and the use of computers

This chapter has only scratched the surface of the multitude of databases and data reviews that are now available. For instance, more than 100 materials databases of many kinds are listed by Wawrousek *et al.* (1989), in an article published by one of the major repositories of such databases. More and more of them are accessible via the internet. The most comprehensive recent overview of "Electronic access to factual materials information: the state of the art" is by Westbrook *et al.* (1995). This highly informative essay includes a 'taxonomy of materials information', focusing on the many different property considerations and property types which an investigator can be concerned with. Special attention is paid to mechanical properties. The authors focus also on the *quality and reliability of data*: quality of source, reproducibility, evaluation status, etc., all come into this, and alarmingly, they conclude that numerous databases on offer today "consist wholly or in part of data that would not even meet the criteria for 'limited use'." They home in on the many on-line databases accessible through STN International, a scientific and technical information network. In fact, there are a number of organisations worldwide, most connected to the internet, which merge a number of independent databases; the Förderverein Werkstoffdokumentation in Germany is another such. ASM International has recently issued a Directory of Materials Property databases (ASM 2000).

One feature discussed by Westbrook *et al.* is the nature of the enquiries that a database can handle... i.e., the quality of its associated software. One example is 'range searching', finding all materials that have values of a particular property in a specified numerical range. This aspect of using databases has recently been examined critically, in two linked papers, by Ashby (1998) and by Bassett *et al.* (1998). For instance, different *categories* of materials have different ranges of value of the thermal expansion coefficient (p. 202). The ranges for ceramics and glasses (small values) do not overlap with the range for polymers (large values). Ashby's comment on this is: "Correlations exist between the values of mechanical, thermal, electrical and other properties which derive from the underlying physics of bonding and packing of atoms. Some of these correlations have a simple theoretical basis and can be expressed as dimensionless groups with much narrower value ranges; they allow a physically based check on property values and *allow some properties to be estimated when values for others are known*" (my italics). The second paper shows how missing data can be estimated by using these principles.

Ashby, some years ago, produced a computerised database called the *Materials Selector* (CMS 1995) which allows the best material to be selected for a particular application with combined criteria (as it might be, least weight and stiffness in a certain range). The ideas which led to this were presented in a book (Ashby 1992). John Rodgers, who operates the CRYSTMET crystallographic database for metals, is now planning to create a database for unfamiliar materials, using the CMS environment and information in CRYSTMET, incorporating a range of calculated or estimated physical property values. It begins to look as though data estimation using information in databases may become a new, distinct activity within materials science and engineering. In fact, the process already has a name – *datamining*. According to John Rodgers, this means "the extraction of implicit, previously unknown and potentially useful information from data" (Rodgers 1999).

Materials selection is as much an art as a rigorous science, and another computational approach to it, based on ideas of artificial intelligence, has been proposed by Arunachalam and Bhaskar (1999). They call their approach "bounded rationality" and exploit it to analyse the background to some notorious disasters based on material failure. We can always learn from failure as well as from success.

REFERENCES

Allen, F. (1998) *Acta Cryst. A* **54**, 758.

Arunachalam, V.S. and Bhaskar, R. (1999) *MRS Bull.* **24** (**10**), 57.

ASM (2000) *Directory of Materials Property Databases*, supplement to *Advanced Materials and Processes*, August.

ASTM (1993) *Manual on the Building of Databases* (ASTM, Philadelphia).

Ashby, M.F. (1992) *Materials Selection in Mechanical Design* (Pergamon Press, Oxford).

Ashby, M.F. (1998) *Proc. Roy. Soc. Lond. A* **454**, 1301.

Bassett, D., Brechet, Y. and Ashby, M.F. (1998) *Proc. Roy. Soc. Lond. A* **454**, 1323.

CMS (1995) *Cambridge Materials Selector* (Granta Design Ltd., Trumpington, Cambridge, England).

Effenberg, G. (2001) Article on data compilations on phase diagrams, in ed. *Encyclopedia of Materials*. Buschow, K.H.J. *et al.* (Pergamon Press, Oxford).

Greenaway, F. (1996) *Science International: A History of the International Council of Scientific Unions* (Cambridge University Press, Cambridge).

Hansen, M. (1936) *Der Aufbau der Zweistofflegierungen* (Springer, Berlin).

Hansen, M. and Anderko, K. (1958) *Constitution of Binary Alloys* (McGraw-Hill, New York).

Kleppa, O. (1994) *J. Phase Equil.* **15**, 240.

Massalski, T.B. (1989) *Metall. Trans.* **20B**, 445.

Murray, J. (1985) Assessment and calculation of the Ti–Cu phase diagram, in *Noble Metal Alloys* ed. Massalski, T.B. *et al.* (The Metallurgical Society of AIME, Warrendale, PA).

Okamoto, H. and Massalski, T.B. (1993) *J. Phase Equil.* **14**, 316.

Rodgers, J.R. (1999) Private communication, 21 September.

Villars, P., Onodera, N. and Iwata, S. (1998) *J. Alloys Comp.* **279**, 1.

Wassermann, G. and Wincierz, P. (1981) *Das Metall-Laboratorium der Metallgesellschaft AG: Chronik und Bibliographie* (Metallgesellschaft, Frankfurt).

Wawrousek, H., Westbrook, J.H. and Grattidge, W. (1989) Data sources of mechanical and physical properties of materials, in *Physik Daten*, vol. **30**(1) (Germany, Karlsruhe Fachinformationszentrum).

Weast, R.C. (1985) Letter dated 7 September.

Westbrook, J.H., Kaufman, J.G. and Cverna, F. (1995) *MRS Bull.* **20**(8), 40.

White, G. (1991) Letter dated 3 September.

Chapter 14
The Institutions and Literature of Materials Science

Chapter 14
The Institutions and Literature of Materials Science

14.1. TEACHING OF MATERIALS SCIENCE AND ENGINEERING

The emergence of university courses in materials science and engineering, starting in America in the late 1950s, is mapped in Section 1.1.1. The number and diversity of courses, and academic departments that host them, have evolved. An early snapshot of the way the then still novel concept of MSE was perceived by educators, research directors and providers of research funds can be found in an interesting book (Roy 1970) in which, for example, a panel reported that a representative of the GE Company "stressed that his company regards the university as a provider of people and not as an institution which supplies all of the solutions to industry's materials problems. The university should train both materials scientists and engineers, should clearly recognise the difference between these two groups, and should provide the basis for interdisciplinary cooperation." Rustum Roy, the editor of that volume, repeatedly called for just such interdisciplinary cooperation on campus; the high point of his campaign was a paper published in 1977 (Roy 1977). He has done much to bring about just such interdisciplinarity at his own university, Pennsylvania State University, which for many years has hosted an interdisciplinary Materials Research Laboratory of the kind whose history is outlined in Section 1.1.3. His role in creating the Materials Research Society was similarly motivated.

The present situation, both in the US and elsewhere, is examined in a recent survey article (Flemings and Cahn 2000). In the United States, the number of core MSE departments (i.e., independent university departments granting bachelor through doctorate degrees) in 1999 was 41. On top of that, 14 departments are still specific to particular categories of materials, and another 41 are either joint with other disciplines that are peripheral to MSE, or are wholly embedded in departments of other disciplines, such as mechanical or chemical engineering. So, merged or embedded departments are as numerous as independent departments. After a sharp peak in 1982, the number of students granted bachelor's degrees in the US specifically in materials or metallurgy declined somewhat, stabilising at $\approx$1200 per annum in the 1990s. The number of faculty members in MSE departments in 1997 was estimated at 625 (Flemings 1999).

In England (excluding Scotland, Wales and Northern Ireland), there were 21 mainline MSE departments in 1998; Fig. 9.4 (Chapter 9) shows plots of student

503

numbers in the US. On the continent of Europe, where institutes and not full departments are the organisational rule, it is much more difficult to pick out those institutes which are properly described as being in the MSE mainline; an attempt by a range of national societies to list appropriate university institutes has led to numbers ranging from 79 in Germany, via 48 in France to only 4 in Sweden... but many of the institutes listed are in fields which are peripheral to, or barely connected with, MSE. In some universities on the continent, a number of institutes are combined into a materials department. To pick just one example, at the eminent Eidgenössische Technische Hochschule in Zürich, Switzerland, the following institutes (or groups) are currently combined: biomechanics; biomedical engineering; metals and metallurgy; metallic high-performance materials (the distinction between these last two is typical of continental modes of organisation); nonmetallic inorganic materials; polymer chemistry; polymer physics; polymer technology; supramolecular chemistry; surface science and technology. Thus, here semiconductors have been hived off to another department.

Fig. 14.1 shows an impressionistic 'ternary diagram' showing the emphasis on three broad fields relevant to MSE at a range of German universities that prepare students in the study of materials. If one thing is crystal clear, it is that there is no one ideal way of teaching MSE laid up in heaven, and the example of the Swiss department indicates that there is much scope for variety.

In spite of statistical problems, two things are clear from a close examination of student numbers in various countries and institutions: MSE courses are burgeoning, and the best mainline departments are going from strength to strength. However, some of the weaker departments/institutes (those with relatively few students) are being forced by resolute academic deans into marriages with quite distinct disciplines – which (experience suggests) can be a precursor of brain death – or being closed down altogether.

Flemings (1999) reflected under the title "What next for departments of materials science and engineering?" A particularly interesting feature of his paper is a comparison of the characteristics and activities of a class of students who graduated with metallurgy degrees from M.I.T. (Flemings's university) in 1951, with those of another class who graduated in MSE in 1991. In each case, statistics were collected 7 years after graduation; not all students responded (See Table 14.1).

The most striking features, apart from the sharp drop in fecundity, are the large numbers of graduates who went on to obtain business qualifications (Masters of Business Administration, MBAs); the fact that in 1958, working in metallurgy and in engineering seems to have been synonymous in the eyes of respondents, but not so in 1998; the drastic fall in the numbers who gave research and development as their current métier, in spite of a sharp rise in those taking advanced degrees; and the fact that, around the age 30, *none* of the 1998 respondents had become university faculty.

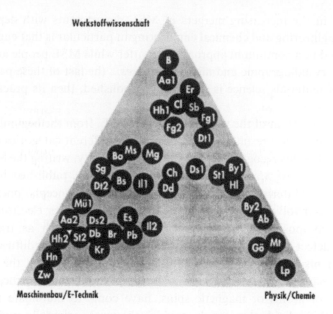

Figure 14.1. Estimated emphasis on three broad fields – Werkstoffwissenschaft = materials science; Maschinenbau/E-technik = mechanical and electrical engineering; Physik-Chemie = physics and chemistry – in MSE education at various German universities from (DGM 1994).

Table 14.1. Particulars of two graduating M.I.T. classes, 7 years after graduation.

	Class of 1951 (%)	Class of 1991 (%)
With advanced degrees	37	64
With MBAs	0	43
Working in metallurgy	89	14
Working in engineering	89	43
University faculty	19	0
In R&D, including faculty	48	14
Married	96	62
Mean number of children per graduate	1.8	0.1

As Flemings points out, compared with the middle of the twentieth century, MSE departments now have to prepare their students for quite different professional lives. The key question that seems to arise from these figures is: Do university departments put too much emphasis on research? And yet, before we conclude that they do, we must remember that it is widely agreed that research is what keeps university faculty alert and able to teach in an up-to-date way. It may well be that what students currently want, and what the health and progress of MSE demands, are two distinct things.

A danger in the increasing mergers of MSE departments with departments of mechanical engineering and chemical engineering in particular is that engineers are in general wedded to a continuum approach to matter while MSE people are concerned with atomic, crystallographic and micro-*structures*... the last of these particularly. If that aspect of materials science is sidelined or abolished, then its practitioners lose their souls.

The key justification of the whole concept of MSE, from the beginning, has been the mutual illumination resulting from research on different categories of materials. The way I worded this recognition in my editorial capacity, writing the Series Preface for the 25 volumes of *Materials Science and Technology*, published between 1991 and 2000, was: "Materials are highly diverse, yet many concepts, phenomena and transformations involved in making and using metals, ceramics, electronic materials, plastics and composites are strikingly similar. Matters such as transformation mechanisms, defect behaviour, the thermodynamics of equilibria, diffusion, flow and fracture mechanisms, the fine structure and behaviour of interfaces, the structures of crystals and glasses and the relationship between these, the statistical mechanics of assemblies of atoms or magnetic spins, have come to illuminate not only the behaviour of the individual materials in which they were originally studied, but also the behaviour of other materials which at first sight are quite unrelated. This continual cross-linkage between materials is what has given rise to Materials Science, which has by now become a discipline in its own right as well as being a meeting place of constituent disciplines... Materials Technology (or Engineering) is the more practical counterpart of Materials Science, and its central concern is the processing of materials, which has become an immensely complex skill...."

Whether I was justified in saying that Materials Science "has by now become a discipline in its own right", is briefly discussed in the last chapter.

The most idiosyncratic of the materials families are polymers and plastics. The mutual illumination between these and the various categories of inorganic crystalline materials has been slow in coming, and this means that teaching polymer science in broad materials science departments and relating the properties of polymers to other parts of the course, has not been easy. Yet things are improving, partly because more and more leading researchers and teachers in polymer physics are converted metallurgists. One of these reformed metallurgists is Edward Kramer, now in the Materials Department at the University of California, Santa Barbara. In a message (private communication, 2000) he pointed to three links from his own experience:

(1) In a semicrystalline polymer, the crystals are embedded in a matrix of amorphous polymer whose properties depend on the ambient temperature relative to its glass transition temperature. Thus, the overall elastic properties of the semicrystalline polymer can be predicted by treating the polymer as a composite material

with stiff crystals embedded in a more compliant amorphous matrix, and such models can even be used to predict the linear viscoelastic properties.

(2) Thermodynamics and kinetics of phase separation of polymer mixtures have benefited greatly from theories of spinodal decomposition and of classical nucleation. In fact, the best documented tests of the theory of spinodal decomposition have been performed on polymer mixtures.

(3) A third topic is the mutual diffusion of different macromolecules in the melt. Here, the original formulation of the interdiffusion problem in metals proved very useful even though the mechanisms involved are utterly different. When a layer of polymer A with a low molecular weight diffuses into a layer of the same polymer with high molecular weight, markers placed at the original interface move towards the low-molecular-weight side, just as in Kirkendall's classical experiments with metals (Section 4.2.2). The viscous bulk flow that drives this marker displacement is equivalent to the vacancy flux in metals.

I shall be wholly convinced of the beneficial conceptual synergy between polymers and other classes of materials when polymer scientists begin to make more extensive use of phase diagrams.

In earlier chapters, especially Chapters 2 and 3, the links of materials scientists to neighbouring concerns such as solid-state physics, solid-state chemistry, mineralogy, geophysics, colloid science and mechanics have been considered, and need not be repeated here. *Suffice it to say that materials scientists and engineers have proved themselves to be very open to the broader world of science.* A good proof of this is the experience of the Research Council in Britain that distributes public funds for research in the physical sciences. It turns out that the committee which judges claims against the funds provided for materials science and engineering (a committee composed mainly of practising materials scientists) awards many grants to departments of physics, chemistry and engineering as well as to mainline MSE departments, whereas the corresponding committees focused on those other disciplines scarcely ever award funds to MSE departments.

14.2. PROFESSIONAL SOCIETIES AND THEIR EVOLUTION

The plethora of professional societies now linked to MSE can be divided into three categories – old metallurgical societies, either unregenerate or converted to broader concerns; specialised societies, concerned with other particular categories of materials or functions; and societies devoted to MSE from the time of their foundation. Beyond this, there are some federations, umbrella organisations that link a number of societies.

All the societies organise professional meetings, and often publish the proceedings in their own journals; many of the larger societies publish multiple journals. Most societies also publish a range of professional books.

14.2.1 Metallurgical and ex-metallurgical societies

There have long been a number of renowned national societies devoted to metals and alloys, some of them more than a century old. They include (to cite just a few examples, using early – not necessarily original – names) the Metallurgical Society of the American Institute of Mining, Metallurgical and Petroleum Engineers, The American Society for Metals, the Institute of Metals in London, the Deutsche Gesellschaft für Metallkunde, the Société Française de Métallurgie, the Indian Institute of Metals, the Japan Institute of Metals. Most of these have now changed their names because, at various times, they have sought to broaden their remit from metals to materials; the Indian and Japanese bodies have not hitherto changed their names. Some bodies have simply resolved to become broader; one has become simply TMS (which represents The Minerals, Metals and Materials Society), another, ASM International. Other societies have broadened by merging with other preexisting societies: thus the Institute of Metals in London first became the Metals Society, which merged with the Iron and Steel Institute to become the Institute of Metals once again, and eventually merged with other societies concerned with ceramics, polymers and rubber to become the Institute of Materials.

The journals published by the various societies have mostly undergone repeated changes of name. Thus, the old *Journal of the Institute of Metals* first split into *Metal Science* and *Materials Technology* and finally reunited as *Materials Science and Technology*. TMS and ASM International joined forces to publish *Metals Transactions*, which recently turned into *Metallurgical and Materials Transactions*; this journal replaced two earlier ones published separately by the two societies, each of these having changed names repeatedly. The German journal published by the Deutsche Gesellschaft für Metallkunde (now the D.G. für Materialkunde, DGM) was and remains the *Zeitschrift für Metallkunde*; most of the papers remain metallurgical and most of them are now in English. (The history of the DGM, "in the mirror of the Zeitschrift für Metallkunde", is interestingly summarized in an anniversary volume, DGM 1994.) The French society has replaced 'metals' with 'materials' in its name, and likewise incorporated the word in the rather lengthy title of its own journal (*Revue de Métallurgie: Science et Génie des Matériaux*). These many name changes must be a librarian's nightmare.

The underlying idea fueling the many changes of names of journals is that by changing the name, societies can bring about a broadening of content. By and large this has not happened, and the journals have remained obstinately metallurgical in character, because when a journal is first published, it quickly acquires a firm identity in the minds of its readers and of those who submit papers to it, and a change of name does not modify this identity. In my view, only a very resolute and proactive editor, well connected through his own scientific work to the scientific community, and with clear authority over his journal, has any hope of gradually bringing about a genuine transformation in the nature of an existing, well-established journal. The alternative, of course, is to start completely new journals, some independent of societies; this alternative strategy is discussed in Section 14.3.

In Europe, a Federation of Materials Societies, FEMS, was established in 1987; it links 19 societies in 17 countries (website: http://www.fems.org). It plays a role in setting up Europe-wide conferences on materials, keeps national societies informed of each other's doings, and seeks to avert timetable conflicts. Further federations feature in the next section.

14.2.2 Other specialised societies

Numerous societies are devoted to ceramics, to glass or to both jointly. The American Ceramic Society is the senior body; the European Ceramic Society is an interesting example of a single body covering a wide but still restricted geographical area. Societies covering polymers (and elastomers sometimes treated as a separate group) are multifarious, both nationally and internationally. Still other specialisms, such as composite materials, carbon and diamond are covered by commercial journals rather than by specialised societies, but even where there is no society to organise conferences in a field, yet independent and self-perpetuating groups of experts arrange such conferences without society support. Semiconductor devices and integrated circuits are mostly covered by societies closely linked to the electrical engineering profession. There are a number of societies, such as the Royal Microscopical Society in Britain, which focus on aspects of materials characterization. Any attempt to list the many specialised professional bodies would be unproductive.

14.2.3 Materials societies ab initio

The first organization to carry the name of materials science was a British club, the Materials Science Club, founded by a group of materials-oriented British chemical engineers in 1963. This group organised broad meetings on topics such as 'materials science in relation to design' and 'biomechanics', and published some of the contributions in its own quarterly *Bulletin*. The Club brought together a very wide range of some hundreds of scientists and engineers from universities, industry and government laboratories, including a proportion of foreign members, awarded medals, and published almost 100 issues of its *Bulletin* before difficulties in organising its affairs without any paid staff eventually brought about its absorption, in the late 1980s, by the Institute of Metals in London, and thereby its extinction. Only one complete set of the *Bulletin* survives, in the library of the City University in London. While it lasted, it was a very lively organization.

Undoubtedly, the key organization created to foster the new concepts of interdisciplinary research on materials is the Materials Research Society, MRS, founded in the US in 1973, after 7 years of exhaustive discussions. It is to be particularly noted that its name carries the words 'Materials Research', not 'Materials Science'. 'Materials Research' avoids specifying which kinds of scientists and engineers should be involved in the society; all that is required that their work should contribute to an understanding and improvement of materials. According to illuminating essays (Roy and Gatos 1993) by two of the founders of the MRS, Rustum Roy and Harry Gatos (whom we have met in Section 10.4.1), from the start the society was to focus on research involving cooperation between different disciplines, of which MSE was to be just one – albeit a vital one. Gatos is forthright in his essay: "The founding and operation of MRS was the culmination of my ten years of frustrated effort in searching for a professional home (old, renovated or new) for the young, homeless materials science. The leaders of the existing materials societies strenuously resisted accepting that materials science existed outside the materials they dealt with, be they metals, ceramics, or polymers. The founders of MRS were just a small but 'driven' minority..." Certainly my own experience of starting Britain's first university department of materials science in 1965 confirms what Gatos (who was at MIT) says about professional societies at that time; when I first attended a meeting of the MRS in 1976, I realised that I had found my primary intellectual home, inchoate though it was in that year. The MRS took some years to reach its first 1000 members, but after that grew rapidly.

There was a further consideration in the minds of the founders, though that has been kept rather quiet in public. In the early 1970s, physicists and chemists working in American industry, especially the many working on aspects of materials, were not made welcome in their professional physics and chemistry societies, which were inclined to ignore industrial concerns. These two groups played a substantial part in bringing the MRS to life; it must also be said immediately that enlightened figures in industry, especially William O. Baker, director of research at Bell Telephone Laboratories, from an early stage supported MRS by word and deed. MRS from the beginning welcomed industrial scientists and topics of close concern to industry. It is thus natural that today, as many as 25% of the ≈12,500 members of MRS (in more than 60 countries) are in industry (as against 63% in academe and 12% in government laboratories) (Rao 2000).

Roy and Gatos, as also Phillips (1995) in her even more recent snapshot of the MRS, all emphasise two features of the society: the major role of volunteer activity by members in taking scientific decisions and making the society work (in its early years, it had no paid staff), and the invention of the principle of simultaneous symposia, organized by members, each on a well-defined, limited topic, that constitutes the main business of the society's annual meetings, a practice, as Roy points out, "now copied almost universally by most disciplinary societies." Several hundred volumes of proceedings of these symposia have been published by 2000. The MRS now has a large, paid headquarters staff, essential for what has become a large and variegated organization.

In addition to the symposium proceedings, MRS publishes a monthly *MRS Bulletin*, and in 1986 it founded an archival research journal, *Journal of Materials Research* (*JMR*), and both are going strong. I have had many occasions in this book to cite expository articles in the *Bulletin*, in particular. The *JMR* is run in an unusual way, typical of the MRS: each submitted paper is sent to one of a panel of principal editors (chosen periodically by the society's council) and he/she reports on the paper to the Editor-in-Chief, who alone communicates with authors. I was one of the first batch of principal editors, and found that this system worked well. An essay on the genesis and principles of this journal, three years afterwards, was published by Kaufmann (1988). *JMR* has only one Achilles' heel: as Roy (1993) pointed out, "the MRS has not been able to involve the polymer community to a major extent; less than 5% of the *JMR* is (in 1993) devoted to polymers." This is a lasting problem for all who seek to foster a broadly based discipline of MSE. However, *JMR* is publishing an increasing proportion of papers on the broad theme of materials processing, and this is a particularly useful service.

Once it was well-established, and mindful of its many foreign members, the MRS encouraged the progressive creation of local MRSs in a number of countries. There are now 10 of these, in Australia, China, Mexico, Argentina, India, Japan, South Korea, Russia, Taiwan and Europe (embodying various European countries, and domiciled in France). Some are more active than others; in particular, the Indian body, MRS-I, publishes its own successful research periodical, *Bulletin of Materials Science*, and the Chinese MRS has organized a succession of major international conferences. Overarching these societies is the International Union of Materials Research Societies; the original MRS has helped a great deal in setting up this federal supervisory body, but in no sense does it dominate it. One example of the help this federation gives to constituent bodies is a major MSE conference held in Bangalore, India, in 1998 (proceedings, IUMRS-ICA 98 1999).

In Japan, the Japan Federation of Materials acts as an umbrella organisation for 18 Japanese materials societies, and very recently, in 2000, it has co-sponsored a new English-language Japanese journal, *Science and Technology of Advanced Materials*, with (among other aims) the laudable editorial objective of "concise presentations, so that interested readers can read an issue from cover to cover."

One primary aim of the MRS, to achieve a breakdown of interdisciplinary barriers, has been well achieved, according to one of the prime godfathers of materials science, the American Frederick Seitz (Fig. 3.19, Chapter 3). In a book primarily devoted to Italian solid-state physics (Seitz 1988), he remarks: "I might say a few words about the 55 odd years in which I have been associated with solid-state physics or, as it is sometimes called in the US, solid-state science, or condensed matter physics or materials science. When I entered the field as a graduate student in the early 1930s the overall field was strongly compartmenta-lised into three divisions which had relatively little interaction... One division was related to work in the field of metallurgy and ceramics... The second division related to research on materials for electrical engineering and electronics,... The third division related to the investigations of what might be called the fundamentalist scientists." Of these three divisions, Seitz says: "While these divisions still exist, the flow of information between them is now much greater than it was and the research groups in each have many common bonds, mainly because of the application of solid-state physics." This is a robust physicist's view of the broadening of materials research.

Of course, many other professional societies have played their part in this successful reaching out between specialisms. As outlined above, the big metallurgical societies have broadened resolutely, and the American Physical Society and American Chemical Society are now much more hospitable to their members in industry than they apparently were 30 years ago.

14.3. JOURNALS, TEXTS AND REFERENCE WORKS

There is now an immense range of scientific journals, broad, narrow and in-between, to serve the great range of materials. The journals published by the many professional societies have encountered increasing competition from the many published by commercial publishers, but those, in turn, are now under severe pressure because of a growing librarians' revolt against subscription prices that rise much faster than general inflation.

14.3.1 Broad-spectrum journals
One classification is of special importance: there is a small minority of materials journals that can be described as *broad-spectrum*, compared with a much larger number which are specialised to a greater or lesser degree. Probably the first broad-

spectrum journal was *Journal of Materials Science*, *JMS*, launched by a commercial publisher in 1966. I was the first chairman of editors, so had a major role in forming policy. My insistence was that there should be several editors with complementary fields of expertise and independent powers of decision over submitted papers, and I encouraged those editors to be proactive (to use a current jargon-word) and seek out key papers on novel topics. This worked well, and the publication of such key papers then encouraged other authors in the same field to steer their papers to *JMS*. The 1969 paper from which Fig. 6.6 (Chapter 6) was reproduced was an example of this successful policy. Since the journal was broad-spectrum *from the beginning* (including, incidentally, polymer physics) that was how it has always been perceived and it has not become specialised, even when the 6 editors had to be replaced by one editor after some years (because of my enhanced academic duties in 1973 that deprived me of time to edit). However, there have been several spin-off mini-journals, including one devoted to the new editor's specialism, biomedical engineering. *JMS* has also always been very international.

Another journal, *Materials Science and Engineering* (*MSE*), was started by another commercial publisher at about the same time as *JMS*. This had only one editor, a metallurgist, from the start, and so in spite of its stated objectives, it remained almost wholly metallurgical for many years. When eventually it became broader under a new editor, it was split into several independent journals with distinct editorial boards, each of them relatively broad-spectrum – in particular, one devoted to functional materials, and another to biomimetics. The main *MSE* remained in being, and has remained largely metallurgical after 35 years.

The MRS archival journal, *Journal of Materials Research*, already mentioned, is another broad-spectrum journal that has worked well, except for its limited polymer content. Here again, the principle of multiple editorship seems to have been an important component of success.

Some older journals, such as *Journal of Physics and Chemistry of Solids*, which has been published for some 60 years and now focuses to some degree on functional materials, have long been broad-spectrum. Others have a broad-spectrum name but in fact are relatively narrowly focused: an example is *Materials Research Bulletin*, which in fact is concerned mostly with the chemistry of inorganic materials. Its subtitle is *an international journal reporting research on the synthesis, structure and properties of materials.* (This journal now has a supplement entitled *Crystal Engineering*.) Likewise, an English-language journal simply called *Advanced Materials* began publication 10 years ago in Germany, and is highly successful; in spite of its comprehensive title, it is wholly focused on materials chemistry, especially processing. In recent years, the archetype of broad spectrum, *Nature*, has begun to pay special attention to papers on materials processing, self-assembly techniques in particular, as the many references to that journal in Chapter 11 testify.

In Russia, after many years of a successful but purely metallurgical journal entitled *Fizika Metallov i Metallovedenie* (the last word representing 'knowledge of materials' and not, as I had supposed, 'metallography' (Rabkin 2000)), a group of influential materials scientists in 1997 started a journal entitled *Materialovedenie*, which word I believe to be the best current Russian form of 'materials science'. In spite of the editors' best efforts, the journal is finding it difficult to break away from a metallurgical focus.

In Japan, as recorded above, a new journal called *Science and Technology of Advanced Materials* has just begun publication.

An interesting, broad-spectrum journal founded in 1997 by Roy is *Materials Research Innovations*; one of its objectives is to bypass normal methods of editorial scrutiny; submitting authors who have published a sufficient number of papers in other, peer-reviewed, journals are assumed, in effect, to have reviewed themselves.

A number of journals devoted wholly to review articles, shading from metallurgy to genuine materials science, are now appearing; the grandfather of this group is *Progress in Materials Science* (which began in 1949 as *Progress in Metal Physics*). Another excellent example is *Materials Science and Engineering – Reports: A Review Journal*.

14.3.2 The birth of Acta Metallurgica

The journal whose genesis is to be described here is of extreme importance in the history of modern physical metallurgy and, later, materials science. Its birth in 1953 coincided with the high point of the 'quantitative revolution' portrayed in Chapter 5, and preceded by a few years the beginning of materials science. It transformed the metallurgical researcher's perception of the discipline and it clearly contributed to the currents of thought that first brought materials science into being in 1958.

Acta Metallurgica owed its birth to a resolute metallurgist, Herbert Hollomon, whom we met in Section 1.1.2 in his capacity as leader of materials research at the General Electric Corporate R&D Center in New York State. According to a history of the journal (Hibbard 1988), an update thereto (Fullman 1996) and private information from Seitz (2000), Hollomon perceived soon after World War 2 that publications from a new post-war surge of research were widely scattered throughout the physical, chemical and metallurgical literature and that there was a "need for a unifying journal in which the fruits of such research could be gathered more effectively." A number of eminent researchers, including among others Frederick Seitz, Harvey Brooks (the founding editor of *Journal of Physics and Chemistry of Solids*), Cyril Stanley Smith (see Section 14.4.1) and Bruce Chalmers, joined in discussions that led, in 1951, to an approach to the American Society of Metals which then offered generous financial support; in this the ASM was later joined by the American Institute of Mining, Metallurgical and Petroleum Engineers. During the next year, a board of governors chaired by Smith was created, and appointed Bruce Chalmers, then a professor in Toronto, Canada (see Section 9.1.1) to be editor. Hollomon was secretary/treasurer of the board of governors.

Acta Metallurgica began publication in the spring of 1953 and at once created a huge impact in the profession with its many rigorous, quantitative papers, long and short. The journal's standards were very high from the beginning, and aspects of physics (such as for instance nuclear magnetic resonance) found their place in the journal from the first volume. Cyril Smith, in his preface to the first issue, memorably remarked: "Now, metallurgy is too broad to be encompassed by a single human mind: it is essential to enlist the interest of the 'pure' scientists, and to increase the number of metallurgists whose connections with production and managerial problems are partially sacrificed in order that they may be more concerned with physics and physical chemistry as a framework for useful metallurgical advance."

By 1967, the flood of short papers had become so great that a separate journal, *Scripta Metallurgica*, was hived off. These Latin titles were intended to symbolise the international character of the journal. Chalmers edited the journal until 1974, when Michael Ashby took over the reins which he held until 1995; at that point a more collegiate editorial structure was instituted. In 1990, the adjective '*metallurgica*' was supplemented by '*materialia*', and in 1996 the journals simply became *Acta materialia* and *Scripta materialia* (some classicist seems to have advised the board of governors, at a late stage, that lower case letters are de rigueur in Latin!)

Acta Metallurgica was unique among journals in having from the beginning a completely independent board of governors which is the formal owner, permanently guaranteed financially by the two leading American metallurgical societies. The initially contracted publisher in Toronto proved to have difficulty in sustaining the printing effort, and when it seemed that the project might be stillborn, Seitz (then chairman of the governing board of the American Institute of Physics) brought in the publishing facilities of that Institute to rescue the situation; much effort was involved in the rescue. By that time, Chalmers had moved to Harvard. However, in 1955 Hollomon met Robert Maxwell, proprietor of Pergamon Press, on an airplane; they took to each other (both were forceful characters to a degree) and Hollomon, who seems to have had quasi-dictatorial powers over the board of governors of *Acta Metallurgica*, insisted that Pergamon Press should take over publication of the journal; it has published it (and its temporary sister publications... like *Materials and Society*) since 1955. However, Pergamon Press has never owned the copyright or the journal itself, and policy decisions have always been taken by the board of governors with input from a very international roster of advisers.

In recent years, under the leadership of a coordinating editor, Subra Suresh, *Acta Materialia* and its letter journal have sought energetically to broaden the remit of the journals, with some success but also some difficulties. In January 2000, Suresh edited a fine 'millenium issue', entitled *Materials science and engineering: current status and future directions*; it included 21 overviews, including excellent treatments of polymers.

14.3.3 Specialised journals

Scientific journals devoted to particular categories of materials, or procedures, become ever more numerous. Some are national, others continental or international in scope; some are highly specific, others somewhere between broad and narrow spectrum; some publish in English or another language only, others accept papers in several languages. All I can usefully do here is to cite a few examples.

An example of a journal hovering between broad and narrow spectrum is *Journal of Alloys and Compounds*, subtitled "an interdiciplinary journal of materials science and solid-state chemistry and physics." One which is more restrictively focused is *Journal of Nuclear Materials* (which I edited for its first 25 years). Ceramics has a range of journals, of which the most substantial is *Journal of the American Ceramic Society*. *Ceramics International* is an example of an international journal in the field, while *Journal of the European Ceramic Society* is a rather unusual instance of a periodical with a continental remit. More specialised journals include *Solid State Ionics: Diffusion and Reactions*, and a new *Journal of Electroceramics*, started in 1997.

Polymer journals are very plentiful and most of them are relatively broad in coverage. Examples – *Polymer* (*the international journal for the science and technology of polymers*), *Progress in Polymer Science* and *New Polymeric Materials*. To repeat a statement made in Chapter 2: "As late as 1960, only four journals were devoted exclusively to polymers – two in English, one in German and one in Russian. Now, however, the field is saturated: a survey in 1994 came up with 57 journal titles devoted to polymers that could be found in the Science Citation Index, and this does not include minor journals that were not cited."

Other examples of specialised journals include *Composites Science and Technology*; a broad journal called *Carbon* and a more specific one, *Diamond and Related Materials*; and *Biomaterials* (incorporating *Clinical Materials*). I have already mentioned the new *Crystal Engineering*, which joins such journals as *Crystal Research and Technology* and in turn was joined in 2001 by *Crystal Growth and Design*. Beyond that, there are the several forms of the classic journal *Acta Crystallographica* (which may have been the first to adopt a Latin title). A whole series of new journals cover computer modelling and simulation of materials: *Computational Materials Science* is one, *Modelling and Simulation in Materials Science and Engineering* is another.

A large group of journals covers various aspects of characterization, including electron microscopy. *Micron* and *Ultramicroscopy* are two of these, *Materials Characterization* (published in association with the International Metallographic Society) is another.

Materials chemistry is now served by a whole range of journals, ranging from the venerable *Journal of Solid-State Chemistry* and *Materials Research Bulletin* (already mentioned) to *Materials Chemistry and Physics* (which, interestingly, now incorporates *The International Journal of the Chinese Society for Materials Science...* which appears to be distinct from the Chinese MRS) and *Journal of Materials Chemistry* (published by the RSC in London) – also *Chemistry of Materials*, published by the ACS. In France, *Annales de Chimie: Science des Matériaux* is an offshoot of a journal originally founded by Lavoisier in 1789 (shortly before he lost his head). *Journal of Materials Synthesis and Processing* is an interesting periodical with somewhat narrower focus.

In this listing of examples, I have excluded straight metallurgical journals and the many devoted to solid-state physics, such as the venerable *Philosophical Magazine* and *Physical Review B*.

14.3.4 Textbooks and reference works

One of the defining features of a new discipline is the publication of textbooks setting out its essentials. In Section 2.1.1, devoted to the emergence of physical chemistry, I pointed out that the first textbook of physical chemistry was not published until 1940, more than half a century after the foundation of the field. Materials science has been better served. In what follows, I propose to omit entirely all textbooks devoted to straight physical metallurgy, of which there have been dozens, say little about straight physics texts, and focus on genuine MSE texts.

As we saw in Chapter 3, the founding text of modern materials science was Frederick Seitz's *The Modern Theory of Solids* (1940); an updated version of this, also very influential in its day, was Charles Wert and Robb Thomson's *Physics of Solids* (1964). Alan Cottrell's *Theoretical Structural Metallurgy* appeared in 1948 (see Chapter 5); although devoted to metals, this book was in many ways a true precursor of materials science texts. Richard Weiss brought out *Solid State Physics for Metallurgists* in 1963. Several books such as *Properties of Matter* (1970), by Mendoza and Flowers, were on the borders of physics and materials science. Another key 'precursor' book, still cited today, was Darken and Gurry's book, *Physical Chemistry of Metals* (1953), followed by Swalin's *Thermodynamics of Solids*.

However, the first text specifically for students of materials science was Lawrence van Vleck's *Elements of Materials Science: An Introductory Text for Engineering Students* (1959), which was very widely used. It appeared only a year after the initiatives at Northwestern University which gave birth to MSE (Section 1.1.1). In 1970, he published *Materials Science for Engineers*. Later, in 1973, the same author brought out *A Textbook of Materials Technology*; in his preface to this, van Vlack says that it was prepared "for those initial courses in materials which need the problem-solving approach of the technologist and the engineer, but which must fit into curricula designed for those who have a minimal background in the sciences." Thus its approach was very different from Morris Fine's book, mentioned next.

In 1964, two competing series of slender volumes appeared: one, the 'Macmillan Series in Materials Science', came from Northwestern: Morris Fine wrote a fine account of *Phase Transformations in Condensed Systems*, accompanied by Marvin Wayman's *Introduction to the Crystallography of Martensite Transformations* and by *Elementary Dislocation Theory*, written by Johannes and Julia Weertman. The second series, edited at MIT by John Wulff, was entitled 'The Structure and Properties of Materials', and included slim volumes on *Structure, Thermodynamics of Structure, Mechanical Behaviour* and *Electronic Properties*.

From the early 1970s onwards, more substantial texts began to appear, notably Arthur Ruoff's *An Introduction to Materials Science* (1972), a book of 700 pages. This was followed by *The Principles of Engineering Materials* (1973) by Craig Barrett, William Nix and Alan Tetelman, then *Metals, Ceramics and Polymers* (1974), 640 pages, by Oliver Wyatt and David Dew-Hughes (the first book, after Cottrell's, by British authors), and then another British book, *Structure and Properties of Engineering Materials* (1977) by Bryan Harris and Anthony Bunsell. In Germany, Erhard Hornbogen brought out *Werkstoffe* (1973). In the Ukraine (while the Soviet Union still existed) an anonymous editor brought out a multiauthor volume (in Russian) entitled *Fizicheskoe Materialovedenie v SSSR* (1986); this is probably the only such book ever to focus on research in one country. In 1982, I.S. Miroshnichenko brought out a specialised book on quenching (of alloys) from the melt. Very recently, Bernhard Ilschner in Lausanne has masterminded a series of texts in materials science in the French language.

A fresh start has been made by Samuel Allen and Edwin Thomas of MIT, with *The Structure of Materials* (1998), the first of a new MIT series on materials. The authors say that "our text looks at one aspect of our field, the structure of materials, and attempts to define and present it in a generic, 'materials catholic' way." They have succeeded, better than others, in integrating some crucial ideas concerning polymers into mainline materials science.

A number of somewhat more specialised texts also began to appear, such as Anderson and Leaver's *Materials Science* (1969); in spite of its broad title, this book by two members of the Electrical Engineering Department at Imperial College, London, was wholly devoted to electrical and magnetic (functional) materials. So was *Electronic and Magnetic Behaviour of Materials* (1967) by Allen Nussbaum of the University of Minnesota.

A good example of a book aimed specifically at processes is Alexander and Brewer's *Manufacturing Properties of Materials* (1963). More recently, there have been some fine texts aimed directly at developing for fledgling engineers a systematic approach for selecting materials during the design process: *Engineering Materials – an Introduction to their Properties and Applications* (1980), by Ashby and Jones, is probably the best example.

There have also been some excellent books and collections of articles written at a popular level. The master of this difficult art was James (J.E.) Gordon, who brought out two immensely successful titles, *The New Science of Strong Materials, or Why You Don't Fall Through the Floor* (1968) and *Structures, or Why Things Don't Fall Down* (1978). The magazine *Scientific American* consecrated the issue of September 1967 entirely to a number of surveys of materials, from a very wide range of perspectives; the lead article was by Cyril Stanley Smith. These articles also came out as a book, *Materials*, published by Freeman. In October 1986, another issue of the same periodical was devoted to materials for economic growth. In 1980, the great French physicist André Guinier (the discoverer of zones in precipitation-hardened light alloys), brought out *La Structure de la Matière, du Ciel Bleu à la Matière Plastique*; this was later translated into English. I have myself for many years contributed 1000-word articles to *Nature* on many aspects of materials science: a selection of 100 of these appeared in 1992 under the title *Artifice and Artefacts*.

A valuable source of up-to-date reviews of many aspects of MSE is a series of books, *Annual Reviews of Materials Science*, published for the last 30 years. There has been one extensive series of high-level multiauthor treatments right across the entire spectrum of MSE, in the form of 25 books collectively entitled *Materials Science and Technology: A Comprehensive Treatment* (1991–2000), masterminded by Peter Haasen, Edward Kramer and myself. There have also been three encyclopedias, *the Encyclopedia of Materials Science and Engineering* (1986), the *Encyclopedia of Advanced Materials* (1994) and the *Encyclopedia of Materials* (2001), which last has appeared in both printed and on-line versions and will receive annual updates.

14.4. MATERIALS SCIENCE IN PARTICULAR PLACES

Recently, at an international conference, during the 'afternoon off' when we were all ambling in the sunshine, a young Algerian student asked me for a 'word of wisdom'. What elderly scholar can resist such a dewy-eyed approach from youth? So I reflected for a moment and then told him: "Remember that there is not really such thing as Algerian science... or British or American science. There is just science, a worldwide collective endeavour. The thing that is invariant is the belief in the importance of a form of internationalism that really works, a pursuit of truth that unites mankind." This last sentence is a formulation that I owe to my wife.

What I said to the young man was both true and untrue. It is quite true that working with, or at least communicating with, one's colleagues worldwide is one of the things that most makes a life in science worth while. Yet what one can do in a particular place depends on the resources and stimulus available, which in turn depend on the traditions and economy of that place. For instance, the traditions of (say) a Middle Eastern country may predispose scientists (and even engineers) there to focus on theoretical work at the expense of experiment. So in that limited sense, there is interest in saying something about how materials science and engineering have developed in different places, and to try to draw some conclusions. That is my objective in this section. I have picked people and institutions on the basis of personal acquaintance in years past, and that is why there is a certain metallurgical bias in my choices, since my personal research was on metals. I have outlined *particular* institutions in the USA, Japan, Australia (with an aside on Germany), Argentina and Russia, and tried to paint brief portraits of the people that brought them into being. I have not attempted the hopeless task of painting a complete portrait of those countries.

Those countries apart, if there were much more space available I could outline research institutions for materials science in the many European countries that possess them, in India, China and Korea, in Canada, Brazil, Israel. The fact that I do not implies no disrespect for the many fine experts in those lands.

14.4.1 Cyril Smith and the Institute for the Study of Metals, Chicago

A number of American research institutions and the people who shaped them have already featured in this book: the creation of the Materials Research Laboratories; Robert Mehl's influence on the Naval Research Laboratory and on Carnegie Institute of Technology; Hollomon's influence on the GE laboratory; Seitz's influence on the University of Illinois (and numerous other places); Carothers and Flory at the Dupont laboratory; the triumvirate who invented the transistor and the atmosphere at Bell Laboratories that made this feat possible; Stookey, glass-ceramics and the Corning Glass laboratory. I would like now to round off this list with an account of a most impressive laboratory that came to grief, and the man who shaped it.

Cyril Stanley Smith (1903–1992) (Fig. 14.2) was a British-born metallurgist who studied at Birmingham University and then emigrated to the United States as a young man, took a doctorate at MIT, and spent 16 years as a successful researcher

Figure 14.2. Portrait of Cyril Stanley Smith in old age (courtesy of MIT museum).

on alloys in an industrial copper-and-brass company; he obtained numerous patents. He became well known for the originality and clarity of his researches, and in 1943 Robert Oppenheimer recruited him to be joint head of the metallurgical effort in the bomb project at Los Alamos. When the War ended in the summer of 1945, he agreed to an invitation from the University of Chicago (which had a highly active president, Robert Hutchings) to create there a novel kind of laboratory devoted to the study of metals in particular, and the solid state more generally. In 1946, the Institute for the Study of Metals opened its doors on the Chicago campus in the same building where in 1942 the world's first nuclear reactor had gone critical. (It is ironic that this earlier project at the time was called 'The Metallurgical Laboratory of the Manhattan District' with the aim of totally confusing anybody who might have been inquisitive.) An account of the first 15 years of the Institute, by one of its members, has recently appeared (Kleppa 1997).

   In April 1946, a few months before the Institute began operation, Smith made public a memorandum detailing the principles on which it was to be founded. (The memorandum is reprinted in Kleppa's paper, and might be described as an elaboration of the ideas concisely set out 7 years later in Smith's preface to the first issue of *Acta Metallurgica*, from which some words are quoted in Section 14.3.2) Indeed, the creation of the Institute and later of *Acta Metallurgica* were two sides of one coin. In his memorandum, Smith saw physicists as the masters of theoretical work on metals, physical chemists as students of reactions and of the associated thermodynamics, while metallurgists would undertake research on matters like diffusion, phase transformations, grain growth, and "similar fields in which a phenomenological approach must precede or accompany the strictly mathematical". He also indicated, fatefully, that "the Institute will maintain close connections with the instructional activities of the university, but it is not intended to establish a separate Department of Metallurgy, and consequently, no degrees in metallurgy will be awarded."

Cyril Smith succeeded in attracting some very distinguished researchers at the beginning, including Charles Barrett (who had worked with Mehl in Pittsburgh), Clarence Zener, Norman Nachtrieb, the eminent crystallographer William Zachariasen, Andrew Lawson, Joseph Burke, Earl Long, the Chinese T'ing-Sui Kê... to mention only a few of the metallurgists, ceramists, physicists and chemists whom Smith had recruited (partly drawing on his acquaintances at Los Alamos). Smith also secured a large group of industrial sponsors, drawing on his industrial past. The Institute published a series of quarterly reports, distributed to the sponsors and some other favoured recipients; these reports in the early years contained entire papers which later appeared in journals, especially *Physical Review* (*Acta Metallurgica* not yet being in existence). Even most of the text of Zener's short but extremely influential book, *Elasticity and Anelasticity of Metals*, published by the University of Chicago Press in 1948, first saw the light of day in a quarterly report. Many topics were unusual, for instance Barrett's work on low-temperature phase transformations and Nachtrieb's on diffusion under hydrostatic pressure (which delivered insights into diffusion mechanisms). The Institute quickly came to be perceived as the leading fundamental research laboratory devoted to metals, and many visitors came; one was Brian Pippard, who in 1955–1956 performed there his famous work on the shape of the Fermi surface in copper (Section 3.1).

Smith himself stimulated many researchers but, though he wrote a celebrated paper on the evolution of microstructure, did not take any graduate students, and so he did not perhaps initially perceive the implications of the fact that large numbers of doctoral students came from the university's physics and chemistry departments to work with some of the permanent Institute staff... but there were no metallurgically trained students to draw on. Some of the Institute staff became closely involved with the physics or chemistry departments, and one even became chairman of the physics department. A consequence of this situation was that Smith could not attract further metallurgists to join the Institute, and junior metallurgists who came for short attachments found that they did not want to stay permanently, even when offered tempting posts. Smith's initial decision not to push for the creation of a metallurgy department proved to be the occasion of the Institute's downfall in the end, because the metallurgists had no sense of belonging to the university as a whole.

In 1955, Smith took a year's sabbatical to pursue his interests in metallurgical history (see, for instance, Section 3.1.2) which led in 1960 to the publication of his *A History of Metallography*. Earl Long became director, but resigned when in 1961 the Institute in effect was taken over by the physicists and chemists and its name was changed to the James Franck Institute (after a German émigré physicist), still its name today. It was a classic organisational coup, and nothing was said about it in the next quarterly report. Interest in metals lapsed almost completely, though Charles Barrett remained for a time. Kleppa, the author of the valedictory article cited above, was the most persistent of the early staff members, and carved out a distinguished place for himself as an expert on experimental thermochemistry of alloys (see an autobiographical paper, Kleppa 1994).

Smith resigned in 1961 and returned to his alma mater, MIT, where he developed his renowned work on the history of metallurgy, drawing on an enormous collection of ancient texts which he had begun to form in the 1930s. It would be fair to say that his courageous conception at Chicago eventually failed, and turned into something entirely different with his departure, because there was no academic home for the metallurgists on the Institute staff and the lack of such a home impeded recruitment and retention of staff. Scientists, like all other people, need a sense of belonging. In the various National Laboratories in America where so much distinguished research on materials goes on today, there is no such problem, but universities are different.

14.4.2 *Kotaro Honda and materials research in Japan*

When, in 1867, the repressive shogun was overthrown, the Meiji (Imperial) Restoration took place and Japan was at last thrown open to the world, the Japanese government soon recognised that Japan had a lot of catching up to do with respect to science and engineering. At once, a number of foreign professors were recruited to teach at Japanese universities, especially from Germany, Britain, France and America. One of those who came was Alfred Ewing, the many-faceted magnetician and engineer whom we met in Chapter 3. He lectured at the physics department at the Imperial University of Tokyo, 1878–1883 and proved effective in instilling an interest in magnetism among the students there. The variegated ways in which the Japanese government located and persuaded such foreign experts to help Japan, and the national differences in the behavior of such experts, are interestingly examined in a book about the 'formation of science' in Japan (Bartholomew 1987).

Honda (1870–1954) was a farmer's son. He had a difficult youth, looked down on by his father and suffering from low self-esteem. His brother talked him out of adopting agriculture as a profession and eventually he went to Tokyo to study physics, where he graduated in 1897. Clearly, Ewing's influence was still felt there, because Honda homed in on magnetism for his initial research. He stayed in Tokyo for 10 years, influenced by Hantaro Nagaoka, an excellent teacher of physics who was interested in metals and magnetostriction, and acquired a doctorate. In 1907, the Ministry of Education awarded him a travelling scholarship and he spent the next 4 years divided between Gustav Tammann in Göttingen (a metallurgist, see Fig. 3.8) and René Du Bois in Berlin (a magnetician). Fig. 14.3 shows a photograph of Honda at this time, as well as a later photograph when he had become famous in Japan. Both photographs suggest how thoroughly he had overcome the handicaps of his childhood. In these years in Europe, he studied the changes in magnetic properties of numerous elements as a function of temperature, and also the periodicity of the atomic susceptibility in a range of metals (Honda 1910). He spent a week with Ewing (who was now in Cambridge), discussing his findings, and received Ewing's praise. Like many Japanese who visited the west, Honda was much influenced by at least one of his teachers, and became as brusque and demanding with his collaborators as was Tammann, and would not brook contradiction. Honda, having overcome his own childhood inferiority, was wholly unimpressed by other people's class or status. According to contemporary records, however, he did not share Tammann's quick temper and was always imperturbable.

Figure 14.3. Portraits of Kotaro Honda as a young man and in middle age (courtesy of Reiner Kirchheim, Göttingen).

On his return home in 1911, Honda was appointed professor of physics at the new Tohoku Imperial University in Sendai, in the north of Japan; this institution had been established only in 1906, when the finance minister twisted the arm of an industrialist who had made himself unpopular because of pollution caused by his copper mines and extracted the necessary funds to build the new university. A provisional institute of physical and chemical research was initiated in 1916, divided into a part devoted to novel plastics and another to metals. This proved to be Honda's lifetime domain; he assembled a lively team of young physicists and chemists. In the same year, Honda invented a high-cobalt steel also containing tungsten and chromium, which had by far the highest coercivity of any permanent-magnet material then known. He called it KS steel, for K. Sumitomo, one of his sponsors, and it made Honda famous.

In 1919, after much politics (the details of which can be found in Bartholomew's book) Honda's group was inaugurated under the name of Iron and Steel Research Institute. Three years later, to broaden its terms of reference, it was rechristened the Research Institute for Iron, Steel and Other Metals (RIISOM). The Institute was wholly focused on intensive research, at all hours of the day and night; Honda was full of scientific ideas and implacable with his colleagues and students. Among many other early successes were a series of improved magnetic alloys, details of which can be found in a survey of Japanese research in magnetism (Chikazumi 1982). Honda succeeded in always maintaining an excellent balance between fundamental and applied concerns. (According to Bartholomew, "the business interests that came to support him thought his work theoretical, but academics thought it applied.") By the end of Honda's reign, Japan had moved a long way from the view expressed in a 1907 editorial, that basic researchers were "eccentrics whose work is a form of dissipation."

The prodigious research output of the Institute often first saw the light of day in the *Science Reports of the Tohoku Imperial University*, and its successors; in my younger days, I received these regularly and found them rivetting reading.

In 1931, Honda, loaded with honors, became president of the university, and the Institute was directed by a successor. The Institute expanded its space, personnel and range of interests, until in 1987, in the words of the current descriptive brochure, "it was reorganised as a countrywide collaborative research institute to meet the rapid progress in materials science and renamed Institute for Materials Research." Its institutional centre has remained in Sendai. Its constituent 31 laboratories range from very pure to very applied science, from crystal physics to irradiation studies, from low-temperature physics to solid-state chemistry under high pressure, from high-purity metals to crystal chemistry, from magnetism to solidification and casting metallurgy – to name just a few. One recent director was Tsuyoshi Masumoto, who made his name in research on metastable alloys and has very recently founded a new broad-spectrum journal, *Science and Technology of Advanced Materials*, mentioned in Section 14.3. A recent director was Hiroyasu Fujimori, another distinguished physicist and metallurgist.

Today, there are many eminent researchers on materials in Japan, alike in universities and in various national research institutes, and latterly in Tsukuba Science City – but the Tohoku Institute has always held a special place, owing to the energy, determination and organising ability of its founder and the habits of work which he instilled in his staff.

14.4.3 *Walter Boas and physics of solids in Australia*

In the 1930s, the world's greatest migration of scientists took place under the lash of Nazism. It has sometimes been asserted that Hitler may have lost the War because of the talent he forced to flee, and that the American development of the atomic bomb that shortened the War so drastically might have been much slower without that migration. Other, less cataclysmic, consequences also flowed from the migration, and this Section is devoted to one of them.

Walter Boas (1904–1982) (Fig. 14.4) was a German physicist of Jewish parentage. In 1922 he entered the Technische Hoschschule in Berlin to study electrical engineering, but two years later he switched to physics because "I wanted a sounder grounding in fundamentals and disliked the large amounts of design work..." These words come from Boas's personal record deposited with the Australian Academy of Sciences in 1973 (with a 1979 addendum). In Berlin, Boas was stimulated by the great physicists who crowded that city, and lived through the heroic days of early wave mechanics. Early in 1927 he had to choose a professor with whom to carry out the small research project required for his diploma (equivalent to a bachelor's degree). After much thought, he approached Professor Richard Becker, who accepted him for a project to check his new theory of metallic creep under stress. At first Boas was not too happy with the task assigned to him and his first tests were unsuccessful. Then, "Becker...came to the laboratory and gave me a proper dressing down in his wise and direct way which had a permanent effect on all my work: 'You must apply yourself with all your love and your whole soul to your project, otherwise no experiment will ever succeed'. How true this is and how often I thought of these words."

Boas praised Becker's personality, as one of the most ethical men he had ever met, with high principles from which he would not deviate and for which he stood up with great courage, even vis-à-vis the dangerous Nazis. When Becker (1887–1955) died in Göttingen, after a life marked by major contributions to nucleation theory and ferromagnetism as well as plasticity and even explosives, his memorial speeches all insisted on his striking honesty and uprightness of character. Heisenberg at the

Figure 14.4. Portrait of Walter Boas (courtesy of CSIRO, Melbourne).

time commented on the youthful rapport which developed between Becker and his students. It is also noteworthy that 6 years after his attention to Boas, Becker did the same for Egon Orowan, who also switched from electrical engineering to physics and studies of plasticity (see Section 3.2.3.2).

In later 1927, Becker saw to it that in spite of the severe Depression, Boas was given a job at the Kaiser-Wilhelm-Institut für Metallkunde in Berlin. Becker told Boas that he was dean of the faculty which had to adjudicate on the 'habilitation' (right to lecture at university) of a young Austrian metallurgist at the Institute, Erich Schmid, and that (in spite of his – Becker's – high ethical standards!) he would tell Schmid that he would agree to his habilitation only if he appointed Boas to a position in his section." He did, and was duly habilitated. On such curious practices can a lifetime career depend. With Schmid, Boas began the experiments on the plasticity of metal crystals (Sections 2.1.6, 4.2.1) that culminated in the publication, in 1935, of their joint book *Kristallplastizität*. Before that, in 1930, Boas obtained his doctorate (as Becker's first doctoral student); his thesis "created a small storm since it was the shortest ever submitted, 15 pages." Boas and Becker remained good friends, and in contact, for the rest of Becker's life.

Boas accompanied Schmid to the University of Fribourg in Switzerland in 1933, when Schmid was granted a chair there, and there they finished their book. Boas was lucky; he left Germany before Hitler took power. Though his parents were not observant Jews, and in fact Boas had been baptised, that would not have saved him from persecution. The Swiss, however, had a rule that prevented any refugee from staying beyond 5 years (which would give him automatic rights of permanent residence) and so Boas had to leave in late 1937. The committee in London that looked after refugees, run by the legendary Esther Simpson who has died only recently, found Boas a temporary haven with Sir William Bragg at the Royal Institution in London. Again with Esther Simpson's help, Boas managed to secure appointment to a 2-year Carnegie lectureship in metallurgy at the University of Melbourne in Australia, and in April 1938 he and his new wife left for Australia where he lived contentedly for the rest of his long life, becoming a citizen in 1944.

Boas's post at the University of Melbourne was made permanent, and for 9 years he taught and supervised research, and helped train metallurgists for war work. A number of noted metallurgists, such as Robert Honeycombe, passed through his hands. Then, after the War, the department was faced by insuperable difficulties and research facilities disappeared. In 1947 he applied for and secured a senior post in the Division of Tribophysics of the Commonwealth Scientific and Industrial Research Organization, CSIRO, also in Melbourne. 'Tribophysics' in the name was a residue of the interests of F.P. Bowden who studied wear and lubrication and sought a more elegant name to denote his interests (Bowden had moved on to Cambridge). Boas's duties were to undertake, and direct, research in the physics of solids. In spite of the tribophysics name, Boas took a broad view of his remit, and studied many aspects of metal physics. CSIRO made no difficulties about his choice of themes; they had an attitude to their senior scientists rather like that of Bell Laboratories... choose the best and give them their heads. How different from the situation today!

In 1949, Boas was appointed to the post of divisional chief after the incumbent retired. He took some persuasion to accept this, because it meant more attention to administration and supervision and would make it difficult for Boas to undertake personal research. But he accepted this and had an enormous influence on a whole generation of Australian physicists and metallurgists, aided in this by his open and affable personality. He remained in charge until 1969; his list of publications could not be large, though he brought out two more didactic books. He became known throughout the world; thus, in 1953 he became a senior adviser to the newly established *Acta Metallurgica*, and was invited to many international scientific occasions as a representative of his adopted country, which he defended fiercely against those who tried to slight it.

A book of scientific articles in celebration of Boas's 75th birthday (Borland et al. 1979) includes a biographical sketch of Boas by J.F. Nicholas. A substantial account of his life can be found in an obituary by Clarebrough and Head (1987).

After his retirement, the character of CSIRO changed utterly, and research in the Division, which repeatedly changed name, became wholly focused on applications. Boas's team fell apart and the scientific atmosphere in the Division became quite different. According to Clarebrough (2000), (one of Boas's most distinguished collaborators and the father of the first differential scanning calorimeter), "the decay of 'Walter's team' is a long story, but the main cause was a change in the politics of science funding in this country. CSIRO was hit first and now it has spread to the universities where funds must be raised for research from industrial sources and basic science is no longer funded." According to a news story in *Nature* (Swinbanks 1996) under the title "Basic research fighting for survival", the bureaucracy of science funding in New Zealand had by that year become 'horrendous' and applicants had to outline the expected outcome of their research. "This has encouraged them to pursue low-risk research rather than long-term fundamental research." Swinbanks went on to point out that Australia was headed in the same direction, but the story of Boas's Division suggests that it is already far down that path.

However, none of what has happened recently can detract from the contribution made by Walter Boas, that eminent physicist of solids, to the scientific life of his adopted and beloved country.

14.4.4 Jorge Sabato and materials science in Argentina

In 1955, being a Spanish-speaking metallurgist, I was invited to spend some weeks in Buenos Aires to deliver a course in elementary modern metallurgy to some members of the mainly youthful staff of the Atomic Energy Commission. The person who invited me, a dynamo of energy and originality, was Jorge Sabato (1924–1983) (Fig. 14.5), an Argentinian metallurgist who had recently joined the Commission's laboratory from local industry. I proved to be just the first of a procession of foreign experts, and later on Sabato organized more ambitious courses for which auditors came from all over South America. The Atomic Energy Commission came to be South America's leading focus of expertise in metallurgical engineering.

While I was in Argentina in 1955, Sabato took me to visit a brand new laboratory in Patagonia, deep in the 'south', near the ski resort of San Carlos de Bariloche. This was, and still is, the Centro Atomico de Bariloche (CAB). It is an institution (formally part of a local university) for research and teaching in physics, ranging from particle physics to solid-state physics. Its origin is one of the most curious in the entire history of academe.

Figure 14.5. Portrait of Jorge Sabato (courtesy of Heraldo Biloni).

In 1948, an Austrian physicist named Ronald Richter came to Argentina and managed to persuade President Perón that he knew the secret of cheap nuclear energy and, more specifically, cheap nuclear bombs. He was exceedingly tight-lipped about the details, but he sufficiently convinced the unscientific politician that he progressively accumulated money, laboratory space and extensive scientific equipment. Like the fisherman's wife of the fairy tale, he kept on coming back to the president's house, the Casa Rosada, to ask for more. In due course he was given exclusive use of an island, Huemul, on the lake of Nahuel Huapi, a beauty spot near Bariloche, a staff of engineers and masses of hardware; no visitors were admitted. Perón was taken for a comprehensive ride, and it was not till early 1951 that a brave 'real' physicist, José Balseiro, was able to persuade Perón to set up a commission of experts to inspect the island. They found that the project 'lacked scientific seriousness'. This phrase comes from a fascinating book that describes the entire history, both scientific and political, of this unserious episode (Mariscotti 1984, 1996).

After the commission reported, it took some time for the government to extract Richter from his island; surprisingly, he was allowed to retire peacefully and (I was informed at the time of my visit) to take up chicken farming near Buenos Aires. So by 1953, there was a mass of virtually unused laboratory equipment on the isle of

Huemul: what was to be done with it? Balseiro had plans for an advanced university institution, to be set up in comparative liberty far from the political quagmire of Buenos Aires, and persuaded Perón to make available a piece of land and some buildings on the shore of Lake Nahuel Huapi, just a mile or two from the island. This became the Centro Atómico de Bariloche, opened in 1954. Balseiro himself was a highly distinguished theoretical physicist. Sabato succeeded in persuading him to include metal physics in the syllabus of the Centro; he was able to convince the suspicious Balseiro that he did mean metal *physics*, not metals techology which would continue to be pursued at the Commission's laboratory in the capital city. A well-known Austrian physicist, Gunther Schöck, came to inaugurate this part of the subject-matter, the first of many foreign scientists who have enjoyed spending periods in Bariloche; that Schöck was an enthusiastic skier played its part in attracting him to this beauty spot. (There are some points of similarity between events in Melbourne and in Bariloche – foreign scientists played a crucial part in both places.) Sabato's diplomacy ensured that the CAB and the Buenos Aires laboratory thereafter worked closely together.

I came back in 1959 to deliver a course of crystallography lectures at the CAB, and by that time the metal physics was well established. It has continued to flourish, and broaden; many papers of note were published, and a succession of international materials symposia have been held there. The CAB director, Balseiro, died young, of cancer, and the latest of a succession of directors is José Abriata, an Argentinian materials scientist. Most observers, I believe, both in South America and beyond, would concur that the Bariloche centre is the most distinguished physics laboratory in South America. Materials science plays an important part there, and credit for that belongs to Jorge Sabato.

14.4.5 *Georgii Kurdyumov and Russian materials science*
The early path of scientists in Russia was not an easy one, especially for those precocious individuals who were far ahead of their contemporaries. In Chapter 3, I mentioned Federov, one of the three co-inventors of the theory of space groups in the late nineteenth century, who found no comprehension at home of his difficult ideas. Here I can also refer to Mikhail Vasilevich Lomonosov (1711–1765), an early Russian polymath, a scientist and littérateur, who took an interest in the whole of science as it was perceived in the Petersburg of his day. He insisted on the crucial bond between chemistry and physics long before Wilhelm Ostwald was born ("a chemist lacking knowledge of physics is like a blind man who seeks by touch"); he took an active part in improving glass technology in his native country; he put paid to the phlogiston theory well before Lavoisier did. These facts come from an as-yet unpublished essay on Lomonosov by a British science historian, Michael Hoare,

very kindly made available to me. Hoare says: " We may without serious reservation identify (Lomonosov) as the first modern physical chemist, materials scientist, mineralogist, ceramist, research administrator and scientific educationalist, and still not nearly exhaust the rollcall of his activities." (In a more recent message, Hoare, 2000, expresses the view that some of the ideas credited to Lomonosov should really be laid at the door of Robert Boyle, a century earlier; their ideas are sometimes hard to disentangle.) Lomonosov had to spend too much of his energy in fighting political battles in the Petersburg Academy of Sciences; he was not the first nor the last to be so beset. But the man with whom I am mainly concerned in this Section did not spend his time in politics, and he had a huge influence on metallurgy and eventually on materials science in his country.

Georgii Vyacheslavovich Kurdyumov (1902–1996) (Fig. 14.6), the son of a priest, was the most famous metallurgist of his generation in the Soviet Union, a man who was not only a great research scientist but also a man of rare human qualities. He and the many people who collaborated closely with him spent decades on a single

Figure 14.6. Portrait of Georgii V. Kurdyumov.

broad cluster of problems; with some scientists, that would be evidence of a lack of imagination, but with Kurdyumov, it showed a deep and persistent curiosity and an ability to ask clear questions and to design crucial experiments that would answer them clearly. It is my great regret that I never had a chance to meet him.

Kurdyumov was educated as a physicist at Abram Ioffe's famous institute in Petersburg (= Petrograd = Leningrad = now, St. Petersburg). Alexander Roytburd, one of Kurdyumov's later collaborators, in a memoir of his revered master (Roytburd 1999), describes the Ioffe Institute as "the cradle of all soviet physics". (The 2000 Nobel Prize winner, Zhores Alferov, is currently director.) Roytburd goes on to claim that "working all his life in physical metallurgy, Kurdyumov kept close ties with the physics community and did a lot for the development of metallurgy in physical metallurgy, which is the fundament of modern materials science".

As a very young scientist, Kurdyumov homed in on the important problem of how and why carbon steel becomes hard when it is quenched from red heat. At an early stage, he established firmly that the hard phase was *martensite*, a distorted form of body-centred cubic iron with dissolved carbon. He studied the crystallography of martensite as a function of carbon content, took his diploma, and then in 1930 he was one of a group of 220 Soviet scientists who were allowed to spend a period abroad. He joined the German metallurgist Georg Sachs in Berlin; Sachs had found out how to grow single crystals of copper-based alloys, and Kurdyumov quickly found out how to convert this skill to the growth of crystals of alloyed austenite. Austenite is the (face-centred cubic) high-temperature phase – which by alloying can be stabilised at room temperature – from which martensite forms during quenching. He then cooled the austenite crystals enough for them to convert to martensite, and studied, by means of X-ray diffraction, the orientation relation between the two phases. This resulted in the celebrated Kurdyumov–Sachs relationship, still much quoted today. He also became convinced that the martensitic transformation (Kurdyumov's term) took place by a form of shear, without diffusion, and thereby initiated 70 years of intensive research on this type of transformation, in steels particularly (but not only in steels).

Back in the Soviet Union, he moved to the Ukraine to help, with his scientist wife, create a research institute in Dniepropetrovsk, where he continued with his researches. He was invited to be director, sought to escape from this fate (he complained that he would be a bad administrator, and that by administering he would lose contact with real science and then become unable to direct scientific work properly) but was persuaded to overcome his scruples. The rest of his long career he both administered (usually more than one institute at once) and remained a unique scientist. During the War, the institute had to move, and after the War, it was moved again, to Moscow, and Kurdyumov with it. While in Moscow, he also created a laboratory of metal physics in Kiev, Ukraine, and directed both the Moscow and the

Kiev institutes (Khandros 1992). In both laboratories, more and more subtleties of the martensite transformation, its crystallography, kinetics and mechanism, were undertaken, and Kurdyumov became internationally famous and established close links with American and Japanese scientists in particular. He and Khandros (Kurdyumov and Khandros 1948) published the first study in depth of a 'thermoelastic' phase transformation, the precursor of many later studies of shape-memory alloys (nowadays familar to many people in the form of 'shape-memory spectacles'), which are based on stress-induced martensitic transformations. I recall reading this paper in translation in 1948 and being astonished at its quality. The foregoing is only a rough outline of a few of the many profound studies that emerged from Kurdyumov's laboratories, resulting in almost 300 papers. Roytburd's (1999) panegyric goes into much greater detail. Kurdyumov remained director of the Moscow laboratory until 1978. His most important achievement was to train a whole generation of research metallurgists to a high standard, and to provide a model for them.

Two further aspects merit notice. Kurdyumov's Moscow laboratory was part of the great network of laboratories administered by the Soviet Academy of Sciences, and his Kiev one belonged to the Ukrainian Academy of Sciences. Throughout the Soviet sphere of influence, and also in China, the science academies were the chief organisers of scientific research – essentially, the academies were, and are, organs of state – whereas in the West, the academies are independent bodies of experts, ready to advise governments but not to administer laboratories. Briefly, just before the First World War, when the US National Academy of Sciences (NAS) was in its formative state, there were voices raised in favour of laboratories to be run by the NAS, but the idea was soon abandoned. Independence from the state was too precious to be thus compromised.

All who knew Kurdyumov well testify to his kindness, consideration of his staff and courage. Roytburd remarks: "In order to be a founder of a new scientific era one has to be a great scientist; to create a science school one has to be a great person. Kurdyumov was this scientist and this person. His personality attracted everyone who met him, but we, his pupils, colleagues and friends especially felt enchanted by him... He was Director of our institute but that was not what defined his authority. He was a spiritual leader because of his vision of the problem, his enthusiasm, his respect for the scientific truth, his belief that the truth is more important than success." Another of Kurdyumov's collaborators, Evgeny Glickman, informed me (Glickman 1999) that "my aunt who worked with Georgii Vyacheslavovich Kurdyumov in Dniepropetrovsk always stressed that at the peak of the Stalin terror, 1937–1938, Georgii Vyacheslavovich Kurdyumov saved her and many colleagues by answering for them at KGB with his own life. By no means was this safe – or typical – behavior at that time."

Throughout Kurdyumov's active career, metallurgy was taught in some places, semiconductors in others, ceramics in others still... there was no materials science. In 1985, encouraged by Gorbachev's perestroika and the opportunity for independent initiative which it brought, one courageous Russian professor, Yu.D. Tretyakov, who had earlier spent some time in Rustum Roy's laboratory at Pennsylvania State University (see the first paragraph of this chapter) and been fired by Roy's ideas, was able to set up Russia's first undergraduate course in materials science (nauk o materialakh) at Moscow's Lomonosov State University, in the form of a five-year programme. There is a degree of chemical emphasis, in view of Tretyakov's background, but on paper, as presented at a meeting at Penn State in 1999 (Tretyakov 1999, see also Tretyakov 2000), it seems a thoroughly well-conceived and balanced programme, and it appears to be attracting much competition for entry. This Section thus ends where it began, with awareness of the name of Mikhail Vasilevich Lomonosov. I would like to think that Professor Tretyakov was also in some measure fired by the personal example of Georgii Vyacheslavovich Kurdyumov.

REFERENCES

Bartholomew, J.R. (1987) *The Formation of Science in Japan: Building a Research Tradition* (Yale University Press, New Haven and London) pp. 63, 186.

Borland, D.W., Clareborugh, L.M. and Moore, A.J.W. (1979) *Physics of Metals: A Festschrift for Dr. Walter Boas on the Occasion of his 75th Birthday* (Department of Mining and Metallurgy, University of Melbourne, and CSIRO, Australia).

Chikazumi, S. (1982) *J. Appl. Phys.* **53**, 7631.

Clarebrough, L.M. (2000) Letter dated 7 August.

Clarebrough, L.M. and Head, A.K. *Walter Boas, 1904–1982, Historical Records of Australian Science* **6**(4) (July) 507.

DGM (1994) *75 Jahre: Die Geschichte der DGM im Spiegel der Zeitschrift für Metallkunde* (Deutsche Gesellschaft für Materialkunde, Oberursel).

Flemings, M.C. (1999) *Annu. Rev. Mater. Sci.* **29**, 1.

Flemings, M.C. and Cahn, R. W. (2000) *Acta Mater.* **48**, 371.

Fullman R.L. (1996) Update to Hibbard's history of *Acta Metallurgica*, unpublished

Glickman, E. (1999) Message dated 25 October.

Hibbard, W.R.H. (1988) A concise history of Acta Metallurgica Inc. and the journals, in, *Robert Maxwell and Pergamon Press*, ed. E. Maxwell (Pergamon Press, Oxford).

Honda, K. (1910) *Ann. Physik* **32**, 1027.

IUMRS–ICA 98 (1999) Proceedings, *Bull. Mater. Sci.* (India) **22**.

Kaufmann, E.N. (1988) *MRS Bulletin* **13** (September) 37.

Khandros, L.G. (1992) Pamphlet on *Georgii Vyacheslavovich Kurdyumov* (Ukrainian Academy of Sciences).

Kleppa, O.J. (1994) *J. Phase Equili.* **15**, 240.
Kleppa, O.J. (1997) *JOM* **49**(1) (January) 18.
Kurdyumov, G.V. and Khandros, L.G. (1948) *Doklady Akad. Nauk SSSR* **66**, 211
Mariscotti, M. (1984, 1996) *El Secreto Atómico de Huemul* (Estudio Sigma, Buenos Aires) esp. p. 271.
Phillips, J.M. (1995) *Adv. Mater.* **7**, 773.
Rabkin, E. (2000) Message, 14 May.
Rao, G. (2000) Message from the MRS, 10 January.
Roy, R. (editor) (1970) *Materials Science and Engineering in the United States* (The Pennsylvania State University Press, University Park and London).
Roy, R. (1977) *Interdisciplinary science on campus – the elusive dream, Chemical Engineering News* (August issue).
Roy, R. and Gatos, H. (1993) *MRS 20th Annniversary, MRS Bulletin* (September) pp. 74, 84
Roytburd, A.L. (1999) *Mater. Sci. Eng. A* **273–275**, 1
Seitz, F. (1988) in *The Origins of Solid-State Physics in Italy, 1945–1960*, ed. G. Giuliani, Societá Italiana di Fisica, Bologna, p. 215.
Seitz, F. (2000) Letter, 7 July.
Swinbanks, D. (1996) *Nature* **379**, 112.
Tretyakov, Yu.D. (1999) website: www.hsms/msu.ru.
Tretyakov, Yu.D. (2000) *MRS Bulletin* **25**(8), 97.

Chapter 15
Epilogue

Chapter 15
Epilogue

The time has come to draw together the threads of what has gone before. MSE is a huge domain; again and again I have had to warn the reader that I could only scratch the surface of some theme in the space available to me, and still I have covered more than 560 pages with a combination of history and depiction.

First, what is materials science? I have gone through my professional life almost without addressing this question explicitly; I have always believed that the right way to address it is by means of what philosophers call an 'ostensive definition', pointing to something and saying "This is it". This inclination was my main reason for accepting, in 1965, the hard labour of creating a new *Journal of Materials Science*; that journal was meant to demonstrate what my novel subject actually was, and I believe it helped to do that. This book is also an essay in ostensive definition. When I had just been appointed professor of materials science at Sussex University, I did write an article under the title 'What is materials science?' (Cahn 1965). Summarising my disquisition, I wrote: "...the materials scientist has to work at several levels of organisation, each of which is under-pinned by the next level. Here, again, he is brother under the skin of the biologist, who does just the same: starting with the cell wall, say, he goes on to study the morphology and economy of the cell as a whole, then the isolated organ (made up of cells), then the organism as a whole." I still hold today that this feature is central to our subject – applied to inanimate and artificial nature by us and to animate nature by biologists – and that the concept of *microstructure* is the most important single defining theme of MSE. To this can be added the slightly broader modern concept of *mesostructure,* a term particularly beloved of modellers and simulators of polymers... the level of organisation in between the atomic/molecular level and macroscopic appearance.

Merton Flemings, a very experienced professor of MSE at MIT, has recently discussed (Flemings 1999) the question: "What next for MSE departments?" He faces, foursquare, the issue whether something can be both a *multidiscipline*, bringing together for use many classical disciplines, and a discipline in its own right. He is sure that MSE is both of these. The path out of the dilemma "is to view the broad engineering study of structure/property/processing/performance relations of materials, with engineering emphasis... as a discipline". That is, he asserts, what mainline, independent MSE departments teach. This fourfold way is depicted in Figure 15.1(a), a little tetrahedron which was first proposed in a 1989 report. Flemings goes so far as to say that "our survival as a discipline and as independent academic departments within the university system depends on how well we succeed in

articulating this paradigm and employing it to contribute to society". Others prefer
to make this little diagram more complicated; thus Shi (1999), a veteran Chinese
materials scientist, is insistent that 'composition' is an equally important variable,
distinct from structure, 'processing' should be linked with 'synthesis', and at the heart
of the whole enterprise he places 'theory and design of materials and processing',
clearly including computer simulation. His view of things is shown in Figure 15.1(b).

One should not be perturbed by different experts' preferences for different kinds
of polyhedra; after all, these are no more than a visual aid to understanding. The key
thing is that different aspects are intimately related... in these figures, every point is
linked to every other point. Each of these aspects, whether they be divided into four
or six categories, needs a familiarity with some of the classical disciplines such as
physics, chemistry, physical chemistry, and with subsidiary not-quite-independent
sciences such as rheology and colloid science.

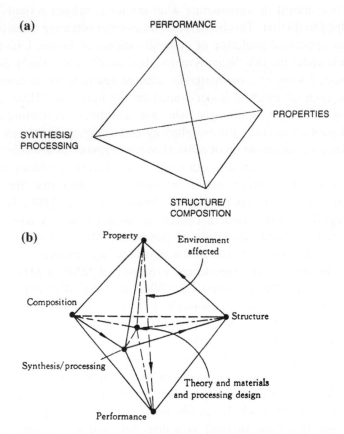

Figure 15.1. (a) The four elements of materials science and engineering, (after Flemings). (b) The
six elements of materials science and engineering (after Shi).

While I entirely agree with both Flemings and Shi about the crucial importance of the components in their diagrams, I persist in my conviction that *microstructure* is the central component that best distinguishes MSE from other disciplines; each chapter of this book demonstrates this centrality. The other components in the diagrams themselves have microstructural features: thus self-assembled materials (a part of processing/synthesis) have carefully controlled microstructure, and composition, because of segregation, varies significantly from point to point – and all this intimately affects properties.

I recall my distinction, in Chapter 2, between emergence (of a discipline) by splitting and emergence by integration, and also my insistence that MSE is a prime example (together with geology) of emergence by integration. This is historically unusual. For instance, in a scholarly study of how chemistry and physics came to be distinct disciplines and then chemistry itself differentiated, Nye (1993) concludes (to simplify drastically) that around 1830 chemistry split decisively from experimental philosophy (or physique générale) by reference to its concern with molecules and their reactions and behaviour, and in doing so left physics behind. It is far harder to reach an acceptable definition of physics than of chemistry, but that has not prevented physicists from driving their discipline forward during the past two centuries. Likewise, we materials scientists practice our mystery whether or not we can define it.

So, nearly half a century after the emergence of the concept, we its practitioners have in materials science and engineering a clearly distinct discipline which in practice doubles up as a multidiscipline, with a substantial number of independent academic departments and research institutes spread around the world, with its own multifarious journals and textbooks, and a large number of professionals, also spread around the world, who call themselves materials scientists and engineers and communicate with each other on that basis. We have a profession to be proud of.

REFERENCES

Cahn, R.W. (1965) *What is materials science? Discovery* (July issue, no page numeration).
Flemings, M. C. (1999) *Annu. Rev. Mater. Sci.* **29**, 1.
Nye, M. J. (1993) *From Chemical Philosophy to Theoretical Chemistry* (University of California Press, Berkeley).
Shi, C. (1999) *Progress in Natural Science* (China) **9**, 2.

Name Index

Subject Index

Corrigenda

§ **3.1.2, page 72.** Alois Joseph Franz Xaver von Beck-Widmanstätter, also known as Widmanstätten, was director of a 'cabinet' of manufactures for the Austrian emperor, not curator of a collection of meteorites. He was previously a printer, which no doubt gave him the impetus to use an etched section of a meteorite as a printing plate. He printed from other meteorites before he made the print from the Elbogen which is illustrated on page 73.

Recent researches on meteorites suggest that the cooling rates of meteorites with very coarse structures were typically some tens of degrees per million years. See C. Narayan and J.I. Goldstein, *Geochim. & Cosmochim. Acta* **49** (1985) 397.

In this same section (pp. 73–74) I discuss Henry Sorby's introduction of reflected-light microscopy of metals, later extended to rock sections examined by transmitted light; here I say that the Swiss geologist (Horace-Benedict) de Saussure ridiculed Sorby for venturing to "look at mountains through a microscope". De Saussure in fact died in 1799, 50 years before Sorby's first work with petrographic sections. What de Saussure did was to ridicule the 'pretensions' of 'natural philosophers' who discoursed on the origins of mountains without ever leaving their armchairs; it clearly seemed to him that looking at rocks through a microscope, if it were achieved, would amount to not leaving an armchair! The episode is discussed by D.W. Humphries in his chapter on Sorby, on page 17 of *The Sorby Centennial Symposium on the History of Metallurgy*, edited by C.S. Smith (Gordon and Breach, New York, 1965).

§ **3.2.1.2, page 110.** Through an unfortunate oversight, I omitted to include here the reference I had intended to some very important early work on dislocation theory by the physicist Jacques Friedel (Paris); Friedel is mentioned in another connection on page 137. He was stimulated to work on dislocations by his stay in Bristol, 1949–1952. His contributions are recorded, together with others, in his book *Les Dislocations* (Gauthier-Villars, Paris, 1956), published in a thoroughly revised and updated English translation by Pergamon Press (Oxford), 1964. The personal background to his researches can be found in Friedel's autobiography (*Graines de Mandarin*, Editions Odile Jacob, Paris, 1994, pp. 169–179).

§ **4.2.1, page 160.** In connection with Figure 4.3, of a spherical copper crystal following oxidation, it was mentioned that the sphere had been electrolytically polished. That technique (which can be simply described as the inverse of electrodeposition) has been vital for the production of strain-free metal surfaces for examination by microscopy; the most important pioneer in developing this technique was the French metallurgist P.A. Jacquet (see Jacquet, *Metallurgical Reviews* **1** (1956) 157).

Single-crystal turbine blades (mentioned at the bottom of page 165) are used not only in jet engines but also in some land-based steam turbines, notably those manufactured by GE.

An early overview of crystal growth can be found in a book by H.E. Buckley, *Crystal Growth* (Chapman and Hall, London, 1951).

§ 4.2.4, page 176. A useful adjunct to the history of crystallography is a *Historical Atlas of Crystallography*, edited by J. Lima-de-Faria (International Union of Crystallography, Dordrecht, 1990).

§ 6.2.2, page 217. Abbe, whose theory of diffraction-limited resolution is cited here, does *not* have an accent on his final letter. It appears that this is a common error.

§ 6.3.1, page 235. Clair Patterson's researches on lead contamination in the atmosphere, discussed in the last paragraph of this section, emerged from his earlier work on the age of the earth on the basis of precision measurements of concentrations of different lead isotopes, some radioactive, work for which he became famous. His definitive paper about this work is in *Geochim. & Cosmochim. Acta* **10** (1956) 230.

§ 7.3, page 281. Magnetic ferrites were first studied in Germany and then France, early in the 20th century, but no materials of commercially usable quality were found. This was first achieved in 1932 by a Japanese team, Drs. Takeshi Takei and Y. Kato, who developed a combination of magnetite and cobalt ferrite; the first patent was taken out in 1935, at the time when the Dutch work was just beginning. An account in English of this important early work in Japan can be found in a paper entitled "The Past, Present and Future of Ferrites", by M. Sugimoto, in *J. Amer. Ceram. Soc.* **82** (1999) 269.

§ 9.1.5, page 357. The seminar proceedings on ultrapure metals published in America in 1961 were preceded by a year by a French (CNRS) symposium on the same subject, centered on the researches of a chemical metallurgy group led by the influential metallurgist Georges Chaudron.

§ 9.4, page 367. Local minima in plots of grain-boundary energy versus misorientation, as seen in Figure 9.9 on page 371, are linked through the concept of DSC lattices (Displacement Shift Complete). A DSC lattice is defined as the coarsest lattice which includes the lattices of the two bounding grains (A and B) as sublattices; such a lattice is found for 'special' misorientations between the bounding grains. The 'less coarse' such an overarching lattice is, the lower the corresponding grain-boundary energy. The inverse of the proportion of lattice points common to A and B is called the sigma (Σ) number for that misorientation. This very influential concept was introduced by Walter Bollmann (*Crystal Defects and Crystal Interfaces*, Springer, Berlin, 1970), building on earlier ideas due to M.L. Kronberg and F.H. Wilson (*Trans. AIME* **185** (1949) 501).

§ 10.6.1, page 414. My statement at the bottom of page 416 that "the term 'icosahedral symmetry' is sometimes used" is inaccurate. That term is *always* used for the type of quasicrystals originally discovered by Shechtman, because these have not one, but six fivefold symmetry axes, like an icosahedron. Later, quasicrystals with only a single five-fold axis, combined with periodic stacking along that axis, were also found. Figure 10.7 does *not* show an icosahedral structure, but rather uniaxial tenfold symmetry.

§ 13.2.1, page 491. The 'Rubber Bible' (outlined on page 493) is now accessible online, with enhanced searching capabilities.

§ 14.3.1, page 512. A promising new broad-spectrum journal, *Nature Materials*, began publication in September 2002.

Epilogue, page 539. An important book by Mary Jo Nye concerning the emergence of disciplines is mentioned on page 541. Of several others which might have been added, I wish to cite a chapter by J. Dupré, "Metaphysical disorder and scientific disunity" in *The Disunity of Science. Boundaries, Contexts and Powers*, edited by P. Galison and D.J. Stump (Stanford University Press, Stanford, 1996) p. 101.

Printed and bound by CPI Group (UK) Ltd, Croydon, CR0 4YY

03/10/2024

01040410-0020